Emerging Nanomaterials in Biomedical Science

nova science publishers

www.novapublishers.com

Nanotechnology Science and Technology

More information about this series can be found at
https://novapublishers.com/product-category/series/nanotechnology-science-and-technology/

Biomedical Devices and Their Applications

More information about this series can be found at
https://novapublishers.com/product-category/series/biomedical-devices-and-their-applications/

Eepsita Priyadarshini
Ajith Manayil Parambil
Editors

Emerging Nanomaterials in Biomedical Science

nova
science publishers
www.novapublishers.com

DOI: https://doi.org/10.52305/SPVQ0543

Library of Congress Cataloging-in-Publication Data

Library of Congress Control Number: 2025940959

ISBN: 979-8-89530-600-0 (Hardcover)
ISBN: 979-8-89530-707-6 (eBook)

Published by Nova Science Publishers, Inc. † New York

We dedicate this book to our families,
whose unwavering support, patience, and encouragement
have been the foundation of our academic journey.
Their love and belief in us have been our greatest source
of strength throughout the making of this book.

Dr. Eepsita Priyadarshini and Dr. Ajith Manayil Parambil

Contents

Preface

In an era marked by rapid technological advancements, the convergence of nanotechnology and biomedicine stands out as one of the most promising frontiers in scientific research and healthcare innovation. "Emerging Nanomaterials in Biomedical Science" is a comprehensive exploration of this dynamic intersection, offering readers an in-depth analysis of the latest developments, applications, and emerging trends in nanomaterials for biomedical applications.

Authored by a distinguished team of experts, this book thoroughly examines cutting-edge synthesis techniques and practical biomedical applications of nanomaterials. Each chapter meticulously details specific nanomaterials, their synthesis, properties, and fabrication methods, emphasizing their crucial roles in advancing various aspects of healthcare, such as nanomedicines, tissue engineering, theranostics, diagnostics, and regenerative medicines.

One of the standout features of this book is its interdisciplinary approach, bridging the gap between nanotechnology and biomedicine to offer a holistic understanding of how nanomaterials are revolutionizing healthcare. This approach ensures that readers not only grasp the fundamental principles of nanoparticle synthesis and characterization but also appreciate the broader implications of these materials in real-world biomedical settings.

Section 1 lays the groundwork by offering a general overview of nanomaterials, focusing on their synthesis, characterization, and fundamental applications in biomedicine. This section sets the stage for subsequent discussions by providing essential background knowledge.

Section 2 delves into the role of nanomaterials in drug delivery, exploring targeted delivery mechanisms, encapsulation, and controlled release strategies. This section highlights nanomaterials' potential to enhance therapeutic interventions' efficacy and precision.

Section 3 examines the application of nanomaterials in bioimaging and diagnosis, detailing principles of biosensors and nanobiosensors, contrast agents for imaging, and the theranostic potential of nanoparticles. These insights underscore the transformative impact of nanotechnology on disease detection and monitoring.

Section 4 focuses on regenerative medicine and tissue engineering, discussing the design of scaffolds using nanomaterials, biomimetic approaches, and stimuli-responsive materials for tissue regeneration. This section showcases the potential of nanomaterials to facilitate tissue repair and regeneration.

Section 5 addresses the critical aspects of biocompatibility and safety assessment of nanomaterials, emphasizing toxicological considerations and ensuring that these materials can be safely integrated into biomedical applications.

Finally, Section 6 explores the ethical issues, challenges, and perspectives surrounding nanomaterials in biomedical science. This section covers regulatory compliance, patient consent, interdisciplinary collaboration, commercialization, and nanomaterials' environmental impact and sustainability.

"Emerging Nanomaterials in Biomedical Science" is an invaluable resource for researchers, students, and professionals seeking to understand the full potential of nanotechnology in healthcare. Whether you are a seasoned researcher looking for the latest advancements or a student exploring the frontiers of nanomedicine, this book provides comprehensive insights and practical guidelines for working with nanomaterials in biomedical settings. With a focus on practical applications and real-world case studies, this book serves as a guide to the revolutionary potential of nanomaterials in creating a more sustainable and advanced healthcare future.

We are excited to present this work and hope it inspires further innovation and collaboration in the ever-evolving field of nanotechnology in biomedicine.

Acknowledgments

We extend our deepest gratitude to Nova Science Publishers, USA, for their unwavering support and belief in the importance of this work. Their commitment to advancing scientific knowledge has been instrumental in bringing "*Emerging Nanomaterials in Biomedical Science*," to fruition.

Our sincere appreciation goes to the dedicated editorial staff at Nova Publishers. Your meticulous attention to detail, constructive feedback, and tireless efforts have been invaluable throughout the entire publication process. We are grateful for your guidance and support, which have significantly enhanced the quality of this book.

We would like to thank all the contributors to this volume. Your expertise, hard work, and passion for nanotechnology and biomedicine are reflected in each chapter. Your contributions have enriched this book, making it a comprehensive resource that bridges the gap between these two dynamic fields.

We also acknowledge the support of various institutions and organizations that have facilitated this project. Your encouragement and resources have been crucial in realizing this vision.

Finally, we are grateful to our families, friends, and colleagues for their unwavering support and understanding. Your encouragement has been a source of strength and inspiration throughout this journey.

Thank you all for your invaluable contributions and support. This book would not have been possible without you. We believe that "*Emerging Nanomaterials in Biomedical Science*" will serve as a significant resource for researchers, students, and professionals in the field, paving the way for future innovations and discoveries.

Acronyms

2-DG	2-Deoxy-D-Glucose
3D	three Dimensional
ABC transporter	ATP binding cassette transporter
AC	Alternate Current
ACE2	Angiotensin-Converting Enzyme 2
AF	Annulus Fibrosus
AFMSC	Amniotic Fluid-derived Mesenchymal Stem Cells
Ag	Silver
AgNPs	Silver Nanoparticles
AIDS	Acquired Immune Deficiency Syndrome
ALI	Acute Lung Injury
AM	Additive Manufacturing
AMD	Age-related Macular Degeneration
AMs	Alveolar macrophages
ATP	Adenosine Triphosphate
Au-IL	Gold-Ionic Liquids
AuNPs	Gold Nanoparticles
BAL	Bronchoalveolar lavage
BBB	Blood-Brain Barrier
BCP	Biphasic Calcium Phosphate
BCRP	Breast Cancer Resistance Protein
bFGF	Basic fibroblast growth factor
BMPs	Bone Morphogenetic Proteins
BMSCs	Bone Marrow Stromal Cells
BTE	Bone Tissue Engineering
C60	Fullerene Nanoparticles
CAD	Computer Aided Design
CAGR	Compound annual growth rate
CAPA	Corrective and Preventive Action
CBN	Carbon black nanoparticles
Cd	Cadmium
CD	Carbon Dot
CD4	Cluster of Differentiation
CDIA	Carbon dot-based immunoassay
cDNA	Complementary DNA
CdS	Cadmium Sulfide
CDSCO	Central Drugs Standard Control Organization

CdSe	Cadmium Selenide
CdTe	Cadmium Telluride
CeO_2	Cerium oxide
CEST	Chemical Exchange Saturation Transfer
CF	Cystic Fibrosis
CFTR	Cystic Fibrosis Transmembrane Conductance Regulator
CHO	Aldehyde group
CLA	Central Licensing Authority
CNP	Carbon-Based Nanomaterials
CNT	Carbon Nanotubes
CNV	Choroidal Neovascularisation
Co	Cobalt Nanoparticles
COOH	Carboxylic Acid
COPD	Chronic Obstructive Pulmonary Disease
CP	Calcium Phosphate
CQD	Carbon Quantum Dots
Cr	Chromium
CRDDS	Control Release
CRET	Chemiluminescence Resonance Energy Transfer
CS-TGA-PF	Chitosan-thioglycolic acid-Pluronic F127
CT	Computed Tomography
CT Scans	Computed Tomography Scan
CuO	Copper Oxide Nanoparticles
CV	Cyclic Voltammetry
CVD	Chemical Vapour Deposition
DDS	Drug Delivery System
DHA	Docosahexaenoic acid
DLS	Dynamic Light Scattering
DMOG	Dimethylallyl Glycine
DNA	Deoxyribose Nucleic Acid
DOX	Doxorubicin
DPPC	Dipalmitoylphosphatidylcholine
DSC	Differential scanning calorimetry
DSPC	Distearoylphosphatidylcholine
ECL	Electrochemiluminescence
ECM	Extracellular Matrix
EDC	1-Ethyl-3-(3-dimethyl aminopropyl) carbodiimide
EGDMA	Ethylene glycol dimethyl acrylate
EIS	Electrochemical Impedance Spectroscopy
ELISA	Enzyme-Linked Immunosorbent Assay
EPA	Environmental Protection Agency
EPA	Eicosapentaenoic acid
EPR	Enhanced Permeability and Retention
ER	Endoplasmic reticulum
ERT	Enzyme Replacement Therapy
ESCRT	Endosomal Sorting Complexes Required for Transport Machinery

ESCs	Embryonic Stem Cells
ETC	Electron Transport Chain
EVs	Extracellular Vesicles
FCNs	Fluorescent carbon nanoparticles
FDA	Food and Drug Administration
FDM	Fused Deposition Modeling
Fe_2O_3	Ferric oxide
Fe_3O4	Iron Oxide
FET	Field-Effect Transistors
FFAs	Free fatty acids
FO	Fish oil
FRET	Fluorescence Resonance Energy Transfer
FRI	Fluorescence Imaging
FTIR	Fourier Transform Infrared Spectroscopy
GA	Geographic Atrophy
GAA	Acid Alpha-Glucosidase
GBN	Graphene-Based Nanoparticle
GBR	Guided Bone Regeneration
GCCNP	Glycol Chitosan-Coated Cerium Oxide Nanoparticles
GCE	Glassy Carbon Electrode
GC-MS	Gas chromatography & mass spectroscopy
GCP	Good Clinical Practice
Gd	Gadolinium
GDF 5	Growth Differentiation Factor 5
Gd-NGQD	Gadolinium Nano Graphene Quantum Dots
GelMA	Gelatin Methacrylate
GI	Gastrointestinal
GIT	Gastrointestinal tract
GMP	Good Manufacturing Practices
GNP	Graphene Nano-Platelets
GNPs	Gold Nanoparticles
GO	Graphene Oxide Nanoparticles
GQD	Graphene Quantum Dots
GUVs	Giant Unilamellar Vesicles
H_2O_2	Hydrogen Peroxide
H_2SO_4	Sulfuric acid
Hap	Hydroxyapatite
HAVIC	Human Aortic Valvular Interstitial Cells
HBV	Hepatitis B Virus
HCl	Hydrochloric Acid
HCV	Hepatitis C Virus
hDPSCs	Human Dental Pulp stem cells
HER-2	Human epidermal growth factor receptor-2
HIV	Human Immunodeficiency Virus
HIV 1	Human Immunodeficiency Virus type 1
HK	Histidine Kinase

HLA	Human leukocyte antigen
HNO_3	Nitric Acid
HPV	Human Papilloma Virus
HSV	Human Herpesvirus 6
HUVECs	Human Umbilical Vein Endothelial Cells
IARC	International Agency for Research on Cancer
IDEs	Investigational Device Exemptions
IL	Interleukin
ILVs	Intraluminal Vesicles
IOLs	Intraocular Lenses
IONP-CPC	Iron Oxide Nanoparticles -Calcium Phosphate Cement
IP	Intellectual Property
IPO	Indian Patent Office
ISCT	International Society for Cell and Gene Therapy
ISEV	International Society for Extracellular Vesicles
ISO	International Standard Organization
IVM	Ivermectin
$KMnO_4$	Potassium Permanganate
LbL	Layer-by-layer
LFIA	Lateral Flow Immuno Assay
LMs	Lipid Micelles
LNPs	Lipid nanoparticles
LSD	Lysosomal Storage Disease
LUVs	Large Unilamellar Vesicles
MAG	Monoacylglycerols
MC	Microbial nanocellulose
MDD	Major depressive disorder
MDR-TB	Multi-drug-resistant *Mycobacterium Tuberculosis*
MeGC	Methacrylate Glycol Chitosan
Me-Ge;	Methacrylated Gelatin
MeHA	Meshacrylated Hyaluronic Acid (Me-HA)
MEMS	Micro Electro Mechanical Systems
MHRA	Medicines and Healthcare products Regulatory Agency
miRNA	microRNAs
MISEV	Minimal Information for Studies of Extracellular Vesicles
MIT	Molecular Imprinting Technology
MLVs	Multilamellar Vesicles
MMP	Mitochondrial Membrane Potential
MMPs	Metalloproteinases
Mn	Manganese
MN test	Micronucleus test
Mn-NGQD	Manganese Nano Graphene Quantum Dots
MNP	Magnetic Nanoparticles
MNPs	Metal-based nanoparticles
Mo	Molybdenum
Mo/NaO	Monoolein and Sodium Oleate

MOFs	Metal-organic frameworks
MRI	Magnetic Resonance Imaging
mRNA	Messenger RNA
MRP2	Multidrug resistance-associated protein 2
MSCs	Mesenchymal stem cell
MVBs	Multivesicular Bodies
MVLs	Multivesicular Liposomes
MWCNTs	Multi-walled carbon nanotubes
NABs	Nucleic Acid Biosensors
$NaHCO_3$	Sodium bicarbonate
nAMD	Neovascular AMD
$NaNo_3$	Sodium nitrate
NCL	National Chemical Laboratory
NEI	National Eye Institute
NEI	National Eye Institute
NGB	Next Generation Bioprinting
NH_2	Ammonium
NH2-IL	Ammonium- Ionic Liquids
NHS	N-hydroxysuccinimide
NIOSH	National Institute for Occupational Safety and Health
NIR	Near-Infrared Spectroscopy
NIRF	Near-Infrared Fluorescence
NLCs	Nanostructured Lipid Carriers
nm/	Nanometers
NMPA	National Medical Products Administration
NMs	Nanomaterials
NNI	National Nanotechnology Initiative
$NO^{\bullet}$	Nitric Oxide
NOX:NADPH	(Nicotinamide adenine dinucleotide phosphate hydrogen) oxidase
NP/ NPs	Nanoparticle/ Nanoparticles
NS	Nanosphere
nSMase2	Neutral Sphingomyelinase 2
NSs	Nanoscaffolds
O^{*}_2	Singlet Oxygen
O_2	Oxygen Molecule
O^{-}_2	Superoxide Radicals
OA	Osteoarthritis
OCP	Octacalcium Phosphate
OCT	Optical Coherence Tomography
OH	Hydroxide
$OH^{\bullet}$	Hydroxyl Radicals
OLVs	Oligolamellar Vesicles
OMVs	Outer Membrane Vesicles
PADs	Peptidyl Arginine Deiminases
PAI	Photo-Acoustic Imaging
PC	Phosphatidylcholine

PCL	polycaprolactone
PCT	Patent Cooperation Treaty
PDDC	R)-(1-(3-(3,4-methoxyphenyl)-2,6-dimethylimidazo[1,2-b] pyridazin-8-yl) pyrrolidin-3-yl)-carbamate
pDNA	Plasmid DNA
PDO	Polydioxanone
PE	Phosphatidylethanolamine
PE	Polyethylene
PEG	Polyethylene glycol
PEG-GO	Percutaneous Endoscopic Gastrostomy
PEG-PLA	Polyethylene Glycol-Block-Polylactic Acid
PEGylated	Polyethylene glycolated
PET	Positron Emission Tomography
PGA	Poly(glycolic acid)
P-gp	P-glycoprotein
PHEMA	Poly-Hydroxyethyl Methacrylate
PL	Photoluminescence
PLA	Poly-Lactic Acid
PLGA	Poly Lactic co-Glycolic Acid
PLGA-PLH-PEG	Poly(D,L-Lactic-Co-Glycolic Acid)-B-Poly(L-Histidine)-B-Poly(Ethylene Glycol)
PMDA	Pharmaceuticals and Medical Devices Agency
PMMA	Polymethyl Methacrylate
PNIPAM	Poly(N-isopropylacrylamide)
POCT	Point-Of-Care Testing Immunochromatographic assays
PP	Polypropylene
PPF	Polypropylene Fumarate
PQS	Pseudomonas Quinolone Signal
PS	Polystyrene Nanoparticles
PTCA	10-perylene tetracarboxylic acid
PTFE	Polytetrafluoroethylene
PtNPs	Platinum-based Nanoparticles
PTT	Partial Thromboplastin Time
PUFAs	Polyunsaturated fatty acids
PVA	Polyvinyl Alcohol
PVC	Polyvinyl Chloride
PVD	Physical Vapour Deposition
PVP	Polyvinylpyrrolidone
QBs	Quantum dot nanobeads
QCT	Quercetin
QD	Quantum Dots
QMS	Quality Management System
QS	Quorum Sensing
QY	Quantum Yields
RA	Rheumatoid arthritis
RBCs	Red Blood Cells

RES	Reticuloendothelial System
RGD	Arginine-Glycine-Aspartic Acid
rGO	reduced Graphene Oxide
RM	Regenerative medicine
RNA	Ribonucleic acid
RNS	Reactive Nitrogen Species
ROS	Reactive Oxygen Species
RP	Rapid prototyping
RPE	Retinal Pigment Epithelium
RR	Response Regulator
RSI	Raman Spectroscopy Imaging
SARS-CoV-2	Severe acute Respiratory Syndrome Coronavirus-2
SCE	Sister Chromatid Exchange
SDS	Sodium dodecyl sulfate
SDS/DDAB	Sodium Dodecyl Sulfate and Didodecyldimethylammonium Bromide
SEDDS	Self-Emulsifying Drug Delivery Systems
SEM	Scanning Electron Microscopy
SERS	Surface-enhanced Raman Scattering
SFF	Solid-Freeform Fabrication
SH_2	Hydrogen Sulphide
SiO_2	Silicon Dioxide
SiO_2	Silicon Oxide
siRNA	Small Interfering Ribonucleic acid
SLA	State Licensing Authority
SLA	Stereolithography
SLNs	Solid Lipid Nanoparticles
SLS	Selective Laser Sintering
SNP	Silver Nanoparticles
SOD	Superoxide Dismutases
SPIONs	Superparamagnetic Iron Oxide Nanoparticles
SPR	Surface-Plasma Resonance
SR	Silicone rubber
SSE	Semisolid Extrusion
SUVs	Small Unilamellar Vesicles
SWCNT	Single Long-Wrapped Graphene Sheet
SWCNTs	Single walled carbon nanotubes
TA	Tannic acid
TCS	Two-Component System
TE	Tissue Engineering
TEM	Transmission electron microscope
Ti	Titanium
TiO_2	Titanium dioxide
TLC	Thin Layer Chromatography
TNFα	Tumour Necrosis Factor alpha
TRAP	Tartrate Resistant Acid Phosphatase

ULVs	Unilamellar Vesicles
US	Ultrasound
USD	United States dollar
USPTO	United States Patent and Trademark Office
UV	Ultraviolet
VEGF	Vascular Endothelial Growth Factor
WC	Watson-Crick
WHO	World Health Organization
XRD	X-ray diffraction
ZnO	Zinc oxide
β-TCP	β tri-calcium phosphate
μ-CDIA	microfluidic carbon dot immunoassay

Section 1

Chapter 1.1

Exploring Nanomaterials: Types and Biomedical Applications

Gurmeet Singh[1]
Deepak Khatak[2]
and Annesha Roy[2,*]
[1]School of Physical Sciences, Jawaharlal Nehru University, New Delhi, India
[2]School of Environmental Sciences, Jawaharlal Nehru University, New Delhi, India

Abstract

Nanomaterials (NMs) are crucial in technological advancements due to their superior performance compared to bulk materials and their tunable physical, chemical, and biological properties. They are categorized based on size, shape, origin, composition, and capping agents. In biomedicine, NMs are significant for their enhanced reactivity, large surface area, and ability to interact with biological molecules at cellular and molecular levels. Notable examples include carbon nanotubes, liposomal, polymeric, metal nanoparticles (NPs) (like silver and gold), and quantum dots. These materials have special advantages, such as enhanced fluorescence, targeted drug release, or increased imaging contrast. NMs can be synthesized via top-down or bottom-up methods using physical, chemical, or biological routes, each providing control over size, shape, and functionalization. NMs enhance biosensing, tissue engineering, imaging, and drug delivery in biomedicine. They ensure targeted and controlled drug release, improve MRI contrast for precise diagnostics, support tissue regeneration, and create highly sensitive and specific biosensors for detecting infections and biomarkers. The diverse properties of NMs promise better patient outcomes and advanced treatments. This chapter provides an overview of the various types of NMs, size-dependent characteristics, and the applications of nanostructures in biomedicine, highlighting the current research status and the main applications of biological nanostructures.

Keywords: nanomaterials, nanotechnology, biosensors, biomedicine, biosensing, drug delivery

---

* Corresponding Author Email: annesharoy28@gmail.com

In: Emerging Nanomaterials in Biomedical Science
Editors: Eepsita Priyadarshini and Ajith Manayil Parambil
ISBN: 979-8-89530-600-0

1. Introduction

Nearly all synthetic substances, including the ones already in use, can gain new properties and capabilities because of nanotechnology, which is considered an enabling technology (Singh et al. 2021). Nanomaterials (NMs) with various dimensions and specific ratios reveal many features through their distinct chemistry and physics. Moreover, one of the primary causes of increased spontaneous effects is the alteration in the surface-to-volume ratio. As the structure gets smaller, the surface-to-volume ratio increases, improving the conformation's chemistry and physics compared to bulk matter (Riley et al. 2021). This technology opens a world of creative possibilities, from state-of-the-art biological applications in lab settings to strong functional and structural applications across several sectors (Ramos et al. 2017).

Novel prospects in biological research and therapeutic applications have been made possible by recent developments in the creation of NMs. The analytical and bioanalytical sciences have seen a significant transformation in the last several decades because of nanostructured materials. The application of this discipline to the widest range of commercial domains, such as biomedicine, food, bioengineering, integrated catalysis, nano-catalysis, fuel, energy storage, and energy conversion, has been considerably facilitated by NMs (Ramos et al. 2017, S C Baetke et al. 2015). Compared to traditional bioanalytical techniques, nano-assays with the help of various NMs provide greater sensitivity, selectivity, and throughput. These NMs can identify biochemical alterations in living cells at the single-molecule level (Qiu et al. 2020). Because of their ultrasmall size, high surface area to volume ratio, adjustable optical emission, improved mechanical capabilities, and superparamagnetic activity, NMs have gained popularity in biomedical applications. They must also be biocompatible and easily manufactured (Bayda et al. 2019). In general, there is great interest in three types of NMs: one-dimensional materials, such as nanowires and rods; two-dimensional materials, such as films, nanosheets, nanobelts, and nanodisks, and zero-dimensional materials, also known as quantum dots, which vary in diameter and shape (Elsabahy et al. 2012). Diagnostic applications include imaging and the use of biosensors in the field of biomedicine. Dental adhesives and DNA vaccine-containing NMs have been widely used in illness prevention. The pharmacokinetics associated with the creation of nanoparticle surfaces contribute to the development of immune-safe biomedicine (Elsabahy et al. 2012, Singh et al. 2019). Nanoscale devices made of NMs are easily able to interact both extracellularly and intracellularly with biomolecules because of their miniscule size. They can monitor cells inside the living system, diagnose illness, and provide medicine in previously unattainable methods since they can efficiently access most body parts (Dey and Goswami 2011).

In this chapter, we have primarily focused on NMs, which are widely used in various biological applications, with particular attention to their uses in medicine, diagnostics, drug/gene delivery, and tissue engineering.

2. Types of Nanomaterials

NMs may be categorized into three primary classes depending on their chemical composition: carbon-based NMs; inorganic NMs (metal oxide, quantum dots); and organic NMs (liposomal and polymeric NMs). The chemical characteristics of their bulk materials are often retained by

NMs, which is advantageous when selecting a particular NM for a biological application. The following NMs are employed in nanomedicine:

2.1. Metal-Based Nanomaterials

One-dimensional nanosized (with size 1 to100 nm) metals are called metallic NMs. Metallic NMs are broadly classified into four groups: metallic NPs (0D), nanotubes, nanorods, metallic nanowires (1D), nanolayers, nano-sheets, nano-plates (2D), and nanoshells with some other nanostructures (3D) (Martínez-Alcalá et al. 2020). Metal precursors are used to create metal NPs. The unique optoelectrical characteristics of these NPs are attributed to constrained surface plasmon resonance (De Zárate et al. 2020). Noble metal and alkali NPs that exhibit a measurable absorption band in the solar electromagnetic spectrum include Au, Cu, and Ag (Karim et al. 2020).

In biological and biomedical applications, metallic NPs such as gold (Au), copper (Cu), silver (Ag), iron oxide (Fe_2O_3), titanium oxide (TiO_2), zinc oxide (ZnO), platinum (Pt), gadolinium (Gd), palladium (Pd) and selenium (Se) are most often utilized (Martínez-Alcalá et al. 2020, Karim et al. 2020). For example, relatively small concentrations of metals like Cu and Ag may be extremely toxic to bacteria (Favier et al. 2018). Metals have a broad range of uses as antimicrobial agents in industry, healthcare, and agriculture in general because of their biocidal effect (Fouda et al. 2019). For the electrochemical detection of biomolecules, several metal oxide NPs, including ZnO, NiO, TiO_2, MnO_2, Fe_2O_3, and Co_3O_4 have been investigated. In addition, hybrids of metal oxides have garnered sufficient interest in this field. CuO-NPs are valuable in a variety of applications, such as superstrong materials, sensors, antibacterial agents, and catalysts, due to their distinct properties (Immanuel et al. 2019).

2.2. Polymeric Nanomaterials

Natural or synthetic polymers are used to form polymeric NMs, which have special properties for drug delivery systems. These characteristics include a higher level of biological compatibility that reduces cytotoxicity, increased and persistent efficacy, limited discharge rates, strong stability, scaling up reproducibility, fewer treatment frequencies, lower toxicity, and the ability to co-administer with other drugs to produce synergistic effects (Vasile et al. 2019). The main polymers utilized to produce drug delivery-specific NMs include poly (methyl methacrylate) and poly lactic-coglycolic acid, poly(alkyl) cyanoacrylates, poly PLA-PEG, and polycaprolactone. Occasionally, PEG-conjugated polymeric NPs are synthesized to provide higher blood or tissue circulation frequencies (Banik et al. 2015).

Dendrimers are the most widely used kind of polymeric NMs. Dendrimers are polymer-based NMs with tremendous promise that have been used to carry drugs and genes; nevertheless, dendrimers containing cations can occasionally have genotoxic consequences (Gardikis et al. 2019). Properties like improved biological compatibility, easy preparation, increased solubility, and increased blood half-life ability allow effective drug and gene delivery via entrapment/encapsulation (Sánchez et al. 2019). Interestingly, it has been shown that conjugating dendrimers which include lipids or PEG molecules might help in reducing dendrimer-associated charges and unwanted cellular reactions, leading to improved characteristics, both physical and biological and perhaps exert positive therapeutic benefits (Otsuka et al. 2003).

2.3. Liposomal Nanomaterials

Liposomes are composed of lipid vesicles with a watery center that are sandwiched by artificial or natural phospholipid bilayers. Hydrophilic medications can be encapsulated in the aqueous solution, whereas hydrophobic and amphiphilic pharmaceuticals can dissolve in the phospholipid bilayers (Bae et al. 2011).

Liposomes are extensively used in the management of cancer, particularly in the administration of anticancer medications and chemotherapies. Liposomes are coupled with poly (D, L-lactic acid)-polyethylene glycol (PLA-PEG) for effective distribution (Mallick et al. 2014). However, the broad biological dispersion and unintended consequences of non-targeted liposomal delivery systems, which mostly affect non-targeted tissues, are a significant disadvantage. The tailored liposomal delivery mechanism prevents the pharmaceuticals from quickly leaving the bloodstream, increases the potency of the encapsulated medications, and provides a consistent release of medications. As a result, they can enhance the medication's efficacy and exert beneficial effects on target tissues. Topical and parenteral administration methods are also viable for liposomal medication delivery. Moreover, they have properties in common with the real cell membrane, which has been used in many pharmacokinetic and biomaterial-cell interaction studies (Zare-Zardini et al. 2015).

2.4. Carbon-Based Nanomaterials

Carbon nanotubes (CNT) that have one or more walls of micrometer lengths and nanometer diameters may be synthesized commercially. The broad range of biomedical uses of carbon nanotubes, including imaging and treatment, are anticipated due to their ease of internalization by live cells (Negri et al. 2020). However, most solvents do not dissolve carbon nanotubes. To make the nanotubes soluble in buffers suitable for biological applications, noncovalent coating of amphiphilic molecules or functionalization of different chemical groups on the nanotube surface are performed (Simon et al. 2019). Carbon nanotubes have a special electromagnetic characteristic that makes them extremely sensitive in a variety of imaging modalities, including near-infrared rays and photoacoustic molecular imaging (Song et al. 2018).

Carbon-based nanostructures come in a variety of shapes and sizes, such as carbon nanotubes, nanospheres, and nanodiamonds. They are entirely composed of carbon (Cha et al. 2013). Carbon nanotubes and other carbon-based NMs are generated from graphene, a supramolecular two-dimensional planar structure. Through various chemical or physical conjugation techniques, small organic compounds, peptides, proteins, and DNA have been loaded into carbon nanotube drug delivery devices (Bianco et al. 2005). CNTs is especially used in biomedicine for biosensing, diagnosis, and treatment. The exceptional capacity of CNTs to function as transporters of molecules of various origins, such as proteins, DNA/RNA, enzymes, and medications has made this possible. Apart from their established ability to transport, it has also been shown that CNTs can efficiently penetrate cells and the blood-tissue barrier (Chou SG et al. 2003, Saliev et al. 2019).

2.5. Quantum Dots (QDs)

QDs are semiconductor-containing nanocrystals that exhibit excellent optical as well as electrical properties that are not present in their bulk counterparts (Alavi et al. 2020). QDs are especially well-suited for use as optical probes because they exhibit longer fluorescence lifetimes than conventional organic fluorescent dyes, higher molar extinction coefficients, less photobleaching, and high quantum yields (QYs). For this reason, QDs are thought to be a useful alternative to organic dyes and optical markers in a variety of bioassay applications (Kadian et al. 2021). The primary technique for synthesizing quantum dots (QDs) is surface passivation of an inner layer, or the core, by depositing a capping layer, or shell, made of an inorganic semiconductor. The bandgap of the core layer is less than that of the shell. QD synthesis normally takes place at high temperature, in a coordinating solvent, using an organometallic shell precursor (such as Zn or Cd) and a core material from the 16th group (commonly S or Se) of periodic table (Kadian et al. 2021, Bailey et al. 2004).

The direct interactions between QDs and bacterial cells further boost the overall antibacterial activity. The generation of reactive oxygen species (ROS), disruption of the cell structure, as well as cytoplasm leakage resulting in DNA binding and changed gene expression are some of the ways that QDs work against bacteria (Gao et al. 2004). The capacity of QDs to pass the blood-brain barrier has allowed them to demonstrate their promise in the biomedical field, particularly in the areas of bioimaging, neuroimaging, theranostics, and medication delivery, particularly to the brain. Because of their excellent optical characteristics and strong biocompatibility, GQDs are now being used as bioimaging markers both *in vitro* and *in vivo*. *In vitro* research demonstrated that QDs can quickly and efficiently label a variety of cell lines, including neural stem cells and neurons (Perini et al. 2020).

3. Routes of Synthesis

NMs can be synthesized in a variety of ways. Through a comprehensive study of the benefits and drawbacks associated with each synthesis technique, scientists can choose the most suitable methodology for generating superior NMs that possess customized characteristics for certain uses. One such approach is shown in Figure 1. In general, there are three different routes for synthesizing NMs which are discussed below (Patil et al. 2021).

3.1. Physical Methods

These methods pertain to the fabrication and creation of NMs by physical procedures and techniques such as mechanical, thermal, and electromagnetic radiation. In these methods, the synthesis starts from bulk materials and under systematic conditions, the NMs get formed. We refer to these as top-down approaches (Satyanarayana et al. 2018). The following methods are part of the physical method.

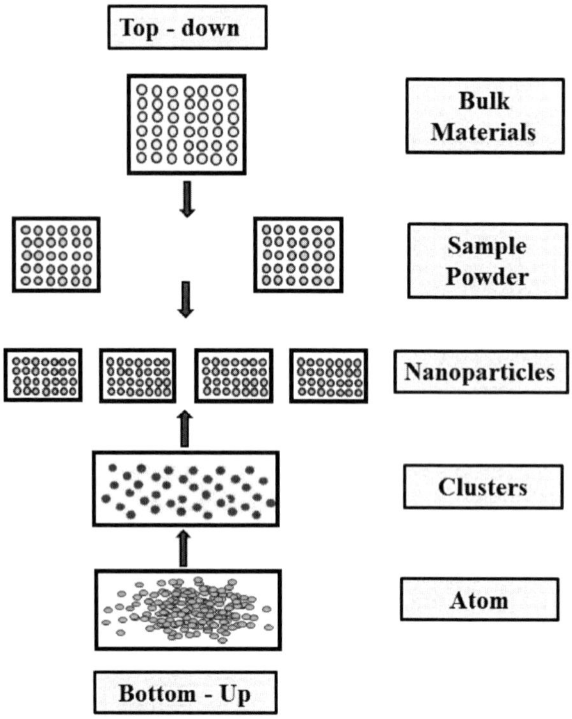

Figure 1. Synthesizing NMs using the top-down and bottom-up approaches.

3.1.1. Physical Vapor Deposition (PVD)

In this method, a solid substance is evaporated using physical processes including heat, vacuum, and momentum and then it is deposited on the thin-film deposition. After that, the vaporized material condenses onto a substrate to create a coating or thin layer. Numerous materials, such as metals, alloys, semiconductors, insulators, and NMs, can be deposited on thin films using this method (Rossnagel et al. 2003).

PVD is used in several areas of biomedicine and catalysis, including implant coating, tissue engineering, medical devices, and biosensor creation etc. (Stuart et al. 2021). PVD coatings have medical advantages such as biocompatibility or wear-resistant coatings for medical equipment. Although there are other coating techniques, PVD offers the benefit of being a practical procedure at lower temperatures in a comparatively inert atmosphere (Lüdecke et al. 2020, Arweiler-Harbeck et al. 2001).

3.1.2. Ion Sputtering

In this method, high-energy ions bombard a target material, ejecting atoms or clusters and forming a vapor that condenses into thin films or NPs. The size and distribution of NPs can be precisely controlled using ion sputtering. Vacuum sputtering lowers the possibility of contamination. A vast array of materials, including metals, alloys, and compounds, can be produced via ion sputtering. Because ion sputtering can happen at comparatively low temperatures, substrate damage is minimized (Kafle et al. 2020). Ion sputtering can produce

NMs for catalysis, optics, and biomedicine applications. Using this method, nanostructured surfaces can be produced for use in biosensors and tissue engineering. Nanoalloys can be synthesized using ion sputtering for use in energy storage, biomedicine, and catalysis. J.A. Lenis et al. have discussed the multilayer Hydroxyapatite - Ag/TiN-Ti coatings for biomedical applications synthesized by sputtering (Lenis et al. 2019).

3.1.3. Laser Ablation

Laser ablation is a process in which the target sample is placed inside a vacuum chamber. The sample is exposed to a high-pulsed laser beam, which causes plasma to develop. Next, the plasma is transformed into a colloidal nanoparticle solution. The second-harmonic group-type laser is commonly used for nanoparticle creation. The type of laser, pulses, solvent type, and pulsing time are factors that influence the final product (Patil et al. 2021). This method has several uses in a variety of domains, such as thin film deposition, surface modification, spectroscopy, microscopy, and materials research.

Laser ablation is used in microscopy to provide high-resolution pictures of biological and material samples. The production of NMs, thin layer deposition, surface patterning and modification, micro-structuring and machining, spectroscopic analysis, and biomedical applications like tissue engineering and drug delivery are a few typical uses for laser ablation (Kim et al. 2017). Andrei I. Pastukhov has synthesized boron NPs by using this technique and has used them for biomedical applications (Pastukhov et al. 2022).

3.2. Chemical Methods

NMs are synthesized chemically using solvents and chemical precursors. Large volumes can be produced with high regularity using these techniques. Below is a discussion of some chemical techniques:

3.2.1. Sol-Gel Method

Various NMs such as metal oxides, ceramic NPs, and solid materials can be synthesized using the wet-chemical Sol-gel method (bottom-up approach). There are four essential processes in the sol-gel route. First, precursors, usually metal alkoxides, are combined to generate a colloidal solution or sol. Subsequently, the sol undergoes hydrolysis and polycondensation processes to form a gel-like diphasic system. After that, the solvent is eliminated by drying, which causes densification and shrinking. To improve the material's mechanical qualities, structural stability, and crystallinity, it is ultimately subjected to a thermal treatment. Finally, this process results in samples with high purity, crystallinity, and uniform-sized particles (Singh et al. 2023). The synthesis of materials with specific qualities may be done effectively using the Sol-Gel process. It has several benefits, including low-temperature requirements, flexibility, simplicity, high purity, homogeneity, and control over particle size and shaping. This makes it possible to precisely manage the qualities of the finished product and produce materials with the necessary features (Navas et al. 2021). Gisela M Luz et al. have prepared bioactive glass NPs by using the Sol-gel method and have shown their biomedical application (Luz et al. 2011).

3.2.2. Precipitation Method

Precipitation synthesis is a popular and flexible method for creating NPs. In this procedure, the precursor materials are dissolved in de-ionized water to create a solution, and then a precipitating agent is added slowly with continuous stirring which promotes the creation of white suspension. After centrifuging the white product for a few minutes at high rpm (~5000 rpm), the product is rinsed three times with distilled water and then once more with pure alcohol. The final product is calcined for three hours at a high temperature (500°C) in an air environment (Ghorbani et al. 2015). A few benefits of the precipitation process are its ease of use and affordability, its high yield and purity, and its controllable particle size and form. It is also quite versatile since it allows for versatility in the selection of precursor and precipitating agents. This method is a desirable choice for a variety of applications since it is scalable for large-scale manufacturing. Julia Mürbe et al. (Mürbe et al. 2008) have discussed manganite NPs for biomedical applications.

3.2.3. Chemical Reduction Method

This method requires little equipment, is easy to operate, and is the most adaptable method for producing NMs. Most chemical reduction techniques follow the two-phase Brust–Schiffrin process, which involves chemical reduction at the oil-water interface. The thiolated molecules are absorbed and stabilized in the organic phase in the subsequent step (Li et al. 2011). In this method, the aqueous solution containing the metal ions is employed as the precursor. The next step is to add chemical-reducing agents to reduce metal ions. The most used reducing agents are sodium borohydride, sodium citrate, sodium ascorbate, hydrazine hydrate, and volatile organic solvents (e.g., Toluene) (Vaseghi et al. 2020). Agglomeration and cluster formation are the results of this process, which is followed by the creation of NPs. The size of the generated NPs may be effectively controlled by using such chemical reduction agents (Pandey et al. 2015).

3.3. Biological Methods

3.3.1. Green Synthesis Using Plants and microorganisms

The biosynthetic route provides a safe, environmental friendly, and biocompatible method for creating NPs for use in biomedical applications (Razavi et al. 2015). Green synthesis focuses on decreasing the use of toxic chemicals (Jadoun et al. 2020). The presence of phytochemicals in leaves, fruits, roots, stems, and seeds has led to the utilization of these parts of plants in the synthesis of different NPs as phytochemicals function as stabilizing and reducing agents (Narayanan et al. 2011). Microorganisms such as bacteria, fungi, actinomycetes, etc. are also considered as potential reducing agents for the synthesis of NMs. Metal NPsmay be easily created *in vivo* by growing certain microorganisms with metallo-regulatory molecules—specifically, proteins and peptides involved in the process of detoxifying metals—under specified circumstances and by feeding them metal cations in the culture medium (Bansal et al. 2012). A few well-documented examples of metal NP formation are zinc sulfide NPs in bacteria, gold NPs in Precambrian algal blooms, algal cells, CdS NPs in bacteria and yeast, and magnetite NPs in bacteria. Most of these NP formations result from natural regulatory processes (Park et al. 2015, Bhattacharya et al. 2005). There have been reports of the effective and speedy extracellular production of Ag, Cu, and Au NPs utilizing broth extracts from various plants, including *Aloe vera, Pelargonium graveolens*, and *Tamarindus indica* (Marchiol 2012).

3.3.2. Enzymatic Synthesis

Developing novel techniques for synthesizing heterogeneous solid supports has become necessary as enzymatic catalysis has changed throughout (Arsalan et al. 2018). To enhance various enzymatic features, such as enhanced activity, thermal stability, and pH tolerance, NMs are quite useful as they allow immobilization of enzymes. (Hulkoti et al. 2014). Extracellular enzymes such as cellobiohydrolase D, glucosidase, acetyl xylan esterase and β-glucosidase are produced by several fungi and are known to be important in the production of NMs. Nitrate reductase, released by fungi, aids in the bioreduction and synthesis of NMs and is widely used in the extracellular production of silver NPs (Ovais et al. 2018).

4. Applications of Nanomaterials

Manufactured nanostructures, nano-based devices and nanomedicine can be a means to fully detect, repair, regulate, monitor, build, defend, and enhance human physiological systems. The use of NMs in biomedicine as shown in Figure 2, includes medication delivery, diagnostics, treatment, photoablation, implants and prostheses, vaccinations, gene therapy, wound dressings, and nanorobotics.

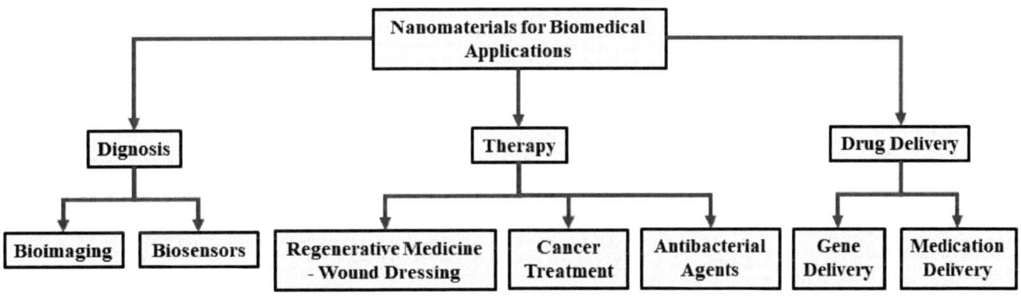

Figure 2. Different applications of nanomaterials in biomedical sciences.

4.1. Drug Delivery Systems

Most studies conducted about nanomedicine focus on using nanomaterials in drug delivery systems (DDS). Targeting and delivering drugs precisely, preserving biocompatibility during application and treatment procedures, and accelerating manufacturing are the objectives of using NPs in drug delivery (De Jong 2008). Drugs can attach themselves to NPs through adsorption or during the' manufacturing process. DDS can enhance many of the pharmacological qualities of traditional (or "free") medications. DDS are made to act as drug reservoirs or to change the pharmacokinetics and biodistribution of the medications they are linked with. For example, mesoporous silica NMs have developed into a multipurpose drug delivery platform that can deliver chemotherapeutic drugs to several kinds of cancer cells both *in vitro* and *in vivo*. Because of their enormous internal surface areas, pore volumes, and variable pore sizes, mesoporous silica NMs-based drug carriers have demonstrated the ability to integrate a high dose of pharmaceuticals into the internal pore system (Lee et al. 2009). Hydrophobic and charged anticancer medications can be packaged effectively and protectively using this carrier, enabling regulated and on-demand delivery. For example, paclitaxel can be administered using TiO_2 NPs functionalized with folic acid and polyethylene glycol (Venkatasubbu et al. 2013).

4.2. Biosensors

In the last decade, nano biosensors have emerged as a key instrument in the field of diagnostics. Nanobiosensors have practical medical benefits, including detecting bacteria or viruses that may cause fatal illnesses. In the field of diagnostics, NMs are essentially used for the early identification of the progression of disorders like cancer. Because they have a higher surface area than volume, NMs are mostly utilized in bioimaging as contrasting agents because their surfaces are engineered with imaging agents (Baptista et al. 2007). Numerous types of NMs, including metal, oxides, semiconductors, and carbon NPs (CNPs), have been employed in developing biosensors. These NPs have many functions in various sensing systems (Spadavecchia et al. 2013). Due to their distinct optical characteristics and simplicity of usage with various biomarkers in aqueous solution, gold NPs have dominated NM-based biosensing applications that have been developed so far (Boisselier et al. 2009). Due to their catalytic qualities, silver NPs (AgNPs) are receiving a great deal of attention in the field of developing innovative biosensors. Liu and Hu have reported a novel H_2O_2 biosensor that uses hybrid sol-gel methods to electrochemically detect myoglobin (Mb) mounted on AgNP-doped carbon nanotube (CNT) film (Liu et al. 2009).

Numerous investigations are being conducted to use carbon nanotubes (CNTs) in electrochemical biosensing due to their high electrocatalytic activity. In addition to being smaller than microelectrodes, nanoscale sensors composed of carbon nanotubes (CNTs) have superior signal-to-noise ratios. Palanisamy et al. developed a glucose biosensor using multi-walled CNTs and zinc oxide composites, which has high repeatability and reproducibility for detecting glucose in urine samples (Palanisamy et al. 2012). Magnetic NPs are also contributing to the development of more sensitive, quick, and affordable biosensors since most biological samples have very low magnetic susceptibility, which lowers the background against which measurements are performed. With exceptional sensitivity, magnetic-based biosensors identify medicines, infections, tumor cells, enzymes, and other proteins (Haun et al. 2011).

4.3. Therapeutic Applications

NMs have the ability to delivery drugs to specific targets and translocate through the bloodstream therefore, engineered NPs are currently the focus of biomedical research (Anderson et al. 2019). NMs are also being developed for the treatment of various diseases. They are appealing even from a clinical standpoint because of this. A tailored cancer vaccine comprising autologous resected tumor cell membranes and *E. Coli* membranes has been produced by Chen et al. (2021) where the vaccine is administered as NPs. Compared to the limited immunogenicity of simply autologous tumor antigens, it has been observed that these hybrid NPs elicit a higher immunity against cancer, with the *E. coli* membrane acting as an adjuvant to increase the immune response (Chen et al. 2021).

4.4. Magnetic Hyperthermia

Hyperthermia refers to the method of using heat to kill the diseased body part. This method is typically used in conjunction with radiation therapy to eradicate cancerous cells while sparing healthy cells. While there are three types of hyperthermia—regional, entire, and local—local hyperthermia is the most chosen since it is considered safer. Magnetic NPs that have been customized by encapsulating them with antibodies targeting the target cells and infusing them into the arteries providing blood to the tumor are utilized in local hyperthermia treatment. Once at their intended location, the NPs impose an alternating magnetic field. By relaxing the magnetic moments of the magnetic NPs, the applied magnetic field helps stabilize them. This equilibrium condition results in the release of heat energy, which is lethal to the target cells (Mao et al. 2016). A few NMs employed for magnetic hyperthermia include Fe-Pt, Fe_2O_3, cobalt, and manganese-doped ferrite NPs. In computerized tomography, contrasting agents such as gold, silver, and bismuth are used because they have longer blood circulation durations and greater X-ray absorption than other commonly employed contrasting agents. Another phenomenon that provides higher contrast when studying soft tissues is magnetic resonance imaging (MRI). In the past, this approach employed gadolinium-based contrast agents, but NP-based contrast agents can provide superior contrast, a longer retention period, and fewer side effects. Iron oxide NPs are typically utilized for this (Naik et al. 2023).

4.5. Photoablation

Photoablation treatment is another application for NMs. There are two kinds of photoablation treatment: photothermal therapy and photodynamic therapy. In photodynamic treatment, light-sensitive materials are utilized that turn harmful when exposed to specific wavelengths of light. TiO_2 NPs are typically employed because of their photo-catalytic activity and biocompatibility. They produce photoactivated holes and electrons when they encounter a certain light wavelength. The electrons go into the conduction band, but the holes remain in the valence band. The outside of the NP accumulates both holes and electrons. Superoxide radicals are created when oxygen molecules combine with electrons, and hydroxyl radicals are created when water molecules react with holes. Ultimately, the targeted sick cells die because of these ROS. However, photothermal therapy is more akin to hyperthermia; it involves using infrared light sources to irradiate NMs, which generate heat and ultimately cause diseased cells to die by apoptosis. The NMs in photothermal therapy comprise titanium oxide and gold (McNamara et al. 2016).

4.6. Medical Robotic

Numerous medical uses, including medication administration, surgery, and identifying dangerous cells, are possible using nanorobots. Additionally, they can penetrate cells and fix damaged DNA, aid in the healing of tissues, organs, and blood clots, dissolve kidney and gallstones, and measure blood glucose levels in humans (Felfoul et al. 2016). Nanotechnology will significantly impact biopsies and operations as these procedures may be performed with greater precision and accuracy thanks to nanorobots. With the use of nanorobotics, even the sick body sections that are challenging for traditional technologies to reach may be readily reached. Nanotechnology has been successfully used to break up blood clots in mouse brains and collect histological specimens from the pig bile duct. Various devices, such as nano knives that manipulate individual axons with the help of external electric fields, are available. Soon, this technology may prove to be very beneficial for neurosurgical applications (Soto et al. 2018).

Conclusion

A vast range of extraordinarily potent items may be created with nanotechnology. NPs are undoubtedly widely used in various industries, including electronics, textiles industries, and in medical sciences. Because there is a possibility that nanotechnology will have some adverse effects on human life, it is difficult to apply it directly in biological sciences. A new area called "nanotoxicology" is emerging, alerting us to the possible risks associated with NPs. Therefore, preclinical research should be done to assess the biosafety of nanotechnology before directly utilizing it in medical applications.

The method used to synthesize NPs is essential for determining their characteristics and applications across several fields. Furthermore, due to its affordability, ease of use, and environmental friendliness, biogenic and bioinspired synthesis has gained a lot of interest recently. Using natural materials (such as plant extracts and microorganisms) in the production process reduces excessive chemical consumption, and new applications demanding biocompatibile NPs are made possible.

Acknowledgments

G. Singh and D. Khatak sincerely thank the University Grant Commission (UGC), India, for awarding the UGC JRF fellowship.

References

Alavi, Mehran, Erfan Jabari, and Esmaiel Jabbari. 2020 "Functionalized carbon-based nanomaterials and quantum dots with antibacterial activity: a review." *Expert Review of Anti-Infective Therapy* 19, no. 1: 35–44. https://doi.org/10.1080/14787210.2020.1810569.

Anderson, Simon D., and Vanessa V. Gwenin. 2019 "Magnetic Functionalized Nanoparticles for Biomedical, Drug Delivery and Imaging Applications." *Nanoscale Research Letters* 14, no. 1. https://doi.org/10.1186/s11671-019-3019-6.

Arsalan, Abdullah, and Hina Younus. 2018 "Enzymes and nanoparticles: Modulation of enzymatic activity via nanoparticles." *International Journal of Biological Macromolecules* 118: 1833–47. https://doi.org/10.1016/j.ijbiomac.2018.07.030.

Arweiler-Harbeck, D., A. Sanders, M. Held, M. Jerman, H. Ehrich, and K. Jahnke. 2001 "Does Metal Coating Improve the Durability of Silicone Voice Prostheses?" *Acta Oto-Laryngologica* 121, no. 5: 643–46. https://doi.org/10.1080/00016480121012.

Bae, Ki Hyun, Hyun Jung Chung, and Tae Gwan Park. 2011 "Nanomaterials for Cancer Therapy and Imaging." *Molecules and Cells/Molecules and Cells* 31, no. 4: 295–302. https://doi.org/10.1007/s10059-011-0051-5.

Baetke, S. C., T. Lammers, and F. Kiessling, 2015 "Applications of nanoparticles for diagnosis and therapy of cancer," *˜the œBritish Journal of Radiology/British Journal of Radiology* 88, no. 1054: 20150207, https://doi.org/10.1259/bjr.20150207.

Bailey, Robert E., Andrew M. Smith, and Shuming Nie. 2004 "Quantum dots in biology and medicine." *Physica. E, Low-Dimensional Systems and Nanostructures* 25, no. 1: 1–12. https://doi.org/10.1016/j.physe.2004.07.013.

Banik, Brittany L., Pouria Fattahi, and Justin L. Brown. 2015 "Polymeric nanoparticles: the future of nanomedicine." *Wiley Interdisciplinary Reviews. Nanomedicine and Nanobiotechnology* 8, no. 2: 271–99. https://doi.org/10.1002/wnan.1364.

Bansal, Vipul, Atul Bharde, Rajesh Ramanathan, and Suresh K. Bhargava. 2012 "Inorganic materials using 'unusual' microorganisms." *Advances in Colloid and Interface Science* 179–182: 150–68. https://doi.org/10.1016/j.cis.2012.06.013.

Baptista, Pedro, Eulália Pereira, Peter Eaton, Gonçalo Doria, Adelaide Miranda, Inês Gomes, Pedro Quaresma, and Ricardo Franco. 2007 "Gold nanoparticles for the development of clinical diagnosis methods." *Analytical and Bioanalytical Chemistry/Analytical & Bioanalytical Chemistry* 391, no. 3: 943–50. https://doi.org/10.1007/s00216-007-1768-z.

Bayda, Samer, Muhammad Adeel, Tiziano Tuccinardi, Marco Cordani, and Flavio Rizzolio. 2019 "The History of Nanoscience and Nanotechnology: From Chemical–Physical Applications to Nanomedicine." *Molecules/Molecules Online/Molecules Annual* 25, no. 1: 112. https://doi.org/10.3390/molecules25010112.

Bhattacharya, Debaditya, and Rajinder K. Gupta. 2005 "Nanotechnology and Potential of Microorganisms." *Critical Reviews in Biotechnology* 25, no. 4: 199–204. https://doi.org/10.1080/07388550500361994.

Bianco, Alberto, Kostas Kostarelos, Charalambos Partidos, and Maurizio Prato. 2005 "Biomedical applications of functionalised carbon nanotubes." *Chemical Communications*, no. 5: 571. https://doi.org/10.1039/b410943k.

Boisselier, Elodie, and Didier Astruc. 2009 "Gold nanoparticles in nanomedicine: preparations, imaging, diagnostics, therapies and toxicity." *Chemical Society Reviews* 38, no. 6: 1759. https://doi.org/10.1039/b806051g.

Cha, Chaenyung, Su Ryon Shin, Nasim Annabi, Mehmet R. Dokmeci, and Ali Khademhosseini.2013 "Carbon-Based Nanomaterials: Multifunctional Materials for Biomedical Engineering." *ACS Nano* 7, no. 4: 2891–97. https://doi.org/10.1021/nn401196a.

Chen, Long, Hao Qin, Ruifang Zhao, Xiao Zhao, Liangru Lin, Yang Chen, Yixuan Lin, et al. 2021 "Bacterial cytoplasmic membranes synergistically enhance the antitumor activity of autologous cancer vaccines." *Science Translational Medicine* 13, no. 601. https://doi.org/10.1126/scitranslmed.abc2816.

Chou SG, Dresselhaus MS, Humphreys E, Chung SY, VanderSande J, Chiang YM et al. 2003 "Raman analysis of carbon nanotube bundles for bio-electronic applications." *Materials Research Society Symposium - Proceedings.* 2003; 737:395-400.

De Jong, None. 2008 "Drug delivery and nanoparticles: Applications and hazards." *International Journal of Nanomedicine*, 133. https://doi.org/10.2147/ijn.s596.

De Zárate, David Ortiz, Carlos García-Meca, Elena Pinilla-Cienfuegos, José A. Ayúcar, Amadeu Griol, Laurent Bellières, Esther Hontañón, Frank E. Kruis, and Javier Martí. 2020 "Green and Sustainable

Manufacture of Ultrapure Engineered Nanomaterials." *Nanomaterials* 10, no. 3: 466. https://doi.org/10.3390/nano10030466.

Dey, D., and T. Goswami. 2011 "Optical Biosensors: A Revolution Towards Quantum Nanoscale Electronics Device Fabrication." *Journal of Biomedicine and Biotechnology* 2011: 1–7. https://doi.org/10.1155/2011/348218.

Elsabahy, Mahmoud, and Karen L. Wooley. 2012 "Design of polymeric nanoparticles for biomedical delivery applications." *Chemical Society Reviews* 41, no. 7: 2545. https://doi.org/10.1039/c2cs15327k.

Favier, Isabelle, Daniel Pla, and Montserrat Gómez. 2018 "Metal-based nanoparticles dispersed in glycerol: An efficient approach for catalysis." *Catalysis Today* 310: 98–106. https://doi.org/10.1016/j.cattod.2017.06.026.

Felfoul, Ouajdi, Mahmood Mohammadi, Samira Taherkhani, Dominic De Lanauze, Yong Zhong Xu, Dumitru Loghin, Sherief Essa, et al. 2016 "Magneto-aerotactic bacteria deliver drug-containing nanoliposomes to tumour hypoxic regions." *Nature Nanotechnology* 11, no. 11: 941–47. https://doi.org/10.1038/nnano.2016.137.

Fouda, Amr, Gomaa Abdel-Maksoud, Mohamed Ali Abdel-Rahman, Salem S. Salem, Saad El-Din Hassan, and Mohamad Abdel-Haleem El-Sadany. 2019 "Eco-friendly approach utilizing green synthesized nanoparticles for paper conservation against microbes involved in biodeterioration of archaeological manuscript." *International Biodeterioration & Biodegradation* 142: 160–69. https://doi.org/10.1016/j.ibiod.2019.05.012.

Gao, Xiaohu, Yuanyuan Cui, Richard M Levenson, Leland W K Chung, and Shuming Nie. 2004 "*In vivo* cancer targeting and imaging with semiconductor quantum dots." *Nature Biotechnology* 22, no. 8: 969–76. https://doi.org/10.1038/nbt994.

Gardikis, Konstantinos, Maria Micha-Screttas, Costas Demetzos, and B. R Steele. 2012 "Dendrimers and the development of new complex nanomaterials for biomedical applications." *Current medicinal chemistry* 19, no. 29: 4913-4928.

Ghorbani, Hamid, Fedos Mehr, Hossein Pazoki, and Behrad Rahmani. 2015 "Synthesis of ZnO Nanoparticles by Precipitation Method." *Oriental Journal of Chemistry/Oriental Journal of Chemistry* 31, no. 2: 1219–21. https://doi.org/10.13005/ojc/310281.

Haun, Jered B., Tae-Jong Yoon, Hakho Lee, and Ralph Weissleder. 2011 "Molecular Detection of Biomarkers and Cells Using Magnetic Nanoparticles and Diagnostic Magnetic Resonance." In *Methods in Molecular Biology*, 33–49. https://doi.org/10.1007/978-1-61779-052-2_3.

Hulkoti, Nasreen I., and T. C. Taranath. 2014 "Biosynthesis of nanoparticles using microbes—A review." *Colloids and Surfaces. B, Biointerfaces* 121: 474–83. https://doi.org/10.1016/j.colsurfb.2014.05.027.

Immanuel, S., Aparna, T. K., Sivasubramanian, R. Chapter 5 - Graphene–Metal Oxide Nanocomposite Modified Electrochemical Sensors. In Graphene-Based Electrochemical Sensors for Biomolecules, Pandikumar, A., Rameshkumar, P., Eds.; Elsevier: 2019, 113-138, https://doi.org/10.1016/C2017-0-02915-4.

Jadoun, Sapana, Rizwan Arif, Nirmala Kumari Jangid, and Rajesh Kumar Meena. 2020 "Green synthesis of nanoparticles using plant extracts: a review." *Environmental Chemistry Letters* 19, no. 1: 355–74. https://doi.org/10.1007/s10311-020-01074-x.

Kadian, Sachin, Sushanta K. Sethi, and Gaurav Manik. 2021 "Recent advancements in synthesis and property control of graphene quantum dots for biomedical and optoelectronic applications." *Materials Chemistry Frontiers* 5, no. 2: 627–58. https://doi.org/10.1039/d0qm00550a.

Kafle, Bhim Prasad. 2020 "Introduction to nanomaterials and application of UV–Visible spectroscopy for their characterization." In *Elsevier eBooks*, 147–98. https://doi.org/10.1016/b978-0-12-814866-2.00006-3.

Karim, Nazmul, Shaila Afroj, Kate Lloyd, Laura Clarke Oaten, Daria V. Andreeva, Chris Carr, Andrew D. Farmery, Il-Doo Kim, and Kostya S. Novoselov. 2020 "Sustainable Personal Protective Clothing for Healthcare Applications: A Review." *ACS Nano* 14, no. 10: 12313–40. https://doi.org/10.1021/acsnano.0c05537.

Kim, Myungjoon, Saho Osone, Taesung Kim, Hidenori Higashi, and Takafumi Seto. 2017 "Synthesis of Nanoparticles by Laser Ablation: A Review." *Kona Powder and Particle Journal/Kona* 34, no. 0: 80–90. https://doi.org/10.14356/kona.2017009.

Lee, Chia-Hung, Shih-Hsun Cheng, Yu-Jing Wang, Yu-Ching Chen, Nai-Tzu Chen, Jeffrey Souris, Chin-Tu Chen, Chung-Yuan Mou, Chung-Shi Yang, and Leu-Wei Lo. 2009 "Near-Infrared Mesoporous Silica Nanoparticles for Optical Imaging: Characterization and *In Vivo* Biodistribution." *Advanced Functional Materials* 19, no. 2: 215–22. https://doi.org/10.1002/adfm.200800753.

Lenis, J. A., G. Bejarano, P. Rico, J. L. Gómez Ribelles, and F. J. Bolívar. 2019 "Development of multilayer Hydroxyapatite - Ag/TiN-Ti coatings deposited by radio frequency magnetron sputtering with potential application in the biomedical field." *Surface & Coatings Technology/Surface and Coatings Technology* 377: 124856. https://doi.org/10.1016/j.surfcoat.2019.06.097.

Li, Ying, Oksana Zaluzhna, Bolian Xu, Yuan Gao, Jacob M. Modest, and YuYe J. Tong. 2011 "Mechanistic Insights into the Brust−Schiffrin Two-Phase Synthesis of Organo-chalcogenate-Protected Metal Nanoparticles." *Journal of the American Chemical Society* 133, no. 7: 2092–95. https://doi.org/10.1021/ja1105078.

Liu, Chuan-Yin, and Ji-Ming Hu. 2009 "Hydrogen peroxide biosensor based on the direct electrochemistry of myoglobin immobilized on silver nanoparticles doped carbon nanotubes films." *Biosensors & Bioelectronics/Biosensors & Bioelectronics (Online)* 24, no. 7: 2149–54. https://doi.org/10.1016/j.bios.2008.11.007.

Lüdecke, Claudia, Jörg Bossert, Martin Roth, and Klaus D. Jandt. 2013 "Physical vapor deposited titanium thin films for biomedical applications: Reproducibility of nanoscale surface roughness and microbial adhesion properties." *Applied Surface Science* 280: 578–89. https://doi.org/10.1016/j.apsusc.2013.05.030.

Luz, Gisela M, and João F Mano. 2011 "Preparation and characterization of bioactive glass nanoparticles prepared by sol–gel for biomedical applications." *Nanotechnology* 22, no. 49: 494014. https://doi.org/10.1088/0957-4484/22/49/494014.

Mallick, Sudipta, and Joon Sig Choi. 2014 "Liposomes: Versatile and Biocompatible Nanovesicles for Efficient Biomolecules Delivery." *Journal of Nanoscience and Nanotechnology* 14, no. 1: 755–65. https://doi.org/10.1166/jnn.2014.9080.

Mao, Xinpei, Jiadi Xu, and Honggang Cui. 2016 "Functional nanoparticles for magnetic resonance imaging." *Wiley Interdisciplinary Reviews. Nanomedicine and Nanobiotechnology* 8, no. 6: 814–41. https://doi.org/10.1002/wnan.1400.

Marchiol, Luca. 2012"Synthesis of metal nanoparticles in living plants." *Italian Journal of Agronomy* 7, no. 3: 37. https://doi.org/10.4081/ija.2012.e37.

Martínez-Alcalá, Isabel, and Maria Pilar Bernal. 2020 "Environmental Impact of Metals, Metalloids, and Their Toxicity," 451–88. https://doi.org/10.1002/9781119487210.ch21.

McNamara, Karrina, and Syed a. M. Tofail. 2016 "Nanoparticles in biomedical applications." *Advances in Physics: X* 2, no. 1: 54–88. https://doi.org/10.1080/23746149.2016.1254570.

Mürbe, Julia, Annett Rechtenbach, and Jörg Töpfer.2008 "Synthesis and physical characterization of magnetite nanoparticles for biomedical applications." *Materials Chemistry and Physics* 110, no. 2–3: 426–33. https://doi.org/10.1016/j.matchemphys.2008.02.037.

Naik, Atharva Yatin, and Priya Sundarrajan. 2023 "Applications of nanomaterials in biomedical science." *Annals of Phytomedicine* 12, no. 1. https://doi.org/10.54085/ap.2023.12.1.54.

Narayanan, Kannan Badri, and Natarajan Sakthivel. 2011 "Green synthesis of biogenic metal nanoparticles by terrestrial and aquatic phototrophic and heterotrophic eukaryotes and biocompatible agents." *Advances in Colloid and Interface Science* 169, no. 2: 59–79. https://doi.org/10.1016/j.cis.2011.08.004.

Navas, Daniel, Sandra Fuentes, Alejandro Castro-Alvarez, and Emigdio Chavez-Angel. 2021 "Review on Sol-Gel Synthesis of Perovskite and Oxide Nanomaterials." *Gels* 7, no. 4: 275. https://doi.org/10.3390/gels7040275.

Negri, Viviana, Jesús Pacheco-Torres, Daniel Calle, and Pilar López-Larrubia. 2020 "Carbon Nanotubes in Biomedicine." *Topics in Current Chemistry* 378, no. 1. https://doi.org/10.1007/s41061-019-0278-8.

Otsuka, Hidenori, Yukio Nagasaki, and Kazunori Kataoka. 2003 "PEGylated nanoparticles for biological and pharmaceutical applications." *Advanced Drug Delivery Reviews* 55, no. 3: 403–19. https://doi.org/10.1016/s0169-409x(02)00226-0.

Ovais, Muhammad, Ali Talha Khalil, Nazar Ul Islam, Irshad Ahmad, Muhamamd Ayaz, Muthupandian Saravanan, Zabta Khan Shinwari, and Sudip Mukherjee. 2018 "Role of plant phytochemicals and microbial enzymes in biosynthesis of metallic nanoparticles." *Applied Microbiology and Biotechnology* 102, no. 16: 6799–6814. https://doi.org/10.1007/s00253-018-9146-7.

Palanisamy, Selvakumar, Srikanth Cheemalapati, and Shen-Ming Chen. 2012 "Enzymatic Glucose Biosensor Based on Multiwalled Carbon Nanotubes-Zinc Oxide Composite." *International Journal of Electrochemical Science* 7, no. 9: 8394–8407. https://doi.org/10.1016/s1452-3981(23)18002-3.

Pandey, Ankur, and Ramachandran Manivannan. 2015 "A Study on Synthesis of Nickel Nanoparticles Using Chemical Reduction Technique." *Recent Patents on Nanomedicine* 5, no. 1: 33–37. https://doi.org/10.2174/1877912305666150417232717.

Park, Tae Jung, Kyoung G. Lee, and Sang Yup Lee. 2015 "Advances in microbial biosynthesis of metal nanoparticles." *Applied Microbiology and Biotechnology* 100, no. 2: 521–34. https://doi.org/10.1007/s00253-015-6904-7.

Pastukhov, Andrei I, Iaroslav B Belyaev, Julia C Bulmahn, Ivan V Zelepukin, Anton A Popov, Irina N Zavestovskaya, Sergei M Klimentov, Sergey M Deyev, Paras N Prasad, and Andrei V Kabashin. 2022 "Laser-ablative aqueous synthesis and characterization of elemental boron nanoparticles for biomedical applications." *Scientific Reports* 12, no. 1. https://doi.org/10.1038/s41598-022-13066-8.

Patil, Nilesh, Rajveer Bhaskar, Vishal Vyavhare, Rahul Dhadge, Vaishnavi Khaire, and Yogesh Patil. 2021 "OVERVIEW ON METHODS OF SYNTHESIS OF NANOPARTICLES." *International Journal of Current Pharmaceutical Research*, 11–16. https://doi.org/10.22159/ijcpr.2021v13i2.41556.

Perini, Giordano, Valentina Palmieri, Gabriele Ciasca, Marco De Spirito, and Massimiliano Papi. 2020 "Unravelling the Potential of Graphene Quantum Dots in Biomedicine and Neuroscience." *International Journal of Molecular Sciences* 21, no. 10: 3712. https://doi.org/10.3390/ijms21103712.

Qiu, Lin, Ning Zhu, Yanhui Feng, Efstathios E. Michaelides, Gaweł Żyła, Dengwei Jing, Xinxin Zhang, Pamela M. Norris, Christos N. Markides, and Omid Mahian. 2020 "A review of recent advances in thermophysical properties at the nanoscale: From solid state to colloids." *Physics Reports* 843: 1–81. https://doi.org/10.1016/j.physrep.2019.12.001.

Ramos, Ana P., Marcos a. E. Cruz, Camila B. Tovani, and Pietro Ciancaglini. 2017 "Biomedical applications of nanotechnology." *Biophysical Reviews* 9, no. 2: 79–89. https://doi.org/10.1007/s12551-016-0246-2.

Razavi, Mehdi, Erfan Salahinejad, Mina Fahmy, Mostafa Yazdimamaghani, Daryoosh Vashaee, and Lobat Tayebi. 2015 "Green Chemical and Biological Synthesis of Nanoparticles and Their Biomedical Applications." In *Springer eBooks*, 207–35. https://doi.org/10.1007/978-3-319-15461-9_7.

Riley, Parand R., and Roger J. Narayan. 2021"Recent advances in carbon nanomaterials for biomedical applications: A review." *Current Opinion in Biomedical Engineering* 17: 100262. https://doi.org/10.1016/j.cobme.2021.100262.

Rossnagel, S. M. 2003 "Thin film deposition with physical vapor deposition and related technologies." *Journal of Vacuum Science & Technology. A. Vacuum, Surfaces, and Films* 21, no. 5: S74–87. https://doi.org/10.1116/1.1600450.

Saliev, Timur. 2019 "The Advances in Biomedical Applications of Carbon Nanotubes." *C* 5, no. 2: 29. https://doi.org/10.3390/c5020029.

Sánchez, Alfredo, Anabel Villalonga, Gonzalo Martínez-García, Concepción Parrado, and Reynaldo Villalonga. 2019 "Dendrimers as Soft Nanomaterials for Electrochemical Immunosensors." *Nanomaterials* 9, no. 12: 1745. https://doi.org/10.3390/nano9121745.

Satyanarayana, T. 2018 "A Review on Chemical and Physical Synthesis Methods of Nanomaterials." *International Journal for Research in Applied Science and Engineering Technology* 6, no. 1: 2885–89. https://doi.org/10.22214/ijraset.2018.1396.

Simon, Juliette, Emmanuel Flahaut, and Muriel Golzio. 2019 "Overview of Carbon Nanotubes for Biomedical Applications." *Materials* 12, no. 4: 624. https://doi.org/10.3390/ma12040624.

Singh, Bhupendra Kumar, Sunwoo Lee, and Kyungsu Na. 2019 "An overview on metal-related catalysts: metal oxides, nanoporous metals and supported metal nanoparticles on metal organic frameworks and zeolites." *Rare Metals/Rare Metals* 39, no. 7: 751–66. https://doi.org/10.1007/s12598-019-01205-6.

Singh, Gurmeet, Arpit Gaur, Priyanka Bisht, and Rabindra Nath Mahato. 2023 "Structural, Morphological, and Magnetocaloric Properties of Nanocrystalline Pr0.7La0.3MnO3." *ECS Journal of SolidState Science and Technology* 12, no. 8: 087002. https://doi.org/10.1149/2162-8777/aceaaa.

Singh, Kshitij Rb, Vanya Nayak, Jay Singh, Ajaya Kumar Singh, and Ravindra Pratap Singh. 2021 "Potentialities of bioinspired metal and metal oxide nanoparticles in biomedical sciences." *RSC Advances* 11, no. 40: 24722–46. https://doi.org/10.1039/d1ra04273d.

Song, Rongguo, Qianlong Wang, Boyang Mao, Zhe Wang, Danli Tang, Bin Zhang, Jingwei Zhang, et al. 2018 "Flexible graphite films with high conductivity for radio-frequency antennas." *Carbon* 130: 164–69. https://doi.org/10.1016/j.carbon.2018.01.019.

Soto, Fernando, and Robert Chrostowski. 2018 "Frontiers of Medical Micro/Nanorobotics: *in vivo* Applications and Commercialization Perspectives Toward Clinical Uses." *Frontiers in Bioengineering and Biotechnology* 6. https://doi.org/10.3389/fbioe.2018.00170.

Spadavecchia, Jolanda, Alexandre Barras, Joël Lyskawa, Patrice Woisel, William Laure, Claire-Marie Pradier, Rabah Boukherroub, and Sabine Szunerits. 2013 "Approach for Plasmonic Based DNA Sensing: Amplification of the Wavelength Shift and Simultaneous Detection of the Plasmon Modes of Gold Nanostructures." *Analytical Chemistry* 85, no. 6: 3288–96. https://doi.org/10.1021/ac3036316.

Stuart, Bryan W., and George E. Stan. 2021 "Physical Vapor Deposited Biomedical Coatings." *Coatings* 11, no. 6: 619. https://doi.org/10.3390/coatings11060619.

Vaseghi, Zahra, and Ali Nematollahzadeh. *2020 "Nanomaterials,"* 23–82. https://doi.org/10.1002/9781119576785.ch2.

Vasile, Cornelia. 2019 "Polymeric nanomaterials: Recent developments, properties and medical applications." *Polymeric nanomaterials in nanotherapeutics*: 1-66. https://doi.org/10.1016/b978-0-12-813932-5.00001-7.

Venkatasubbu, G. Devanand, S. Ramasamy, V. Ramakrishnan, and J. Kumar. 2013 "Folate targeted PEGylated titanium dioxide nanoparticles as a nanocarrier for targeted paclitaxel drug delivery." *Advanced Powder Technology* 24, no. 6: 947–54. https://doi.org/10.1016/j.apt.2013.01.008.

Zare-Zardini, Hadi, Farzad Ferdowsian, Hossein Soltaninejad, Adel Ghorani Azam, Safoura Soleymani, Masoud Zare-Shehneh, Mahtab Mofidi, Roya Rafati, and Leila Ebrahimi. 2015 "Application of Nanotechnology in Biomedicine: A Major Focus on Cancer Therapy " *Journal of Nano Research* 35: 55–66. https://doi.org/10.4028/www.scientific.net/jnanor.35.55.

Chapter 1.2

Nanotechnology in Healthcare: Synthesis, Characterization, and Utilization of Nanoparticles

R. Venkatesh[1,*]
S. Dhasneem[2]
L. Vadivel Kannan[3]
and S. Selvakumar[4]
[1]Department of Physics, PSNA College of Engineering and Technology (Autonomous), Dindigul, Tamil Nadu, India
[2]Department of Physics, Christian College of Engineering and Technology, Oddanchatram, Dindigul, Tamil Nadu, India
[3]Department of Mechanical Engineering, PSNA College of Engineering and Technology (Autonomous), Dindigul, Tamil Nadu, India
[4]Department of Science and Humanities, Thamirabharani Engineering College, Tirunelveli, Tamil Nadu, India

Abstract

This book chapter extensively explores the synthesis, characterization, and multifaceted uses of nanoparticles within the realm of biomedicine. Beginning with an in-depth analysis of various synthesis techniques, including physical, chemical, and biological methodologies, the chapter provides insights into the mechanisms underlying nanoparticle formation and their implications for biomedical applications. Crucially, it examines the importance of nanoparticle characterization through advanced analytical techniques such as electron microscopy, spectroscopy, and surface analysis, elucidating the crucial role of these methods in tailoring nanoparticles for specific biomedical purposes. The chapter discusses the basics of nanotechnology in medicine, explaining how tiny particles interact with cells and molecules. It delves into how cells absorb these particles, their intracellular transportation, and the subsequent physiological reactions within the body. This understanding helps create better drug delivery systems, diagnostics, and treatments. Moreover, the chapter explores various nanostructures employed in biomedical applications, ranging from metallic and polymeric nanoparticles to liposomes, dendrimers, and quantum dots. Each nanostructure is examined in terms of its unique physicochemical properties, surface modifications, and biomedical functionalities, highlighting their

* Corresponding Author's Email: lakatesh@gmail.com

In: Emerging Nanomaterials in Biomedical Science
Editors: Eepsita Priyadarshini and Ajith Manayil Parambil
ISBN: 979-8-89530-600-0

potential in applications such as cancer therapy, imaging, biosensing, and regenerative medicine. This chapter is a helpful guide for anyone interested in nanomedicine. It covers how nanoparticles are made, studied, and used in medicine. Its goal is to speed up progress in creating new nanotechnology solutions for medical problems and making people healthier.

Keywords: nanoparticles, synthesis techniques, characterization, nanotechnology, biomedicine, cellular interactions

1. Introduction

It delves into the transformative potential of nanoparticles in medical science. These microscopic structures, usually between 1 and 100 nanometers, exhibit exceptional characteristics that render them highly promising for various biomedical applications (Mele et al. 2018). This chapter extensively examines nanoparticles' synthesis, characterization, and diverse applications within biomedicine.

Synthesis techniques are essential for customizing nanoparticles to fit specific biomedical requirements. Methods like chemical synthesis, physical vapor deposition, and biological synthesis are examined, showcasing their pros and cons. Knowing these techniques is vital for creating nanoparticles with desired biomedical properties. Advanced tools such as electron microscopes and spectroscopy are employed by scientists to understand nanoparticle structures and enhance their suitability for medical applications.

The chapter further delves into the multifaceted applications of nanoparticles in biomedicine, spanning diagnostics, therapeutics, and regenerative medicine. Nanoparticles serve as powerful tools for early disease detection through imaging techniques such as magnetic resonance imaging (MRI), computed tomography (CT), and fluorescence imaging (Bulte et al. 2008). Additionally, nanoparticles offer precision drug delivery systems, allowing for precise transport of therapeutics to diseased tissues while minimizing side effects. Furthermore, nanoparticles hold promise in regenerative medicine by facilitating tissue engineering and promoting cellular regeneration.

2. Synthesis of Nanoparticles

Making nanoparticles involves different methods, each designed for creating nanoparticles with certain traits. Methods like chemical reduction turn metal ions into nanoparticles, while the sol-gel method changes colloidal solutions into gels to make metal oxide nanoparticles. Additionally, precipitation, microemulsion, hydrothermal synthesis, electrochemical deposition, and biological synthesis are employed.

Each method has its strengths in controlling nanoparticle size, shape, and properties. These techniques are crucial in fields like medicine, electronics, and environmental science, pushing progress and innovation across various industries forward.

2.1. Physical Synthesis Methods

Physical synthesis methods are pivotal in the creation of nanoparticles through direct physical processes, offering precise control over their size, shape, and composition (Jagdeo et al. 2023). These techniques, which include vapor condensation, laser ablation, and lithography, are instrumental in tailoring nanoparticles for various biomedical applications.

Vapor condensation techniques involve controlled evaporation and subsequent condensation of materials in specific atmospheres. This process yields nanoparticles with uniform size distributions and crystalline structures (Choi et al. 2002). By manipulating parameters, such as temperature and pressure, researchers can finely tune the properties of the resulting nanoparticles. This precision makes vapor condensation particularly suitable for synthesizing metallic and semiconductor nanoparticles optimized for biomedical applications.

Laser ablation methods harness the power of laser irradiation to ablate target materials, generating nanoparticles with controlled characteristics. This technique allows for the precise manipulation of nanoparticle properties by adjusting laser parameters such as intensity, duration, and wavelength (Riabinina et al. 2012). Laser ablation offers advantages in terms of scalability and versatility, enabling the production of nanoparticles composed of diverse materials for diverse biomedical purposes.

Lithography-based methods, including electron beam lithography and nanoimprint lithography, enable the precise patterning of nanoparticles on substrates. These techniques facilitate the development of advanced biomedical devices and sensors by allowing researchers to engineer intricate nanostructures with tailored functionalities. Electron beam lithography offers sub-nanometer resolution, enabling the creation of complex nanopatterns with exceptional precision (Scuderi et al. 2016). On the other hand, Nanoimprint lithography offers high-throughput production capabilities, making it suitable for large-scale nanopatterning applications.

Physical synthesis methods help researchers create nanoparticles with specific properties for medical uses. By controlling factors like size, shape, and composition, these techniques support biomedical research and the creation of new medical technologies to enhance human health.

2.2. Chemical Synthesis Methods

Chemical synthesis methods are integral to the production of nanoparticles, which employ chemical reactions to convert precursor materials into nanoparticles. These methods are prized for their scalability, reproducibility, and versatility, making them widely favored for generating nanoparticles with diverse compositions and functionalities. Common chemical synthesis approaches include sol-gel synthesis, precipitation, and hydrothermal methods.

Sol-gel synthesis involves the reaction of metal alkoxides with water and subsequent condensation to produce nanoparticles within a sol-gel matrix. This method enables the fabrication of nanoparticles, particularly silica, and metal oxide nanoparticles, tailored for biomedical applications. By carefully controlling reaction conditions such as pH, temperature, and solvent composition, researchers can achieve precise control over nanoparticle size, morphology, and surface chemistry. Sol-gel synthesis offers advantages such as tunable porosity and biocompatibility, rendering it appropriate for applications ranging from drug delivery to tissue engineering (Ibrahim et al. 2022).

Precipitation methods rely on the controlled mixing of precursor solutions to induce the formation and enlargement of nanoparticles. Researchers can regulate the resulting nanoparticles' size, shape, and composition by adjusting reaction temperature, stirring rate, and pH. Precipitation methods are versatile and relevant to a broad spectrum of materials such as metals, metal oxides, organic nanoparticles (Wang et al. 2013). This versatility makes precipitation methods well-suited for producing nanoparticles for biomedical uses, like enhancing medical imaging contrast and transporting drugs.

Hydrothermal methods leverage elevated temperature and pressure conditions to promote nanoparticle formation from aqueous precursors. This method facilitates the synthesis of crystalline nanoparticles with controlled properties, including size, shape, and crystallinity (Li et al. 2016). Hydrothermal synthesis offers advantages such as facile scale-up and control over nanoparticle morphology, making it suitable for producing nanoparticles for biomedical applications, including tissue engineering scaffolds and therapeutic carriers.

2.3. Biological Synthesis Methods

Biological synthesis methods represent a fascinating intersection of biotechnology and materials science, leveraging the inherent capabilities of living organisms to create nanoparticles with tailored characteristics (Kumar et al. 2013). These approaches are gaining traction due to their eco-friendly nature and sustainability, as they minimize the need for harsh chemicals and energy-intensive processes typically associated with conventional nanoparticle synthesis methods.

Microbial synthesis involves harnessing the metabolic activities of microorganisms such as bacteria, fungi, and algae. These organisms possess unique enzymatic systems that reduce metal ions from their soluble forms to form nanoparticles. By controlling the environmental conditions and adjusting parameters such as pH, temperature, and nutrient availability, researchers can fine-tune the resulting nanoparticles' size, shape, and composition (Wypij et al. 2020). This method offers advantages such as scalability, cost-effectiveness, and the ability to produce nanoparticles in a controlled manner.

Plant-mediated synthesis takes advantage of the diverse array of phytochemicals present in plant extracts. These phytochemicals act as reducing and stabilizing agents, facilitating the conversion of metal ions into nanoparticles (Majeed et al. 2019). Plant-mediated synthesis offers several benefits, including biocompatibility and the potential for large-scale production using readily available plant materials. The resulting nanoparticles exhibit properties suitable for various biomedical uses, such as drug delivery, imaging, and tissue engineering.

Biomolecule-mediated synthesis represents a more intricate approach involving using specific biomolecules such as proteins and DNA to guide nanoparticle formation. In protein-templated synthesis, proteins serve as templates or scaffolds, directing the nucleation and growth of nanoparticles with exact management of their dimensions and configuration (Sharma et al. 2014). This method enables the fabrication of complex nanostructures and functional materials tailored for specific applications. Similarly, DNA-directed assembly exploits the programmable nature of DNA molecules to organize nanoparticles into predetermined patterns and structures. By harnessing the principles of molecular self-assembly, researchers can engineer sophisticated nanomaterials with unprecedented precision and functionality.

3. Characterization of Nanoparticles

Characterization of nanoparticles involves analyzing their physical, chemical, and structural properties to understand their behavior and potential applications. Techniques like transmission electron microscopy (TEM), scanning electron microscopy (SEM), and atomic force microscopy (AFM) provide high-resolution imaging of nanoparticle size, shape, and surface morphology. Additionally, spectroscopic methods such as UV-Vis spectroscopy, X-ray diffraction (XRD), and Fourier-transform infrared spectroscopy (FTIR) offer insights into nanoparticle composition, crystal structure, and chemical bonding. Other characterization techniques include dynamic light scattering (DLS) for measuring nanoparticle size distribution, surface area analysis, and zeta potential measurement to assess nanoparticle stability and surface charge. These characterization methods are essential for quality control, optimizing synthesis processes, and tailoring nanoparticles for specific applications like drug delivery, catalysis, and nanoelectronics.

3.1. Electron Microscopy

Electron microscopy techniques (Figure 1), particularly transmission electron microscopy (TEM) and scanning electron microscopy (SEM), are indispensable tools in the field of nanotechnology, offering unparalleled insights into the morphology, structure, and surface characteristics of nanoparticles (Tsapis et al. 2016).

Table 1. Characteristics analysis of Transmission Electron Microscope (TEM) and Scanning Electron Microscope (SEM)

Characteristics	Transmission Electron Microscopes (TEM)	Scanning Electron Microscopes (SEM)
Electron beam	Broad beam	Narrow and precise beam
Image captured	Internal structure	Topographical and surface
Clarity	Higher clarity	Lower clarity
Enlargement	Up to fifty millionfold	Up to two million
Image dimension	2 Dimension	3 Dimension
Sample thickness	Ultra-thin samples	Thin and thick samples
Penetrates the specimen	Yes	No
Sample limitations	More limiting	Less limiting
Preparation of samples	More preparation is required	Less preparation is required
Cost	More expensive	Less expensive
Speed	Slower	Faster
Operation	More complicated and requires training	Easy to use

Transmission electron microscopy (TEM) passes a beam of electrons through a thin specimen, typically a sample containing nanoparticles. As the electrons interact with the atoms within the specimen, they scatter, producing a magnified image that reveals intricate details of the nanoparticle's morphology, size, and crystallinity (Utsunomiya et al. 2003). TEM offers exceptionally high spatial resolution and can resolve features at the nanometer scale. By analyzing TEM images, researchers can precisely determine the size and shape of nanoparticles and their internal structure and composition. Additionally, techniques such as selected area electron diffraction (SAED) can be employed with TEM to characterize the crystalline nature of nanoparticles, providing valuable information about their atomic arrangement and crystallographic orientation (Walck et al. 2020).

Scanning electron microscopes (SEM) utilize the fine and centered electron ray to capture topographical and surface images of samples (Table 1). With lower resolution but magnifications of up to 2,000,000 times, SEMs provide detailed views of both thin and thick samples with less restrictive sample preparation. They are comparatively less expensive, faster, and easier to operate, making them accessible for various research and industrial applications. Transmission electron microscopes (TEM) employ a broad electron beam to penetrate samples, capturing internal structural images with higher resolution and magnifications of up to 50,000,000 times. TEMs require ultra-thin sample preparation, making them more restrictive and costly. Additionally, their operation is more complicated and demands extensive training due to their intricate functionality.

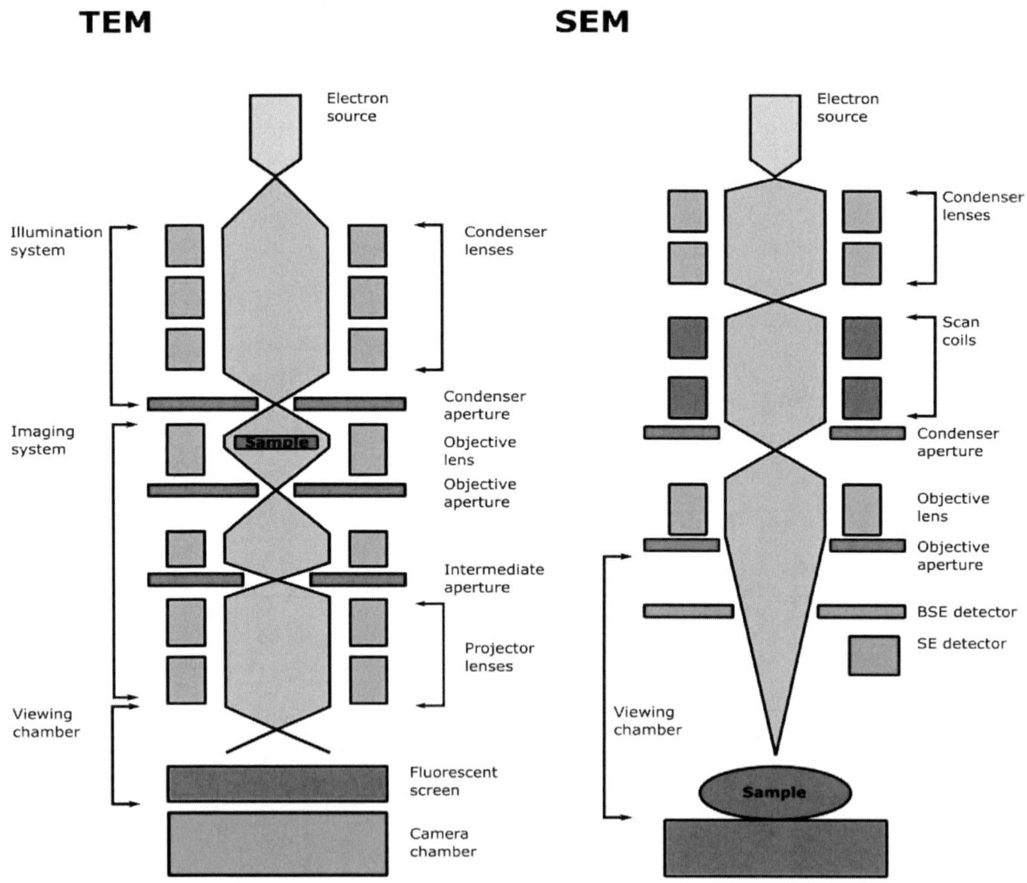

Image Source: https://bitesizebio.com/29197/electron-microscopy-techniques/.

Figure 1. Schematic Diagram of Transmission Electron Microscope (TEM) and Scanning Electron Microscope (SEM).

Scanning electron microscopy (SEM) provides three-dimensional imaging of nanoparticle surfaces by scanning a focused electron beam of electrons across the sample's surface. When the electrons interact with the specimen, they produce secondary and backscattered electrons, which are detected and used to construct an image of the sample's topography. SEM offers excellent depth of field and resolution, making it ideal for visualizing nanoparticle surface morphology, roughness, and spatial distribution (Nurit et al. 2022). By adjusting parameters such as accelerating voltage and beam current, researchers can optimize SEM imaging to highlight specific surface properties and characteristics of the nanoparticles under investigation.

3.2. Spectroscopy

Spectroscopic techniques play a pivotal role in the characterization of nanoparticles, offering valuable insights into their optical, vibrational, and chemical properties. Among these techniques, UV-Vis spectroscopy, Fourier-transform infrared spectroscopy (FTIR), and Raman spectroscopy are widely employed to elucidate different nanoparticle behavior and composition aspects.

UV-Vis spectroscopy is a versatile tool for analyzing the optical behavior of nanoparticles. It involves measuring the absorption and dispersion of light by nanoparticles across a range of wavelengths in the ultraviolet and visible regions of the electromagnetic spectrum. This technique provides information about the plasmonic properties of nanoparticles, including their surface plasmon resonance (SPR), resulting from the synchronized oscillation of free electrons in the nanoparticle's metallic core (Perkampus et al. 1992). By monitoring changes in the UV-Vis spectra, researchers can gain insights into nanoparticle size, shape, aggregation state, and surface chemistry. UV-Vis spectroscopy is particularly useful for studying noble metal nanoparticles, such as gold and silver, which exhibit strong plasmonic effects that can be tuned by controlling their size and shape.

Fourier-transform infrared spectroscopy (FTIR) is employed to analyze nanoparticle chemical composition and surface functionalization. FTIR spectroscopy detects molecular vibrations associated with specific chemical bonds, providing information about the functional groups on the nanoparticle surface. Researchers can identify surface ligands, capping agents, and adsorbed molecules through the FTIR spectra of nanoparticles with standard spectra of known compounds (Gooch et al. 2011). This technique is invaluable for characterizing the surface chemistry of nanoparticles and understanding how they influence their stability, reactivity, and interactions with biological systems.

Raman spectroscopy offers complementary insights into the composition and structure of nanoparticles based on the scattering of monochromatic light. When incident light interacts with nanoparticles, it undergoes inelastic scattering, resulting in frequency shifts that are characteristic of the vibrational modes of the nanoparticle's constituent molecules. By analyzing the Raman spectra, researchers can identify specific chemical species, detect impurities, and investigate nanoparticle phase transitions (Galinetto et al. 2018). Raman spectroscopy detects subtle variations in molecular structure and conformation, making it a powerful tool for studying nanomaterials with complex compositions and structures.

Spectroscopic techniques such as UV-Vis, FTIR, and Raman spectroscopy are essential for characterizing nanoparticles' optical, vibrational, and chemical properties. By combining these techniques with other analytical methods, researchers can comprehensively understand nanoparticle behavior and tailor their properties for various applications in catalysis, sensing, imaging, and drug delivery.

It provides a comparative overview of UV-Vis analysis, FTIR analysis, and Raman analysis, highlighting their respective principles, instrumentation, types of information obtained, sensitivity, sample requirements, spatial resolution, advantages, and disadvantages (Table 2). Each technique offers unique strengths and is suitable for different analytical purposes depending on the nature of the specimen and the precise information needed.

Table. 2 Analyzing features of various spectroscopic techniques

Features	UV-Vis Spectroscopy	Fourier Transform Infrared Spectroscopy (FTIR)	Raman Spectroscopy
Principle	Measures absorption of UV and visible light by electronic transitions.	Measures absorption of infrared light by molecular vibrations.	Measures scattering of monochromatic light by molecular vibrations.
Instrumentation	Uses a UV-Vis spectrophotometer.	Uses an interferometer and Fourier transformation to obtain the spectrum.	Uses a laser and a spectrometer to analyze scattered light.
Type of Information	Provides information on the concentration of chromophores in electronic transitions.	Provides information on molecular vibrations functional groups.	Provides information on molecular vibrations crystal structures.
Sensitivity	Highly sensitive to electronic transitions; detects low concentrations.	More sensitive to polar functional groups (OH, NH, CO).	Less sensitive to polar functional groups, more to non-polar groups.
Sample Requirements	Can analyze liquids, solids, and gases; minimal sample preparation.	Analyzes solids, liquids, gases, minimal sample preparation.	Analyzes solids, liquids, gases, minimal sample preparation.
Spatial Resolution	Not applicable; measures bulk properties.	Generally lower unless coupled with microscopy techniques.	Can achieve higher spatial resolution.
Advantages	Rapid analysis, quantitative measurements, wide application range.	Rapid analysis identifies the functional group's ease of use.	Analyzes crystalline material's structural information, less affected by water.
Disadvantages	Limited structural information requires chromophores for absorption.	Prone to water vapor interference, it is less effective with strong fluorescence.	Prone to fluorescence interference, lower sensitivity to certain groups.

3.3. Surface Analysis

Surface analysis techniques are crucial for understanding nanoparticle surfaces' chemical composition and topographical features, which are crucial in defining their characteristics and interactions with the surrounding environment. Among these techniques, X-ray photoelectron spectroscopy (XPS) and atomic force microscopy (AFM) are widely used to provide detailed information about surface composition and morphology at the nanoscale.

X-ray photoelectron spectroscopy (XPS) is a powerful surface-sensitive technique that analyzes materials' chemical composition and elemental states, including nanoparticle surfaces. XPS operates by irradiating the sample surface with X-rays, which eject photoelectrons from the outermost atomic layers. By measuring the kinetic energies of these photoelectrons, researchers can determine the binding energies associated with different elements and chemical species on the nanoparticle surface. This information allows for identifying surface elements, oxidation states, and chemical bonding configurations. XPS is particularly useful for characterizing functionalized nanoparticles, identifying surface ligands, and monitoring chemical changes induced by surface modifications or environmental interactions (Qu et al. 2013).

Atomic force microscopy (AFM) offers precision imaging of nanoparticle surfaces with nanometer-scale spatial resolution. In AFM, a pointed probe tip affixed to a cantilever is scanned over the sample surface while monitoring the deflection of the cantilever. By maintaining a constant interaction between the probe tip and the specimen surface, AFM measures the topographical surface changes with exceptional precision (Mehlhorn et al. 2016). AFM can also map surface properties such as mechanical properties, adhesion forces, and electrical conductivity at the nanoscale. Additionally, AFM techniques such as scanning Kelvin probe microscopy (SKPM) enable the characterization of surface potential variations and charge distribution on nanoparticle surfaces. AFM is invaluable for studying nanoparticles' surface morphology, roughness, and structural features, providing valuable insights into their physical properties and surface interactions (Rohwerder et al. 2007).

The XPS and AFM constitute powerful tools for surface analysis, offering complementary details regarding the chemical composition and topographical characteristics of nanoparticle surfaces. By combining these techniques with other analytical methods, researchers can comprehensively understand surface structure-function relationships in nanoparticles and tailor their surface properties for specific applications in catalysis, sensing, nanoelectronics, and biomedicine.

4. Nanoparticles in Biomedical Applications

Nanoparticles hold immense potential in biomedical applications due to their unique properties. They can be designed to aim for particular cells or tissues, making them promising candidates for drug delivery systems. Encapsulating drugs within nanoparticles can enhance drug stability, prolong circulation time, and improve drug uptake at the target site while reducing side effects. Additionally, nanoparticles can be modified with biomolecules like antibodies or peptides for targeted imaging or therapy of diseases like cancer. Moreover, they can serve as contrast agents in imaging techniques like MRI, CT scans, and fluorescence microscopy, aiding in the early detection and monitoring of diseases. With ongoing research and development, nanoparticles continue revolutionizing biomedical fields, offering new possibilities for diagnosis, treatment, and personalized medicine.

4.1. Nanoparticles in Drug Delivery

Nanoparticles have garnered significant attention as advanced carriers for drug delivery, revolutionizing medicine by offering many benefits over conventional drug formulations. Among the key benefits of nanoparticle drug delivery systems is their ability to encapsulate drugs within biocompatible materials. This encapsulation not only enhances the stability and solubility of the drugs but also protects them from degradation and premature clearance in the body (Madkour et al. 2019). By shielding drugs within nanoparticles, researchers can prolong their circulation time in the bloodstream, thereby improving their bioavailability and therapeutic efficacy.

Nanoparticle carriers can be designed to aim at particular tissues, cells, or even tiny parts inside cells, reducing unwanted effects and making them safer for the body. Adding targeting molecules like antibodies or peptides onto nanoparticle surfaces can send drugs exactly where needed, leaving healthy tissues alone. This targeted method improves how well drugs work, letting higher doses reach the right spot while causing fewer side effects elsewhere in the body.

Nanoparticle drug delivery systems have the benefit of releasing drugs slowly and steadily, keeping therapeutic levels stable for a long time. Scientists can control how fast drugs are released to meet specific treatment requirements by adjusting factors like nanoparticle size, shape, and composition. This controlled release not only helps patients stick to their treatment plans but also ensures drugs work well by keeping their levels right for longer periods.

Conventional drug delivery systems (DDS) typically involve administering medication in a manner where the drug rapidly enters the bloodstream, often resulting in fluctuations in drug levels and potential side effects. Nanomedicine-based control release drug delivery systems (CRDDS) employ nanoparticles to encapsulate and deliver drugs to specific target sites in a controlled manner. Through controlled-release drug delivery systems, these nanoparticles can release the drug gradually, allowing for sustained therapeutic effects while minimizing adverse reactions. This innovative approach holds promise for enhancing drug efficacy, improving patient outcomes, and reducing healthcare costs.

Several nanoparticle-based drug delivery systems have been developed, each with unique characteristics and applications. Liposomes, composed of phospholipid bilayers, are versatile carriers capable of encapsulating hydrophobic and hydrophilic medications within their watery core or lipid envelope. Polymeric nanoparticles, fabricated from biodegradable polymers such as polylactic-co-glycolic acid (PLGA) or polyethylene glycol (PEG), offer tunable properties and sustained release profiles. Dendrimers are intricately branched macromolecules with precisely defined structures that allow exact management of drug loading and release rates (Zhang et al. 2019).

Table 3. Parameters analysis of Conventional Drug Delivery System (DDS) and Nanomedicine-based Control Release Drug Delivery System (CRDDS)

Parameters	Conventional Drug Delivery System (DDS)	Nanomedicine-based Control Release Drug Delivery System (CRDDS)
Size of Carriers	Micro to macro-sized carriers (e.g., tablets, capsules)	Nano-sized carriers (e.g., liposomes, nanoparticles)
Drug Loading Efficiency	Limited capacity	High capacity
Release Kinetics	Often immediate release	Controlled and sustained release
Targeting Capability	Limited targeting ability	Enhanced targeting
Drug Stability	Stability concerns over time	Improved stability
Biocompatibility	Generally good	Enhance biocompatibility
Drug Solubility	Limited options for poorly soluble drugs	Improved solubility
Route of Administration	Oral, parenteral, topical, etc.	Diverse options, often specific to nanocarrier
Side Effects	Potential for systemic side effects	Reduced systemic side effects
Manufacturing Complexity	Relatively straightforward	Complex synthesis and formulation
Cost	Generally lower	Higher
Clinical Translation	Established in clinical practice	Advancing towards clinical use
Examples	Tablets, capsules, patches	Liposomal doxorubicin, nanoparticle-based vaccines

Extensive research has focused on nanoparticle-based drug delivery systems for treating various diseases, such as cancer, infectious diseases, inflammatory disorders, and neurological conditions. The comparison of conventional drug delivery systems (DDS) to nanomedicine-based control release drug delivery systems (CRDDS) involves outlining key features and characteristics of each approach (Table 3).

4.2. Nanoparticles in Imaging

Nanoparticles possess extraordinary physical properties that render them invaluable as contrast agents in biomedical imaging, enabling researchers and clinicians to visualize biological structures and processes with unprecedented detail and precision. Leveraging these unique properties, nanoparticles can be tailored and functionalized to function as adaptable contrast agents in various imaging methods, such as optical, magnetic, and acoustic techniques.

Table 4. Different types of nanoparticles in imaging systems

Aspects	Magnetic Resonance Imaging (MRI)	Surface-enhanced Raman Scattering (SERS)	Quantum Dots	Carbon Nanotubes
Type of Nanoparticles	Iron oxide nanoparticles, gadolinium-based contrast agents, etc.	Gold or silver nanoparticles	Semiconductor nanoparticles (e.g., CdSe, CdTe)	Single - walled carbon nanotubes (SWCNTs), Multi-walled carbon nanotubes (MWCNTs)
Imaging Mechanism	Alters local magnetic fields affect T1 or T2 relaxation times	Enhances Raman signals due to plasmonic effects	Fluorescence emission based on quantum confinement	Contrast agents due to high surface area, near-infrared absorbance
Applications	Structural and functional imaging, tumor detection, angiography	Biosensing, molecular imaging, detection of biomolecules	Bioimaging, cellular tracking, sentinel lymph node mapping	Contrast-enhanced imaging, drug delivery, biosensors
Advantages	High-resolution imaging, minimally invasive, penetration into deep tissues	High sensitivity, molecular specificity, multiplexing	Bright and stable fluorescence, tunable emission wavelengths	High surface area, near-infrared absorbance, biocompatibility
Surface Functionalization	PEGylation, targeting ligands (antibodies, peptides)	Raman reporter molecules, biocompatible coatings	Ligand exchange for targeting, surface passivation	Functionalization with targeting ligands, biomolecules
Challenges	Longitudinal relativity, renal clearance, biocompatibility	Signal reproducibility, background interference	Toxicity concerns, stability in biological environments	Dispersion stability, potential toxicity, batch-to-batch variability
Recent Advances	Functionalized nanoparticles for targeted imaging	Engineered substrates for improved signal enhancement	Improved quantum yield, biocompatible coatings	Functionalized for specific applications, enhanced solubility
Future Directions	Integration with molecular imaging techniques	Development of portable and sensitive SERS probes	Enhanced stability, integration into theragnostic platforms	Biomedical applications, theragnostic capabilities

In optical imaging, nanoparticles play a pivotal role as contrast agents due to their distinctive ability to interact with light. Metallic nanoparticles, such as gold nanorods and silver nanoparticles, exhibit strong plasmonic effects that can enhance optical signals, making them ideal for techniques like fluorescence imaging and surface-enhanced Raman scattering (SERS). By functionalizing these nanoparticles with targeting ligands and fluorescent dyes, researchers can achieve highly specific and sensitive imaging of diseased tissues and cells (Gupta et al. 2017). Gold nanoparticles, for example, are frequently utilized in molecular imaging and cancer diagnostics, enabling precise localization and visualization of tumor cells.

Magnetic nanoparticles offer unparalleled capabilities for magnetic resonance imaging (MRI), a non-invasive imaging modality widely used in clinical diagnostics. Nanoparticles composed of iron oxide and gadolinium-based compounds serve as contrast agents in MRI by influencing the relaxation rates of adjacent water molecules, thereby enhancing contrast between different tissues. Iron oxide nanoparticles are favored for their biocompatibility and ability to generate strong T2-weighted MRI contrast (Orooji et al. 2022). These nanoparticles are employed for imaging various anatomical structures and pathological conditions, including tumors, cardiovascular diseases, and neurodegenerative disorders.

Moreover, nanoparticles like quantum dots and carbon nanotubes demonstrate outstanding optical characteristics that facilitate fluorescent and photoacoustic imaging. Quantum dots, semiconductor nanocrystals with tunable emission spectra, are used in fluorescence imaging to visualize cellular processes and track biomolecules with high sensitivity and multiplexing capabilities. On the other hand, carbon nanotubes possess unique photoacoustic properties, converting absorbed light into acoustic signals that can be detected with high resolution. This makes them promising candidates for multimodal imaging applications, combining the strengths of optical and acoustic imaging modalities.

This comprehensive outline of the various types of nanoparticles used in imaging, their specific imaging mechanisms, applications, advantages, challenges, recent advances, and future directions for each category of MRI, SERS, Quantum Dots, and Carbon Nanotubes (Table. 4). This structured approach provides a clear comparison of how different nanoparticles contribute to advancing imaging technologies in medical and biological sciences.

4.3. Nanoparticles in Biosensing

Nanoparticles serve as versatile sensing platforms for detecting and quantifying various biomolecules, pathogens, and environmental contaminants, offering exceptional sensitivity, selectivity, and multiplexing capabilities. By functionalizing nanoparticles with specific recognition elements, such as antibodies, aptamers, and molecular probes, researchers can design highly sensitive biosensors for various analytical applications, spanning healthcare, environmental monitoring, and food safety.

Gold nanoparticles are among the most widely used nanomaterials in biosensing applications, owing to their unique optical properties and facile surface functionalization. In colorimetric assays, gold nanoparticles undergo a visible color change upon binding to target analytes, enabling rapid and visual detection of proteins, nucleic acids, and small molecules. Additionally, gold nanoparticles are employed in surface plasmon resonance (SPR) sensors (Koch et al. 2006), where changes in the refractive index near the nanoparticle surface are detected as shifts in the SPR wavelength, providing real-time and label-free monitoring of biomolecular interactions with high sensitivity.

Quantum dots and up-conversion nanoparticles offer powerful fluorescence-based detection platforms for biomolecular targets characterized by their intense and tunable emission spectra. These nanoparticles exhibit exceptional photostability and brightness, enabling multiplexed detection of multiple analytes simultaneously. By conjugating quantum dots or up-conversion nanoparticles with specific recognition molecules, researchers can develop biosensors capable of detecting proteins, nucleic acids, and pathogens with high sensitivity and specificity. Moreover, the unique optical properties of quantum dots enable precise quantification and imaging of biomolecules in complex biological samples.

Magnetic nanoparticles represent another class of versatile sensing platforms, particularly in magnetic biosensors and biosensing arrays. These nanoparticles are functionalized with recognition molecules and employed with magnetic detection techniques to enable rapid and quantitative analysis of biological samples. Magnetic biosensors offer minimal sample preparation, high throughput, and portability, making them suitable for point-of-care diagnostics and field-based applications (Kurlyandskaya et al. 2022).

4.4. Nanoparticles in Regenerative Medicine

Nanoparticles are at the forefront of regenerative medicine, revolutionizing approaches to tissue engineering, stem cell therapy, and wound healing by harnessing their unique properties to modulate cellular behavior and promote tissue regeneration (Hashemi et al. 2022). When incorporated into scaffolds, hydrogels, and matrices, these tiny structures offer myriad benefits that accelerate healing processes and support tissue regeneration *in vitro* and *in vivo*.

Tiny metal particles like silver and titanium dioxide have natural germ-killing abilities, which are useful for healing wounds. When these particles are added to bandages or scaffolds, they help support the wound, stop infections, and speed up healing. Silver particles, for instance, can kill germs like bacteria, fungi, viruses, and heal tissue.

Biodegradable polymeric nanoparticles are another cornerstone of regenerative medicine, providing a versatile platform for administering growth factors, cytokines, and therapeutic agents to injured tissues (Couvreur et al. 2018).

Researchers can use biocompatible polymers to trap bioactive molecules, forming controlled-release systems. These systems can encourage cell growth, blood vessel formation, and tissue healing injury. These nanoparticles hold onto therapeutic substances and release them slowly, directly to the injured area, reducing side effects elsewhere in the body.

Nanofibers and nanocomposites mimic natural structures, aiding cell behavior like adhesion and migration. Electro-spun nanofibers, for example, create scaffold-like structures resembling natural tissues, promoting cell attachment and tissue integration. Adding nanoparticles boosts mechanical strength, surface properties, and biological activity, making scaffolds customized for tissue engineering needs.

In tissue engineering and regenerative medicine, nanoparticles are increasingly recognized for their potential to manipulate cellular processes intricately. They possess the ability to tune cellular signaling pathways finely, alter gene expression profiles, and direct stem cell differentiation. This capability is largely attributed to their surface modifications and encapsulation of bioactive molecules, allowing them to interact at a molecular level with cells and influence their behavior during tissue regeneration endeavors (Raja et al. 2022). Moreover, nanoparticles are pivotal in advancing the fabrication of complex tissues and organs tailored for transplantation. By enhancing the mechanical properties of scaffolds, controlling the release of growth factors, and mimicking the native extracellular matrix environment, nanoparticles facilitate the creation of functional tissue constructs. This breakthrough holds promise in mitigating the critical shortage of donor organs and tissues, offering potential solutions by developing engineered tissues that can integrate seamlessly with the recipient's body. Thus, nanoparticles represent a transformative toolset in pursuing innovative therapies and personalized regenerative treatments.

Conclusion and Future Perspectives

In summary, this chapter comprehensively investigates nanoparticle synthesis, characterization, and applications in biomedicine. This chapter catalyzes advancements in nanomedicine research and technology by delving into the intricate mechanisms governing nanoparticle formation, highlighting the significance of advanced characterization techniques, and showcasing the diverse spectrum of biomedical applications.

In the chapter, different ways to make nanoparticles, from traditional chemistry to new biological methods have been discussed. Knowing these methods helps make nanoparticles with the right properties for medical uses. It also highlights the importance of using tools like electron microscopy and spectroscopy to understand the structure and function of nanoparticles. By using these tools, researchers can figure out how nanoparticles behave and make them work better for medical purposes. The chapter emphasized how nanoparticles are promising in solving important healthcare problems and transforming the field. They can be used for precise drug delivery, imaging, regenerative medicine, and biosensing, offering new ways to diagnose diseases sooner, deliver treatments better, and help tissues heal.

In the future, advancing nanoparticle technology holds immense promise for revolutionizing medicine, but significant research is still necessary to overcome existing challenges and fully harness its potential. One crucial area of focus is nanoparticle safety; ensuring that these tiny particles do not cause unintended harm to the body is paramount. Researchers are investigating how nanoparticles interact with biological systems and how their long-term effects can be mitigated or eliminated. Another major hurdle is scalability. While nanoparticles show incredible potential at a laboratory scale, translating this effectiveness to large-scale medical applications remains formidable. This requires technological advancements and innovative manufacturing processes that can produce nanoparticles efficiently and consistently.

Addressing these challenges necessitates collaboration among diverse fields of expertise. Nanoparticle research intersects with disciplines ranging from chemistry and materials science to biology and medicine. By pooling their knowledge and ideas, experts can develop novel strategies to improve the safety and effectiveness of nanoparticles in medical applications. Furthermore, continuous exploration of new therapeutic avenues is essential. Nanoparticles offer versatile capabilities, such as targeted drug delivery, imaging enhancements, and even therapeutic interventions at the molecular level. Exploring these possibilities could lead to breakthroughs in treating diseases more effectively and with fewer side effects than current methods allow.

The chapter provides a comprehensive roadmap for understanding and utilizing nanoparticles in medicine. If successfully navigated, this path can potentially transform healthcare globally. By making treatments more precise, efficient, and accessible, nanoparticle technology could significantly improve health outcomes worldwide, marking a pivotal advancement in medical science.

References

Bulte, Jeff. W. M., and Michel M. J. Modo. "Introduction: The Emergence of Nanoparticles as Imaging Platform in Biomedicine." *Nanoparticles in Biomedical Imaging*, 2008, 1–5. https://doi.org/10.1007/978-0-387-72027-2_1.

Choi, C. J., O. Tolochko, and B. K. Kim. "Preparation of Iron Nanoparticles by Chemical Vapor Condensation." *Materials Letters* 56, no. 3 (October 2002): 289–94. https://doi.org/10.1016/s0167-577x(02)00457-3.

Couvreur, P., L. Grislain, V. Lenaerts, F. Brasseur, P. Guiot, and A. Biernacki. "Biodegradable Polymeric Nanoparticles as Drug Carrier for Antitumor Agents." *Polymeric Nanoparticles and Microspheres*, January 18, 2018, 27–94. https://doi.org/10.1201/9781351075893-2.

Galinetto, Pietro, Benedetta Albini, Marcella Bini, and Maria Cristina Mozzati. "Raman Spectroscopy in Zinc Ferrites Nanoparticles." *Raman Spectroscopy*, April 18, 2018. https://doi.org/10.5772/intechopen.72864.

Gooch, Jan W. "Fourier-Transform Infrared Spectroscopy." *Encyclopedic Dictionary of Polymers*, 2011, 323–24. https://doi.org/10.1007/978-1-4419-6247-8_5263.

Gupta, S., A. Banaszak, T. Smith, and N. Dimakis. "Molecular Sensitivity of Metal Nanoparticles Decorated Graphene-family Nanomaterials as Surface-enhanced Raman Scattering (SERS) Platforms." *Journal of Raman Spectroscopy 49*, no. 3 (December 27, 2017): 438–51. https://doi.org/10.1002/jrs.5318.

Hashemi, Zahra Sadat, Mahlegha Ghavami, Saeed Khalili, and Seyed Morteza Naghib. "The Emerging Role of Exosome Nanoparticles in Regenerative Medicine." *Nano pharmaceuticals in Regenerative Medicine,* March 29, 2022, 67–93. https://doi.org/10.1201/9781003153504-5.

Ibrahim, Sundus S., and Ziad T. Khodair. "Synthesis, Characterization, and Antifungal Activity of Cd2SnO4 Nanoparticles Prepared by Sol-Gel Technique." *Neuro Quantology 20*, no. 2 (April 1, 2022): 137–42. https://doi.org/10.14704/nq.2022.20.2.nq22081.

Jagdeo, Kailas R. "Physical Methods for Synthesis of Nanoparticles." *Nano chemistry*, January 4, 2023, 66–76. https://doi.org/10.1201/9781003081944-4.

Koch, Karl-Wilhelm. "Surface Plasmon Resonance." *Encyclopedic Reference of Genomics and Proteomics in Molecular Medicine,* 2006, 1832–35. https://doi.org/10.1007/3-540-29623-9_4880.

Kumar, Anu, Kuldeep Kaur, and Sarika Sharma. "Synthesis, Characterization and Antibacterial Potential of Silver Nanoparticles by Morus Nigra Leaf Extract." *Indian Journal of Pharmaceutical and Biological Research 1*, no. 04 (December 31, 2013): 16–24. https://doi.org/10.30750/ijpbr.1.4.4.

Kurlyandskaya, Galina V., ed. Sensors and Biosensors Related to Magnetic Nanoparticles. *MDPI*, 2022. https://doi.org/10.3390/books978-3-0365-4616-2.

Li, Jianlin, Qingliu Wu, and Ji Wu. "Synthesis of Nanoparticles via Solvothermal and Hydrothermal Methods." *Handbook of Nanoparticles*, 2016, 295–328. https://doi.org/10.1007/978-3-319-15338-4_17.

Madkour, Loutfy H. "Nanoparticle and Polymeric Nanoparticle-Based Targeted Drug Delivery Systems." *Nucleic Acids as Gene Anticancer Drug Delivery Therapy*, 2019, 191–240. https://doi.org/10.1016/b978-0-12-819777-6.00013-5.

Majeed, Shahnaz, and Muhammad Samad. "Plant Mediated Synthesis of Silver Nanoparticles and Its Antibacterial Effect." *Journal of Pure and Applied Microbiology* 13, no. 2 (June 30, 2019): 1267–72. https://doi.org/10.22207/jpam.13.2.69.

Mehlhorn, Heinz. "Atomic Force Microscopy (AFM)." *Encyclopedia of Parasitology*, 2016, 240–240. https://doi.org/10.1007/978-3-662-43978-4_4501.

Mele, Elisa. "Introduction: Smart Materials in Biomedicine." *Smart Nanoparticles for Biomedicine*, 2018, 1–13. https://doi.org/10.1016/b978-0-12-814156-4.00001-x.

Nurit, Taitel-Goldman. "Minerals Observed by Scanning Electron Microscopy (SEM), Transmission Electron Microscopy (TEM) and High-Resolution Transmission Electron Microscopy (HRTEM)." *Electron Microscopy*, August 17, 2022. https://doi.org/10.5772/intechopen.102477.

Orooji, Farnaz, and Russell Butler. "Applying Unet for Extraction of Vascular Metrics from T1-Weighted and T2-Weighted MRI." December 20, 2022. https://doi.org/10.1101/2022.12.18.520922.

Perkampus, Heinz-Helmut. "Recent Developments in UV-VIS Spectroscopy." *UV-VIS Spectroscopy and Its Applications,* 1992, 81–130. https://doi.org/10.1007/978-3-642-77477-5_5.

Qu, Jun, and Harry M. Meyer. "X-Ray Photoelectron Spectroscopy (XPS)." *Encyclopedia of Tribology*, 2013, 4133–38. https://doi.org/10.1007/978-0-387-92897-5_1222.

Raja, N. S., and Manish Dash. "Application of Nanoparticles in Soft Tissue Engineering." *Application of Nanoparticles in Tissue Engineering*, 2022, 33–65. https://doi.org/10.1007/978-981-16-6198-3_3.

Riabinina, Daria, Mohamed Chaker, and Joëlle Margot. "Dependence of Gold Nanoparticle Production on Pulse Duration by Laser Ablation in Liquid Media." *Nanotechnology 23*, no. 13 (March 16, 2012): 135603. https://doi.org/10.1088/0957-4484/23/13/135603.

Rohwerder, Michael, and Florin Turcu. "High-Resolution Kelvin Probe Microscopy in Corrosion Science: Scanning Kelvin Probe Force Microscopy (SKPFM) versus Classical Scanning Kelvin Probe (SKP)." *Electrochimica Acta 53*, no. 2 (December 2007): 290–99. https://doi.org/10.1016/j.electacta.2007.03.016.

Scuderi, Mario, Marco Esposito, Francesco Todisco, Daniela Simeone, Iolena Tarantini, Luisa De Marco, Milena De Giorgi, et al. "Nanoscale Study of the Tarnishing Process in Electron Beam Lithography-Fabricated Silver Nanoparticles for Plasmonic Applications." *The Journal of Physical Chemistry C 120*, no. 42 (October 13, 2016): 24314–23. https://doi.org/10.1021/acs.jpcc.6b03963.

Sharma, Garima, Ashish Sharma, Riju Bhavesh, Jongbong Park, Bilguun Ganbold, Ju-Suk Nam, and Sang-Soo Lee. "Biomolecule-Mediated Synthesis of Selenium Nanoparticles Using Dried Vitis Vinifera (Raisin) Extract." *Molecules 19*, no. 3 (February 27, 2014): 2761–70. https://doi.org/10.3390/molecules19032761.

Tsapis, Nicolas. "Imaging Polymer Nanoparticles by Means of Transmission and Scanning Electron Microscopy Techniques." *Polymer Nanoparticles for Nanomedicines*, 2016, 205–19. https://doi.org/10.1007/978-3-319-41421-8_8.

Utsunomiya, Satoshi, and Rodney C. Ewing. "Application of High-Angle Annular Dark Field Scanning Transmission Electron Microscopy, Scanning Transmission Electron Microscopy-Energy Dispersive X-Ray Spectrometry, and Energy-Filtered Transmission Electron Microscopy to the Characterization of Nanoparticles in the Environment." *Environmental Science & Technology 37*, no. 4 (January 16, 2003): 786–91. https://doi.org/10.1021/es026053t.

Walck, Scott D. "Recipes for Consistent Selected Area Electron Diffraction Results: Part 2: Recording the SAED Pattern." *Microscopy Today 28*, no. 3 (May 2020): 54–59. https://doi.org/10.1017/s1551929520000887.

Wang, Benyang, Qifeng Wei, and Shiliang Qu. "Synthesis and Characterization of Uniform and Crystalline Magnetite Nanoparticles via Oxidation-Precipitation and Modified Co-Precipitation Methods." *International Journal of Electrochemical Science* 8, no. 3 (March 2013): 3786–93. https://doi.org/10.1016/s1452-3981(23)14431-2.

Wypij, Magdalena, and Patrycja Golińska. "Mechanism of Microbial Synthesis of Nanoparticles." *Microbial Nanotechnology*, May 25, 2020, 100–116. https://doi.org/10.4324/9780429276330-6.

Zhang, Xingzhen, Lixia Zhao, Guangxi Zhai, Jianbo Ji, and Anchang Liu. "Multifunctional Polyethylene Glycol (PEG)-Poly (Lactic-Co-Glycolic Acid) (PLGA)-Based Nanoparticles Loading Doxorubicin and Tetrahydrocurcumin for Combined Chemoradiotherapy of Glioma." *Medical Science Monitor 25* (December 19, 2019): 9737–51. https://doi.org/10.12659/msm.918899.

Chapter 1.3

Limitations of Current Technologies and Advantages of Nanomaterials in Biomedical Sciences

Deepa Thakur[1,*]
Chirag Porwal[2]
Aakansha Suchitta[3]
and Rahul Sharma[4,†]
[1]Department of Materials Science and Engineering, Uppsala University, Uppsala, Sweden
[2]Department of Mechanical and Materials Engineering, Indian Institute of Technology Mandi, Kamand, Himachal Pradesh, India
[3]Optics and Photonic Center, Indian Institute of Technology Delhi, Hauz Khas, New Delhi, India
[4]Department of Physics and Astronomy, Uppsala University, Uppsala, Sweden

Abstract

Conventional techniques of drug delivery are a deficit of target selectivity, resulting in systemic side effects and reduced effectiveness. Many drugs exhibit poor solubility and stability, resulting in inadequate absorption and therapeutic effect. Current diagnostic technologies (e.g., MRI and CT scans) have limited resolution, making the detection of early-stage diseases or small pathological changes challenging. Traditional chemotherapies and radiation therapies can damage healthy tissues along with cancerous ones, causing severe side effects. The widespread use of antibiotics has prompted the appearance of resistant bacterial strains, complicating infection management. With great promise in biomedical sciences, nanotechnology is a fast-developing field that can be engineered to deliver drugs directly to specific cells, minimizing adverse consequences and maximizing therapeutic impact. Nanomaterials exhibit unique physiochemical properties, such as enhanced absorption and reactivity, greater surface area, increased molar extinction coefficients, and quantum effects. Nanoparticles can enhance the solubility and stability of drugs, ensuring better absorption and prolonged circulation time. Additionally, they enhance the resolution of imaging techniques, allowing for the detection of early-stage diseases and small-scale abnormalities. Nanoparticles with inherent antimicrobial properties (e.g., silver nanoparticles) can combat resistant bacterial strains and reduce infection rates. This chapter highlights the limitations of current technologies and the

* Corresponding Author's Email: deepathakur81374@gmail.com
† Corresponding Author's Email: rahuluppsala@gmail.com

In: Emerging Nanomaterials in Biomedical Science
Editors: Eepsita Priyadarshini and Ajith Manayil Parambil
ISBN: 979-8-89530-600-0

significant advancement that nanotechnology has made in the last few decades for biomedical applications. By addressing the limitations of current biomedical technologies, nanomaterials offer significant advancements in precision, efficacy, and safety, clearing the path for more effective treatments and diagnostic tools in the biomedical sciences.

Keywords: nanomaterials, gene therapy, diagnostic techniques, therapeutic methods, regenerative medicines

1. Introduction

Biomedical techniques consist of the application of engineering and technology in the healthcare sector. This area is constantly growing and workson advancements to upgrade human healthcare. Biomedical techniques include research related to organs-on-chips, wearable medical devices, surgical robotics, nanorobots, artificial intelligence in medical imaging, and medicinal chemistry. Revolutionary advancements in medical imaging and diagnosis techniques are reshaping medical practices. Progress in robotics, bionics, and prosthetics has significantly revolutionized the healthcare industry. The precision of medical imaging technology is increasing, reducing the reliance on invasive diagnostic procedures. As new medical and scientific discoveries emerge, the role of biomedical engineers will be pivotal in translating these innovations into advancements that benefit healthcare and humanity.

1.1. Introduction to Nanomaterials in Biomedical Science

The research in nanomaterials is at its peak and is used in multiple applications, including biomedical engineering. A nanomaterial possesses one or more size dimensions in the range of 1–100 nm and can have a variety of shapes and morphologies (Thakur et al. 2021; 2022; 2023; 2024). Nanomaterials possess high strength, toughness, high diffusivity, optical properties, advantageous sintering capabilities, and many more features due to the quantum confinement effect. The progress in nanoscience resulted in various applications, such as tissue-engineered scaffolds, bio-imaging, gene therapy, brain and nervous system, heart disorder, ocular disease, cancer therapy, biosensors, implant engineering, cell culture, and drug delivery applications, etc. (Anjum et al. 2021). Figure 1a shows the schematic portrayal of the utilization of nanoparticles in various healthcare applications. Figure 1b shows the key parameters to ensure the suitability of nanomaterials for biomedical applications.

The field of nanomaterial-engineered biomedicines has various challenges, including compatibility of nanomaterials with living organs and toxicity. This requires a detailed investigation of the interaction of the nanostructured materials with biosystems using contemporary characterization methods. Recently, a class of nanostructures in mesoporous form was found useful for supplying therapeutic drugs to tumor cells. These demonstrate very little drug leakage into the living cells. Mesoporous nanostructures are blessed with enhanced porosity and surface functionality. The mini-sized pore controls the amount of drug release and its efficient loading at the targeted site (Sawy et al. 2021).

Magnetic nanoparticles have played a big role in biomedical science, specifically imaging therapies, such as X-ray, computed tomography (CT) scans, magnetic resonance imaging

(MRI), etc. Those nanoparticles that can be manipulated by a magnetic field come under the category of magnetic nanoparticles. In general, magnetic nanoparticles include two parts: a magnetic element, such as iron (Fe), nickel (Ni), and cobalt (Co), and a chemical functional group. Magnetic nanoparticles have various potential properties, such as X-ray absorption, high efficiency of energy transmission, and generation of reactive oxygen constituents (Yoon et al. 2017). Additionally, magnetic nanoparticles are advantageous for cancer therapy and diagnosis. Iron-oxide-based magnetic nanoparticles are utilized in MRI (Morales et al. 2012). Magnetic CoPt nanomaterials are employed as an MRI contrast agent to detect implanted brain stem cells (Meng et al. 2011). Nanoparticles, such as iron oxide, magnetite, maghemite, and gold, are useful in magnetic fluid hyperthermia. The nanoparticles are introduced at the tumor's location, and a high-frequency magnetic field is applied. The heat generated by these nanoparticles increases the temperature of the tumor up to 40-46°C, which helps kill cancer cells (Sharifi, Shokrollahi, and Amiri 2012; Javidi et al. 2015; Javidi et al. 2014). Magnetic nanoparticles can also cure cancer by combining hyperthermia and drug-controlled release (Estelrich et al. 2015).

Modern medical science demands highly sensitive, quick response, dynamic, and reliable diagnostic devices. Sensors based on nanostructures are considered highly suitable matches for the same. Despite the various advantages and ongoing research in nanomaterials for biological applications, nanoparticles' safety remains a primary constraint that restricts their application. The challenges and limitations of nanoparticles and current biomedical techniques are discussed in the following section.

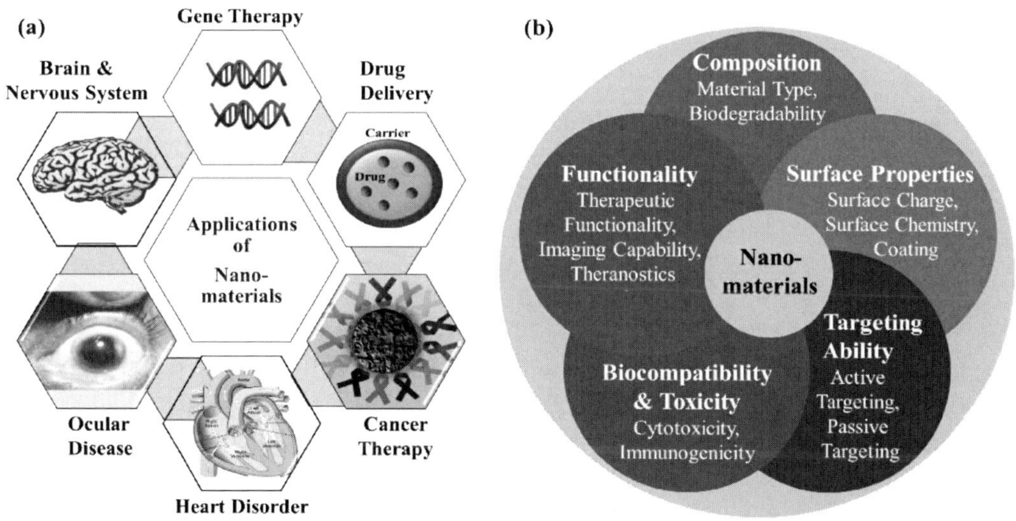

Figure 1. (a) The schematic shows the application of nanomaterials for various biomedical applications. (Schematic prepared using Creative Commons) (b) Key parameters to ensure the suitability of nanomaterials for biomedical applications.

2. Limitations of Current Biomedical Techniques

Before the clinical trials of human drugs, preclinical tests were performed on animals but rarely translated into secure and potent drugs for humans. About 90% of proposed drugs are rejected in clinical trials due to inefficacy and toxicity to humans (Mak, Evaniew, and Ghert 2014).

Despite spending time and resources on biomedicine research, many challenges still remain (Bergström and Lindmark 2019). Animal physiology is different from humans, in terms of drug efficacy, making it difficult to predict the human response toward drug exposure (Council 2007). Novel approach based on nanomaterials, can help develop biomedical techniques which are faster, more efficient, and cost-effective.

2.1. Conventional Drug Delivery Systems

Classical drug delivery systems include medicines in the form of tablets, capsules, syrups, ointments, etc. The aim of the drug delivery system is to release the active chemicals to get a desired therapeutic response in the body. However, the traditional drug delivery systems suffer from inferior bioavailability, poor solubility, and specificity and are not able to release drugs at a desired rate. In the absence of an efficient drug delivery system, the complete therapeutic process can be rendered purposeless. Conventional drugs such as tablets, capsules, and syrups are abolished from anatomy rapidly. The concentration of the drug increases in the body when it is taken, and then it starts decaying exponentially. Hence, drugs taken via conventional methods do not remain in the body long enough to produce a desirable therapeutic effect. One solution to the above problem is to take multiple doses within a day. In several cases, adverse effects of the drugs have been seen on the body. Enormous attempts have been made in the last few decades to upgrade drug delivery systems with the utility of nanomaterials to upgrade targeted drug delivery.

2.2. Imaging and Diagnostic Techniques

Medical imaging provides a visual portrayal of desired tissues and organs of the living body, helping to observe the abnormality in the body. Various healthcare imaging technologies are available, such as X-rays, CT scans, MRI, ultrasound, sonography, etc. These advanced imaging and diagnostic techniques are beneficial for the detection of various diseases such as myocardial diseases, cancer, neurological disorders, heart problems, abdominal illnesses, fractures in the bone, and many more. The biggest advantage of the imaging method is its invasive procedure. These imaging techniques do not require any blood or other fluid or organ to be removed from the body. Ultrasound scans, use sound waves to generate pictures of internal structures, generally abdominal, heart, blood vessels, and fetuses in the uterus. The most significant merit of ultrasound scans is their ability to visualize real-time movement. It is a safer imaging technique without the risk of ionizing radiation.

MRI is among the most recognized biomedical imaging techniques. In MRI, the machine is cylindrical, and a magnetic field is emitted. This generated magnetic field compels atoms in the living body to align in a specific direction. After turning off the frequencies, the atoms return to their positions, and a radio signal is emitted. This emitted signal is processed by the computer to convert this into images. One of the primary benefits of MRIs is that they do not involve any radiation exposure, making it generally safe for patients.

However, in addition to the advantages in diagnosis, these imaging techniques have some associated risks. The X-ray machine is a common diagnosis tool but involveshigh-energy X-rays that pass through the body. Hence, its repetitive exposure can be harmful to health. MRI machines require helium for cooling which makes it costly. Additionally, people with metal implants cannot be imaged due to high magnetic fields.

CT scans are yet another very useful technique. CT scans combine X-ray techniques with computer processing to capture imaging of body parts such as bones, blood vessels, and organs. In CT scans, X-ray beams move around the body, and computer software develops 3D images of the targeted body part. CT scans expose patients to ionizing radiation. Excessive ionizing radiation exposure increases the risk of deoxyribonucleic acid (DNA) damage and cancer. An incredible amount of effort has been made to boost the resolution and sensitivity of the imaging techniques. However, high cost, unavailability, maintenance, and limited resources restrict regular usage of advanced imaging tools. (Hussain et al. 2022).

2.3. Surgical and Therapeutic Methods

Surgery is used when a body part needs to be repaired, removed, or replaced by cutting and opening the body. Surgeries generally carry risks as they are invasive. The practical risks can be postsurgical pain, bruising, malignant hyperthermia, nerve damage, internal bleeding, septic shock, and even death (Ahmed et al. 2017). Risk is higher in emergency surgery, open surgery, organ transplant surgery, and long surgeries. Surgery needs an extremely experienced surgeon who can perform the task precisely. Generally, open surgeries need long recovery time and intensive care (Dharap, Barbaniya, and Navgale 2022; Kim, Ryu, and Kim 2022; Billroth et al. 2017). Robot-assisted surgery is a new technique that combines a machine's precision with human surgeons' expertise. Robotic arms and a high-definition vision system that projects a magnified three-dimensional image of the surgical spot to be operated upon are employed in robotic surgery. Robotic surgery enhances the safety of procedures and lowers the risk of failure (Reddy et al. 2023).

2.4. Regenerative Medicine and Tissue Engineering

The desire to completely repair or regenerate damaged tissues or organs and restore their functions has been a long-held human aspiration (Han et al. 2020). The advancement of regenerative medicine and tissue engineering makes this dream come true. Regenerative medicine should support cell growth and nutrition transportation. This field has been developing for more than four decades now. Multiple milestones in tissue regeneration have been achieved, but many basic issues still demand more advancement. This includes limited scaffold materials, biocompatibility, slow and incomplete tissue regeneration, and immune rejection issues. Regenerating a functioning heart remains a significant bioengineering challenge. The heart's pumping action necessitates mechanical forces to compress a chamber full of blood with defined inlets and outlets. It is well-established that mechanical weight plays a crucial role in the evolution and structuring of cardiac tissue. Similarly, creating a synthetic scaffold that mimics the properties of the native cornea is still challenging. Moreover, the regeneration of injuries to the core and outlying nervous system, whether traumatic or degenerative, continues to be a major hurdle. The problem can be resolved by collaborating with the expertise of biomaterials scientists and medical doctors (Han et al. 2020; Ikada 2006). Marcelo Muñoz and the group developed a sprayable nanotherapeutic, based on a multiarmed peptide grafted nanogold (Munoz et al. 2022). The *in vitro* and in-vivo testing was performed on an infarcted myocardial surface andthe study was conducted on the mouse heart. The spray-on treatment was found to enhance the cardiac function of the mouse by 2.4 folds. The schematic of this study is shown in Figure 2a. Cardiovascular disease is the second most frequent reason for death in Canada (Public Health Agency of Canada, n.d.). Heart attack, medically called myocardial infarction, takes place due to coronary blood flow blockage. It is among the top disease that causes the maximum number of annual deaths ("The Top 10 Causes of Death," 2020). These reports make the study related to nanoparticles for heart-related diseases extremely important.

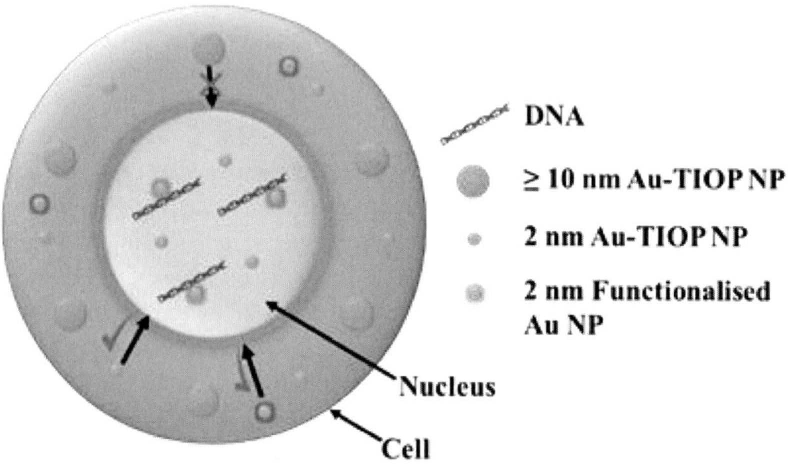

Figure 2. Schematic showing the distribution and localization of different sized (2 nm and 10 nm) tiopronin-covered Au nanoparticles (Au-TIOP NPs) in cancer cells. (Schematic diagram is inspired from Huo et al. 2014).

2.5. Gene Therapy and Genetic Engineering

With the advancement in biological science, gene therapy that corrects mutated genes is possible. At present, gene therapy is chiefly practiced in research laboratories only. It is important to optimize and design the delivery vehicles called vectors. Vector is used to efficiently release genes at specific sites and must be biocompatible. Most studied species in gene therapy are viruses because they can invade cells easily and incorporate their genetic material. Nevertheless, there is significant worry about intensified immune responses and genetic handling, particularly in germline cells. Recent studies showed several limitations in the use of vectors. Many scientists have approved gene therapy for somatic cells, especially for cystic fibrosis and Duchenne muscular dystrophy (Kohn, Chen, and Spencer 2023; Gonçalves and Paiva 2017). Shuaidong Huo and the group explored size-dependent properties of gold nanoparticles for intracellular distribution in MCF-7 breast cancer cells (Huo et al. 2014). It was found that the nanoparticles with a size <10 nm were capable of invading the nucleus, and larger nanoparticles were observed only in the cytoplasm. The authors devised a method to control gene expression by directly introducing triplex-forming oligonucleotides into the nucleus with the help of ultrasmall 2 nm gold nanoparticles (Huo et al. 2014). Figure 2 represents the distribution and localization of tiopronin (TIOP)-covered Au nanoparticles in MCF-7 cancer cells. It was observed that the 2-6 nm Au-NPs functionalised with triplex-forming oligonucleotide was effectively reducing c-myc RNA and c-myc protein. This shows the importance of size dependency on activity.

Gene therapy is a powerful methodology for future healthcare engineering, however needs to follow bioethics. Few researchers have employed the CRISPR-Cas9 technology to customize embryonic cells genetically and HIV resistance by inserting the CCR5 (Tebas et al. 2014) (Callaway 2016). Their genetic analysis revealed that four among 26 embryos were effectively altered. However, bioethics suggests conducting animal trials before human trials (Gonçalves and Paiva 2017).

3. Advantages of Nanomaterials in Biomedical Applications

Nanomaterials have emerged as a transformative force in biomedicine and diagnostics research, providing novel solutions and significant advancements across various domains. Some key advantages are discussed in this section.

3.1. Enhanced Drug Delivery Systems

It is possible to achieve precise drug delivery to targeted cells or tissues, which enhances therapeutic efficacy by utilizing nanoparticles as drug carriers. Antibodies, peptides, or other ligand molecules can be affixed to the surfaces of nanoparticles, which helps in easy identification and binding to specific biomarkers found on targeted cells. The engineered nanomaterials offer control and sustain the release of therapeutic agents, which maintain optimal drug levels over longer periods. Further, these nanomaterials act as preservative agents protecting drug degradation or premature clearance from the body. For example, researchers have demonstrated that polymeric nanoparticles can protect drugs from enzymatic degradation, enhance their half-life and ensure sustained therapeutic levels (Begines et al. 2020; Gothwal, Khan, and Gupta 2016). Further, Rosenholm et al. have reported that stimuli-responsive nanoparticles can release their payload in response to specific external conditions such as pH, temperature, or external magnetic fields. These conditions allow for good control over the rate of drug release (Rosenholm, Sahlgren, and Lindén 2011). Biodegradable nanoparticles that degrade slowly and release drugs gradually enable a sustained therapeutic effect. Nanoparticles or engineered nanostructures can encapsulate hydrophobic drugs, further improving their solubility. Clinical diagnostics and point-of-care therapeutics also allow better dispersion in biological fluids and enhance the solubility and stability of therapeutic agents (Gothwal, Khan, and Gupta 2016).

3.2. Advancement in Imaging and Diagnostic Techniques

Nanomaterials are immensely revolutionizing bioimaging and diagnostic techniques. They offer high resolution and enhanced sensitivity. Quantum dots, for instance, are semiconductor nanocrystals that exhibit unique optical properties, including size-tunable fluorescence and exceptional brightness. These properties make quantum dots highly suitable for imaging applications, enabling the visualization of biological processes at the molecular and cellular levels (Goldman and Mattoussi 2005). As discussed earlier, magnetic nanoparticles enhance the contrast of MRI images, allowing for the detection of tumors, inflammation, and other pathological conditions with high sensitivity.

Liquid biopsy, a non-invasive method for detecting cancer biomarkers in blood samples, has also benefited from nanomaterial advancements. Nanoparticles can capture and detect circulating tumor cells, cell-free DNA, and exosomes, providing valuable information for early cancer detection and monitoring treatment responses (Tong et al. 2017). Due to their plasmonic properties, gold nanoparticles can be used in photoacoustic imaging, where laser-induced ultrasound waves provide high-resolution images of tissues. Combining photoacoustic imaging with other modalities, such as MRI or CT, further improves diagnostic capabilities.

3.3. Precision in Surgical and Therapeutic Methods

Nanoparticles can be employed in targeted therapies that deliver therapeutic agents directly to infected tissues, minimizing damage to healthy tissues. For example, in photothermal therapy, gold nanoparticles can be targeted to tumor cells and, upon exposure to near-infrared light, generate localized heat that destroys cancer cells without harming surrounding healthy tissues (Mehrmohammadi et al. 2013). Nanorobots are tiny devices powered by nanomaterials and hold promise for performing precise surgical tasks at the cellular level. These nanorobots can navigate the bloodstream, reach specific sites, and perform tasks such as removing blood clots or delivering drugs with high precision (Ba, Teijeiro, and Rivas 2013). In radiation therapy, nanoparticles can act as radiosensitizers, enhancing cancer cells' sensitivity to radiation. This helps reduce the need for high radiation doses, substantially minimizing damage to healthy tissues (Mehrmohammadi et al. 2013).

3.4. Innovations in Regenerative Medicine and Tissue Engineering

Regenerative medicine and tissue engineering need superior scaffold materials with enhanced biocompatibility, which can be offered by nanomaterials as these provide a conducive environment for cell growth and differentiation. Such a favorable environment for cellular growth and tissue regeneration can be provided by nanofibers, nanotubes, nanocomposites, and hydrogels, which can easily mimic the natural extracellular matrix (Jun et al. 2018). For example, electrospun nanofibers can create highly porous and fibrous scaffolds that promote cell adhesion, proliferation, and differentiation. These nanofiber scaffolds can be functionalized with growth factors or peptides to promote tissue regeneration (Chakrapani 2023). Hydroxyapatite nanoparticles incorporated into scaffolds for bone tissue engineering, reported by Mo et. al. (Gothwal, Khan, and Gupta 2016), and polyethylene glycol (PEG) coated nanoparticles for an improved immune response are some prominent advances (Soo et al. 2016). Moreover, engineered nanomaterials can release immunomodulatory agents that suppress or modulate the immune response, reducing the risk of rejection. This approach is particularly valuable in organ transplantation and tissue grafting (Song, Musetti, and Huang 2017).

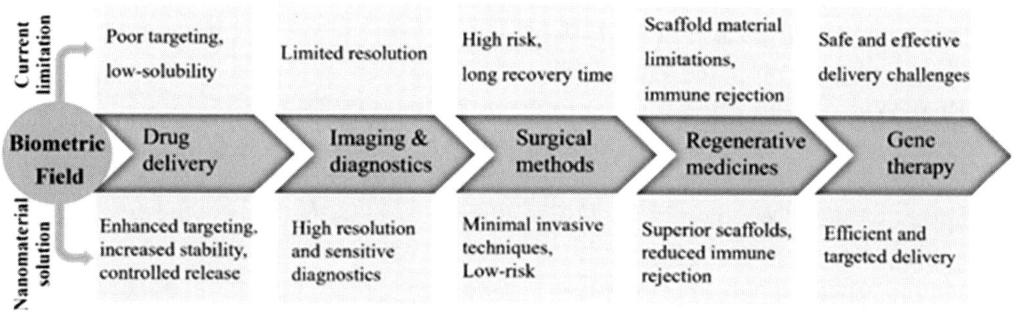

Figure 3. Transformative Potential of Nanomaterials.

3.5. Breakthroughs in Gene Therapy and Genetic Engineering

Recent years have witnessed substantial advancements in the field of gene therapy and genetic engineering as evidenced by CRISPR-Cas9 technology (A. Doudna and Charpentier 2014), CAR-T Cell Therapy (Sterner and Sterner 2021), and spinal muscular atrophy (SMA) treatment (Mercuri et al. 2020), to name a few. Gene therapy and genetic engineering aim to achieve an efficient and targeted delivery system for genetic material. Nanomaterials significantly contribute to making some of the major improvements and advancements in this field. For example, nanomaterial-based drug delivery systems are known to reduce immunogenicity associated with gene therapy. Shielding therapeutic cargo from immune recognition has been realized by encapsulating genetic material within nanoparticles. Genetic engineering encompasses viral and non-viral vectors that deliver genetic material into target cells. Viral vectors are effective but pose risks such as immunogenicity and insertional mutagenesis. On the other hand, non-viral nanocarriers are safer options. Liposomes, polymeric nanoparticles, and dendrimers are non-viral options that offer safer alternatives for delivering genes to target cells (Nayerossadat, Maedeh, and Ali 2012). Further, lipid nanoparticles (LNPs) are also being used to deliver messenger RNA (mRNA) for gene therapy and vaccine applications. LNPs protect mRNA from early degradation, facilitating its delivery to cells (Sahay et al. 2013). Additionally, surface modifications, such as PEGylation or affinity pre-concentrations of aptamers, can further reduce the immunogenicity of nanoparticles and improve selective binding (Shiraishi and Yokoyama 2019). In a study conducted by Patricia Mora-Raimundo et al. (Mora-Raimundo et al. 2021), nanoparticles were used to deliver SOST small interfering RNA (siRNA) to reduce the expression of the SOST gene, aiming to enhance bone formation. Figure 3a demonstrates that SOST expression is higher in ovariectomized mice (OVX) compared to non-ovariectomized control mice. However, when injected, nanoparticles carrying SOST siRNA and osteostatin (OST-siRNA) significantly reduced SOST expression by ~50% after 2 weeks and 60% after 3 weeks. The systemic administration of osteostatin and siRNA-loaded nanoparticles effectively decreased the target gene's expression, confirming the knockdown activity of the injected system. The synthesized nanoparticles successfully protected the siRNA and peptide from enzymatic degradation, facilitated their transport to target cells, and achieved the desired therapeutic effect (Mora-Raimundo et al. 2021).

4. Case Studies and Clinical Applications

4.1. Real-World Examples of Nanomaterials in Drug Delivery

In clinical point-of-care diagnostics, numerous cancer-based studies have employed molecular biomarkers or biosensors for early-stage detection of diseases and drug delivery applications. Doxil, the first Food and Drug Administration (FDA)-approved nanodrug, is one of the finest examples of nanomaterials in drug delivery applications. It has been particularly effective in treating Kaposi's sarcoma, ovarian cancer, and multiple myeloma. The chemical formulation of Doxil consists of a chemotherapeutic agent known as doxorubicin encapsulated within a liposome-based nanoparticle. This liposomal encapsulation substantially enhances the efficacy of this drug by improving its pharmacokinetic behavior and reducing its toxicity. The polyethylene glycol (PEG) coating on the liposome is robust enough to prevent identification and destruction by the immune system, allowing prolonged circulation time in the bloodstream. Patients receiving Doxil experience fewer side effects compared to those receiving conventional doxorubicin. This is primarily due to the reduced cardiotoxicity caused by the liposomal delivery system (Barenholz 2012; Gabizon, Shmeeda, and Barenholz 2003). Recent studies have confirmed its efficacy and safety profile, making it fundamental in oncological nanomedicine and drug therapeutics (Senapati et al. 2018).

Abraxane is another recent important breakthrough in nanomedicine. It is widely known for its effective treatment of metastatic breast cancer. Abraxane uses albumin-bound nanoparticles (a well-known chemotherapy drug) to deliver paclitaxel. The nano-formulation enhances the solubility and bioavailability of paclitaxel and helps deliver drugs to the tumor cells efficiently. In clinical trials, patients treated with Abraxane have shown higher response rates and improved progression-free survival. Their use has recently expanded to lung and pancreatic cancer, demonstrating their versatility (Moreno-aspitia, Perez, and Moreno-aspitia 2005; Socinski et al. 2012).

4.2. Success Stories in Imaging and Diagnostic Methods

Bioimaging capabilities and medical diagnostics techniques such as advanced MRI techniques, optical coherence tomography (OCT), liquid biopsies, wearable biosensors, genome sequencing, proteomics, CRISPR-Cas9-based diagnostics platforms (as mentioned in section 2.5) have direct links with nanoparticle research. Biocompatible nanoparticles are extensively being used in imaging and diagnostics. Ferumoxytol (an iron oxide nanoparticle) enhances contrast in imaging lymph nodes and liver lesions in clinical settings. This is helpful for patients who have renal insufficiency and cannot tolerate gadolinium (Reimer and Balzer 2003; Bashir et al. 2015).

Similarly, quantum dots are also helpful in imaging cancer cells and tracking cellular processes. They are successfully used to label HER2 receptors on breast cancer cells, causing precise imaging and targeted therapy. This approach can significantly improve the accuracy of cancer diagnostics soon (Alivisatos 2004; Gambhir and Weiss 2005). Quantum dots are also used in multiplexed diagnostics to detect multiple biomarkers and simultaneously enhance accuracy and speed (Huang et al. 2020).

4.3. Advances in Surgical Techniques and Regenerative Medicines

Researchers in clinical diagnostics have dedicatedly contributed towards advanced surgical techniques and regenerative medicines and the current research can witness robotic surgeries, advanced endoscopic and single port-laparoscopic surgeries with reduced invasiveness and higher chances of faster recovery. A recent clinical study on diabetic foot ulcers showed that nanofiber dressings can accelerate wound closure and reduce infection rates compared to traditional wound care methods. The nanofibers' ability to provide a moist environment and sustained drug release was crucial to these outcomes. Bioactive molecules and stem cells in nanofiber dressings have improved therapeutic efficacy (Shahid et al. 2022). Nanoparticles are also being used to improve bone regeneration.

Hydroxyapatite nanoparticles closely mimic the natural mineral component of bone, making them ideal for bone grafting and tissue engineering applications. In a clinical trial, hydroxyapatite nanoparticles were combined with stem cells and implanted into patients with bone defects precisely at the defect site. The use of hydroxyapatite nanoparticles resulted in significant bone regeneration and integration with the surrounding tissue, showing the nanoparticles' potential for enhancing bone repair and regeneration (Dorozhkin and Epple 2002; Liu et al. 2017). Recent studies are being performed to develop multifunctional nanoparticles that can simultaneously deliver osteogenic factors and antimicrobial agents to prevent infections while promoting bone growth (Jeong et al. 2024).

4.4. Impacts on Gene Therapy

Gene therapy has also benefited from nanomaterial research. LNPs are being used as drug-delivery agents for mRNA vaccines. These nanoparticles protect the mRNA from degradation. They facilitate cellular uptake, assisting in the synthesis of therapeutic proteins. One recent example is the successful development of COVID-19 mRNA vaccines (Pfizer-BioNTech and Moderna). These vaccines showed high efficacy in fighting COVID-19 and controlled the pandemic. This shows the use of lipid nanoparticles in treating various genetic disorders (Jackson and Kester 2020; Pardi et al. 2018). Now, researchers are trying to use mRNA vaccines for personalized cancer immunotherapy, which also shows promising preliminary results (Song, Musetti, and Huang 2017).

Gold nanoparticles (GNPs) are also better carriers for delivering CRISPR-Cas9 gene editing components. GNPs have better biocompatibility and enhance cellular uptake, making them a suitable candidate for gene therapy. In a recent groundbreaking study, GNPs delivered CRISPR-Cas9 components to correct a genetic mutation in Duchenne Muscular Dystrophy (DMD) mouse models. The treated mice showed improvement in muscle functioning and their structure. This demonstrates the potential of GNPs in the treatment of human genetic disorders (Lee et al. 2017; Nelson and Gersbach 2016). This technology is advancing day by day to target other genetic disorders. For example, Cystic fibrosis and sickle cell anemia are under preclinical trials (Gonçalves and Paiva 2017).

5. Challenges and Future Directions

Nanobiotechnology is progressing, and many researchers are working hard to improve healthcare. Yet, there are various challenges and limitations, which are explained below:

5.1. Current Limitations and Challenges

The global population demands advanced healthcare with high precision and early-stage disease detection. In developing countries like India, Pakistan, Sri Lanka, etc., where most of the population resides in rural areas, the first challenge is to build hospitals everywhere that are equipped with basic diagnostic facilities. Conventional drug delivery systems often suffer from the lack of target specificity, low solubility and stability, resulting in systemic side effects and reduced therapeutic efficacy. Imaging and diagnostic technologies, despite significant advancements, continue to face challenges such as limited resolution and sensitivity, which hinder the early detection of diseases. Surgical and therapeutic methods involve significant risks and necessitate highly skilled practitioners, whereas regenerative medicine and tissue engineering grapple with issues like scaffold material limitations, biocompatibility, and immune rejection. Although gene therapy shows promise, it is hampered by challenges related to the safe and effective delivery of genetic material.

The existing biomedical techniques can be upgraded with the assistance of nanomaterials. Nanomaterials have various advantages, but the toxicity risk and detrimental effects should be studied before human trials (Moshfegh 2009; Turchanin et al. 2011; Albalawi et al. 2021). The manufacturing, scalability, and low-cost availability of advanced diagnosis equipment are also challenges that must be addressed.

5.2. Future Trends and Innovations

Nanomaterials hold immense transformative potential in overcoming the limitations of current biomedical technologies. Their unique physicochemical properties make them ideal candidates for improving drug delivery systems. Nanoparticles can be precisely engineered to target selective cells or tissues, decreasing off-target impacts and maximizing therapeutic virtue. They also enhance the solubility and stability of hydrophobic drugs, ensuring better absorption and prolonged circulation time. Nanomaterials, such as quantum dots and magnetic nanoparticles, offer unparalleled resolution and sensitivity in imaging and diagnostics. This enables the early detection of diseases and precise imaging of biological processes. Developing non-invasive diagnostic procedures using nanomaterials reduces the need for biopsies and invasive surgeries, significantly improving patient comfort and safety.

The nanomedicine market is estimated to be ~ USD 350.8 billion in 2025 (Albalawi et al. 2021). Nanomaterials can pass the blood-brain barrier, making them viable candidates for biomedicine and drug delivery (Moshfegh 2009; Turchanin et al. 2011; Albalawi et al. 2021). Robotic surgery and gene therapy are the future of healthcare. The future directions and challenges are summarised in Table 1.

Table 1. Future directions and challenges

Biomedical Field	Future Directions	Challenges
Drug Delivery	Personalized and targeted therapies	Ensuring biocompatibility, scaling up production
Imaging and Diagnostics	Multi-modal imaging techniques	Reducing costs, improving accessibility
Surgical Methods	Robotic and minimally invasive surgeries	Training and infrastructure requirements
Regenerative Medicine	Advanced scaffold materials, organ regeneration	Biocompatibility, immune response management
Gene Therapy	Safe and efficient gene editing techniques	Regulatory and ethical considerations, long-term safety

Conclusion

The future of nanomaterials in biomedical science is exceptionally promising. As research and development in nanotechnology continue to progress, the potential for nanomaterials to revolutionize healthcare becomes more apparent. They offer solutions to long-standing challenges in drug delivery, diagnostics, surgery, regenerative medicine, and gene therapy, making treatments more effective, precise, and safe.

However, for nanomaterials to achieve widespread clinical adoption, several challenges mustbe addressed. These include ensuring the biocompatibility and safety of nanomaterials, scaling up manufacturing processes, and reducing costs to make advanced treatments accessible to a broader population. Rigorous clinical trials and preclinical evaluations are essential to establish the efficacy and ensure the safety of nanomaterial-based solutions and associated therapies. Continued research and interdisciplinary collaboration will be crucial in realizing the full potential of nanomaterials, thus leading to improved and favorable patient outcomes and assist in progressive advancement of the global healthcare system.

Acknowledgments

Deepa Thakur and Rahul Sharma acknowledge Uppsala University; Chirag Porwal thanks the Indian Institute of Technology Mandi, and Aakansha Suchitta acknowledges the Indian Institute of Technology Delhi for the required facilities.

References

"The Top 10 Causes of Death." 2020. *World Health Organization.*

Ahmed, Faransa A, Shimaa A Khalaf, Safaa R Mahmoud, and Ibrahim A Ibrahim. 2017. "Risk Factors of Surgical Site Infection among Infants: Descriptive Study." *Assiut Scientific Nursing Journal* 5 (12): 108–16. https://doi.org/10.21608/asnj.2017.60621.

Albalawi, Fahad, Mohd Zobir Hussein, Sharida Fakurazi, and Mas Jaffri Masarudin. 2021. "Engineered Nanomaterials: The Challenges and Opportunities for Nanomedicines." *International Journal of Nanomedicine* 16: 161–84. https://doi.org/http://doi.org/10.2147/IJN.S288236.

Alivisatos, Paul. 2004. "The Use of Nanocrystals in Biological Detection." *Nature Biotechnology* 22 (1): 47–52. https://doi.org/10.1038/nbt927.

Anjum, Sumaira, Sara Ishaque, Hijab Fatima, Wajiha Farooq, Christophe Hano, Bilal Haider Abbasi, and Iram Anjum. 2021. "Emerging Applications of Nanotechnology in Healthcare Systems: Grand Challenges and Perspectives." *Pharmaceuticals* 14 (8): 707. https://doi.org/10.3390/ph14080707.

Ba, Manuel, Antonio Teijeiro, and Jose Rivas. 2013. "Magnetic Nanoparticle-Based Hyperthermia for Cancer Treatment." *Reports of Practical Oncology and Radiotherapy* 8: 397–400. https://doi.org/10.1016/j.rpor.2013.09.011.

Barenholz, Yechezkel Chezy. 2012. "Doxil® — The First FDA-Approved Nano-Drug: Lessons Learned." *Journal of Controlled Release* 160 (2): 117–34. https://doi.org/10.1016/j.jconrel.2012.03.020.

Bashir, Mustafa R, Lubna Bhatti, Daniele Marin, and Rendon C Nelson. 2015. "Emerging Applications for Ferumoxytol as a Contrast Agent in MRI." *Journal of Magnetic Resonance Imaging* 41 (4): 884–98. https://doi.org/10.1002/jmri.24691.

Begines, Belén, Tamara Ortiz, María Pérez-Aranda, Guillermo Martínez, Manuel Merinero, Federico Argüelles-arias, and Ana Alcudia. 2020. "Polymeric Nanoparticles for Drug Delivery: Recent Developments and Future Prospects." *Nanomaterials* 10 (7): 1403. https://doi.org/10.3390/nano10071403.

Bergström, Fredrik, and Bo Lindmark. 2019. "Accelerated Drug Discovery by Rapid Candidate Drug Identification." *Drug Discovery Today* 24 (6): 1237–41. https://doi.org/10.1016/j.drudis.2019.03.026.

Billroth, Theodor, Johan Von Mikulicz, Thomas Morton, and Charles Mcburney. 2017. "Challenges of Surgery in the 21st Century." *Cirugía y Cirujanos.* 85 (1): 1–3. https://doi.org/10.1016/j.circen.2017.01.001.

Callaway, Ewen. 2016. "Second Chinese Team Reports Gene Editing in Human Embryos." *Nature* 8. https://doi.org/10.1038/nature.2016.19718.

Chakrapani, Gayathri. 2023. "Functionalization of Electrospun Nanofiber for Biomedical Application." *Journal of Applied Polymer Science*, no. 140: e53906. https://doi.org/10.1002/app.53906.

Council, National Research. 2007. "Toxicity Testing in the 21st Century: A Vision and a Strategy." In *National Academies Press.*

Dharap, Satish B, Priya Barbaniya, and Shantanu Navgale. 2022. "Incidence and Risk Factors of Postoperative Complications in General Surgery Patients." *Cureus* 14 (11): e30975. https://doi.org/10.7759/cureus.30975.

Dorozhkin, Sergey V, and Matthias Epple. 2002. "Biological and Medical Significance of Calcium Phosphates." *Angewandte Chemie - International Edition* 41: 3130–46. https://doi.org/https://doi.org/10.1002/1521-3773(20020902)41:17<3130:AID-ANIE3130>3.0.CO;2-1.

Doudna, A., Jennifer, and Emmanuelle Charpentier. 2014. "The New Frontier of Genome Engineering with CRISPR-Cas9." *Science* 346 (6213): 1258096. https://doi.org/10.1126/science.1258096.

Estelrich, Joan, Elvira Escribano, Josep Queralt, and Maria Antònia Busquets. 2015. "Iron Oxide Nanoparticles for Magnetically Guided and Magnetically Responsive Drug Delivery." *International Journal of Molecular Sciences* 16 (4): 8070–8101. https://doi.org/10.3390/ijms16048070.

Gabizon, Alberto, Hilary Shmeeda, and Yechezkel Barenholz. 2003. "Pharmacokinetics of Pegylated Liposomal Doxorubicin Review of Animal and Human Studies." *Clinical Pharmacokinetics* 42 (5): 419–36. https://doi.org/10.2165/00003088-200342050-00002.

Gambhir, S S, and S Weiss. 2005. "Quantum Dots for Live Cells, *in Vivo* Imaging, and Diagnostics." *Science* 307 (5709): 538–44. https://doi.org/10.1126/science.1104274.

Goldman, Ellen R, and Hedi Mattoussi. 2005. "Quantum Dot Bioconjugates for Imaging, Labelling and Sensing." *Nature Materials* 4 (6): 435–46. https://doi.org/doi.org/10.1038/nmat1390.

Gonçalves, Giulliana Augusta Rangel, and Raquel de Melo Alves Paiva. 2017. "Gene Therapy: Advances, Challenges and Perspectives." *Einstein* 15 (3): 369–75. https://doi.org/10.1590/S1679-45082017RB4024.

Gothwal, Avinash, Iliyas Khan, and Umesh Gupta. 2016. "Polymeric Micelles: Recent Advancements in the Delivery of Anticancer Drugs." *Pharmaceutical Research* 33: 18–39. https://doi.org/10.1007/s11095-015-1784-1.

Han, Fengxuan, Jiayuan Wang, Luguang Ding, Yuanbin Hu, and Wenquan Li. 2020. "Tissue Engineering and Regenerative Medicine: Achievements, Future, and Sustainability in Asia." *Frontiers in Bioengineering and Biotechnology* 8 (83): 1–35. https://doi.org/10.3389/fbioe.2020.00083.

Huang, Lei, Shulin Tian, Wenhao Zhao, Ke Liu, and Xing Ma. 2020. "Multiplexed Detection of Biomarkers in Lateral- Flow Immunoassays." *Analyst* 145 (8): 2828–40. https://doi.org/10.1039/c9an02485a.

Huo, Shuaidong, Shubin Jin, Xiangdong Xue, Keni Yang, Anil Kumar, Paul C Wang, Jinchao Zhang, et al. 2014. "Ultrasmall Gold Nanoparticles as Carriers for Nucleus-Based Gene Therapy Due to Size-Dependent." *ACS Nano* 8 (6): 5852–62. https://doi.org/10.1021/nn5008572.

Hussain, Shah, Iqra Mubeen, Niamat Ullah, Syed Shahab, Ud Din, Bakhtawar Abduljalil Khan, Muhammad Zahoor, Riaz Ullah, Farhat Ali Khan, and Mujeeb a Sultan. 2022. "Modern Diagnostic Imaging Technique Applications and Risk Factors in the Medical Field: A Review." *BioMed Research International* 2022 (1): 5164970. https://doi.org/10.1155/2022/5164970.

Ikada, Yoshito. 2006. "Challenges in Tissue Engineering." *Journal of Royal Society Interface* 3 (10): 589–601. https://doi.org/10.1098/rsif.2006.0124.

Jackson, Nicholas A C, and Kent E Kester. 2020. "The Promise of MRNA Vaccines: A Biotech and Industrial Perspective." *Npj Vaccines* 5 (1): 11. https://doi.org/10.1038/s41541-020-0159-8.

Javidi, M, M Heydari, A Karimi, M Haghpanahi, M Navidbakhsh, and A Razmkon. 2014. "Evaluation of the Effects of Injection Velocity and Different Gel Concentrations on Nanoparticles in Hyperthermia Therapy." *J Biomed Phys Eng* 4 (4): 151–62. https://doi.org/PMC4289522.

Javidi, Mehrdad, Morteza Heydari, Mohammad Mahdi Attar, Mehrdad Javidi, Morteza Heydari, Mohammad Mahdi Attar, Mohammad Haghpanahi, and Alireza Karimi. 2015. "Cylindrical Agar Gel with Fluid Flow Subjected to an Alternating Magnetic Field during Hyperthermia." *International Journal of Hyperthermia* 31 (1): 33–39. https://doi.org/10.3109/02656736.2014.988661.

Jeong, Hyewoo, Hayeon Byun, Jinkyu Lee, Yujin Han, Seung Jae Huh, and Heungsoo Shin. 2024. "Enhancement of Bone Tissue Regeneration with Multi-Functional Nanoparticles by Coordination of Immune, Osteogenic, and Angiogenic Responses." *Advanced Healthcare Materials*, 2400232. https://doi.org/10.1002/adhm.202400232.

Jun, Indong, Hyung-seop Han, James R Edwards, and Hojeong Jeon. 2018. "Electrospun Fibrous Scaffolds for Tissue Engineering: Viewpoints on Architecture and Fabrication." *International Journal of Molecular Sciences* 19 (3): 745. https://doi.org/10.3390/ijms19030745.

Kim, Jihye, Hwan Ryu, and Tae-hwan Kim. 2022. "Early Reoperation Rates and Its Risk Factors after Instrumented Spinal Fusion Surgery for Degenerative Spinal Disease: A Nationwide Cohort Study of 65,355 Patients." *Journal of Clinical Medicine Article* 11 (12): 3338. https://doi.org/10.3390/jcm11123338.

Kohn, Donald B, Yvonne Y. Chen, and Melissa J. Spencer. 2023. "Successes and Challenges in Clinical Gene Therapy." *Gene Therapy* 30 (10): 738–46. https://doi.org/10.1038/s41434-023-00390-5.

Lee, Kunwoo, Michael Conboy, Hyo Min Park, Fuguo Jiang, Hyun Jin Kim, Mark A Dewitt, Vanessa A Mackley, et al. 2017. "Nanoparticle Delivery of Cas9 Ribonucleoprotein and Donor DNA *in Vivo* Induces Homology-Directed DNA Repair." *Nature Biomedical Engineering* 1 (11): 889–901. https://doi.org/10.1038/s41551-017-0137-2.

Liu, Mei, Xin Zeng, Chao Ma, Huan Yi, Zeeshan Ali, Xianbo Mou, Song Li, Yan Deng, and Nongyue He. 2017. "Injectable Hydrogels for Cartilage and Bone Tissue Engineering." *Bone Research* 5 (1): 17014. https://doi.org/10.1038/boneres.2017.14.

Mak, Isabella W Y, Nathan Evaniew, and Michelle Ghert. 2014. "Lost in Translation: Animal Models and Clinical Trials in Cancer Treatment." *Am J Transl Res* 6 (2): 114–18. https://doi.org/PMCID: PMC3902221.

Mehrmohammadi, Mohammad, Soon Joon Yoon, Douglas Yeager, and Stanislav Y Emelianov. 2013. "Photoacoustic Imaging for Cancer Detection and Staging." *Current Molecular Imaging* 2 (1): 89–105.

Meng, Xiaoting, Hugh C. Seton, Le T. Lu, Ian A. Prior, Nguyen T. K. Thanh, and Bing Song. 2011. "Magnetic CoPt Nanoparticles as MRI Contrast Agent for Transplanted Neural Stem Cells Detection." *Nanoscale* 3 (3): 753–1276. https://doi.org/10.1039/c0nr00846j.

Mercuri, Eugenio, Maria Carmela Pera, Mariacristina Scoto, Richard Finkel, and Francesco Muntoni. 2020. "Spinal Muscular Atrophy — Insights and Challenges in the Treatment Era." *Nature Reviews Neurology* 16 (12): 706–15. https://doi.org/10.1038/s41582-020-00413-4.

Morales, De Marı P, Ingrid B Bo, Johannes T Heverhagen, Davide Prosperi, and Wolfgang J Parak. 2012. "Biological Applications of Magnetic Nanoparticles." *Chem Soc Rev* 41 (11): 4306–34. https://doi.org/10.1039/c2cs15337h.

Mora-Raimundo, Patricia, Daniel Lozano, Manuel Benito, Francisca Mulero, Miguel Manzano, and María Vallet-regí. 2021. "Osteoporosis Remission and New Bone Formation with Mesoporous Silica Nanoparticles." *Advanced Science* 8 (16): 2101107. https://doi.org/10.1002/advs.202101107.

Moreno-aspitia, Alvaro, Edith A Perez, and Alvaro Moreno-aspitia. 2005. "Nanoparticle Albumin-Bound Paclitaxel (Abi-007): A Newer Taxane Alternative in Breast Cancer." *Future Oncol.* 1 (6): 755–62. https://doi.org/10.2217/14796694.1.6.755.

Moshfegh, A Z. 2009. "Nanoparticle Catalysts." *Journal of Physics D: Applied Physics* 42 (23): 233001. https://doi.org/10.1088/0022-3727/42/23/233001.

Munoz, Marcelo, Cagla Eren Cimenci, Keshav Goel, Maxime Comtois-bona, Mahir Hossain, Christopher Mctiernan, Matias Zun, et al. 2022. "Nanoengineered Sprayable Therapy for Treating Myocardial Infarction." *ACS Nano* 16 (3): 3522−3537. https://doi.org/10.1021/acsnano.1c08890.

Nayerossadat, Nouri, Talebi Maedeh, and Palizban Abas Ali. 2012. "Viral and Nonviral Delivery Systems for Gene Delivery." *Advanced Biomedical Research* 1 (2): 27. https://doi.org/10.4103/2277-9175.98152.

Nelson, Christopher E, and Charles A Gersbach. 2016. "Engineering Delivery Vehicles for Genome Editing." *Annual Review of Chemical and Biomolecular Engineering* 7: 637–62. https://doi.org/10.1146/annurev-chembioeng-080615-034711.

Pardi, Norbert, Michael J Hogan, Frederick W Porter, and Drew Weissman. 2018. "MRNA Vaccines — a New Era in Vaccinology." *Nature Publishing Group* 17 (4): 261–79. https://doi.org/10.1038/nrd.2017.243.

Public Health Agency of Canada. n.d. "Six Types of Cardiovascular Disease." https://doi.org/http://www.phac-aspc.gc.ca/cd-mc/cvd-mcv/cvd-mcv-eng. php.

Reddy, Kavyanjali, Pankaj Gharde, Harshal Tayade, Mihir Patil, Lucky Srivani Reddy, and Dheeraj Surya. 2023. "Advancements in Robotic Surgery: A Comprehensive Overview of Current Utilizations and Upcoming Frontiers." *Cureus* 15 (12): e50415. https://doi.org/10.7759/cureus.50415.

Reimer, Peter, and Thomas Balzer. 2003. "Ferucarbotran (Resovist): A New Clinically Approved RES-Specific Contrast Agent for Contrast-Enhanced MRI of the Liver: Properties, Clinical Development, and Applications." *European Radiology* 13: 1266–76. https://doi.org/10.1007/s00330-002-1721-7.

Rosenholm, Jessica M, Cecilia Sahlgren, and Mika Lindén. 2011. "Multifunctional Mesoporous Silica Nanoparticles for Combined Therapeutic, Diagnostic and Targeted Action in Cancer Treatment." *Current Drug Targets,* 12 (8): 1166–86. https://doi.org/10.2174/138945011795906624.

Sahay, Gaurav, William Querbes, Christopher Alabi, Ahmed Eltoukhy, Sovan Sarkar, Christopher Zurenko, Emmanouil Karagiannis, et al. 2013. "Efficiency of SiRNA Delivery by Lipid Nanoparticles Is Limited by Endocytic Recycling." *Nature Biotechnology* 31 (7). https://doi.org/10.1038/nbt.2614.

Sawy, Amany M, Ahmed Barhoum, Sara A Abdel, Salwa M El-hallouty, Wafaa G Shousha, Ahmed A Maarouf, and Ahmed S G Khalil. 2021. "Insights of Doxorubicin Loaded Graphene Quantum Dots: Synthesis, DFT Drug Interactions, and Cytotoxicity." *Materials Science & Engineering C* 122: 111921. https://doi.org/10.1016/j.msec.2021.111921.

Senapati, Sudipta, Arun Kumar Mahanta, Sunil Kumar, and Pralay Maiti. 2018. "Controlled Drug Delivery Vehicles for Cancer Treatment and Their Performance." *Signal Transduction and Targeted Therapy* 3 (1): 7. https://doi.org/10.1038/s41392-017-0004-3.

Shahid, Abdus, Abdur Rahim, Mohammad Asaduzzaman, Chowdhury Mohammad, and Abul Kashem. 2022. "Development of Antibacterial Nanofibrous Wound Dressing and Conceptual Reaction Mechanism to Deactivate the Viral Protein by Nigella Sativa Extract." *Advances in Traditional Medicine* 22 (2): 283–91. https://doi.org/10.1007/s13596-020-00538-3.

Sharifi, Ibrahim, H Shokrollahi, and S Amiri. 2012. "Ferrite-Based Magnetic Nanofluids Used in Hyperthermia Applications." *Journal of Magnetism and Magnetic Materials* 324 (6): 903–15. https://doi.org/10.1016/j.jmmm.2011.10.017.

Shiraishi, Kouichi, and Masayuki Yokoyama. 2019. "Toxicity and Immunogenicity Concerns Related to PEGylated-Micelle Carrier Systems: A Review." *Science and Technology of Advanced Materials* 20 (1): 324–36. https://doi.org/10.1080/14686996.2019.1590126.

Socinski, Mark A, Igor Bondarenko, Nina A Karaseva, Anatoly M Makhson, Igor Vynnychenko, Isamu Okamoto, Jeremy K Hon, et al. 2012. "Weekly Nab-Paclitaxel in Combination with Carboplatin Versus Solvent-Based Paclitaxel Plus Carboplatin as First-Line Therapy in Patients with Advanced Non – Small-Cell Lung Cancer: Final Results of a Phase III Trial." *Journal of Clinical Oncology* 30 (17): 2055–62. https://doi.org/10.1200/JCO.2011.39.5848.

Song, Wantong, Sara N Musetti, and Leaf Huang. 2017. "Biomaterials Nanomaterials for Cancer Immunotherapy." *Biomaterials* 148: 16–30. https://doi.org/10.1016/j.biomaterials.2017.09.017.

Soo, Jung, Qingguo Xu, Namho Kim, Justin Hanes, and Laura M Ensign. 2016. "PEGylation as a Strategy for Improving Nanoparticle-Based Drug and Gene Delivery." *Advanced Drug Delivery Reviews* 99: 28–51. https://doi.org/10.1016/j.addr.2015.09.012.

Sterner, Robert C, and Rosalie M Sterner. 2021. "CAR-T Cell Therapy: Current Limitations and Potential Strategies." *Sterner and Sterner Blood Cancer Journal* 11 (4): 69. https://doi.org/10.1038/s41408-021-00459-7.

Tebas, Pablo, David Stein, Winson W Tang, Ian Frank, Shelley Q Wang, Gary Lee, D Ph, et al. 2014. "Gene Editing of CCR5 in Autologous CD4 T Cells of Persons Infected with HIV." *The New England Journal of Medicine* 370 (10): 901–10. https://doi.org/10.1056/NEJMoa1300662.

Thakur, Deepa, Ankita Mathur, Moolchand Sharma, Aditi Halder, and Rahul Vaish. 2021. "Photocatalytic and Wettability Behavior of Regenerative Bio-Inspired CuxO (X=1,2)." *Materials Research Bulletin* 144 (July): 111489. https://doi.org/10.1016/j.materresbull.2021.111489.

Thakur, Deepa, Chirag Porwal, Vishal Singh, Viswanath Balakrishnan, and Rahul Vaish. 2024. "2D Transition Metal Dichalcogenides: Synthesis Methods and Their Pivotal Role in Photo, Piezo, and Photo-Piezocatalytic Processes." *Separation and Purification Technology* 337: 126462. https://doi.org/10.1016/j.seppur.2024.126462.

Thakur, Deepa, Moolchand Sharma, Viswanath Balakrishnan, and Rahul Vaish. 2022. "Reusable Piezocatalytic Water Disinfection Activity of CVD-Grown Few-Layer WS2 on Sapphire Substrate." *Environmental Science: Nano* 9 (2): 805–14. https://doi.org/10.1039/d1en00926e.

Thakur, Deepa, Pawan Kumar, Arjun Barwal, Deep Jariwala, and Erich Stach. 2023. "Site-Specific Optical Encryption via Nanoscale Integration of Carbon on Monolayer WS2." *Carbon* 214: 118339. https://doi.org/10.1016/j.carbon.2023.118339.

Tong, Bing, Yan Xu, Jing Zhao, Minjiang Chen, Jia Xing, and Wei Zhong. 2017. "Prognostic Significance of Circulating Tumor Cells in Non-Small Cell Lung Cancer Patients Undergoing Chemotherapy." *Oncotarget* 8 (49): 86615–24. https://doi.org/10.18632/oncotarget.21255.

Turchanin, Andrey, Dirk Weber, Christian Kisielowski Matthias B€uenfeld, Mikhail V. Fistul, Thomas Weimann Konstantin B. Efetov, Rainer Stosch, Joachim Mayer, and Armin G€olzh€auser. 2011. "Conversion of Self-Assembled Monolayers into Nanocrystalline Graphene: Structure and Electric." *ACS Nano* 5 (5): 3896–3904. https://doi.org/10.1021/nn200297n.

Yoon, Hong Yeol, Sangmin Jeon, Dong Gil You, Jae Hyung Park, Ick Chan Kwon, Heebeom Koo, and Kwangmeyung Kim. 2017. "Inorganic Nanoparticles for Image-Guided Therapy." *Bioconjugate Chem* 28 (1): 124–34. https://doi.org/10.1021/acs.bioconjchem.6b00512.

Section 2

Chapter 2.1

Nanomaterials for Drug Delivery

Vennila Selvaraj[1,*]
Suresh Sagadevan[2]
and Gurunathan Karuppasamy[1]
[1]Nanofunctional Materials Laboratory, Department of Nanoscience and Technology, Science Campus, Tamilnadu, India
[2]Nanotechnology and Catalysis Research Centre, University of Malaya, Kuala Lumpur, Malaysia

Abstract

The exciting field of nanomedicine uses nanoscale materials to revolutionize healthcare. These nanomedicines have enormous potential for the delivery of drugs. When drugs are delivered precisely to the target place, nanomedicine offers promising approaches to treating chronic diseases. From chemotherapy to immunotherapy, nanoparticles have become a versatile weapon in the fight against various diseases. This chapter describes the latest advances in nanomedicine and drug delivery systems. We explore how scientists use the power of nanomaterials to improve the effectiveness of existing and new nanomedicines. The world of insignificant nanotechnology, where materials are made at theatomic and molecular level, has a key impact on our daily lives. This technology has transformed various industries from fighting diseases to providing targeted drugs. Nanotechnology is changing our quality of life and shaping the world economy. The chapter concludes with a vision of future trends and prospects in this dynamic field.

Keywords: nanomedicine, nanomaterials, drug delivery system, types of nanoparticles, nanotechnology

1. Introduction

Nanotechnology is a rapidly evolving field of materials and devices that are designed and manufactured in tiny quantities of nanometers (nm) measured in a billionth of a meter ("The National Institutes of Health Roadmap for Medical Research Website: Http://Www.

* Corresponding Author's Email: vennilaphysics93@gmail.com

In: Emerging Nanomaterials in Biomedical Science
Editors: Eepsita Priyadarshini and Ajith Manayil Parambil
ISBN: 979-8-89530-600-0

Nihroadmap.Nih.Gov. Accessed 14 November 2007." 2008). This field has enormous potential for medicine, especially in developing nanoparticles (NPs). Nanomedicine is a branch of nanotechnology that focuses on creating new and improved tools for diagnosing and treating diseases. These applications range from medical imaging and drug delivery to tissue regeneration (Version 2024). The unique size of NPs is like biomolecules such as proteins and DNA and allows interactions at the cell level. Although traditionally defined as materials within the 1-100 nm range, nanomedicine proposes extending this size limit to 1000 nm in some applications. Nanotechnology is expected to have a value of $528 billion in medicine by 2019 and will continue to grow strongly over the next few years (Manuscript 2014). Advancements in engineered nanomaterials are opening doors for more precise disease diagnosis and treatment. Nanotechnology offers solutions to challenges faced with conventional drug delivery. From difficulties with distribution throughout the body to limitations within cells, nanotech can address these issues through targeted approaches. This includes delivering drugs to specific cells, transporting them to organelles within those cells, and other innovative methods. To translate these promising possibilities into clinical practice, the National Nanotechnology Initiative (NNI) was established by the US National Science and Technology Council (NSTC) in 2000 ("National Nanotechnology Initiative - Leading to the Next Industrial Revolution" 2000). The NNI set clear goals and challenges for the field. This initiative has fueled recent research efforts to improve nanotechnology, with nanoparticles (NPs) being a major area of exploration and progress. NPs have the potential to significantly improve the stability and solubility of the drugs they carry, allowing them to pass through cell membranes more easily and circulate in the body for longer periods. This can ultimately enhance treatment safety and effectiveness (Kou et al. 2018) (Blanco, Shen, and Ferrari 2015).

In this book chapter, nanomaterials offer a revolutionary approach to drug delivery. Their unique properties can overcome the limitations of conventional therapies, leading to more targeted, effective, and safer treatments. As research progresses, nanomaterials are poised to play a pivotal role in shaping the future of medicine.

2. Types of Nanomaterials for Drug Delivery

Several classes of nanomaterials are showing promise in drug delivery, each offering unique benefits and considerations for use. Let's explore some of the most widely studied types of nanoparticles.

2.1. Polymeric Nanoparticles

These engineered materials are crafted from synthetic polymers, designed to be compatible with the body, and break down naturally over time. They act as microscopic shields, encapsulating drugs and protecting them from premature breakdown in the body. This allows for controlled release of the drug, delivering it steadily over a desired timeframe. Some common examples of these synthetic polymeric nanoparticles include polyacrylamide, polyacrylate, and chitosan (Tiyaboonchai 2003).

2.1.1. Biodegradable and Synthetic Polymers Used in Nanoparticle Fabrication

The field of drug delivery has seen a surge in interest in natural, biodegradable polymers. This focus on biocompatible and naturally degradable materials offers a promising alternative to conventional dosage forms. Researchers are particularly interested in these polymers due to their potential for improved bioavailability, reduced toxicity, and targeted drug delivery. These advancements pave the way for a new generation of medicines – nanomedicines (Idrees et al. 2020). The concept of targeted drug delivery, likened to "magic bullets" by Paul Ehrlich over a century ago, is finally becoming a reality. In 1996, Kumar and Banker's vision of submicron drug delivery systems took a major step forward. Liposomes and micro/nanoparticles have emerged as key players in the field of drug delivery. However, liposomes face limitations like stability issues, inconsistent production, and limited drug-loading capacity, particularly for larger molecules. As a result, polymeric nanoparticle drug delivery systems are emerging as a powerful alternative to conventional methods (Tiyaboonchai 2003).

2.1.2. Benefits of Biocompatibility, Water Solubility, and Modifiable Surfaces for Targeting

These tiny particles act like cautious carriers within your body. Their biocompatible design ensures they don't cause irritation or immune system attacks. Water solubility allows them to seamlessly travel through your bloodstream and bodily fluids, reaching their target zone effortlessly. Most importantly, scientists can equip these particles with customizable surface markers. These markers act like address labels, specifically directing the particles towards diseased cells, like a guided missile finding its target. This targeted delivery minimizes harm to healthy cells while maximizing the impact on diseased ones. By combining these three properties, scientists have created a powerful tool for drug delivery, gene therapy, and various other medical applications (Large et al. 2019).

Nanoparticles (NPs) can be engineered for "active targeting" by decorating their surfaces with specific molecules. These surface modifications allow NPs to interact with molecules overexpressed by diseased cells or tissues (Ahmad et al. 2022). This approach improves drug delivery efficiency by guiding NPs toward their intended targets (Mitchell et al. 2021). Various molecules can be used for surface modification, including small proteins, peptides, antibodies, aptamers, and oligos saccharides. These modifications enhance targeting and can play a crucial role in reducing toxicity, especially for materials like silver nanoparticles (Large et al. 2019). For instance, attaching human albumin to NPs can simultaneously decrease their toxicity and promote targeting to specific sites (Rea et al. 2014). Creating these targeted NPs is a multi-step process requiring careful characterization at each stage. Techniques like transmission electron microscopy (TEM), scanning electron microscopy (SEM), Fourier transform infrared spectroscopy (FTIR), dynamic light scattering (DLS), and zeta potential analysis are commonly used to assess the shape, size, chemical composition, and surface charge of the NPs. This detailed understanding ensures the successful development of targeted nanoparticles for optimized drug delivery (Terracciano et al. 2015) (Silvestri et al. 2019).

2.2. Liposome Delivery

Liposomes are microscopic spheres with a unique structure. Imagine tiny bubbles formed by a double layer of fatty molecules (phospholipids), like those in cell membranes. These spheres enclose a watery core, making them ideal carriers for various drugs. One of the major advantages of liposomes is their biocompatibility – they are non-toxic and naturally degradable to the body. This characteristic makes them a powerful tool for drug delivery. Liposomes offer several benefits. They can protect drugs from degradation in the body, ensuring they remain potent until they reach their target. Liposomes can bypass barriers that hinder some drugs from reaching their target cells or tissues. Tailoring liposome surfaces allows for targeted drug delivery to diseased areas, reducing potential side effects in healthy tissues. Overall, liposomes represent a significant advancement in drug delivery, offering improved therapeutic efficacy and reduced systemic toxicity (Guimarães, Cavaco-Paulo, and Nogueira 2021). As shown in Figure 1(a) the general form of Liposomes the unique structure of liposomes allows them to act as versatile carriers for a wide range of drugs due to their ability to accommodate molecules with varying solubility. Imagine these liposomes like microscopic packaging with different compartments, Water-soluble (hydrophilic) drugs can be encapsulated within the watery core of the liposome. Fat-soluble (hydrophobic) drugs can be embedded within the liposomes' fatty layer (phospholipid bilayer). Molecules with both water and fat-devoted properties (amphiphilic) can reside at the interface between the watery core and the fatty layer. This ability to accommodate drugs with different properties makes liposomes a highly adaptable drug delivery platform. As shown in Figure 1(b), liposomes can be constructed from various types of lipids, offering flexibility for specific drug delivery needs.

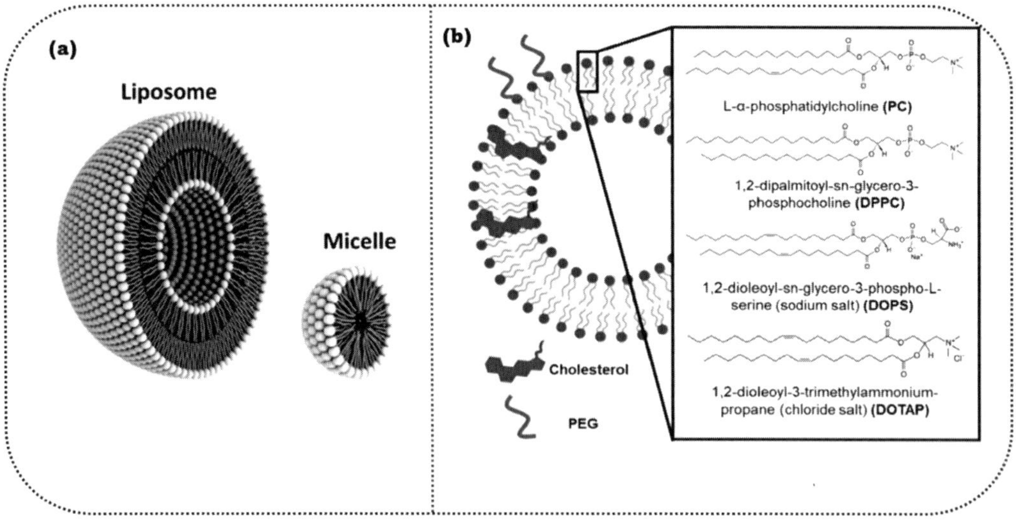

Figure 1. (a) General form of Liposomes. (b) Liposomes can be constructed from various lipids, offering flexibility for specific drug delivery needs. Reproduced with permission from Gonzalez et al. Copyright 2020, ACS. (Gonzalez Gomez and Hosseinidoust 2020).

2.2.1. Structure and Composition of Liposomes

Liposomes come in various structures based on their organization and number of membrane layers. Here is a breakdown of the main categories:

Unilamellar Vesicles (ULVs) have a single, concentric bilayer membrane surrounding an aqueous core. ULVs can be further classified by size: Small Unilamellar Vesicles (SUVs), ranging from 30-100 nm, are the smallest type of liposome (some studies report SUVs up to 200 nm) (Fan, Marioli, and Zhang 2021). Large Unilamellar Vesicles (LUVs), larger than 100 nm, offer more space for cargo. (Some studies report LUVs up to 500 nm). Giant Unilamellar Vesicles (GUVs) are the biggest ULVs, exceeding 1000 nm in size (Pattni, Chupin, and Torchilin 2015). Oligolamellar Vesicles (OLVs), with 2-5 concentric bilayer membranes, OLVs exhibit an onion-like structure. Multilamellar Vesicles (MLVs) have the most concentric bilayers, typically more than 5, creating a more complex layered structure (Kim, Kim, and Kim 1993). Multivesicular Liposomes (MVLs): Unlike MLVs, MVLs possess a unique "honeycomb" structure. These structures have many separate fluid-filled compartments, with each compartment surrounded by a single, thin membrane made of two layers. This classification system helps researchers choose the most suitable liposome type for their specific needs, considering factors like size, cargo capacity, and desired release profile (Liu, Chen, and Zhang 2022).

2.3. Inorganic Nanoparticles

Researchers are excited about the potential of inorganic nanoparticles to transform drug delivery. These microscopic particles, smaller than 100 nanometers wide, offer distinct advantages over traditional methods. A key benefit is their ability to place in specific cells or tissues (Nasirzadeh, Nazarian, and Hayat 2016). This is achieved by attaching targeting molecules to the nanoparticle surface. These molecules bind to receptors on the desired cells, acting like a guided missile for drug delivery. This targeted approach can increase drug effectiveness and minimize side effects on healthy cells (Unnikrishnan et al. 2023). Inorganic nanoparticles can also be designed to release drugs slowly over time. By controlling the degradation rate of the nanoparticles, researchers can create a sustained release system for medicines that need to be delivered over a long period or have a short lifespan in the body (Asad, Jacobsen, and Teleki 2022).

This technology rapidly evolves, but hurdles must be overcome before widespread clinical use. One challenge is ensuring the safety of these nanoparticles, as some may be toxic to cells (Afzal et al. 2022). Additionally, the manufacturing cost needs to be reduced. Despite these challenges, continued research holds the promise of significant advancements in drug delivery merits to inorganic nanoparticles (Marcum Zachary A, Sevick MA 2013).

2.3.1. Inorganic Nanoparticles for Drug Delivery

Recent advancements in inorganic nanoparticle synthesis and modification have significantly propelled their use in biomedicine (Unnikrishnan et al. 2023). These nanoparticles offer a unique platform for integrating multiple functionalities within a single structure (Gao et al. 2004). For instance, gold nanoparticles can be precisely controlled in size and shape, and their surfaces can be decorated with antibodies (Gao et al. 2004). This allows for targeted imaging and light-mediated destruction of cancer cells using near-infrared light for deeper tissue penetration. Similar progress has been achieved with polymer-coated superparamagnetic iron oxide nanoparticles. By attaching various components like fluorescent molecules, cancer-targeting agents, anticancer drugs, or siRNA to the polymer coating, these multifunctional nanoparticles can not only target tumors but also be visualized within the body using both magnetic resonance imaging (MRI) and fluorescence imaging techniques (Lee et al. 2007).

3. Advantages of Nanomaterials in Drug Delivery

Recent breakthroughs in drug delivery have emerged with the development of nanocarrier systems. These carriers provide controlled release of therapeutic drugs to target specific diseased cells. Nanomaterials (NMs) and nanoparticles (NPs) are microscopic materials typically sized between 1 and 100 nanometers (L. Cheng, Jiang, and Wang 2013), Nanomaterials (NMs) exhibit a remarkable combination of unique properties in terms of physics, chemistry, mechanics, electricity, and interaction with living systems. This arises from their drastically reduced size and the corresponding increase in surface area compared to their bulk counterparts. Importantly, the high surface area to volume ratio in NMs can give rise to quantum confinement effects, which further tailor their properties in distinctive ways (Singh 2010). Using NMs and nanostructures as drug delivery systems (DDS) has garnered significant interest in nanomedicine (Allen et al. 2012). These nanocarriers can encapsulate or chemically conjugate drugs, allowing for controlled release at target tissues (Hami 2021). Nanomaterials' minuscule size and expansive surface area allow them to enter cells and interact with biomolecules more effectively. This paves the way for researchers to harness nanotechnology in drug delivery systems, potentially improving drug absorption, bioavailability, and stability. These advancements could address the shortcomings faced by traditional drug delivery methods.

Nanoparticles possess several key attributes that make them advantageous drug delivery vehicles. These include high stability, substantial cargo capacity, and the ability to encapsulate hydrophilic and hydrophobic therapeutic moieties. Additionally, nanoparticles can be formulated for administration via diverse routes, including oral and inhalation delivery. Notably, controlled (sustained) release of the drug payload from the nanoparticle matrix can be achieved through design strategies. Collectively, these properties of nanoparticles offer the potential to improve drug bioavailability, reduce dosing frequency, and potentially address medication non-adherence (Gelperina et al. 2005).

3.1. Enhanced Solubility and Bioavailability of Drugs

Poor aqueous solubility is a major hurdle in drug development, hindering a molecule's therapeutic potential. This necessitates strategies to improve bioavailability, a critical factor for successful drug action. Low-solubility drugs present significant challenges during formulation and development, impacting their pharmacokinetic and pharmacodynamic profiles, including distribution, protein binding, and absorption. Orally administered drugs, constituting many pharmaceutical forms, particularly require good water solubility for effective delivery. In the quest for new medicines, screening programs identify a substantial number (approximately 40%) of new chemical entities (NCEs) with limited solubility and bioavailability, posing a formidable obstacle in the pharmaceutical formulation landscape (Bhalani et al. 2022) (Yadava et al. 2015).

3.2. Improved Targeting and Delivery to Specific Tissues

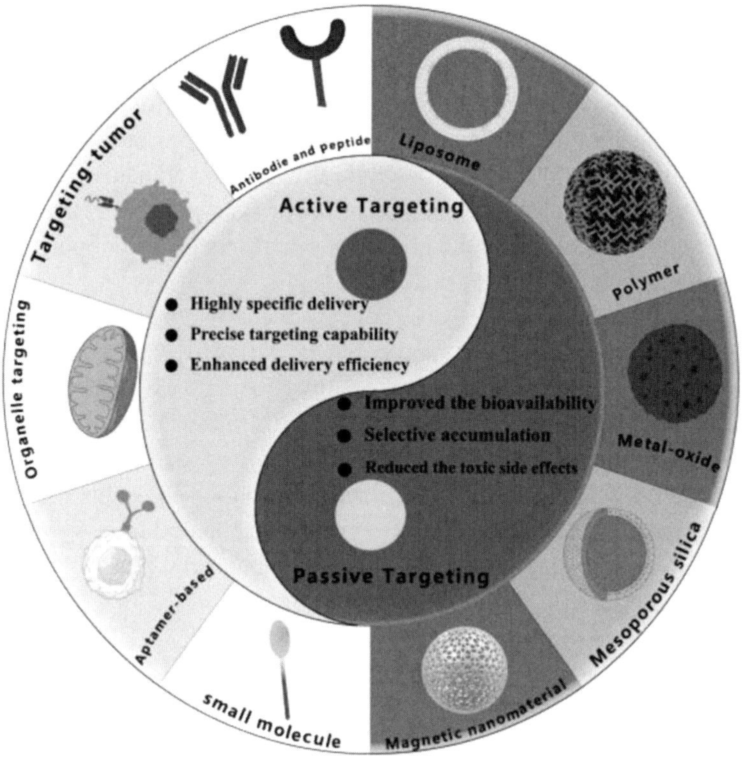

Figure 2. Passive and active targeting mechanisms (Adapted from Li et al. 2023).

Nanoparticle-mediated targeted drug delivery offers a precise and efficacious approach for delivering therapeutic agents directly to tumor cells or tissues, minimizing exposure to healthy cells (Glasgow and Chougule 2015). This strategy leverages nanocarriers, accumulating at tumor sites due to the enhanced permeability and retention (EPR) effect (Dang and Guan 2020). A crucial component of these systems is the targeting moiety, specifically designed to bind with receptors or biomolecules overexpressed on the target cells (Di Stefano 2023). Consequently, targeted drug delivery facilitates personalized medicine due to its potential for reduced drug dosage, enhanced therapeutic efficacy, and minimized side effects, further exploration of nanoparticle-based carriers (Chandra, Barick, and Bahadur 2011). Nanocarriers offer a revolutionary approach to drug delivery by improving bioavailability and efficacy through various targeting mechanisms (Chamundeeswari, Jeslin, and Verma 2019). By acting as delivery vehicles, these carriers can influence how drugs move through the body and reach their target tissues, ultimately enhancing their ability to enter cells. Drugs are directed towards specific tumor sites, significantly improving therapeutic effects and minimizing side effects on healthy tissues. Nanotechnology enhances drug stability and bioavailability, leading to prolonged therapeutic action and potentially reducing dosing frequency and drug amount required. Figure 2 explains Nanocarriers, due to their inherent size and properties, passively accumulate in tumor sites by exploiting the Enhanced Permeability and Retention (EPR) effect (Pushpalatha, Selvamuthukumar, and Kilimozhi 2017). Tumors create a unique environment for drug delivery. Leaky blood vessels and poor lymphatic drainage allow nanoparticles, or nanocarriers, to seep out of the bloodstream and accumulate within the tumor. We can further enhance this process by equipping nanocarriers with targeting molecules like antibodies or peptides. These molecules recognize and bind to specific receptors found in high numbers on tumor cells, leading to a more precise drug delivery. The beauty of nanocarriers lies in their versatility. By adjusting their size, shape, and surface properties, we can control how they release their cargo and how well they target specific tissues. This allows them to interact precisely with biological molecules, improving their stability and how long they circulate in the bloodstream. The ability to integrate multiple drugs and therapeutic approaches within a single nanocarrier platform holds immense promise for synergistic oncotherapy. This allows for the co-delivery of drugs with complementary mechanisms of action, potentially leading to improved therapeutic outcomes (Li et al. 2023).

The latest breakthroughs in targeted drug delivery focus on creating new "homing beacons" (like antibodies, peptides, and aptamers) and specialized transport systems (such as liposomes, polymers, and metal oxides) tailored for precise delivery using nanoparticles. Optimizing these components fosters stronger interactions with tumor cells, ultimately improving drug delivery efficiency and specificity. Nanocarrier-originated aim drug delivery strategies effectively concentrate drugs within tumor tissue, minimizing distribution to healthy organs and tissues. This targeted approach translates to reduced systemic exposure and associated side effects, improving patient tolerability and treatment outcomes (Shah et al. 2021).

4. Potential Applications of Targeted Therapy in Areas Such as Cancer Treatment and Infectious Diseases

Current cancer treatment strategies have shifted towards targeted therapies that exploit specific molecular aberrations in cancer cells. In contrast to predictable chemotherapy, which intends entirely fast splitting cells, targeted treatments act on precise particles critical for cancer cell growth and survival. This selectivity minimizes damage to healthy tissues and reduces the incidence of severe side effects.

4.1. Concept of Targeted Drug Delivery Using Nanoparticles

Targeted drug delivery represents a paradigm shift in pharmaceutical intervention, aiming for spatiotemporal control of therapeutic delivery. This approach offers distinct advantages compared to conventional methods, by directing the drug payload to diseased cells, targeted delivery minimizes exposure of healthy tissues, thereby mitigating the potential for side impacts (Patra et al. 2018). Localized drug gathering at the point site enhances the therapeutic effect by achieving higher effective concentrations (X. Cheng, Xie, and Sun 2023). Targeted delivery may enable the utilization of potently efficacious drugs that would otherwise be too toxic for systemic administration due to unwanted side effects on healthy tissues. Nanoparticulate systems, typically ranging from 1-100 nm in size (compared to human hair at roughly 80,000 nm), leverage their minute size for efficient cellular and tissue penetration, composing them perfect carriers intended for targeted drug delivery claims (Di Stefano 2023).

4.2. Assessing the Potential of Nanocarriers in Therapeutics

Nanomaterials hold immense possibilities for revolutionizing drug delivery due to their exclusive characteristics, which can overcome challenges faced by traditional methods. Here's a look at key factors considered when evaluating nanomaterials for this application.

4.2.1. In Vivo Assessment of Targeting and Safety

Scientists use specially designed procedures or existing research models to test how well a treatment targets the intended area and assess any safety risks. These assessments encompass distribution, clearance kinetics, hematological parameters, serum biochemistry, and histopathological analysis (Pandey et al. 2023). Figure 3 depicts the intricate nature of some of these models. These studies investigate nanoparticle localization and accumulation patterns within tissues and organs. Radiolabelling techniques can track the nanoparticles in both live and deceased animals. Nanoparticles leave the body, and what happens to them over time following exposure helps us understand how quickly they are eliminated (Pandey et al. 2023). Evaluating changes in blood chemistry and cell counts in living organisms (*in vivo*) after exposure to nanoparticles can further reveal potential toxic effects (Siegrist et al. 2019). Histopathological examination of exposed cells, tissues, and organs provides expensive perceptions keen on the extent of cytotoxicity prompted by the nanoparticles (Sayes, Reed, and Warheit 2007). Following the *in vivo* assessment, a detailed discussion of conventional methods for evaluating nanoparticle cytotoxicity, such as proliferation assays, apoptosis assays, necrosis assays, oxidative stress assays, and DNA damage assays, will be presented (Pandey et al. 2023).

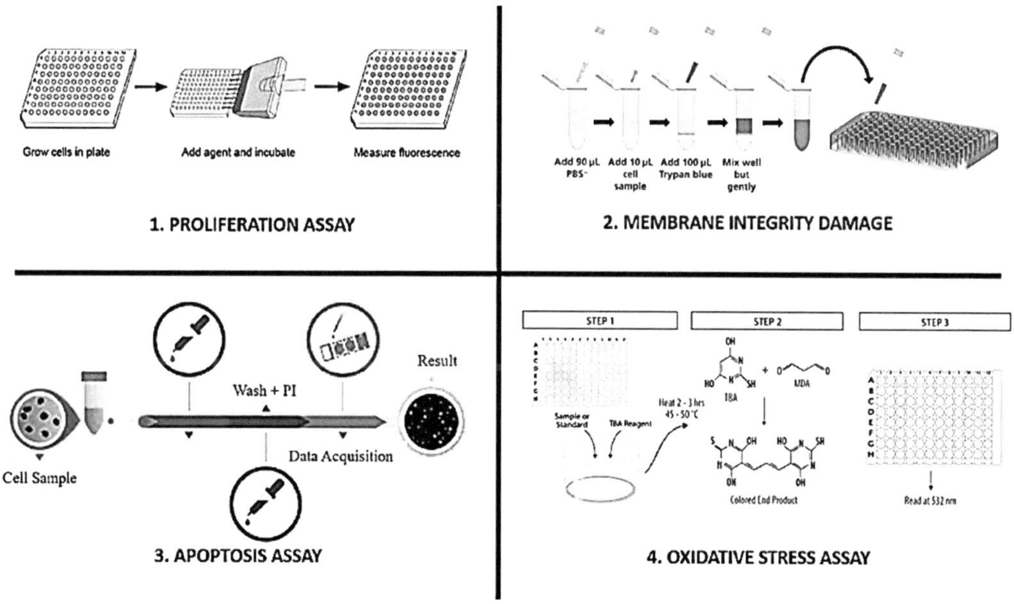

Figure 3. Established experimental models are employed to evaluate targeting efficacy (Adapted from Pandey et al. 2023).

4.2.2. In Vivo Imaging in Drug Discovery: A Window into Treatment Development of Oral Drug Delivery

The complex interplay between the digestive system (GI tract), the medication itself, and its delivery form significantly influences how well a drug is absorbed after being taken orally. To address the limitations in our knowledge of how the gut absorbs substances, the COST Action on Understanding Gastrointestinal Absorption-related Processes (UNGAP)' was initiated in 2017. This initiative aims to deepen our knowledge of intestinal drug absorption and pave the way for future advancements in this area. Studying the human gastrointestinal (GI) tract directly (*in vivo*) is crucial to understanding how our digestive system interacts with medications taken by mouth (orally administered dosage forms). This approach provides deep insights into the mechanisms behind how GI functions and drug performance are linked (Senekowitsch et al. 2022).

Researchers have employed a method called intraluminal sampling to investigate factors affecting intestinal absorption of orally administered drugs. Though invasive and labor-intensive, this technique has become increasingly used over the past two decades. After taking oral dosage forms, it has provided valuable insights into drug concentration profiles within the stomach and small intestine. Scientists have investigated how the drug's characteristics and the way it's delivered affect the amount of drug present within the lumen, which ultimately influences how much reaches the bloodstream (Diakidou et al. 2009). Intraluminal sampling has become a valuable tool in recent years for investigating key questions about drug performance in the gut. These questions include how food and antacids impact medication effectiveness, the distinction between increased drug availability (solubilization) and exceeding its solubility limit (supersaturation), and the role of stomach fluid movement in drug and formulation behavior throughout the digestive system. Interpreting intraluminal concentration data can be complex. The gut environment is constantly changing, with multiple processes like breakdown (disintegration), dissolving (dissolution), exceeding solubility (supersaturation), and reformation (precipitation) happening simultaneously. Despite these challenges, significant advancements in recent decades have provided a clearer picture of drug and formulation behavior in the upper digestive tract. Research is ongoing to further understand drug behavior further down the digestive tract, including the lower small intestine and colon (Lemmens et al. 2020) (Jain and Thareja 2019).

4.2.3. Overcoming Drug Resistance with Combined Delivery Systems

The development of drug resistance is a major hurdle in effective medical treatment. Combination therapy has emerged as a promising solution that targets multiple pathways and enhances treatment efficacy. This approach is particularly valuable in cancer research, where it can overcome multidrug resistance. Studies have shown that combined drug delivery systems can not only reduce the required dosage and minimize side effects but also maintain treatment effectiveness and hinder the development of resistance (Wang and Huang 2020). Researchers, including Zamora-Mera et al. investigated a combination therapy for magnetic hyperthermia treatment using chitosan nanoparticles (CSNPs). These CSNPs were designed to carry both ferrofluid and the chemotherapy drug 5-Fluorouracil (5-FU). The nanoparticles were tested on healthy and cancerous cells, revealing promising results. The magnetic hyperthermia treatment significantly reduced the viability of cancer cells with minimal impact on healthy cells. Interestingly, treatment with just ferrofluid-loaded CSNPs led to rapid cancer cell regrowth. However, combining magnetic hyperthermia with 5-FU loaded CSNPs resulted in less 5-FU release, suggesting a potential interaction between the therapies. This study highlights the potential of using a combined approach for drug delivery in cancer treatment. (Zamora-Mora et al. 2017).

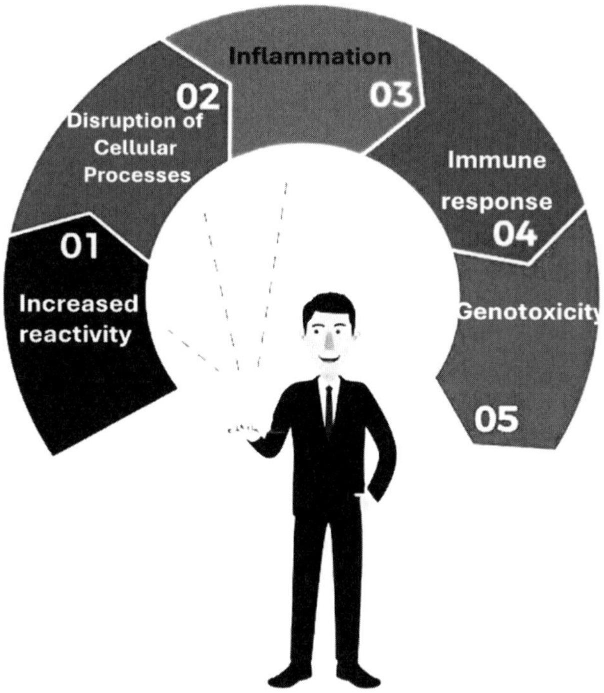

Figure 4. Some potential toxicity concerns of some nanomaterials.

5. Challenges and Future Prospects

5.1. Potential Toxicity Concerns of Some Nanomaterials

Nanotechnology offers exciting advancements in various fields, but the unique properties of nanomaterials raise concerns about potential toxicity. Due to their small size, these materials can bypass biological barriers and enter cells, unlike their larger counterparts. This ability can lead to several potential health risks (Varala et al. 2023)

- Increased reactivity: Nanoparticles have a high surface area-to-volume ratio, making them more chemically reactive. This can lead to the generation of harmful free radicals within cells, causing oxidative stress and damage.
- Disruption of Cellular Processes: Nanoparticles can interfere with essential cellular functions by interacting with proteins and organelles. This disruption can lead to cell death or impaired function (Ezike et al. 2023).
- Inflammation and immune response: The body's immune system may recognize nanoparticles as foreign invaders, triggering inflammation and potentially autoimmune reactions.

- Genotoxicity: Certain nanomaterials may damage DNA, potentially leading to mutations and increasing the risk of cancer. How harmful the effects are dependent on several things. This includes the material, how much and what shape it is in, its surface texture, and how you encountered it (breathing it in, swallowing it, or getting it on your skin). While research is ongoing, the long-term health effects of nanomaterial exposure are still under investigation (Varala et al. 2023).

5.2. Need for Further Research on Biocompatibility and Long-Term Effects

Nanodrugs hold immense promise for targeted therapy and improved drug delivery. However, their unique properties necessitate thoroughly evaluating their interaction with biological systems. This scientific note highlights the critical need for further research on the biocompatibility and long-term effects of nanodrugs (Jahan et al. 2023). The unique properties of nanomedicines, including their miniscule size and extensive surface area, raise concerns about unintended interactions with biological molecules and cellular structures. These interactions could trigger unwanted immune reactions or toxic effects within the body. Material properties include the composition, surface chemistry, and degradation profile of nanodrugs, which can significantly impact their biocompatibility. The long-term persistence of non-biodegradable nanomaterials raises concerns about bioaccumulation and potential toxicity (Ditri 2017).

5.2.1. Research Directions

- Advanced *in vitro* and *in vivo* models: Developing more sophisticated models that better mimic human physiology is essential for the accurate assessment of nano-drug biocompatibility and long-term effects.
- Standardization of testing methods: Establishing standardized protocols for biocompatibility testing of nano drugs will facilitate data comparison and drive forward the development of secure and well-tolerated nanomedicines.
- Nanotoxicology research: We need to study how nanomaterials affect living things (nanotoxicology) to understand the interactions and develop ways to lessen potential dangers.

Conclusion

In conclusion, the application of nanoparticles in nanomedicine presents a paradigm shift in drug delivery. By leveraging their unique properties, nanoparticles can be engineered for targeted release of therapeutics to specific cells and tissues. This targeted approach promises improved treatment efficacy and reduces systemic toxicity in various diseases, including cancer and infectious pathologies. However, challenges persist regarding the long-term safety of certain nanomaterials. Further investigations are warranted to elucidate the biocompatibility profiles of these materials before widespread clinical translation. In essence, nanomedicine offers exciting prospects for the future of medicine, but meticulous safety evaluation is paramount for responsible clinical implementation.

References

Afzal, Obaid, Abdulmalik S. A. Altamimi, Muhammad Shahid Nadeem, Sami I. Alzarea, Waleed Hassan Almalki, Aqsa Tariq, Bismillah Mubeen, et al. 2022. "Nanoparticles in Drug Delivery: From History to Therapeutic Applications." *Nanomaterials* 12 (24): 1–27. https://doi.org/10.3390/nano12244494.

Ahmad, Faheem, Mounir M. Salem-Bekhit, Faryad Khan, Sultan Alshehri, Amir Khan, Mohammed M. Ghoneim, Hui Fen Wu, Ehab I. Taha, and Ibrahim Elbagory. 2022. "Unique Properties of Surface-Functionalized Nanoparticles for Bio-Application: Functionalization Mechanisms and Importance in Application." *Nanomaterials* 12 (8). https://doi.org/10.3390/nano12081333.

Allen, Theresa M, Pieter R Cullis, Theresa M Allen, and Pieter R Cullis. 2012. "*Drug Delivery Systems : Entering the Mainstream*" 1818 (2004). https://doi.org/10.1126/science.1095833.

Asad, Shno, Ann Christin Jacobsen, and Alexandra Teleki. 2022. "Inorganic Nanoparticles for Oral Drug Delivery: Opportunities, Barriers, and Future Perspectives." *Current Opinion in Chemical Engineering* 38: 100869. https://doi.org/10.1016/j.coche.2022.100869.

Bhalani, Dixit V., Bhingaradiya Nutan, Avinash Kumar, and Arvind K. Singh Chandel. 2022. "Bioavailability Enhancement Techniques for Poorly Aqueous Soluble Drugs and Therapeutics." *Biomedicines* 10 (9). https://doi.org/10.3390/biomedicines10092055.

Blanco, Elvin, Haifa Shen, and Mauro Ferrari. 2015. "Principles of Nanoparticle Design for Overcoming Biological Barriers to Drug Delivery." *Nature Biotechnology* 33 (9): 941–51. https://doi.org/10.1038/nbt.3330.

Chamundeeswari, Munusamy, John Jeslin, and Madan Lal Verma. 2019. "Nanocarriers for Drug Delivery Applications." *Environmental Chemistry Letters* 17 (2): 849–65. https://doi.org/10.1007/s10311-018-00841-1.

Chandra, Sudeshna, K. C. Barick, and D. Bahadur. 2011. "Oxide and Hybrid Nanostructures for Therapeutic Applications." *Advanced Drug Delivery Reviews* 63 (14–15): 1267–81. https://doi.org/10.1016/j.addr.2011.06.003.

Cheng, Liang-chien, Xiumei Jiang, and Jing Wang. 2013. "*Cells*," 3547–69. https://doi.org/10.1039/c3nr34276j.

Cheng, Xiaoxiao, Qirong Xie, and Yang Sun. 2023. "Advances in Nanomaterial-Based Targeted Drug Delivery Systems." *Frontiers in Bioengineering and Biotechnology* 11 (April): 1–27. https://doi.org/10.3389/fbioe.2023.1177151.

Dang, Yu, and Jianjun Guan. 2020. "Nanoparticle-Based Drug Delivery Systems for Cancer Therapy." *Smart Materials in Medicine* 1 (February): 10–19. https://doi.org/10.1016/j.smaim.2020.04.001.

Diakidou, Amalia, Maria Vertzoni, Konstantinos Goumas, Erik Söderlind, Bertil Abrahamsson, Jennifer Dressman, and Christos Reppas. 2009. "Characterization of the Contents of Ascending Colon to Which Drugs Are Exposed after Oral Administration to Healthy Adults." *Pharmaceutical Research* 26 (9): 2141–51. https://doi.org/10.1007/s11095-009-9927-x.

Ditri, Emily L.Zale and Joseph W. 2017. "乳鼠心肌提取 HHS Public Access." *Physiology & Behavior* 176 (1): 139–48. https://doi.org/10.1007/978-1-4939-8916-4.

Ezike, Tobechukwu Christian, Ugochukwu Solomon Okpala, Ufedo Lovet Onoja, Chinenye Princess Nwike, Emmanuel Chimeh Ezeako, Osinachi Juliet Okpara, Charles Chinkwere Okoroafor, et al. 2023. "Advances in Drug Delivery Systems, Challenges and Future Directions." *Heliyon* 9 (6): e17488. https://doi.org/10.1016/j.heliyon.2023.e17488.

Fan, Yuchen, Maria Marioli, and Kelly Zhang. 2021. "Analytical Characterization of Liposomes and Other Lipid Nanoparticles for Drug Delivery." *Journal of Pharmaceutical and Biomedical Analysis* 192: 113642. https://doi.org/10.1016/j.jpba.2020.113642.

Gao, Xiaohu, Yuanyuan Cui, Richard M. Levenson, Leland W. K. Chung, and Shuming Nie. 2004. "*In Vivo* Cancer Targeting and Imaging with Semiconductor Quantum Dots." *Nature Biotechnology* 22 (8): 969–76. https://doi.org/10.1038/nbt994.

Gelperina, Svetlana, Kevin Kisich, Michael D. Iseman, and Leonid Heifets. 2005. "The Potential Advantages of Nanoparticle Drug Delivery Systems in Chemotherapy of Tuberculosis." *American Journal of Respiratory and Critical Care Medicine* 172 (12): 1487–90. https://doi.org/10.1164/rccm.200504-613PP.

Glasgow, Micah D. K., and Mahavir B. Chougule. 2015. "Recent Developments in Active Tumor Targeted Multifunctional Nanoparticles for Combination Chemotherapy in Cancer Treatment and Imaging." *Journal of Biomedical Nanotechnology* 11 (11): 1859–98. https://doi.org/10.1166/jbn.2015.2145.

Gonzalez Gomez, Azucena, and Zeinab Hosseinidoust. 2020. "Liposomes for Antibiotic Encapsulation and Delivery." *ACS Infectious Diseases* 6 (5): 896–908. https://doi.org/10.1021/acsinfecdis.9b00357.

Guimarães, Diana, Artur Cavaco-Paulo, and Eugénia Nogueira. 2021. "Design of Liposomes as Drug Delivery System for Therapeutic Applications." *International Journal of Pharmaceutics* 601 (April). https://doi.org/10.1016/j.ijpharm.2021.120571.

Hami, Zahra. 2021. "A Brief Review on Advantages of Nano-Based Drug Delivery Systems." *Annals of Military and Health Sciences Research* 19 (1): 1–6. https://doi.org/10.5812/amh.112274.

Idrees, Humaira, Syed Zohaib Javaid Zaidi, Aneela Sabir, Rafi Ullah Khan, Xunli Zhang, and Sammer Ul Hassan. 2020. "A Review of Biodegradable Natural Polymer-Based Nanoparticles for Drug Delivery Applications." *Nanomaterials* 10 (10): 1–22. https://doi.org/10.3390/nano10101970.

Jahan, Nazish, Noor Ul Huda, Arooj Fatima, and Huma Shamshad. 2023. "Recent Advances and Future Challenges of Nano-Based Drug Delivery System." *Nano and Medical Materials* 3 (1): 1–12. https://doi.org/10.59400/nmm.v3i1.121.

Jain, Akhlesh K., and Suresh Thareja. 2019. "*In Vitro* and *in Vivo* Characterization of Pharmaceutical Nanocarriers Used for Drug Delivery." *Artificial Cells, Nanomedicine and Biotechnology* 47 (1): 524–39. https://doi.org/10.1080/21691401.2018.1561457.

Kim, Taehee, Joon Kim, and Sinil Kim. 1993. "Extended-Release Formulation of Morphine for Subcutaneous Administration." *Cancer Chemotherapy and Pharmacology* 33 (3): 187–90. https://doi.org/10.1007/BF00686214.

Kou, Longfa, Yangzom D. Bhutia, Qing Yao, Zhonggui He, Jin Sun, and Vadivel Ganapathy. 2018. "Transporter-Guided Delivery of Nanoparticles to Improve Drug Permeation across Cellular Barriers and Drug Exposure to Selective Cell Types." *Frontiers in Pharmacology* 9 (JAN): 1–16. https://doi.org/10.3389/fphar.2018.00027.

Large, Danielle E., Jonathan R. Soucy, Jacob Hebert, and Debra T. Auguste. 2019. "Advances in Receptor-Mediated, Tumor-Targeted Drug Delivery." *Advanced Therapeutics* 2 (1): 1–27. https://doi.org/10.1002/adtp.201800091.

Lee, Jae Hyun, Yong Min Huh, Young Wook Jun, Jung Wook Seo, Jung Tak Jang, Ho Taek Song, Sungjun Kim, et al. 2007. "Artificially Engineered Magnetic Nanoparticles for Ultra-Sensitive Molecular Imaging." *Nature Medicine* 13 (1): 95–99. https://doi.org/10.1038/nm1467.

Lemmens, Glenn, Joachim Brouwers, Jan Snoeys, Patrick Augustijns, and Tim Vanuytsel. 2020. "Insight into the Colonic Disposition of Celecoxib in Humans." *European Journal of Pharmaceutical Sciences* 145 (January): 105242. https://doi.org/10.1016/j.ejps.2020.105242.

Li, Jianmin, Qingluo Wang, Guoyu Xia, Nigela Adilijiang, Ying Li, Zhenqing Hou, Zhongxiong Fan, and Jinyao Li. 2023. "Recent Advances in Targeted Drug Delivery Strategy for Enhancing Oncotherapy." *Pharmaceutics* 15 (9). https://doi.org/10.3390/pharmaceutics15092233.

Liu, P., G. Chen, and J. Zhang. 2022. "A Review of Liposomes as a Drug Delivery System : Current." *Molecules* 27 (4): 1372.

Manuscript, Author. 2014. "*NIH Public Access*" 48 (5): 872–97. https://doi.org/10.1002/anie.200802585. Nanomedicine.

Marcum Zachary A, Sevick MA, Handler SM. 2013. "基因的改变NIH Public Access." *Jama* 309 (20): 2105–6. https://doi.org/10.1016/j.cocis.2014.03.004.Inorganic.

Mitchell, Michael J., Margaret M. Billingsley, Rebecca M. Haley, Marissa E. Wechsler, Nicholas A. Peppas, and Robert Langer. 2021. "Engineering Precision Nanoparticles for Drug Delivery." *Nature Reviews Drug Discovery* 20 (2): 101–24. https://doi.org/10.1038/s41573-020-0090-8.

Nasirzadeh, Keyvan, Shahram Nazarian, and Seyed Mohammad Gheibi Hayat. 2016. "Inorganic Nanomaterials: A Brief Overview of the Applications and Developments in Sensing and Drug Delivery." *Journal of Applied Biotechnology Reports* 3 (2): 395–402.

"National Nanotechnology Initiative - Leading To the Next Industrial Revolution." 2000. *Microscale Thermophysical Engineering*. https://doi.org/10.1080/108939500501481 60.

Pandey, Ramendra Pati, Jasmina Vidic, Riya Mukherjee, and Chung Ming Chang. 2023. "Experimental Methods for the Biological Evaluation of Nanoparticle-Based Drug Delivery Risks." *Pharmaceutics* 15 (2): 1–17. https://doi.org/10.3390/pharmaceutics15020612.

Patra, Jayanta Kumar, Gitishree Das, Leonardo Fernandes Fraceto, Estefania Vangelie Ramos Campos, Maria Del Pilar Rodriguez-Torres, Laura Susana Acosta-Torres, Luis Armando Diaz-Torres, et al. 2018. "Nano Based Drug Delivery Systems: Recent Developments and Future Prospects." *Journal of Nanobiotechnology* 16 (1): 1–33. https://doi.org/10.1186/s12951-018-0392-8.

Pattni, Bhushan S., Vladimir V. Chupin, and Vladimir P. Torchilin. 2015. "New Developments in Liposomal Drug Delivery." *Chemical Reviews* 115 (19): 10938–66. https://doi.org/10.1021/acs.chemrev.5b00046.

Pushpalatha, R., S. Selvamuthukumar, and D. Kilimozhi. 2017. "Nanocarrier Mediated Combination Drug Delivery for Chemotherapy – A Review." *Journal of Drug Delivery Science and Technology* 39: 362–71. https://doi.org/10.1016/j.jddst.2017.04.019.

Rea, Ilaria, Nicola M. Martucci, Luca De Stefano, Immacolata Ruggiero, Monica Terracciano, Principia Dardano, Nunzia Migliaccio, et al. 2014. "Diatomite Biosilica Nanocarriers for SiRNA Transport inside Cancer Cells." *Biochimica et Biophysica Acta - General Subjects* 1840 (12): 3393–3403. https://doi.org/10.1016/j.bbagen.2014.09.009.

Sayes, Christie M., Kenneth L. Reed, and David B. Warheit. 2007. "Assessing Toxicology of Fine and Nanoparticles: Comparing *in Vitro* Measurements to *in Vivo* Pulmonary Toxicity Profiles." *Toxicological Sciences* 97 (1): 163–80. https://doi.org/10.1093/toxsci/kfm018.

Senekowitsch, Stefan, Philipp Schick, Bertil Abrahamsson, Patrick Augustijns, Thomas Gießmann, Hans Lennernäs, Christophe Matthys, et al. 2022. "Application of *In Vivo* Imaging Techniques and Diagnostic Tools in Oral Drug Delivery Research." *Pharmaceutics* 14 (4). https://doi.org/10.3390/pharmaceutics 14040801.

Shah, Afzal, Saima Aftab, Jan Nisar, Muhammad Naeem Ashiq, and Faiza Jan Iftikhar. 2021. "Nanocarriers for Targeted Drug Delivery." *Journal of Drug Delivery Science and Technology* 62 (January): 102426. https://doi.org/10.1016/j.jddst.2021.102426.

Siegrist, Stefan, Emre Cörek, Pascal Detampel, Jenny Sandström, Peter Wick, and Jörg Huwyler. 2019. "Preclinical Hazard Evaluation Strategy for Nanomedicines." *Nanotoxicology* 13 (1): 73–99. https://doi.org/10.1080/17435390.2018.1505000.

Silvestri, Brigida, Paolo Armanetti, Gennaro Sanità, Giuseppe Vitiello, Annalisa Lamberti, Gaetano Calì, Alessandro Pezzella, et al. 2019. "Silver-Nanoparticles as Plasmon-Resonant Enhancers for Eumelanin's Photoacoustic Signal in a Self-Structured Hybrid Nanoprobe." *Materials Science and Engineering C* 102 (February): 788–97. https://doi.org/10.1016/j.msec.2019.04.066.

Singh, Surya. 2010. "*Nanomedicine – Nanoscale Drugs and Delivery Systems*." https://doi.org/10.1166/jnn.2010.3617.

Stefano, Antonio Di. 2023. "Nanotechnology in Targeted Drug Delivery." *International Journal of Molecular Sciences* 24 (9): 10–11. https://doi.org/10.3390/ijms24098194.

Terracciano, Monica, Mohammad Ali Shahbazi, Alexandra Correia, Ilaria Rea, Annalisa Lamberti, Luca De Stefano, and Hélder A. Santos. 2015. "Surface Bioengineering of Diatomite Based Nanovectors for Efficient Intracellular Uptake and Drug Delivery." *Nanoscale* 7 (47): 20063–74. https://doi.org/10.1039/c5nr05173h.

"The National Institutes of Health Roadmap for Medical Research Website: Http://Www.Nihroadmap.Nih.Gov. Accessed 14 November 2007." 2008, no. November 2007: 4408. https://doi.org/10.1111/j.1445-2197.2007.04408.x.

Tiyaboonchai, Waree. 2003. "Chitosan Nanoparticles : A Promising System for Drug Delivery." *Naresuan University Journal* 11 (3): 51–66.

Unnikrishnan, Gayathri, Anjumol Joy, M. Megha, Elayaraja Kolanthai, and M. Senthilkumar. 2023. *Exploration of Inorganic Nanoparticles for Revolutionary Drug Delivery Applications: A Critical Review. Discover Nano*. Vol. 18. Springer US. https://doi.org/10.1186/s11671-023-03943-0.

Varala, Ravi, Vijay Kotra, Anil Kumar Kanuri, Mahesh Reddy Burra, and Shaik Nyamathullah. 2023. "Nano Drug Delivery-Benefits, Limitations and Future Perspective." *Nano and Medical Materials* 3 (2): 244. https://doi.org/10.59400/nmm.v3i2.244.

Version, Document. 2024. "*Long-Term Monitoring for Nanomedicine Implants and Drugs*." https://doi.org/10.1038/nnano.2015.341.

Wang, Hairui, and Yongzhuo Huang. 2020. "Combination Therapy Based on Nano Codelivery for Overcoming Cancer Drug Resistance." *Medicine in Drug Discovery* 6: 100024. https://doi.org/10.1016/j.medidd.2020.100024.

Yadava, Sunil K., Jitendra B. Naik, Jayesh S. Patil, Vinod J. Mokale, and Ruby Singh. 2015. "Enhanced Solubility and Bioavailability of Lovastatin Using Stabilized Form of Self-Emulsifying Drug Delivery System." *Colloids and Surfaces A: Physicochemical and Engineering Aspects* 481: 63–71. https://doi.org/10.1016/j.colsurfa.2015.04.026.

Zamora-Mora, Vanessa, Mar Fernández-Gutiérrez, Álvaro González-Gómez, Beatriz Sanz, Julio San Román, Gerardo F. Goya, Rebeca Hernández, and Carmen Mijangos. 2017. "Chitosan Nanoparticles for Combined Drug Delivery and Magnetic Hyperthermia: From Preparation to *in Vitro* Studies." *Carbohydrate Polymers* 157: 361–70. https://doi.org/10.1016/j.carbpol.2016.09.084.

Chapter 2.2

Innovative Nanovehicles: The Synergy of Liposome and Carbon Dot-Based Systems in Drug Delivery and Diagnostics

Twinkle Rout[1]
Audifax Kpeno[2]
Soumyashree Rout[3]
Bhagyashree Nanda[2]
and Arun Kumar Pradhan[4,*]

[1]Department of Surgical Oncology, IMS and SUM Hospital, Siksha 'O' Anusandhan (Deemed to be University), Kalinga Nagar, Bhubaneswar, Odisha, India
[2]Department of Psychiatry, IMS and SUM Hospital, Siksha 'O' Anusandhan (Deemed to be University), Kalinga Nagar, Bhubaneswar, Odisha, India
[3]Department of Neurology, IMS and SUM Hospital, Siksha 'O' Anusandhan (Deemed to be University), Kalinga Nagar, Bhubaneswar, Odisha, India
[4]Center for Biotechnology, Siksha 'O' Anusandhan (Deemed to be University) Kalinga Nagar, Bhubaneswar, Odisha, India

Abstract

Nanotechnology has revolutionized drug delivery systems by providing exact regulations to improve therapeutic effectiveness, control drug release and reduce side effects. These advanced nano-vehicles offer novel solutions for targeted therapy and precise imaging. This chapter explores the synergistic potential of liposome and carbon dot (CD)-based systems, highlighting their unique properties and complementary advantages. Liposomes serve as versatile carriers for controlled and targeted drug delivery because of their biocompatibility, biodegradability, and ability to encapsulate hydrophilic as well as hydrophobic drugs. CDs are characterized by their exceptional fluorescence, biocompatibility, and ease of surface functionalization, they offer robust platforms for bioimaging and diagnostics. Combining these two nanomaterials creates hybrid nano-vehicles that enhance therapeutic efficacy and diagnostic accuracy. Such hybrid systems leverage the controlled release capabilities of liposomes and the superior imaging properties of CDs, paving the way for innovative applications in theranostics, multi-modal

* Corresponding Author's Email: arunpradhan@soa.ac.in

In: Emerging Nanomaterials in Biomedical Science
Editors: Eepsita Priyadarshini and Ajith Manayil Parambil
ISBN: 979-8-89530-600-0

imaging, and enhanced drug delivery. This chapter underscores the promising future of liposome and CD-based nano-vehicles, emphasizing their potential to transform modern medicine through improved treatment outcomes and diagnostic precision.

Keywords: nanoparticle, nano-vehicle, liposome, carbon dot, bioimaging

1. Introduction

Nanotechnology has opened doors for transformative advancements in the medical field, particularly in drug delivery and diagnostics. Nano-vehicles, due to their nanoscale size and unique properties, offer unprecedented opportunities for precise therapeutic and diagnostic applications (Malik et al. 2023). Among various nano-vehicles, liposome and carbon dot (CD)-based systems have shown significant attention due to their distinct and complementary characteristics. The synergy of liposome and CDs-based systems represent a substantial advancement in developing nano-vehicles for medical applications. This hybrid approach leverages the strengths of both materials, addressing some of the limitations inherent in each system when used independently (Mavroidi et al. 2023). By combining the controlled release capability of liposomes with the fluorescent tracking ability of CDs, hybrid nano-vehicles can deliver drugs more efficiently and monitor their distribution and release in real-time. This improves therapeutic efficacy and allows control over drug dosing. Furthermore, integration of CDs into liposome structures provides robust imaging capabilities, enabling the simultaneous delivery of therapeutic agents and real-time imaging of disease sites (Bi et al. 2019). This dual functionality is precious in theranostics, where treatment and diagnosis are combined. Surface modifications of hybrid nano-vehicles customized to target specific tissues or cell types minimize unintended side-effects and lower the risk of systemic toxicity. Moreover, the capability to encapsulate a diverse array of therapeutic agents, coupled with the versatile surface chemistry of CDs, makes these hybrid systems applicable in different medical fields, including infectious diseases, regenerative medicine and oncology (Ruman et al. 2020). The synergy of liposome and CD-based systems opens new avenues for research and innovation. Scientists can explore novel combinations and modifications to enhance the performance of these nano-vehicles further, potentially leading to breakthroughs in personalized medicine. The combination of liposome and CD-based systems mark a substantial advancement in the design and use of nano-vehicles. This innovative method shows significant potential in enhancing the safety, effectiveness, and precision of drug delivery and diagnostics, ultimately resulting in better health-care solutions. (Ruman et al. 2020) (Hamad et al. 2024).

Nanotechnology, through its manipulation of matter at atomic, molecular, and supramolecular levels, has brought revolutionary advancements across various domains, notably in medicine. In healthcare, nanotechnology employs nanoscale materials and devices for disease detection, treatment, and prevention, offering unparalleled prospects to elevate healthcare outcomes (Malik et al. 2023). Nanomaterials, which generally measure between 1 and 100 nanometers, exhibit unique chemical, biological, and physical properties that notably differ from those of larger-scale materials. Due to their high surface area-to-volume ratio, they frequently exhibit heightened reactivity, rendering them well-suited for various biomedical applications. At the nanoscale level, quantum effects can change these materials' optical, electrical, and magnetic characteristics, allowing for novel applications in imaging and diagnostics. The control over the size and shape of nanomaterials enables the creation of customized solutions for specific medical challenges. In brief, nanotechnology has greatly improved drug delivery systems, facilitating targeted and controlled release of therapeutic agents (Malik et al, 2023) (Altammar et al. 2023).

Thus, harnessing the ability to manipulate nanoscale materials creates opportunities for precise, efficient, and personalized medical treatments. As research and development in this field continue to advance, nanotechnology holds the promise of transforming healthcare, improving patient outcomes, and addressing some of the most challenging medical issues of our time (Hamad et al. 2024) (Malik et al. 2023) (Altammar et al. 2023).

Nano-carriers are submicron-sized vehicles designed to target specific cells or tissues in the body to deliver drugs, genes, and other therapeutic agents. The unique characteristics of nano-carriers, such as their small size, high surface area-to-volume ratio, and ability to have their surfaces modified with different ligands, make them ideal for addressing the limitations of traditional drug delivery systems. In this chapter, we have briefly discussed the mechanism of drug encapsulation and release via liposome and CD based nano-vehicles.

2. Types of Nanocarriers

a) Liposomes are spherical vesicles made up of one or more phospholipid bilayers. They possess the ability to encapsulate both hydrophilic and hydrophobic drugs, offering protection from degradation and allowing for controlled release profiles (Nsairat et al. 2022).
b) Polymeric Nanoparticles: These are made from biodegradable polymers like PLGA (polylactic-co-glycolic acid) and can be designed to degrade at specific rates, releasing their payload over a prolonged period (Makadia et al. 2011).
c) Dendrimers: These are extensively branched, tree-like structures centered on a core. Their well-defined, multivalent surfaces enable precise drug loading and targeting capabilities (Mittal et al. 2021).
d) Solid Lipid Nanoparticles (SLNs): These comprise of solid lipids and surfactants, forming a stable matrix for drug encapsulation and controlled release (Munir et al. 2024).
e) Carbon Dots: These are tiny, carbon-based nanoparticles with excellent biocompatibility and versatile surface functionalization options, useful for imaging and drug delivery applications (Khayal et al. 2021).
f) Nanotubes and Nanowires: These structures offer high surface area and the potential for high drug loading. Their unique shapes can facilitate cellular uptake and targeted delivery (Yusuf et al. 2023).

2.1. Advantages of Liposome-Based Nanovehicles

Among the various types of nanocarriers, liposome-based nano-vehicles offer several distinctive advantages. Liposomes are made of natural phospholipids, making them highly biodegradable and biocompatible, minimizing the risk of toxicity and adverse immune reactions. They can encapsulate various therapeutic agents, including hydrophilic, hydrophobic, and amphiphilic drugs (Nakhaei et al. 2023). This versatility allows for the delivery of a broad spectrum of drugs. Encapsulation within liposomes can protect drugs from degradation due to various environmental factors like pH, light, and enzymatic activity, thus enhancing their stability and shelf-life. It can be designed to release its payload in a controlled manner through passive diffusion or external triggers such as pH, temperature, or enzymes, which helps maintain therapeutic drug levels over extended periods. By encapsulating toxic drugs within liposomes, the exposure of healthy tissues to these drugs can be minimized, reducing systemic toxicity and improving patient tolerance (Nakhaei et al. 2023).

3. Liposome-Based Nanovehicles

As discussed, liposomes can encapsulate a variety of therapeutic agents, making them versatile nano-vehicles for delivery of drugs (Nsairat et al. 2022). The basic structure of a liposome includes:

a) Lipid Bilayer: The essential structure of liposomes is the lipid bilayer, which mimics cell membranes. Each bilayer comprise of amphipathic phospholipids, with hydrophilic heads facing the aqueous exterior and interior environments, while the hydrophobic tails are oriented inwards, forming a stable barrier. This configuration effectively enables liposomes to enclose hydrophilic drugs in their aqueous core and hydrophobic drugs within the lipid bilayer (Nsairat et al. 2022).
b) Aqueous Core: The interior region of liposome contains an aqueous core where hydrophilic drugs can be entrapped. This core is surrounded by the lipid bilayer, providing a protected environment for the encapsulated therapeutic agents (Nsairat et al. 2022).
c) Surface Modifications: To enhance stability, biocompatibility, & targeting capabilities of liposomes, their exteriors can be altered using different molecules, such as polyethylene glycol (PEG), ligands, or antibodies. PEGylation, refers to the attachment of PEG chains, and extends the circulation time of liposomes in the bloodstream by decreasing opsonization and subsequent removal by the reticuloendothelial system (RES). Target ligands or antibodies can be attached to these surface functionalization layers of liposome surface to enable active targeting of specific cells or tissues, increasing therapeutic efficacy and reducing side effects (Nsairat et al. 2022).

The composition of liposomes is crucial in determining their physical properties, stability, and drug delivery capabilities (Nsairat et al. 2022) (Nakhaei et al. 2023). Based on the composition, they are categorized into the following types:

a) Phospholipids: Natural phospholipids commonly include phosphatidylcholine (PC) and phosphatidylethanolamine (PE), derived from egg yolk or soybeans. Synthetic versions, such as dipalmitoylphosphatidylcholine (DPPC) and distearoylphosphatidylcholine (DSPC), are often used for their distinct properties and purity. Phospholipids can vary in acyl chain length and degree of saturation, affecting the liposome's fluidity, permeability, and stability (Nsairat et al. 2022) (Nakhaei et al. 2023).

b) Cholesterol: Cholesterol is frequently incorporated into the lipid bilayer to enhance liposome stability by modulating membrane fluidity and reducing permeability. The presence of cholesterol helps prevent the leakage of encapsulated drugs and provides structural integrity to the liposome (Nakhaei et al. 2023).
c) Surface Modifiers: PEGylation refers to attaching PEG chains onto the surface of liposomes, enhancing biocompatibility and circulation time. Surface modification with antibodies or specific ligands (e.g., folate, transferrin) enables targeted delivery to definite cells or tissues, improving therapeutic outcomes (Nakhaei et al. 2023).
d) Encapsulated Agents: Liposomes can encapsulate a broad spectrum of therapeutic agents, including nucleic acids (DNA, RNA), small molecule drugs, proteins, and peptides. The nature of the encapsulated agent (hydrophilic or hydrophobic) determines its location within the liposome. In generalhydrophilic agents/drugs remain in the aqueous core and hydrophobic substances remain within the lipid bilayer (Nsairat et al. 2022) (Nakhaei et al. 2023).

The structure and composition of liposome-based nano-vehicles are carefully designed to optimize their stability, biocompatibility, and drug delivery capabilities, making them powerful tools in modern medicine (Nsairat et al. 2022).

4. Mechanisms of Drug Encapsulation

The encapsulation of drugs within liposome-based nano-vehicles can occur through various methods, each tailored to the nature and characteristics of the drug and liposome (Gonzalez et al. 2019). Key mechanisms include:

a) Passive Encapsulation: Techniques like reverse-phase evaporation, thin-film hydration, and solvent injection are frequently employed to achieve passive encapsulation (Gonzalez et al. 2019).
b) Active Loading/Remote Loading): This method is particularly effective for weakly basic drugs. It involves creating a transmembrane gradient (pH gradient or ion gradient) across the liposome membrane. Drugs are loaded into pre-formed liposomes via diffusion driven by the gradient. On entering the liposome structure, the drugs get protonated or complexed and thereby remain trapped in the liposome. Active loading typically results in higher encapsulation efficiency and drug-to-lipid ratios compared to passive loading (Nakhaei et al. 2023) (Gonzalez et al. 2019).
c) Electrostatic Encapsulation: Polyanionic or polycationic drugs can be encapsulated through electrostatic interactions with oppositely charged lipids in the liposome membrane. This method can enhance drug loading efficiency and stability within the liposome.
d) Lipid Film Hydration: A thin lipid film is hydrated using an aqueous solution containing the drug, forming multilamellar vesicles (MLVs). Techniques like sonication or extrusion can then reduce the size and polydispersity of the liposomes,

producing small unilamellar vesicles (SUVs) or large unilamellar vesicles (LUVs) (Gonzalez et al. 2019).

5. Mechanisms of Drug Release

The release of encapsulated drugs from liposome-based nano-vehicles can be regulated by various mechanisms, which depend on factors such as the composition of the liposome, the encapsulated drug, and the environmental conditions (Sriwidodo et al. 2022). Key mechanisms of drug release include:

a) Diffusion: Drugs can passively diffuse out of the liposome through the lipid bilayer. The diffusion rate depends on the drug's lipophilicity, fluidity, and concentration of cholesterol in the lipid bilayer (Sriwidodo et al. 2022).
b) Liposome Degradation: Liposomes can undergo degradation triggered by environmental factors like pH, temperature, or the presence of enzymes. For example, in acidic environments (tumor tissues or intracellular lysosomes) liposomes get destabilized triggering drug release (Gonzalez et al.2019) (Sriwidodo et al. 2022).
c) Fusion with Cell Membranes: Liposomes can fuse with the cell membranes, releasing their contents directly into the cytoplasm. This mechanism is particularly relevant for delivering drugs to target cells, enhancing cellular uptake and therapeutic efficacy (Sriwidodo et al. 2022)
d) Triggered Release: External stimuli can initiate controlled drug release from liposomes. For example, thermosensitive liposomes release drugs at elevated temperatures, such as those in hyperthermic treatment zones. Likewise, pH-sensitive liposomes release drugs in acidic environments, such as tumor tissues or endosomes. Ultrasound waves can induce cavitation and disrupt liposome membranes, facilitating drug release. Photosensitive liposomes release drugs upon exposure to specific wavelengths of light, allowing precise spatial and temporal control (Sriwidodo et al. 2022) (Franco et al. 2021).

Enzyme Induced Release: Enzyme approachable liposomes are engineered to release their contents when exposed to enzymes, such as matrix metalloproteinases (MMPs) found in tumor microenvironments. These enzymes cleave specific bonds in the liposome membrane, causing its destabilization and subsequent release of the encapsulated drug (Franco et al. 2021).

6. Applications of Liposome Based Nano-Vehicles in Drug Delivery and Gene Therapy

Liposome-based nano-vehicles offer significant potential for improving drug delivery and gene therapy. Liposome-based nano-vehicles have revolutionized drug delivery due to their biocompatibility, ability to encapsulate a wide range of therapeutic agents, and versatility in surface modification (Franco et al. 2021).

a) Cancer Therapy: Liposomes can encapsulate chemotherapeutic agents, improving their transport to tumor sites and minimizing systemic toxicity. An example is Doxil®, a liposomal formulation of doxorubicin used in the therapy of several forms of cancer. It increases the circulation time and accumulates preferentially in tumors via the enhanced permeability and retention (EPR) effects (Nikolova et al. 2022) (Fulton et al. 2023). Moreover, liposomes have the feasibility to co-encapsulate multiple drugs, allowing for simultaneous delivery and synergistic effects. This is particularly useful in overcoming drug resistance in cancer therapy (Fulton et al. 2023).
b) Anti-inflammatory Treatments: Liposomes can deliver anti-inflammatory drugs directly to sites of inflammation by reducing the side effects associated with the systemic administration. For instance, liposomal corticosteroids have been utilized in treating conditions such as rheumatoid arthritis and inflammatory bowel diseases (Kapoor et al. 2014).
c) Antimicrobial Therapy: Liposomal formulations can improve the delivery and efficacy of antimicrobial agents. Liposomes can enhance drug penetration into bacterial biofilms or infected tissues and reduce toxicity. Ambisome® (liposomal formulation of amphotericin B) treats fungal infections with reduced nephrotoxicity (Faustino et al. 2020).
d) Ophthalmic Drug Delivery: Liposomes are also capable of delivering drugs to the eye, providing sustained release and improved absorption. This particularly benefits treating chronic conditions like glaucoma or dry eye syndrome (Mishra et al. 2011).
e) Pulmonary Drug Delivery: Liposomal formulations can be inhaled to treat various respiratory diseases. Besides enhancing drug deposition in the lungs, they allow sustained release, improving treatment efficacy for chronic obstructive pulmonary disease (COPD), asthma, and pulmonary infections (Mehta et al. 2020).

6.1. Applications in Gene Therapy

Gene therapy refers to the technique wherein genetic disorders are treated by delivering nucleic acids (DNA/ RNA) to target cells thereby modifing or correcting faulty genes (Torres-Vanegas et al. 2021). Liposome-based nano-vehicles offer several advantages for gene delivery (Figure 1):

a) Delivery of DNA and RNA: Liposomes can encapsulate plasmid DNA (pDNA), small interfering RNA (siRNA), or messenger RNA (mRNA), shielding them from degradation and aiding their delivery into targeted cells. For instance, liposomal delivery of siRNA can silence specific genes involved in disease processes, offering therapeutic potential for conditions like cancer, viral infections, and genetic disorders (Jarallah et al. 2023).
b) Targeted Gene Delivery: By modifying the surface of liposomes with ligand molecules such as antibodies or peptides, specific binding to target cells is facilitated, thereby improving the precision and effectiveness of gene therapy. This targeted strategy reduces off-target effects and lowers the likelihood of immune reactions (Nsairat et al. 2023).

c) Enhanced Transfection Efficiency: Liposomes help facilitate the cellular uptake of nucleic acids and their escape from endosomes, improving transfection efficiency. Cationic liposomes, possessing a positive charge, can interact with the negatively charged nucleic acids to form complexes that facilitate their cellular entry via endocytosis (Liu et al. 2020).
d) Therapeutic Gene Editing: Liposomes can deliver components of gene-editing systems, such as CRISPR/Cas9, enabling precise modification of genetic sequences within target cells. This approach helps treat genetic diseases by correcting mutations at the DNA level (Yin et al. 2023).
e) mRNA Vaccines: Liposomes are used to formulate mRNA vaccines, such as those developed for the COVID-19 pandemic. These vaccines deliver mRNA encoding viral antigens, which host cells translate into proteins, eliciting an immune response. Liposomal mRNA vaccines offer rapid development, scalability, and strong immunogenicity (Gote et al. 2023).

7. Carbon Dot-Based Nano-Vehicles

CDs, also known as carbon quantum dots (CQDs), are a class of nanomaterials that have attracted significant interest due to their unique optical properties, biocompatibility, and versatility. CD nano-vehicles present a promising platform for drug delivery and bioimaging applications. Their facile synthesis, tunable properties, and biocompatibility make them versatile tools in nanomedicine. Several methods can be employed to synthesize CDs, each with their advantages and specific characteristics (Khayal et al. 2021). The primary methods for synthesis of CD are:

a) Top-Down Approaches:
   - Arc Discharge: This method involves generating CD through the arc discharge between graphite electrodes submerged in a liquid medium, resulting in the release of carbonaceous soot from which CD can be extracted (Khayal et al. 2021, Mansuriya et al. 2021).
   - Laser Ablation: A laser beam is focused on a carbon target submerged in a solvent, producing CD by ablating the carbon material (Khayal et al. 2021).
   - Electrochemical Oxidation: CDs are produced by applying an electrical current to graphite electrodes in an electrolyte solution, causing oxidation and formation of carbon dots (Khayal et al. 2021).

b) Bottom-Up Approaches:
   - Hydrothermal/Solvothermal Synthesis: Organic precursors (e.g., glucose, citric acid) are heated in a sealed vessel at a very high temperature and pressure, leading to carbonization and formation of CDs (Khayal et al. 2021, Mansuriya et al. 2021).
   - Microwave-Assisted Synthesis: Organic precursors are irradiated with microwaves, generating heat that induces carbonization and the formation of CD. This technique allows rapid synthesis of CDs (Mansuriya et al. 2021) (Behi et al. 2021).

- Combustion/Synthetic Methods: Combustion of organic precursors or pyrolysis under controlled conditions is also used in the synthesis of CDs. For example, burning candle soot or natural precursors like orange juice can yield CDs (Khayal et al. 2021) (Mansuriya et al. 2021) (Behi et al. 2021)

c) Other Methods:
   - Chemical Oxidation: Strong acids or oxidizing agents are used to break down larger carbon structures (e.g., carbon black) into smaller CDs (Behi et al. 2021).
   - Template-Assisted Synthesis: CDs are formed using templates/substrates such as mesoporous silica to control size and morphology, followed by template removal (Behi et al. 2021).

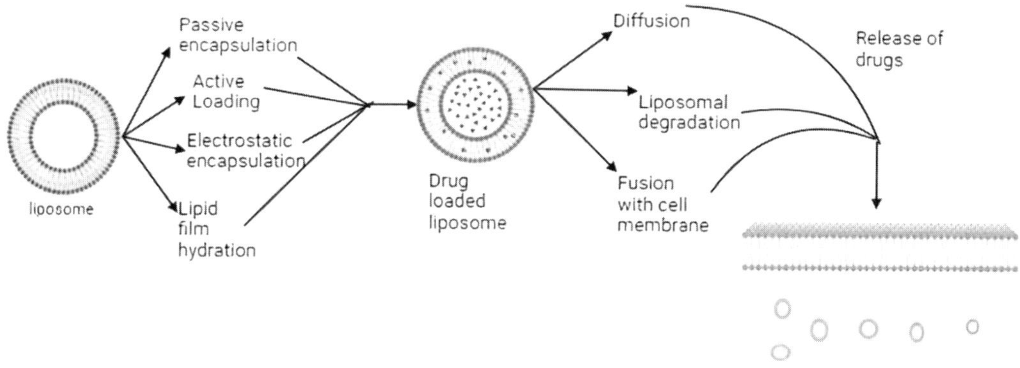

Figure 1. Different liposomal drug encapsulation and delivery mechanism.

8. Properties of CD

CD possess several intriguing properties that make them suitable for various applications in drug delivery and bioimaging.

a) Optical Properties:
   - Photoluminescence: CDs exhibit strong and tunable photoluminescence, emitting light of various colors when excited by a suitable wavelength. This property is attributed to quantum confinement effects and surface defects (Ozyurt et al. 2023).
   - Fluorescence Stability: CDs have excellent photostability compared to traditional organic dyes, which is ideal for long-term imaging and sensing applications (Ozyurt et al. 2023).

b) Biocompatibility: CDs are non-toxic and biocompatible, making them suitable for biomedical applications. Their biocompatibility can be further enhanced through surface modification with biocompatible polymers or functional groups (Hui et al. 2023).

c) Chemical Stability: CDs exhibit chemical stability, resist photobleaching, and maintain their properties under various environmental conditions. They are stable in different pH ranges, which is advantageous for biomedical applications (Hui et al. 2023).

d) Size and Surface Area: CDs typically have sizes below 10 nm, providing a very high surface area to volume ratio. This enhances their interaction with biological molecules and allows for efficient drug loading. The small size enables them to penetrate biological barriers and accumulate in target tissues or cells (Hui et al. 2023).
e) Surface Functionalization: The surface of CD can be easily functionalized with various chemical groups (e.g., carboxyl, hydroxyl, and amine), enhancing their solubility, stability, and ability to conjugate with drugs or target ligands. Surface functionalization allows for the customization of CDs for specific applications s such as targeted drug delivery or imaging (Hui et al. 2023) (Alaghmandfard et al. 2021).
f) Electrical Conductivity: CDs exhibit excellent electrical conductivity, which is beneficial in applications requiring conductive materials, such as sensors and electronic devices (Ozyurt et al. 2023).

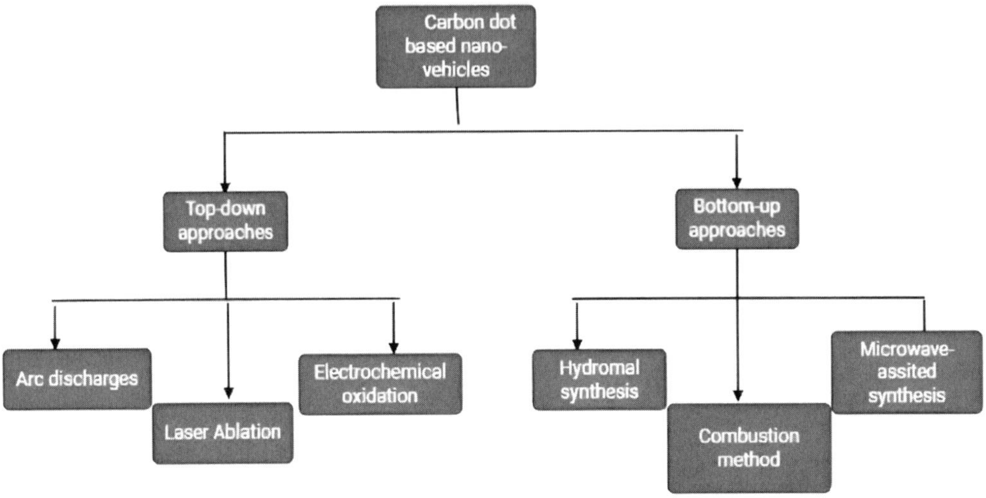

Figure 2. Different CDs-based nano-vehicles mechanism for drug delivery.

9. Applications of CDs in Drug Delivery and Bioimaging

CDs are ideal for drug delivery because of their distinct properties and bioimaging applications. They offer the feasibility of surface functionalization by specific ligand allowing site-specific delivery of drugs. The tissue/cell specific drug release helps enhance therapeutic efficacy and reduces side effects. Moreover, CD can co-deliver multiple therapeutic agents, enabling combinational therapies that synergistically improve the outcome of treatment (Figure 2) (Calabrese et al. 2021).

a) Bioimaging: CDs are excellent fluorescent probes for imaging biological tissues and cells due to their strong and tunable photoluminescence. They can be engineered to emit various colors, enabling multicolor imaging and tracking different biological processes. Their biocompatibility and stability make CD suitable for *in vivo* imaging applications, such as tracking drug biodistribution or visualizing tumors (Calabrese et al. 2021) (Phan et al. 2022).

b) Fluorescence Properties of Carbon Dots (CDs): CDs are well-known for their exceptional fluorescence properties, which are crucial for imaging and sensing applications. These properties are influenced by various factors such as the CD size, surface functional groups, and synthesis method. CDs display strong photoluminescence, with emission wavelengths that can be adjusted by modifying the synthesis parameters and surface plasmon (Phan et al. 2022). CDs generally emit in the blue-green range, however, red and near-infrared (NIR) emissions can also be achieved by modulating surface parameters. CDs show excitation-dependent fluorescence, where the emission wavelength shifts based on the excitation wavelength and exhibit high quantum yields (QY), sometimes up to 80%. This property allows for multi-wavelength imaging. Additionally, QY can be enhanced by surface passivation and functionalization with specific organic or inorganic groups. CDs are more photostable than traditional organic dyes, maintaining their fluorescence intensity over prolonged excitation periods. This makes them ideal for long-term imaging applications. The fluorescence properties of CDs are retained without significant toxicity, which is crucial for biomedical imaging applications (Qureshi et al. 2024).

c) Functionalization and surface modification: Functionalization involves adding functional groups or molecules to the surface of CDs. Functional groups on the CD surface (e.g., amine, carboxyl, and hydroxyl) can further influence their fluorescence properties and enable conjugation with other molecules for targeted imaging. Functionalization and surface modification of CDs are crucial steps to enhance their properties, improve their stability, and tailor them for specific applications. These modifications can significantly impact their biocompatibility, solubility, targeting ability, and functionality in various biomedical and technological applications. This process can be categorized into two main strategies namely, passive and active functionalization (Shabbir et al. 2023).

  1. Passive Functionalization: This involves the amalgamation of functional groups during synthesis of CDs. Common precursors used in synthesis naturally introduce functional groups such as carboxyl (-COOH), hydroxyl (-OH), and amine (-NH2) onto the CD surface, based on the selected organic precursors contaning the desired functional groups. For example, by using citric acid carboxyl groups can be introduced. Passivating agents such as PEG or other polymers are used during synthesis to improve biocompatibility and fluorescence properties (Liu et al. 2020).
  2. Active Functionalization: This approach involves post-synthesis modification of CDs to introduce specific functional groups or molecules. It either involves conjugation of biomolecules (e.g., antibodies, peptides, or DNA) or other functional entities (e.g., drugs, targeting ligands) through covalent bonding ornon-covalent attachment of molecules through electrostatic interactions. Active functionalization is useful for temporary modifications or reversible applications (Mahor et al. 2021).

10. Synergistic Potential of Liposome and Carbon Dot-Based Systems

The integration of liposomes and CDs presents a promising frontier in nanotechnology, particularly for their applications in drug delivery, Theranostics, and bioimaging. Combining these two nanomaterials leverages their advantages while overcoming limitations, enhancing properties and functionalities. Liposomes are biocompatible, biodegradable, and have low toxicity, making them best for drug delivery systems (Bartkowski et al. 2024). CDs are fluorescent nanoparticles known for their excellent photostability, high water solubility, and low toxicity. They possess unique optical properties, such as strong luminescence and the ability to act as photosensitizers. CDs provide strong and stable fluorescence, allowing real-time tracking and monitoring of the nano-vehicles. This is particularly useful for imaging-guided therapy and ensuring precise delivery to target sites. The hybrid system is engineered for multifunctional roles, such as simultaneous drug delivery and bioimaging. This capability is critical for theranostics applications, where diagnostics and therapeutics are combined in a single platform. Functionalization of CDs and liposomes with targeting ligands enables specific binding to target cells, enhancing the efficacy of drug delivery. Additionally, the system can respond to external stimuli (e.g., pH, temperature, light), enabling site-specific release of therapeutics (Bartkowski et al. 2024).

11. Applications of Hybrid Liposome-Carbon Dot Nanovehicles

Key applications of hybrid nano-vehicles (combining liposome and CD) include theranostics and targeted drug delivery with controlled release mechanisms (Li et al. 2016).

a) *Integrated Diagnosis and Treatment*: Hybrid liposome-CD systems can simultaneously deliver therapeutic agents and enable diagnostic imaging. CDs inherent fluorescence properties allow real-time tracking of drug distribution and localization within the body, ensuring precise delivery to target sites while monitoring treatment efficacy. Additionally, the strong and stable fluorescence of CDs allow the use of non-invasive techniques providing detailed insights into the biodistribution and cellular uptake of the nano-vehicles. This capability is crucial for assessing the therapeutic response and making necessary adjustments to the treatment regimen (Kaurav et al. 2023).
b) *Personalized Medicine:* The real-time tracking of delivered drug enables tailoring treatment plans according to individual patients' response. This personalized strategy improves treatment effectiveness while reducing potential side effects. CDs can be functionalized to detect specific biomarkers associated with diseases like cancer. The hybrid system can then deliver targeted therapies while monitoring biomarker levels, facilitating early identification and prompt intervention (Alghamdi et al. 2022).
c) *Cancer Theranostics:* Hybrid liposome-CD nano-vehicles can be engineered to target tumor cells specifically, by using ligands or antibodies that recognize tumor-specific antigens. This targeted approach reduces off-target effects and enhances the therapeutic index. CDs can act as photosensitizers in photodynamic therapy, generating reactive oxygen species (ROS) upon light activation to kill cancer cells.

The liposome component can encapsulate chemotherapeutic agents, synergizing photodynamic therapy (PDT) with conventional chemotherapy (Alwattar et al. 2021).

d) *Targeted Drug Delivery and Controlled Release:* The hybrid organization of liposome-CD allows for encapsulating hydrophilic and hydrophobic drugs within the liposomes as well as on the surface of CDs. This strategy increases the drug loading capacity and enables the delivery of combination therapies. Liposomes protect the encapsulated drugs from premature degradation and clearance, while CDs stabilize the liposomal structure, enhancing overall stability and bioavailability of nano-vehicles (Alavi et al. 2017).

e) CDs can be functionalized with targeting ligands such as peptides, antibodies, or small molecules that specifically bind to receptors overexpressed on target cells (e.g., cancer cells). This targeting capability ensures that the hybrid nano-vehicles deliver their payload specifically to the desired site, minimizing systemic toxicity and improving therapeutic efficacy. The targeted nano-vehicles are taken up by cells through receptor-mediated endocytosis, ensuring efficient intracellular delivery of therapeutic agents (Alwattar et al. 2021) (Alavi et al. 2017). The hybrid system can be engineered to respond to various external stimuli, such as pH, temperature, or light. For instance, in the acidic microenvironment of tumor cells, pH-sensitive liposomes can release their payload selectively at the tumor site. Similarly, light-triggered release mechanisms can be utilized for site-specific drug delivery. The integration of CDs into the liposomal structure can provide a sustained release profile, ensuring extended therapeutic effect and decreased frequency of administration (Alavi et al. 2017).

f) *Overcoming Drug Resistance:* The hybrid system's capacity to co-deliver multiple drugs with different mechanisms of action can help overcome drug resistance in cancer therapy. Further, therapeutic treatment of cancer cells through various pathways reduces the likelihood of development of resistant. In addition, CDs can be designed to inhibit efflux pumps in drug-resistant cells, enhancing the intracellular accumulation of chemotherapeutic agents and improving treatment (Szumilak et al. 2021).

12. Case Studies and Recent Advancements on Hybrid Liposome-CD Nano-Vehicles

Hybrid liposome-CD nano-vehicles are at the forefront of nanomedicine, demonstrating substantial potential in theranosticss and targeted drug delivery (Szumilak et al. 2021). Recent case studies and advancements highlight their versatility and efficacy in various biomedical applications, some of which are discussed below.

12.1. Cancer Treatment and Imaging

- Study on Breast Cancer: In a study conducted by Park et al. hybrid liposome-CD nano-vehicles were analyzed for breast cancer treatment. The liposomes encapsulated doxorubicin, a chemotherapeutic agent, while CDs served as the fluorescent counterpart. The hybrid system demonstrated efficient tumor targeting and reduced systemic toxicity (Table 1). The fluorescence from CDs allows real-time tracking of nano-vehicles, confirming their accumulation at the tumor site and providing valuable insights into biodistribution and pharmacokinetics (Park et al. 2002).

- Theranostics Approach in Melanoma: Researchers developed a hybrid system incorporating liposomes loaded with a photosensitizer and folic acid functionalized CD for targeting melanoma cells. Upon light activation, the CD induced ROS generation that led PDT of cancer cells, while liposomal provided pay-load chemotherapy. This combined approach resulted in significant tumor regression and minimal side effects (Itoo et al. 2022).

12.2. Antimicrobial Therapy

In a study on bacterial infections, hybrid l nano-vehicles were used to deliver antibiotics to the site of infection. CD were functionalized with antibodies specific to bacterial surface proteins, ensuring targeted delivery, while. liposomes encapsulated the antibiotic vancomycin. The developed system showed enhanced antibacterial efficacy, reduced resistance development and allowed tracking of infection through fluorescence imaging (Ferreira et al. 2021).

12.3. Neurodegenerative Diseases

Parkinson's disease treatment: A novel hybrid system was developed for delivering neuroprotective agents to the brain. These were well functionalized with transferrin for crossing the blood-brain barrier (BBB), while liposomes encapsulated dopamine to replenish deficient neurotransmitters in Parkinson's disease (Table 1). This approach demonstrated improved delivery of drugs to the brain, enhanced treatment results, & allowed for monitoring the treatment through non-invasive imaging techniques (Cheng et al. 2022).

13. Recent Advancements of Nanoliposomes

a) Enhanced Theranostics: The design of novel hybrid nano-vehicles has led to the advancement of multifunctional nanoplatforms that can simultaneously perform drug delivery, imaging, and therapy. For instance, incorporating magnetic nanoparticles into the hybrid system has enabled magnetic resonance imaging (MRI) alongside fluorescence imaging, providing complementary diagnostic information and improving treatment precision (Seleci et al. 2017).
b) Stimuli-Responsive Systems: (pH and Temperature Sensitivity) Recent studies have focused on developing stimuli-responsive hybrid systems that assist in release of drugs in response to specific environmental triggers (Rahim et al. 2021).
c) Improved Biocompatibility and Stability: Researchers have explored various surface modifications to enhance the biocompatibility and stability of hybrid liposome-CD nano-vehicles. PEG coating, for instance, has been employed to increase circulation

time in the bloodstream and reduce immune recognition, resulting in more efficient delivery to target sites (Mohanty et al. 2020).

d) Personalized Medicine Approaches: By tailoring the surface functionalization and payload according to individual patient profiles, hybridnano-vehicles can provide highly specific and effective treatments, paving the way for personalized healthcare solutions (Alghamdi et al. 2022).

Table 1. Summary of different drugs used in the treatment of various diseases

Drugs Name	Cancer disease	Neurodegenerative disease	Mechanism of action	Reference
Doxorubicin	Breast cancer	-	Inhibits cancer cell growth by blocking enzyme topoisomerase 2	(Bulbake et al. 2017, Gardikis et al. 2010)
Curcumin	-	Alzheimer disease	Stops the amyloid-protein aggregation	(Utomo et al. 2021)
Levodopa	-	Parkinson disease	Precursor to dopamine	(Salat et al. 2013)
Daunorubicin	Blood cancer	-	Inhibitor of DNA topoisomerase II	(Lowenberg et al. 2009)
Quercetin	-	Huntington disease	Inhibitor of intestinal glucose absorption, insulin secretion, etc.	(Islam et al. 2021, Chakraborty et al. 2014)
Paclitaxel	Gastric and ovarian cancer	-	Inhibitor of angiogenesis	(Hau et al. 1999, Ye et al. 2013)

Conclusion and Limitations

Innovative nano-vehicles, leveraging the exceptional properties of liposomes and CDs, represent a paradigm shift in drug delivery and diagnostics. These advanced systems offer unparalleled opportunities to improve medical treatments' precision, and efficacy. Integrating multifunctional capabilities into a single platform paves the way for developing next-generation therapeutic and diagnostic tools, which are more responsive to the complex needs of modern medicine. As research and development in nanotechnology continues to evolve, the potential for these innovative nano-vehicles to transform healthcare becomes increasingly evident. Their impact will likely extend beyond conventional therapies, influencing areas such as regenerative medicine and immunotherapy. Ultimately, the synergy of liposome and CD based systems signifies a promising frontier in the quest for more effective, personalized, and less invasive medical solutions, heralding a new era in the treatment & diagnosis of diseases.

Despite the several advantages of liposome-CD hybrid system, there are several n challenges that limits their application. Liposomes in specific suffer from stability issues, short circulation times, rapid clearance by the reticuloendothelial system, and complexities in large-scale manufacturing. In addition, their limited drug loading capacity and formulation intricacies also hinder therapeutic efficacy and commercial viability. CDs, on the other hand, grapple with concerns regarding long-term toxicity and inconsistent surface chemistry. Their small size restricts drug loading capacity, while synthesis procedure suffers from reproducibility and scalability issues. Furthermore, CDs biodistribution and clearance mechanisms are not fully understood, leading to unintended accumulation in non-target tissues. Therefore, addressing these limitations is essential and of paramount importance to allow effective clinical translation of these nano-vehicles.

Acknowledgments

We extend our sincere thanks to the Institute for the invaluable support throughout the development of this chapter. We are also grateful to our colleagues for their insightful feedback and encouragement.

References

Alaghmandfard A, Sedighi O, Rezaei NT, Abedini AA, Khachatourian AM, Toprak MS, Seifalian A. Recent advances in the modification of carbon-based quantum dots for biomedical applications. *Materials Science and Engineering: C.* 2021 Jan 1; 120:111756.

Alavi M, Karimi N, Safaei M. Application of Various Types of Liposomes in Drug Delivery Systems. *Adv Pharm Bull.* 2017 Apr;7(1):3-9. doi: 10.15171/apb.2017.002. Epub 2017 Apr 13.

Alghamdi MA, Fallica AN, Virzì N, Kesharwani P, Pittalà V, Greish K. The Promise of Nanotechnology in Personalized Medicine. *J Pers Med. 2022* Apr 22;12(5):673. doi: 10.3390/jpm12050673.

Altammar KA. A review on nanoparticles: characteristics, synthesis, applications, and challenges. *Front Microbiol.* 2023 Apr 17; 14:1155622. doi: 10.3389/fmicb.2023.1155622.

Alwattar JK, Mneimneh AT, Abla KK, Mehanna MM, Allam AN. Smart Stimuli-Responsive Liposomal Nanohybrid Systems: A Critical Review of Theranostics Behavior in Cancer. *Pharmaceutics.* 2021 Mar 8;13(3):355. doi: 10.3390/pharmaceutics13030355.

Bartkowski M, Zhou Y, Nabil Amin Mustafa M, Eustace AJ, Giordani S. CARBON DOTS: Bioimaging and Anticancer Drug Delivery. *Chemistry–A European Journal.* 2024 Jan 11: e202303982.

Behi M, Gholami L, Naficy S, Palomba S, Dehghani F. Carbon dots: a novel platform for biomedical applications. *Nanoscale Adv.* 2021 Dec 13;4(2): 353-376.doi: 10.1039/d1na00559f.

Bi H, Xue J, Jiang H, Gao S, Yang D, Fang Y, Shi K. Current developments in drug delivery with thermosensitive liposomes. *Asian J Pharm Sci.* 2019 Jul;14(4):365-379. doi: 10.1016/j.ajps.2018.07.006. Epub 2018 Oct 31.

Bulbake. U, Doppalapudi. S, Kommineni. N. et al. Liposomal formulations in clinical use: An updated review, *Pharmaceutics,* 2017, Mar 27;9(2):12. doi:10.3390/pharmaceutics9020012.

Calabrese G, De Luca G, Nocito G, Rizzo MG, Lombardo SP, Chisari G, Forte S, Sciuto EL, Conoci S. Carbon Dots: An Innovative Tool for Drug Delivery in Brain Tumors. *Int J Mol Sci.* 2021 Oct 29;22(21):11783. doi: 10.3390/ijms222111783.

Chakraborty. J, Singh. R, Dutta. D, Naskar. A, Rajamma. U et al. Quercetin improves behavioral deficiencies, restores astrocytes and microglia and reduces serotonin metabolism in 3- nitropropionicacis-induced rat model of Huntingtons disease, *CNS Neurosci Ther,* 2014 20(1):10-19, doi:10.1111/cns.12189.

Cheng G, Liu Y, Ma R, Cheng G, Guan Y, Chen X, Wu Z, Chen T. Anti-Parkinsonian Therapy: Strategies for Crossing the Blood-Brain Barrier and Nano-Biological Effects of Nanomaterials. *Nanomicro Lett.* 2022 Apr 15;14(1): 105.doi:10.1007/s40820-022-00847-z.

Faustino C, Pinheiro L. Lipid Systems for the Delivery of Amphotericin B in Antifungal Therapy. *Pharmaceutics.* 2020 Jan 1;12(1):29. doi: 10.3390/pharmaceutics12010029.

Ferreira M, Ogren M, Dias JNR, Silva M, Gil S, Tavares L, Aires-da-Silva F, Gaspar MM, Aguiar SI. Liposomes as Antibiotic Delivery Systems: A Promising Nanotechnological Strategy against Antimicrobial Resistance. *Molecules.* 2021 Apr 2;26(7):2047. doi: 10.3390/molecules26072047.

Franco MS, Gomes ER, Roque MC, Oliveira MC. Triggered Drug Release from Liposomes: Exploiting the Outer and Inner Tumor Environment. *Front Oncol.* 2021 Mar 16; 11:623760. doi: 10.3389/fonc.2021. 623760.

Fulton MD, Najahi-Missaoui W. Liposomes in Cancer Therapy: How Did We Start and Where Are We Now. *Int J Mol Sci.* 2023 Apr 1;24(7):6615. doi: 10.3390/ijms24076615.

Gardikis. K, Tsimplouli. C, Dimas. K, Micha-Screttas. K, et al. New Chimeric advanced drug delivery nano systems(chi-aDDnSs) as doxorubicin carriers, *Int. J. Pharma.,* 2010, Dec 15;402(1-2):231-7. DOI: 10.1016/j.ijpharm.2010.10.007.

Gonzalez Gomez A, Syed S, Marshall K, Hosseinidoust Z. Liposomal Nanovesicles for Efficient Encapsulation of Staphylococcal Antibiotics. *ACS Omega.* 2019 Jun 21;4(6):10866-10876. doi: 10.1021/acsomega.9b00825.

Gote V, Bolla PK, Kommineni N, Butreddy A, Nukala PK, Palakurthi SS, Khan W. A Comprehensive Review of mRNA Vaccines. *Int J Mol Sci.* 2023 Jan 31;24(3):2700. doi: 10.3390/ijms24032700.

Hamad I, Harb AA, Bustanji Y. Liposome-Based Drug Delivery Systems in Cancer Research: An Analysis of Global Landscape Efforts and Achievements. *Pharmaceutics.* 2024 Mar 14;16(3):400. doi: 10.3390/pharmaceutics16030400.

Hau. D. H, Xue. L, Young. L. J, Burke. P. A, Cheung. A. T, Paclitaxel (Taxol): an inhibitor of angiogenesis in a highly vascularized transgenic breast cancer, *Cancer Biother Radiopharm,* 1999 Feb;14(1):31-6, doi:10.1089/cbr.1999.14.31.

Hui S. Carbon dots (CDs): basics, recent potential biomedical applications, challenges, and future perspectives. *Journal of Nanoparticle Research.* 2023 Apr;25(4):68.

Islam. M. S, Quispe. C, Hossain. R, Islam. M. T, et al. Neuropharmacological effects of quercetin: A literature-based review, *Front Pharmacol,* 2021 Jun 17:12:665031. Doi: 10.3389/fphar.2021.665031.

Itoo AM, Paul M, Padaga SG, Ghosh B, Biswas S. Nanotherapeutic Intervention in Photodynamic Therapy for Cancer. *ACS Omega.* 2022 Dec 6;7(50):45882-45909. doi: 10.1021/acsomega.2c05852.

Jarallah SJ, Aldossary AM, Tawfik EA, Altamimi RM, Alsharif WK, Alzahrani NM, As Sobeai HM, Qamar W, Alfahad AJ, Alshabibi MA, Alqahtani SH, Alshehri AA, Almughem FA. GL67 lipid-based liposomal formulation for efficient siRNA delivery into human lung cancer cells. *Saudi Pharm J.* 2023 Jul;31(7): 1139-1148. doi: 10.1016/j.jsps.2023.05.017.

Kapoor B, Singh SK, Gulati M, Gupta R, Vaidya Y. Application of liposomes in treatment of rheumatoid arthritis: quo vadis. *Scientific World Journal.* 2014 Feb 4; 2014:978351. doi: 10.1155/2014/978351.

Kaurav H, Verma D, Bansal A, Kapoor DN, Sheth S. Progress in drug delivery and diagnostic applications of carbon dots: a systematic review. *Front Chem.* 2023 Jul 14; 11:1227843. doi: 10.3389/fchem.2023.1227843.

Khayal A, Dawane V, Amin MA, Tirth V, Yadav VK, Algahtani A, Khan SH, Islam S, Yadav KK, Jeon BH. Advances in the Methods for the Synthesis of Carbon Dots and Their Emerging Applications. *Polymers* (Basel). 2021 Sep 20;13(18):3190. doi: 10.3390/polym13183190.

Li Z, Ye E, Lakshminarayanan R, Loh XJ. Recent advances of using hybrid nanocarriers in remotely controlled therapeutic delivery. *Small.* 2016 Sep;12(35):4782-806.

Liu C, Zhang L, Zhu W, Guo R, Sun H, Chen X, Deng N. Barriers and Strategies of Cationic Liposomes for Cancer Gene Therapy. *Mol Ther Methods Clin Dev.* 2020 Jul 31; 18:751-764. doi: 10.1016/j.omtm.2020.07.015.

Liu J, Li R, Yang B. Carbon dots: a new type of carbon-based nanomaterial with wide applications. *ACS Central Science.* 2020 Dec 14;6(12):2179-95.

Lowenberg. B, Ossenkoppele, g., J, Putten. W. V, et al. High-Dose daunorubicin older patients with acute myeloid leukemia, *N Engl J Med,* 2009 Sept 24;361(13):1235-48., doi:10.1056/NEJMoa0901409.

Mahor A, Singh PP, Bharadwaj P, Sharma N, Yadav S, Rosenholm JM, Bansal KK. Carbon-Based Nanomaterials for Delivery of Biologicals and Therapeutics: A Cutting-Edge Technology. *Journal of Carbon Research* 2021, 7, 19.

Makadia HK, Siegel SJ. Poly Lactic-co-Glycolic Acid (PLGA) as Biodegradable Controlled Drug Delivery Carrier. *Polymers (Basel).* 2011 Sep 1;3(3):1377-1397. doi: 10.3390/polym3031377.

Malik S, Muhammad K, Waheed Y. Emerging Applications of Nanotechnology in Healthcare and Medicine. *Molecules.* 2023 Sep 14;28(18):6624. doi: 10.3390/molecules28186624.

Malik S, Muhammad K, Waheed Y. Nanotechnology: A Revolution in Modern Industry. *Molecules.* 2023 Jan 9;28(2):661. doi: 10.3390/molecules28020661.

Mansuriya BD, Altintas Z. Carbon Dots: Classification, Properties, Synthesis, Characterization, and Applications in Health Care-An Updated Review (2018-2021). *Nanomaterials (Basel).* 2021 Sep 27;11(10):2525. doi: 10.3390/nano11102525.

Mavroidi B, Kaminari A, Sakellis E, Sideratou Z, Tsiourvas D. Carbon Dots-Biomembrane Interactions and Their Implications for Cellular Drug Delivery. *Pharmaceuticals (Basel).* 2023 Jun 2;16(6):833. doi: 10.3390/ph16060833.

Mehta PP, Ghoshal D, Pawar AP, Kadam SS, Dhapte-Pawar VS. Recent advances in inhalable liposomes for treatment of pulmonary diseases: concept to clinical stance. *Journal of drug delivery science and technology.* 2020 Apr 1; 56:101509.

Mishra GP, Bagui M, Tamboli V, Mitra AK. Recent applications of liposomes in ophthalmic drug delivery. *J Drug Deliv.* 2011; 2011:863734. doi: 10.1155/2011/863734.

Mittal P, Saharan A, Verma R, Altalbawy FMA, Alfaidi MA, Batiha GE, Akter W, Gautam RK, Uddin MS, Rahman MS. Dendrimers: A New Race of Pharmaceutical Nanocarriers. *Biomed Res Int.* 2021 Feb 15; 2021:8844030. doi: 10.1155/2021/8844030.

Mohanty A, Uthaman S, Park IK. Utilization of Polymer-Lipid Hybrid Nanoparticles for Targeted Anti-Cancer Therapy. *Molecules.* 2020 Sep 23;25(19):4377. doi: 10.3390/molecules25194377.

Munir M, Zaman M, Waqar MA, Khan MA, Alvi MN. Solid lipid nanoparticles: a versatile approach for controlled release and targeted drug delivery. *Journal of Liposome Research.* 2024 Apr 2;34(2):335-48.

Nakhaei P, Margiana R, Bokov DO, Abdelbasset WK, JadidiKouhbanani MA, Varma RS, Marofi F, Jarahian M, Beheshtkhoo N. Liposomes: Structure, Biomedical Applications, and Stability Parameters with Emphasis on Cholesterol. *Front BioengBiotechnol.* 2021 Sep 9; 9:705886. doi: 10.3389/fbioe.2021.705886. Retraction in: *Front Bioeng Biotechnol.* 2023 Sep 04; 11:1285118. doi: 10.3389/fbioe.2023.1285118.

Nikolova MP, Kumar EM, Chavali MS. Updates on responsive drug delivery based on liposome vehicles for cancer treatment. *Pharmaceutics.* 2022 Oct 15;14(10):2195.

Nsairat H, Alshaer W, Odeh F, Esawi E, Khater D, Al Bawab A, El-Tanani M, Awidi A, Mubarak MS. Recent advances in using liposomes for delivery of nucleic acid-based therapeutics. *OpenNano.* 2023 May 1; 11:100132.

Nsairat H, Khater D, Sayed U, Odeh F, Al Bawab A, Alshaer W. Liposomes: structure, composition, types, and clinical applications. *Heliyon.* 2022 May 13;8(5): e09394. doi: 10.1016/j.heliyon. 2022.e09394.

Ozyurt D, Al Kobaisi M, Hocking RK, Fox B. Properties, synthesis, and applications of carbon dots: A review. *Carbon Trends.* 2023 Jun 8:100276.

Park JW. Liposome-based drug delivery in breast cancer treatment. *Breast Cancer Res.* 2002;4(3):95-9. doi: 10.1186/bcr432. Epub 2002 Apr 18.

Phan LMT, Cho S. Fluorescent Carbon Dot-Supported Imaging-Based Biomedicine: A Comprehensive Review. Bioinorg Chem Appl. 2022 Apr 10; 2022:9303703. doi: 10.1155/2022/9303703. Retraction in: *Bioinorg Chem Appl. 20*23 Dec 6; 2023:9796150. doi: 10.1155/2023/9796150.

Qureshi ZA, Dabash H, Ponnamma D, Abbas MKG. Carbon dots as versatile nanomaterials in sensing and imaging: Efficiency and beyond. *Heliyon.* 2024 May 22;10(11): e31634. doi: 10.1016/j.heliyon.2024.e31634.

Rahim MA, Jan N, Khan S, Shah H, Madni A, Khan A, Jabar A, Khan S, Elhissi A, Hussain Z, Aziz HC, Sohail M, Khan M, Thu HE. Recent Advancements in Stimuli Responsive Drug Delivery Platforms for Active and Passive Cancer Targeting. *Cancers (Basel).* 2021 Feb 7;13(4): 670.doi:10.3390/cancers13040670.

Ruman U, Fakurazi S, Masarudin MJ, Hussein MZ. Nanocarrier-Based Therapeutics and Theranosticss Drug Delivery Systems for Next Generation of Liver Cancer Nanodrug Modalities. *Int J Nanomedicine.* 2020 Mar 3; 15:1437-1456. doi: 10.2147/IJN.S236927.

Salat. D, Tolosa. E, Levopoda in the treatment of Parkinsons disease: Current status and new developments, *J Parkinsons Dis*, 2013 Jan 1;3(3):255-69. Doi:10.3233/JPD-130186.

Seleci M, Ag Seleci D, Scheper T, Stahl F. Theranostics Liposome-Nanoparticle Hybrids for Drug Delivery and Bioimaging. *Int J Mol Sci.* 2017 Jul 2;18(7):1415. doi: 10.3390/ijms18071415.

Shabbir H, Csapó E, Wojnicki M. Carbon quantum dots: the role of surface functional groups and proposed mechanisms for metal ion sensing. *Inorganics.* 2023 Jun 20;11(6):262.

Sriwidodo, Umar AK, Wathoni N, Zothantluanga JH, Das S, Luckanagul JA. Liposome-polymer complex for drug delivery system and vaccine stabilization. *Heliyon.* 2022 Feb 12;8(2):e08934. doi: 10.1016/j.heliyon.2022.e08934.

Szumilak M, Wiktorowska-Owczarek A, Stanczak A. Hybrid Drugs-A Strategy for Overcoming Anticancer Drug Resistance? *Molecules.* 2021 Apr 29;26(9):2601. doi: 10.3390/molecules26092601.

Torres-Vanegas JD, Cruz JC, Reyes LH. Delivery Systems for Nucleic Acids and Proteins: Barriers, Cell Capture Pathways and Nanocarriers. *Pharmaceutics.* 2021 Mar 22;13(3):428. doi: 10.3390/pharmaceutics 13030428.

Utomo. R. Y, Asawa. Y, Okada. S, et al. Development of curcumin-based amyloid beta aggregation inhibitors for Alzheimers disease using the SAR matrix approach, *Bioorg Med Chem*, 2021, Sept 15:46:116357. doi: 10.1016/j.bmc.2021.116357.

Ye. L, He. J, Hu. J, Dong. Q, Wang. H, Fu. F, Tian. J, Antitumor effect and toxicity of Lipusu in rat ovarian cancer xenografts, *Food Chem Toxicol,* 2013, Feb:52:200-6, doi: 10.1016/j.fct.2012.11.004.

Yin X, Harmancey R, McPherson DD, Kim H, Huang SL. Liposome-Based Carriers for CRISPR Genome Editing. *Int J Mol Sci.* 2023 Aug 16;24(16):12844. doi: 10.3390/ijms241612844.

Yusuf A, Almotairy ARZ, Henidi H, Alshehri OY, Aldughaim MS. Nanoparticles as Drug Delivery Systems: A Review of the Implication of Nanoparticles' Physicochemical Properties on Responses in Biological Systems. *Polymers (Basel).* 2023 Mar 23;15(7):1596. doi: 10.3390/polym15071596.

Chapter 2.3

Extracellular Vesicles: Emerging Biomimetic Nanocarriers for Therapeutic Agents in Infectious Diseases

Prince Sebastian[1,*]
Preena Sebastian[2]
and Pavan Yadav[2]
[1]Department of Animal Biology, School of Life Sciences, University of Hyderabad, Hyderabad, Telangana, India
[2]Aster RV Hospital, J. P. Nagar, Bengaluru, Karnataka, India

Abstract

Extracellular Vesicles (EVs) are small nanosized biomimetic vesicles that mimic parental cellular physiology. They are endocytic in origin and secreted by all eukaryotic cells. Various studies suggest EVs with therapeutic payloads have several advantages over conventional synthetic therapeutic carriers. Because of their low immunogenicity and toxicity, stability in circulation, drug release stability, extending drug half-life, tissue-specific targeting, biocompatibility, and high penetration in biological barriers, EVs have gained popularity as therapeutic delivery vehicles. This chapter highlights therapeutic strategies of loading and target-specific delivery in bacterial, viral, and parasitic infections. Finally, we discuss the future perspective of clinical translation and the challenges of EV-mediated drug delivery in infectious diseases.

Keywords: extracellular vesicles, drug delivery, infectious diseases, therapeutic strategies, biomimetic vesicles

1. Introduction

Infectious diseases are majorly classified as bacterial, viral, parasitic or fungal. Pathogenic microorganisms, including bacteria, viruses, parasites or fungi, are the sources of infection. It

* Corresponding Author's Email: princecell1@gmail.com

In: Emerging Nanomaterials in Biomedical Science
Editors: Eepsita Priyadarshini and Ajith Manayil Parambil
ISBN: 979-8-89530-600-0

is one of the major contributors to global mortality and morbidity and poses a threat to public health. These diseases can spread from one person to another, either directly or indirectly. Vaccines and chemotherapeutics are the major preventive strategies for infectious diseases. Vaccine development is time-consuming and challenging, especially against antimicrobial-resistant pathogens (Costanzo and Roviello, 2023). So, the immediate remedy for infectious disease treatment is drug development. In contrast to viral diseases, bacterial, parasitic or fungal diseases are difficult to control with vaccination because of their structural and immunological complexity (Osterloh 2022; Birnbaum and Craft 2011). Antibiotics, anti-parasitic or anti-fungal chemotherapeutics are primary disease control and prevention sources. Indeed, efficient drugs and a versatile platform for their delivery are necessary to prevent or treat emerging infections and drug resistance issues. EVs substantially contribute to the targeted delivery of drugs and offer an effective formulation for clinical translation. EVs produced by the prokaryotic and eukaryotic cells are heterogeneous and are classified according to their biogenesis pathway. However, there is an ongoing debate over the nomenclature criteria of EVs (Witwer and Théry 2019). A variety of cell-derived vesicles were named in literature, including calcifying matrix, microparticles, nanoparticles, microvesicles, exovesicles, endosomes, prostasomes, secretomes, argosomes, tolerosomes, exosomes, and exosome-like vesicles (Colombo, Raposo, and Théry 2014; Witwer and Théry 2019). Cell-derived EVs are heterogeneous vesicles of 30-1000 nm in size and are the choice of drug carriers in the recent era of therapeutics (Couch et al. 2021; Colombo, Raposo, and Théry 2014).

The history of EVs started after 1945 and were considered cellular membrane debris by Chargaff (Chargaff, 1945). Later, in 1967, Peter Wolf demonstrated in electron microscopy that it could be a material derived from platelets named "platelet dust" (Wolf 1967). In 1974, Nunez et al. imaged multivesicular bodies and postulated their membrane fusion (Nunez et al. 1974). Raposo et al. (1996) proved that EVs have an immunological function (Raposo et al. 1996). After that, many studies have highlighted their role in intracellular communication (Tkach and Théry 2016). Since EVs are involved in intracellular communication and have immunological functionality, Zitvogel et al. explored them for an EV-based vaccine approach in 1998 (Zitvogel et al. 1998). Numerous studies on the physiological function of EVs and their potential use as therapeutics have started with this work, which has also opened new therapeutic avenues. Many extensive studies have focused on EVs' potential for therapy in the last several years, especially at the start of 2010. The ISEV (International Society for Extracellular Vesicles) was formed in 2011, and in 2014-18, Minimal Information for Studies of Extracellular Vesicles (MISEV) implemented certain guidelines. ISEV and ISCT (International Society for Cell and Gene Therapy) provide the guidelines for translating innovative EV research into deliverable therapeutics.

EVs are biomimetic membrane vesicles with a lipid bilayer and ligands that can interact specifically with target cells. Thus, EVs promise target-specific delivery of drugs (Burnouf, Agrahari, and Agrahari 2019). Moreover, EVs carry many bioactive molecules and are involved in cellular communications. The ability to shuttle various biomolecules to target cells has made EVs potentially bioactive therapeutic carriers (Uddin et al. 2024). The EVs-based next-generation therapeutics are the need of the hour to control re-emerging and drug-resistant infectious agents. The efficiency of the drugs is determined by many factors, such as their half-life, stability, biocompatibility, target-specific delivery, biological barrier crossing capability, etc., and EVs-based therapeutics satisfy these requirements (Burnouf, Agrahari, and Agrahari 2019; György et al. 2015). This chapter highlights and discusses therapeutic strategies of EVs in various infectious diseases and the future scope of EV-based therapeutics in infectious disease control and treatment.

2. EV-Based Therapeutic Strategies for Infectious Diseases

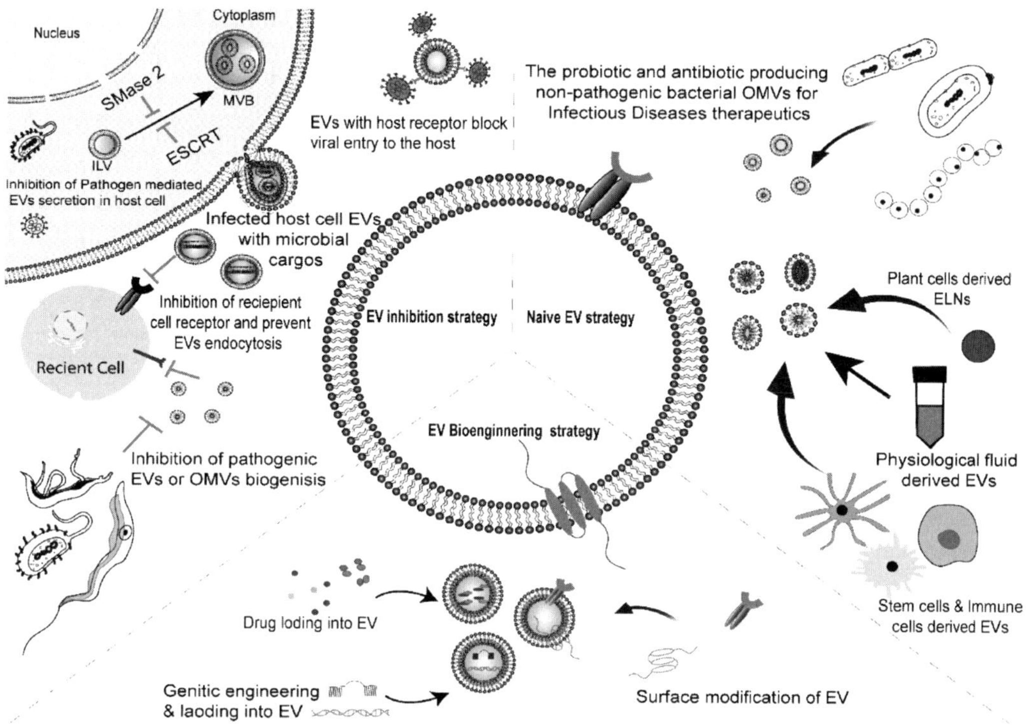

Figure 1. EV mediated therapeutic strategies are represented (i) The inhibition strategy is a major strategy to inhibit the biogenesis of EV, secretion of pathogens and pathogen-infected host cells to prevent or treat the progress of EV-mediated infection. (ii)Naïve EV strategy, the naïve stem cells, immune cells, plant cells, probiotic or non-pathogenic bacteria, and physiological fluids (amniotic, serum, and milk, etc.) are the sources of EVs used as therapeutics for various infectious diseases. (iii) Bioengineering strategy of EVs: the naturally derived EVs are loaded with drug molecules through physical sonication, freeze/thaw cycles, electroporation, extrusion, and mixing. The chemical process is performed through saponin or other transfection reagents. The intrinsic and extrinsic modification of the EVs and attachment of various drug molecules on their surface and delivery to the target. It includes endogenous and exogenous modification of genes through genetic engineering and is used as therapeutics against various infectious agents.

EV-based therapy is a growing trend in nanobiotechnology, and many diseases can be treated with EV therapeutic strategies. It opens an avenue to tackle drug resistance and barrier issues in the specific and efficient delivery of drugs. EVs are cell-free therapy and therapeutics delivery vehicles (Fais et al. 2016). Remarkably, EVs are messenger molecules used by prokaryotic, eukaryotic, and parasitic pathogens. It has been determined that bacterial EVs, also known as outer membrane vesicles (OMVs), are essential mediators in the communication of bacteria and biofilm formation (Kulp and Kuehn, 2010). Interestingly, intracellular pathogens control the host cells and secrete the EVs with microbial molecules that progress the diseases

(Regev-Rudzki, Michaeli, and Torrecilhas, 2021). For example, EV biogenesis is something that viruses can take over to aid in spreading their viral particles and transcription factors (Ramakrishnaiah et al. 2013). So, the pathogenic EV biogenesis inhibition strategy is gaining attention in infectious disease treatment. Another important strategy is that naive cell-derived EVs are potential candidates against infectious agents. Moreover, plant-derived EVs and physiological fluid-derived EVs are suitable in infectious disease therapeutics (Y. Wang et al. 2022a; Cong et al. 2022; Keshtkar et al. 2022; S. Qu et al. 2022; Dai et al. 2024; Y. Wang et al. 2020; García-Martínez et al. 2019). Bioengineered EVs have gained an accelerative momentum among EV-based therapeutics. In this approach, cell-derived EVs are modified or altered for the treatment. Thus, Trojan horse-based delivery of anti-infective medications may be possible with superior specificity and efficacy. Here, different therapeutic strategies, such as EV inhibition strategy, Naive EVs therapeutic strategy, and bioengineering of EVs therapeutic strategy in various infectious diseases, are discussed (Figure 1).

2.1. EVs Inhibition Strategy

EVs regulate the progress of infections by mediating the transfer of specific components of pathogens into the host cells (Gurung et al. 2011; Eliaz et al. 2017; Dong, Filho, and Olivier 2019; Coakley, Maizels, and Buck 2015; Hoy et al. 2014). The EVs with pathogenic cargos of bacteria, viruses, and parasites are enriched in the infected host cells, and they regulate host cells for their infectivity and pathogenicity (McNamara and Dittmer 2020; Regev-Rudzki, Michaeli, and Torrecilhas 2021). These pathogenic cargo-packaged EVs are involved in cellular communication of the host and further progress the infection (Stotz, Brotherton, and Inal 2022). The intracellular secretion of EVs majorly through ESCRT (Endosomal Sorting Complexes Required for Transport Machinery) can follow dependent or independent pathways. Parasite *T. vaginalis*, VPS32, an ESCRTIII complex member, aids in EV biogenesis and molecular cargo sorting (Salas et al. 2022). Similarly, in the parasite, *Trypanosome brucei* exosome production is compromised by gene silencing of the VPS36 (Eliaz et al. 2017). So, it is a prime strategy to suppress the secretion or uptake of pathogen cargo carrying EVs to control the disease progress (Hao et al. 2021). In the ESCRT-dependent pathway, ESCRT forms the multivesicular bodies (MVBs) with Intraluminal Vesicles (ILVs) and secretes EVs into the extracellular space of the cells. MVBs-ILVs are produced via sphingomyelinase hydrolysis mediated Neutral Sphingomyelinase 2 (nSMase2) in the ESCRT-independent pathway (Hessvik and Llorente 2018). The virus hijacks the EV biogenesis pathways to replicate and spread in the host. For example, The HIV-1 virus hijacks the ESCRT-dependent pathway for the efficient replication of the virus (Meng et al. 2015). Human Herpesvirus 6 (HSV) promotes the formation of MVB, indicating the formation of exosomes and virion release (Mori et al. 2008). Rab27a is another crucial factor in the exosome biogenesis pathway, and many viruses hijack the Rab pathway allowing the spread of virus and its pathogenic components. The expression of Rab27a increased in cases of Cytomegalovirus and HIV-1 infection and proved that the interaction promotes the biogenesis of exosomes (Aydin et al. 2021; Ostrowski et al. 2010; Fraile-Ramos et al. 2010). Inhibiting these pathways prevents the secretion of EVs, and through that, pathogenesis could be controlled. A recent study in murine HIV model demonstrated that the inhibition of the enzyme nSMase2 by using an inhibitor (R)-(1-(3-(3,4-methoxyphenyl)-2,6-dimethylimidazo[1,2-b] pyridazin-8-yl) pyrrolidin-3-yl)-carbamate

(PDDC) reversed the major depressive disorder (MDD) and cognitive impairment (CI) in EcoHIV-infected mice. Moreover, the treatment normalized the sphingomyelin and ceramide levels by regulating increased cortical brain nSMase2 activity (X. Zhu et al. 2022). Hepatitis C virus (HCV) replication was prevented by the inhibition of the EV biogenesis (Aydin et al. 2021). The inhibition of ACE2-expressing exosomes with non-structural protein 6 (NSP6) of SARS-CoV-2 reduced the key factors for exosome biogenesis proteins such as tetraspanin-CD63 and ACE2. It prevented the virus's entry into the healthy bystander cells (Lv et al. 2024). The splenic dendritic cells derived from EVs reduce the host brain's *T. gondii* parasite-induced cyst development and protected it from disease Suzuki et al. 2010). The EV secretion in parasitic nematodes was inhibited by ivermectin (IVM), an anthelminthic drug (Loghry et al. 2020). The anti-fungal inhibitor, named turbinmicin, inhibits the vesicle-mediated trafficking pathway and prevents the extracellular vesicle-mediated biofilm formation; interestingly, it shows a far-ranging anti-fungal activity (Zhao et al. 2021). The *Schistosoma japonicum* worms produce EVs that promote egg production in adult worms by pairing miRNAs. A calcium-dependent cysteine protease inhibitor, calpain, prevents EV-mediated miRNA cargo loading and reduces egg production (Kumagai et al. 2022). Bacterial EV inhibitors will act as novel pharmacological agents against infectious agents and regulate EV biogenesis, activity and EV-mediated inflammation control (Chen et al. 2022). For example, *P. aeruginosa* secretes the Pseudomonas Quinolone Signal (PQS) that helps in EV biogenesis and quorum sensing of bacteria (Mashburn-Warren et al. 2008; 2009). A class of calcium-activated enzymes, Peptidyl Arginine Deiminases (PADs), engage in the bacterial OMV biogenesis process (Gavinho et al. 2020; Kosgodage et al. 2019). PAD inhibitors such as Cl-amidine, BB-Cl-amidine AMF30a, and GSK199 effectively lower OMV release in Gram-negative and positive bacteria, increasing antibiotic resistance within *S. aureus* and *E. coli*, revealing new approaches for pharmacological OMV-inhibition and antibacterial resistance prevention (Kosgodage et al. 2019).

Another inhibition strategy is blocking EV propagation by inhibiting the recipient cell endocytosis pathway and uptake processes, such as membrane fusion and phagocytosis (Tian et al. 2014). For example, Dynasore is the inhibitor of the clathrin-dependent pathway. It mediates the uptake of EVs from infected B cells with Epstein–Barr virus into uninfected B cells (Nanbo et al. 2013). The Human Interferon (IFN)-stimulated gene of the Angiotensin Converting Enzyme 2 (hACE2) receptor is recognized by SARS-CoV-2 (Severe acute Respiratory Syndrome Coronavirus-2) for virus entry. IFN-γ-induced viral receptor upregulation of exosomes was used to prevent viral entry, which competitively blocks the viral entry by utilizing the hACE2 receptor (J. Zhang et al. 2021). The FDA-approved oxytetracycline hampers the Ebola virus (EBOV) infection and reduces the secretion of Ebola VP40-carrying EVs (Pleet et al. 2016). Tetherin, another inhibitor, suppresses EVs with EBOV glycoprotein (Nehls et al. 2019). So, the inhibition of EV biogenesis is a promising therapeutic strategy to prevent pathogenic EVs in the infection progression of various infectious diseases.

2.2. Naive EVs Therapeutics

A few bacterial and mammalian EVs are found to have inherent therapeutic activity (Timár et al. 2013; Kesimer et al. 2009), which can make them potential anti-infectious therapeutic agent (Kadurugamuwa and Beveridge 1996) or serve as a model for new biomimetic systems (Hu et al. 2013). The bacteria-secreting OMVs are an effective defensive tactic when vying for environmental niches with other bacterial species. For example, the OMVs of *Myxobacteria* or *Pseudomonas* species contain hydrolytic factors meant to kill prey microbes (Z. Li, Clarke, and Beveridge 1998; Kadurugamuwa and Beveridge 1996). Recent studies demonstrated that these autolytic OMVs with antibiotics from the antibiotic-producing bacteria additionally priming with drugs exhibited synergistic bactericidal effects (Kadurugamuwa and Beveridge 1996; Colzi et al. 2015). Furthermore, the non-pathogenic or probiotic bacteria OMVs are drastically used as efficient drug carriers that are well-tolerated, have low endotoxicity, and feasible large-scale production in the fermenter is possible (Goes et al. 2020; Gujrati et al. 2014; Schulz et al. 2018). Another important study demonstrated that the vaginal symbiotic bacteria *Lactobacillus spp.* (*L. crispatus* BC3 and *L. gasseri* BC12) secrete EVs, inhibiting HIV-1 by diminishing the viral and host cell interaction by hindering Env protein exposure (Ñahui Palomino et al. 2019). The EVs produced from autologous cells may significantly advance in developing personalized medicines with low adverse effects and superior therapeutic efficacy and selectivity. For example, EVs derived from monocytes impede the *in vitro* growth of *S. aureus* by β2 integrin-dependent aggregate formation (Timár et al. 2013). *M. tuberculosis* H37Rv infected J774A.1 macrophage EVs tremendously decreased the bacterial load in the infected mice lungs (García-Martínez et al. 2019). Another study found that virus recognizing and binding sialic acid in the vesicle surface allowed tracheobronchial cells of humans to produce EV-like particles, which is effective against the influenza virus (Kesimer et al. 2009). Mesenchymal stem cells (MSCs) are a probable source of EVs used for therapeutics and are safer, potent and cell-free-mimetic therapeutics (Alcayaga-Miranda, Cuenca, and Khoury 2017). BM-MSC-derived EVs are effective against pneumonia-causing bacterial Acute Lung Injury (ALI) by lowering lung inflammation and increasing lung protein permeability in a mouse model of endotoxin-induced *E. coli* (Abreu, Weiss, and Rocco 2016; Y. G. Zhu et al. 2014). Hepatitis C virus RNA replication was directly suppressed by miRNAs found in umbilical-MSC-derived exosomes, including let-7f, miR-199a, miR-145, and miR-221 (Qian et al. 2016). Similarly, the macrophage exosomes containing the miR-29 family prevent the anti-viral effect of hepatitis C virus infection (Zhou et al. 2016; Tu et al. 2010). The micro RNAs containing exosomes from umbilical cord-MSCs help suppress HIV infection (Z. Zhang et al. 2013). The BM-MSCs derived exosomes against respiratory viruses, especially the SARS-CoV-2, and these exosomes reduce pro-inflammatory cytokines and prevent cytokine storms (Sengupta et al. 2020; Popowski et al. 2021). In the immunocompetent mouse model of severe refractory neutrophilic asthma induced by *Aspergillus* hyphal extract, BM-MSCs-derived EVs reduced allergic airway inflammation (Cruz et al. 2015). The fungal infections, especially *Candida albicans,* can be treated with TLR4-primed and IL-17^{+} primed mesenchymal stem cells derived exosomes (Ghasemian 2021). The EVs derived from peritoneal M2 macrophage prevent the cytokine storm in infectious diseases by inhibiting the pro-inflammatory inducing pathways such as JAK-STAT, NF-κB and p38 and reduce the inflammatory cytokine production (Y. Wang et al. 2022b). Hepatitis B-infected hepatocytes EVs show anti-viral activity by NKG2D ligand expression in macrophages and IFN-γ production via NK cell activation (Kouwaki et al. 2016). Various viral mRNAs and microRNAs packed HEp-2 cells derived EVs with STING (stimulator of IFN genes) stimulation, silencing HSV genes and preventing infection (Kalamvoki, Du, and Roizman 2014). EVs from umbilical cord mesenchymal stem cells of humans mitigate the hyperinflammatory response in COVID-19 infection (Nasiri et al. 2024). The lemon-derived exosome-like nanoparticles (ELNs) could enhance the gut survivability and bile resistance of probiotics *Streptococcus thermophilus* (STH) and *Lactobacillus rhamnosus*, thereby impeding the growth of pathogenic *Clostridium difficile* (Lei et al.2020). The ELNs from ginger prevent the pathogenic bacterial infection of *Porphyromonas gingivalis* by inhibiting its Heme-binding protein 35 (HBP35) by binding on the surface (Sundaram et al. 2019). Furthermore, ginger-originated ELNs with rlcv-miR-rL1-28-3p and aly-miR396a-5p could reduce lung inflammation, expression of the SARS-CoV-2 Nsp12 and genes of spike protein (Teng et al. 2021). Besides mammalian cells, plant cells also have the ability to produce EVs (Woith and Melzig 2019; Cai et al. 2018; Halperin and Jensen 1967; Regente et al. 2009). The plant cell-derived EVs or ELNs are not well-explored areas but have potential as promising biotherapeutics to prevent or treat infectious diseases. Interestingly, physiological fluids-derived EVs contribute a major role in infectious disease treatment. The acellular perinatal amniotic fluid-derived EVs (Zofin) show tremendous healing efficiency and safety in moderate and severe COVID-19. The treatment improved the inflammatory biomarkers such as inflammatory IL-6, TNF- α, CRP, and lymphocyte count (Bellio et al. 2021; Chanda et al. 2024a). Further, AFMSCs (Amniotic Fluid-derived Mesenchymal Stem Cells) derived exosomes are in different phases of clinical trials for treating moderate and severe COVID-19 infection (NCT04384445, NCT04657406) associated adverse effects. The milk-derived exosomes are a better replacement for synthetic carriers of drugs and are widely used to deliver drugs against infectious agents specifically and safely. Another study addressed that the human milk-derived exosomes from healthy donors are packed with micro RNAs that have an enormous capacity to inhibit transmission of HIV-1 (Zahoor et al. 2020a). The exosomes derived from goat milk show an inhibitory effect on the dengue (DNEV) virus by inhibiting viral genome replication and maturation by targeting NS3 expression (Yenuganti et al. 2022). The naïve cell-derived EVs, physiological fluids or stem cells-derived EVs or exosomes serve as potential therapeutics with immune modulatory activity to suppress many pathogenic agents and infections.

2.3. Bioengineering of EVs Therapeutics

The alteration or modifications of the EVs are termed bioengineering, boosting the therapeutic potential. It is either the drug's encapsulation, molecular modification, or attachment to the surface molecules (Nasiri Kenari, Cheng, and Hill 2020a). Another strategy is the bioengineering of artificial exosomes with drug molecules or the genetic engineering and secretion of EVs with loaded nucleic acids (Antimisiaris et al. 2018). For example, stem cells were bioengineered to overexpress the ACE2 to block SARS-CoV-2 virus entry competitively (El-Shennawy et al. 2022; Inal 2020). There are two types of drug loading in EVs namely (1) Endogenous methods, wherein the cells that produce EVs also modify structural protein or nucleic acid components or load drugs into the vesicles before isolation, and (2) Exogenous methods, wherein drug molecules are loaded into EVs after its purification (Nasiri Kenari, Cheng, and Hill 2020b; Fuhrmann et al. 2015). Exogenous loading techniques include the attachment of therapeutic molecules to the surface of EV and allowing compounds to diffuse into the vesicle through the temporary opening of the EV membrane. This is caused by mechanical sonication, electroporation, passive incubation or chemically by saponin treatment. The exogenous loading has less complexity than the endogenous method (Luan et al. 2017; De Jong et al. 2019). Bioengineered EVs are potentially used in infectious disease treatments. The macrophage-derived exosomes were encapsulated with the antibiotic linezolid, and the effect on the mouse model of methicillin-resistant *Staphylococcus aureus* (MRSA) infection was examined. It shows that infection is suppressed compared to free antibiotics (Yang et al. 2018). The therapeutic or prophylactic potential of Quercetin (QCT) loaded EVs in SARS-CoV-2 reduces the inflammation, and ACE2 expressing EVs has a high affinity to the SARS-CoV-2 virus spike protein which hampers the viral entry to the host cell (Raghav et al. 2022). The exosome-loaded LL-37 (LL-37-TM-exo and TM-LL-37-exo) was used to treat Zika virus infection, and it significantly suppresses the viral infection and reduces the broad-spectrum anti-viral effect as well as the virus-mediated testicular and sperm injury (Wang et al. 2024). A recent approach of EVs as a delivery platform for CRISPR-Cas 9 ribonucleoprotein, and it has been successfully evaluated *in vitro* and *in vivo* to inhibit the HSV1 virus. Further, a neuron-targeting peptide RVG suppresses viral replication in neuronal tissue. However, the stringent safety evaluation and enhancement of CRISPR-Cas 9 enrichment was strongly suggested in the study (Wan et al. 2024). Similarly, *in vitro,* along with *in vivo* study reported that engineered EVs can effectively mediate functional CRISPR/Cas9 delivery, resulting in the removal of integrated viral DNA fragments and episomal cDNA. This offers a novel therapeutic approach to treat chronic hepatitis B Virus (HBV) infection (Zeng et al. 2024). The cancer-causing HPV (Human Papilloma Virus)16-E7 protein specifically interacted with a single-chain variable fragment (scFvs). The HPV16-E7 interacts with the tumor suppressor pRb protein, and the dysregulation of this interaction heavily hampers the HPV16-E7-mediated cell proliferation. The Nefmut/anti-HPV16-E7 scFv chimeric product-packed EVs interact with HPV16-E7 to inhibit the growth of cells expressing HPV16-E7 (Ferrantelli et al. 2019). A recent study revealed that a natural flavonoid known as α-mangostin loaded in the milk-derived exosomes cleared 99% of bacteria in macrophages and similarly improved the gut bacterial infection in the *in vivo* model (Qu 2022a et al. 2022). However, stringent evaluation and clinical trials needs to be enforced to implement, monitor and thereby prevent any unforeseen safety issues.

3. EVs-Mediated Drug Delivery Is the Next-Generation Treatment for Infectious Diseases: Future, Scope, and Challenges

In recent years, exosomes have been increasingly explored as platforms for vaccines and drug delivery carriers (Devhare and Ray 2017). Recently, EVs have gained considerable attention in therapeutics as potential next-generation treatment options and delivery carriers for infectious diseases. For instance, clinical trials have been initiated to use EVs for treating respiratory complications associated with SARS-CoV-2 infection (Table 2).

The primary sources of therapeutic exosomes can be derived from dendritic cells, adipose tissue-derived MSCs, T cell-derived exosomes, autologous tumor cells, or mesenchymal stem cells (MSCs), and these cells derived EVs could either be engineered or are naïve form (Perocheau et al. 2021; Deffune, Prudenciatti, and Moroz 2020; Devhare and Ray 2017). Interestingly, acellular products derived exosomes are another noteworthy therapeutics against SARS-CoV-2 in clinical trials derived from human amniotic fluid (Chanda et al. 2024b; Devhare and Ray 2017). Milk-derived exosomes are another important acellular fluid-derived exosome in infectious diseases therapeutics and delivery vehicles that have received more attention in the last few years (Yenuganti et al. 2022; Zahoor et al. 2020; Qu et al. 2022). EV-based therapeutics are in clinical trials for various infectious diseases. However, it must address many questions regarding its safety in clinical translation as therapeutics. For example, the source of the EVs, their proper characterization and conformation of their composite are important because the reliability and reproducibility of EV therapeutics depend on them. So many factors, such as the source of cell, culture conditions, purification, characterization methods, and biological interaction in the system, are unavoidable in EV therapeutics. Here, we discuss the details of the comparative benefits of EVs with other nanocarriers and the significant challenges in their purification, manufacturing strategies in large-scale production, and clinical translation for infectious diseases therapeutics.

3.1. Comparative Benefits of Extracellular Vesicles Over Other Nanocarriers in Infectious Diseases Therapy

The EV-based drug carriers are efficient, target-specific, and intrinsically compatible. On the other hand, it can deliver a sufficient density of loaded cargo over long distances without any immune system hindrance by the camouflage activity of the natural EVs (Parada et al. 2021). Target tissue specificity and multistage therapeutic delivery are other potential credentials that EVs have, and they are not easily achieved by synthetic carriers (Witwer and Wolfram, 2021). Unlike other nanocarriers, EVs have more encouraging properties like decreased immunogenicity and complement activation and might act as a tremendous innate immune system against pathogens (Fuhrmann, Herrmann, and Stevens 2015; György et al. 2015). Recently, the unconventional liposome derived from a prokaryotic membrane loaded with antibacterial molecules acted as a bacterio-mimetics and served better than the conventional liposomes. It opens an avenue for low endotoxic and well-tolerated drug carriers from bacterial OMVs (Colzi et al. 2015; Gujrati et al. 2014). An additional benefit of EVs or OMVs with other nanocarriers is that they can genetically modify cells or bacteria to secrete or load molecular drugs or immunomodulators naturally for the therapy (Fuhrmann, Neuer, and Herrmann 2017). The mammalian EVs and bacterial OMVs are better alternatives for many chemotherapeutic-loaded nanocarriers because of their inherent antibacterial activity with outstanding biocompatibility (Timár et al. 2013; Kesimer et al. 2009). Moreover, natural vesicles have better barrier-crossing ability and tissue homing, which is impossible in case of synthetic carriers (Fuhrmann, Neuer, and Herrmann 2017).

3.2. Challenges and Limitations in Utilizing Extracellular Vesicles for Infectious Disease Treatment

EVs are heterogeneous vesicle populations, making the purification and downstream scale-up process difficult. The biological functionality is achieved only by good manufacturing and the downstream process. The scale-up process enhances the chances of cell death because of the large cell density. It induces apoptosis and cell impurities in the culture, negatively impacting the EVs' downstream functionality. The cellular microenvironment, especially cellular physiology, media composition, growth factors, and other supplementary macro and micronutrients and pH are other paramount factors in EVs' isolation and downstream functionality. Additionally, the scale-up process changes these factors and might alter the phenotype of the source cell line, which is one of the biggest challenges in EV therapeutics (Devhare and Ray 2017). Even though the EVs are biocompatible and immunologically feasible, the safety validations, especially the bacteria-derived OMVs, are more critical than mammalian EVs because of the LPS-mediated endotoxicity (Jahromi et al. 2021). The recent alarming report of a case in Nebraska by theFood and Drug Administration (FDA) on their website states that using exosomes containing unapproved products causes severe adverse effects (FDA, 2020, https://www.fda.gov/vaccines-blood-biologics/consumers-biologics/consumer-alertregenerative-medicine-products-including-stem-cells-and-exosomes, Retrieved on 9th July 2024). Therefore, refined purification and analytical techniques with reliable and cost-effective large-scale production lead to the successful translation of EV therapeutics. The reproducibility, large-scale production as well as good manufacturing practices are still significant issues in industrial translation, even if it passes the clinical trials (Cecchin et al. 2023). The artificial EVs that mimic the natural EVs address sterility and mass production to certain extent. Combining EVs with liposomes packed with drug molecules is one such invention (Piffoux et al. 2018). A promising strategy in EV-based therapeutics involves implanting bioengineered EV-producing cells directly in the body, potentially enhancing EV production with fewer safety concerns (Kojima et al. 2018). According to guidelines from various national and international societies on stem cells and EVs, rigorous safety assessments and well-designed clinical trials are necessary for advancing EV-based therapies (Börger et al. 2020). In addition, standardized purification protocols along with stringent monitoring and analysis during clinical trials are crucial in addressing challenges and improving innovative EV therapeutic applications.

Conclusion

Infectious diseases heavily contribute to global mortality and morbidity of diseases index and hence are a global burden. Moreover, emerging pathogens and microbial resistance are alarming threats to public health. The new prophylactic or treatment strategy is unavoidable in addressing the current problems. EVs are small heterogeneous vesicle that potentially contribute to infectious disease therapeutics. Various therapeutic strategies, such as inhibition of microbial and pathogen-infected host cell-derived EVs, prevent the progress and survival of the pathogens. On the other hand, naive cell-derived and bioengineered EVs from various sources effectively treat many infectious diseases. Recent studies emphasize the increase in investments in EV-based pharmaceutical development and clinical trials of EV therapeutics against infectious diseases. Although EV-based therapeutics have gained rapid growth in nanobiotechnology, they are still in their infancy and need to address certain uncertainties regarding safety. Different approaches are being developed to maximize the therapeutic efficacy of EVs. Refining purification and analytical techniques enable secure and fruitful clinical trials and improve scale-up strategies by overcoming limitations. Despite these, EVs could serve as potential therapeutics and therapeutic carriers to prevent or treat many infectious diseases, and there is a need to unlock their full potential.

Table 2. Extracellular vesicle-mediated infectious diseases therapeutics in clinical trials

Disease or condition	EV origin	Therapeutic Application	Reference (NCT Number)
COVID-19	COVID-19 specific T cell	To prevent SARS-CoV-2 Associated Pneumonia	NCT04389385
SARS-CoV-2 pneumonia	MSCs (Mesenchymal Stem Cells)	To prevent SARS-CoV-2 Associated Pneumonia	NCT04602442
Coronavirus Infection	Allogenic adipose MSCs	To prevent the adverse effects of coronavirus Pneumonia	NCT04276987
SARS-CoV-2 related pneumonia	MSCs	To prevent SARS-CoV-2-related Pneumonia	NCT04491240
COVID-19	Amniotic epithelial stem cells	To hamper the induced SARS-CoV-2-associated hyperactivation and other adverse effects.	NCT04657406
SARS-CoV-2 pneumonia	MSCs	To control SARS-CoV-2 pneumonia and associated adverse effects	NCT05216562
COVID-19, ARDS (acute respiratory distress syndrome)	BMSCs (Bone marrow Mesenchymal Stem Cells)	To prevent ARDS, cytokine storm and hypoxia	NCT04491240
COVID-19	BMSCs	Dyspnea and post-viral syndrome treatment	NCT05116761
COVID-19, ARDS	BMSCs	To prevent COVID-19 or ARDS adverse effects	NCT05354141
SARS-CoV-2 associated pneumonia, ARDS	BMSCs	Prevention of SARS-CoV-2 related pneumonia and ARDS	NCT04493242
Carbapenem-resistant gram-negative bacilli infection	MSCs/progenitor-cell-derived exosomes	To prevent pulmonary infections	NCT04544215
COVID-19	UCMSC-EV (Umbilical Cord Mesenchymal Stem Cells)	To prevent COVID-19-associated pneumonia	NCT05787288
COVID-19	UCMSC-EV	To prevent chronic post COVID-19 cough	NCT05808400
COVID-19, ARDS	UCMSC and Placenta	To prevent COVID-19 adverse effects and ARDS	NCT05387278
COVID-19, ARDS	Perinatal mesenchymal stem cells (PMSC)	To prevent COVID-19 pneumonia and ARDS	NCT04798716
COVID-19	MSC-Exosomes	Immunomodulatory effect in COVID-19 infection	NCT05191381
COVID-19	AFMSC (Amniotic-fluid mesenchymal stem cells)	COVID-19 long haulers	NCT04384445
COVID-19	AFMSC	Mild or moderate COVID-19 prevention	NCT04657406
COVID-19	CD24 expressing 293-TREx™ exosomes from HEK-293 cells derived	Mild or moderate COVID-19 prevention	NCT04969172
SARS-CoV-2 infection	Exosomes from T-REx™-293 cells engineered to express CD24 at high levels	Safety therapy and prevention of adverse events	NCT04747574
COVID-19	CD24 expressing 293-TREx™ derived from HEK-293 cells	To prevent moderate or severe adverse effects of COVID-19	NCT04902183
COVID-19; Novel Coronavirus Pneumonia; ARDS	MSCs	To control adverse events after the viral infection	NCT04798716

Acknowledgments

PS acknowledges the University of Hyderabad, Telangana, for the institutional support and the Indian Council for Medical Research (ICMR), New Delhi, for the ICMR-SRF fellowship.

References

Abreu, Soraia C., Daniel J. Weiss, and Patricia R. M. Rocco. 2016. "Extracellular Vesicles Derived from Mesenchymal Stromal Cells: A Therapeutic Option in Respiratory Diseases?" *Stem Cell Research and Therapy*. BioMed Central. https://doi.org/10.1186/s13287-016-0317-0.

Alcayaga-Miranda, Francisca, Jimena Cuenca, and Maroun Khoury. 2017. "Antimicrobial Activity of Mesenchymal Stem Cells: Current Status and New Perspectives of Antimicrobial Peptide-Based Therapies." *Frontiers in Immunology*. Frontiers Media SA https://doi.org/10.3389/fimmu.2017.00339.

Aydin, Yucel, Ali Riza Koksal, Venu Reddy, Dong Lin, Hanadi Osman, Zahra Heidari, Sadeq Mutlab Rhadhi, William C. Wimley, Mansour A. Parsi, and Srikanta Dash. 2021. "Extracellular Vesicle Release Promotes Viral Replication during Persistent Hcv Infection." *Cells* 10 (5). https://doi.org/10.3390/cells10050984.

Bellio, Michael A., Cassie Bennett, Alissa Arango, Aisha Khan, Xiumin Xu, Cesar Barrera, Vincent Friedewald, and Maria Ines Mitrani. 2021. "Proof-of-Concept Trial of an Amniotic Fluid-Derived Extracellular Vesicle Biologic for Treating High Risk Patients with Mild-to-Moderate Acute COVID-19 Infection." *Biomaterials and Biosystems* 4 (December). https://doi.org/10.1016/j.bbiosy.2021.100031.

Birnbaum, Ron, and Noah Craft. 2011. "Innate Immunity and Leishmania Vaccination Strategies." *Dermatologic Clinics*. https://doi.org/10.1016/j.det.2010.08.014.

Börger, Verena, Simon Staubach, Robin Dittrich, Oumaima Stambouli, and Bernd Giebel. 2020. "Scaled Isolation of Mesenchymal Stem/Stromal Cell-Derived Extracellular Vesicles." *Current Protocols in Stem Cell Biology* 55 (1). https://doi.org/10.1002/cpsc.128.

Burnouf, Thierry, Vibhuti Agrahari, and Vivek Agrahari. 2019. "Extracellular Vesicles as Nanomedicine: Hopes and Hurdles in Clinical Translation." *International Journal of Nanomedicine* 14:8847–59. https://doi.org/10.2147/IJN.S225453.

Cai, Qiang, Lulu Qiao, Ming Wang, Baoye He, Feng-Mao Lin, Jared Palmquist, Sienna-Da Huang, and Hailing Jin. 2018. "Plants Send Small RNAs in Extracellular Vesicles to Fungal Pathogen to Silence Virulence Genes." *Science*. Vol. 360. https://www.science.org.

Cecchin, Riccardo, Zach Troyer, Ken Witwer, and Kevin V. Morris. 2023. "Extracellular Vesicles: The next Generation in Gene Therapy Delivery." *Molecular Therapy*. Cell Press. https://doi.org/10.1016/j.ymthe.2023.01.021.

Centre for Biologics Evaluation and Research, US Food and Drug Administration (2020) Consumer Alert on Regenerative Medicine Products Including Stem Cells and Exosomes. https://www.fda.gov/vaccines-blood-biologics/consumers-biologics/consumer-alert-regenerative-medicine-products-including-stem-cells-and-exosomes

Chanda, D., Del Rivero, T., Ghimire, R., More, S., Mitrani, M. I., Bellio, M. A., & Channappanavar, R. (2024). Acellular Human Amniotic Fluid-Derived Extracellular Vesicles as Novel Anti-Inflammatory Therapeutics against SARS-CoV-2 Infection. *Viruses*, *16*(2), 273. https://doi.org/10.3390/v16020273

Chanda, Debarati, Tania Del Rivero, Roshan Ghimire, Sunil More, Maria Ines Mitrani, Michael A. Bellio, and Rudragouda Channappanavar. 2024a. "Acellular Human Amniotic Fluid-Derived Extracellular Vesicles as Novel Anti-Inflammatory Therapeutics against SARS-CoV-2 Infection." *Viruses* 16 (2). https://doi.org/10.3390/v16020273.

Chargaff, E. J. (1945). *Cell structure and the problem of blood coagulation*. 160(1), 351–359.

Chen, Jianwei, Hongfang Zhang, Siqi Wang, Yujie Du, Bin Wei, Qiang Wu, and Hong Wang. 2022. "Inhibitors of Bacterial Extracellular Vesicles." *Frontiers in Microbiology*. Frontiers Media SA https://doi.org/10.3389/fmicb.2022.835058.

Coakley, Gillian, Rick M. Maizels, and Amy H. Buck. 2015. "Exosomes and Other Extracellular Vesicles: The New Communicators in Parasite Infections." *Trends in Parasitology*. Elsevier Ltd. https://doi.org/10.1016/j.pt.2015.06.009.

Colombo, Marina, Graça Raposo, and Clotilde Théry. 2014. "Biogenesis, Secretion, and Intercellular Interactions of Exosomes and Other Extracellular Vesicles." *Annual Review of Cell and Developmental Biology*. https://doi.org/10.1146/annurev-cellbio-101512-122326.

Colzi, Ilaria, Anna N. Troyan, Brunella Perito, Enrico Casalone, Riccardo Romoli, Giuseppe Pieraccini, Nataša Škalko-Basnet, et al. 2015. "Antibiotic Delivery by Liposomes from Prokaryotic Microorganisms: Similia Cum Similis Works Better." *European Journal of Pharmaceutics and Biopharmaceutics* 94 (July):411–18. https://doi.org/10.1016/j.ejpb.2015.06.013.

Cong, Minghui, Shenyu Tan, Simin Li, Lina Gao, Luqi Huang, Huang Ge Zhang, and Hongzhi Qiao. 2022. "Technology Insight: Plant-Derived Vesicles—How Far from the Clinical Biotherapeutics and Therapeutic Drug Carriers?" *Advanced Drug Delivery Reviews*. Elsevier B.V. https://doi.org/10.1016/j.addr.2021.114108.

Costanzo, Vincenzo, and Giovanni N. Roviello. 2023. "The Potential Role of Vaccines in Preventing Antimicrobial Resistance (AMR): An Update and Future Perspectives." *Vaccines*. MDPI. https://doi.org/10.3390/vaccines11020333.

Couch, Yvonne, Edit I. Buzàs, Dolores Di Vizio, Yong Song Gho, Paul Harrison, Andrew F. Hill, Jan Lötvall, et al. 2021. "A Brief History of Nearly EV-Erything – The Rise and Rise of Extracellular Vesicles." *Journal of Extracellular Vesicles*. John Wiley and Sons Inc. https://doi.org/10.1002/jev2.12144.

Cruz, Fernanda F., Zachary D. Borg, Meagan Goodwin, Dino Sokocevic, Darcy E. Wagner, Amy Coffey, Mariana Antunes, et al. 2015. "Systemic Administration of Human Bone Marrow-Derived Mesenchymal Stromal Cell Extracellular Vesicles Ameliorates Aspergillus Hyphal Extract-Induced Allergic Airway Inflammation in Immunocompetent Mice." *Stem Cells Translational Medicine* 4 (11): 1302–16. https://doi.org/10.5966/sctm.2014-0280.

Dai, Cunchun, Qingjun Xu, Lin Li, Ying Liu, and Shaoqi Qu. 2024. "Milk Extracellular Vesicles: Natural Nanoparticles for Enhancing Oral Drug Delivery against Bacterial Infections." *ACS Biomaterials Science and Engineering*. American Chemical Society. https://doi.org/10.1021/acsbiomaterials.3c01824.

Devhare, Pradip B., and Ratna B. Ray. 2017. "A Novel Role of Exosomes in the Vaccination Approach." *Annals of Translational Medicine*. AME Publishing Company. https://doi.org/10.21037/atm.2016.12.75.

Dong, George, Alonso Lira Filho, and Martin Olivier. 2019. "Modulation of Host-Pathogen Communication by Extracellular Vesicles (EVs) of the Protozoan Parasite Leishmania." *Frontiers in Cellular and Infection Microbiology*. Frontiers Media SA https://doi.org/10.3389/fcimb.2019.00100.

Eliaz, Dror, Sriram Kannan, Hadassa Shaked, Gil Arvatz, Itai Dov Tkacz, Lior Binder, Hiba Waldman Ben-Asher, et al. 2017. "Exosome Secretion Affects Social Motility in Trypanosoma Brucei." *PLoS Pathogens* 13 (3). https://doi.org/10.1371/journal.ppat.1006245.

El-Shennawy, Lamiaa, Andrew D. Hoffmann, Nurmaa Khund Dashzeveg, Kathleen M. McAndrews, Paul J. Mehl, Daphne Cornish, Zihao Yu, et al. 2022. "Circulating ACE2-Expressing Extracellular Vesicles Block Broad Strains of SARS-CoV-2." *Nature Communications* 13 (1). https://doi.org/10.1038/s41467-021-27893-2.

Fais, Stefano, Lorraine O'Driscoll, Francesc E. Borras, Edit Buzas, Giovanni Camussi, Francesco Cappello, Joana Carvalho, et al. 2016. "Evidence-Based Clinical Use of Nanoscale Extracellular Vesicles in Nanomedicine." *ACS Nano*. American Chemical Society. https://doi.org/10.1021/acsnano.5b08015.

Ferrantelli, Flavia, Claudia Arenaccio, Francesco Manfredi, Eleonora Olivetta, Chiara Chiozzini, Patrizia Leone, Zulema Percario, et al. 2019. "The Intracellular Delivery of Anti-HPV16 E7 ScFvs through Engineered Extracellular Vesicles Inhibits the Proliferation of HPV-Infected Cells." *International Journal of Nanomedicine* 14:8755–68. https://doi.org/10.2147/IJN.S209366.

Fraile-Ramos, Alberto, Victoria Cepeda, Edo Elstak, and Peter van der Sluijs. 2010. "Rab27a Is Required for Human Cytomegalovirus Assembly." *PLoS ONE* 5 (12). https://doi.org/10.1371/journal.pone.0015318.

Fuhrmann, Gregor, Andrea Serio, Manuel Mazo, Rekha Nair, and Molly M. Stevens. 2015. "Active Loading into Extracellular Vesicles Significantly Improves the Cellular Uptake and Photodynamic Effect of Porphyrins." *Journal of Controlled Release* 205 (January):35–44. https://doi.org/10.1016/j.jconrel.2014.11.029.

Fuhrmann, Gregor, Anna Lena Neuer, and Inge K. Herrmann. 2017. "Extracellular Vesicles – A Promising Avenue for the Detection and Treatment of Infectious Diseases?" *European Journal of Pharmaceutics and Biopharmaceutics*. Elsevier B.V. https://doi.org/10.1016/j.ejpb.2017.04.005.

Fuhrmann, Gregor, Inge K. Herrmann, and Molly M. Stevens. 2015. "Cell-Derived Vesicles for Drug Therapy and Diagnostics: Opportunities and Challenges." *Nano Today*. Elsevier B.V. https://doi.org/10.1016/j.nantod.2015.04.004.

García-Martínez, Mariano, Luis Vázquez-Flores, Violeta D. Álvarez-Jiménez, Jessica Castañeda-Casimiro, Miguel Ibáñez-Hernández, Luvia Enid Sánchez-Torres, Jorge Barrios-Payán, et al. 2019. "Extracellular Vesicles Released by J774A.1 Macrophages Reduce the Bacterial Load in Macrophages and in an Experimental Mouse Model of Tuberculosis." *International Journal of Nanomedicine* 14:6707–19. https://doi.org/10.2147/IJN.S203507.

Gavinho, Bruno, Bruna Sabatke, Veronica Feijoli, Izadora Volpato Rossi, Janaina Macedo da Silva, Ingrid Evans-Osses, Giuseppe Palmisano, Sigrun Lange, and Marcel Ivan Ramirez. 2020. "Peptidylarginine Deiminase Inhibition Abolishes the Production of Large Extracellular Vesicles from Giardia Intestinalis, Affecting Host-Pathogen Interactions by Hindering Adhesion to Host Cells." *Frontiers in Cellular and Infection Microbiology* 10 (September). https://doi.org/10.3389/fcimb.2020.00417.

Ghasemian, Seyedeh Ommolbanin. 2021. "Application of Exosomes-Derived Mesenchymal Stem Cells in Treatment of Fungal Diseases: From Basic to Clinical Sciences." *Frontiers in Fungal Biology*. Frontiers Media SA https://doi.org/10.3389/ffunb.2021.736093.

Goes, Adriely, Philipp Lapuhs, Thomas Kuhn, Eilien Schulz, Robert Richter, Fabian Panter, Charlotte Dahlem, et al. 2020. "Myxobacteria-Derived Outer Membrane Vesicles: Potential Applicability against Intracellular Infections." *Cells* 9 (1). https://doi.org/10.3390/cells9010194.

Gujrati, Vipul, Sunghyun Kim, Sang Hyun Kim, Jung Joon Min, Hyon E. Choy, Sun Chang Kim, and Sangyong Jon. 2014. "Bioengineered Bacterial Outer Membrane Vesicles as Cell-Specific Drug-Delivery Vehicles for Cancer Therapy." *ACS Nano* 8 (2): 1525–37. https://doi.org/10.1021/nn405724x.

Gurung, Mamata, Dong Chan Moon, Chi Won Choi, Jung Hwa Lee, Yong Chul Bae, Jungmin Kim, Yoo Chul Lee, et al. 2011. "Staphylococcus Aureus Produces Membrane-Derived Vesicles That Induce Host Cell Death." *PLoS ONE* 6 (11). https://doi.org/10.1371/journal.pone.0027958.

György, Bence, Michelle E. Hung, Xandra O. Breakefield, and Joshua N. Leonard. 2015. "Therapeutic Applications of Extracellular Vesicles: Clinical Promise and Open Questions." *Annual Review of Pharmacology and Toxicology*. Annual Reviews Inc. https://doi.org/10.1146/annurev-pharmtox-010814-124630.

Halperin, Walter, and William A Jensen. 1967. "U Ltrastructural Changes during Growth and Embryogenesis in Carrot Cell Cultures 1." *J. Ultrastructure Research*. Vol. 18.

Hao, Y., Song, H., Zhou, Z., Chen, X., Li, H., Zhang, Y., Wang, J., Ren, X., & Wang, X. (2021). Promotion or inhibition of extracellular vesicle release: Emerging therapeutic opportunities. *Journal of controlled release: official journal of the Controlled Release Society*, *340*, 136–148. https://doi.org/10.1016/j.jconrel.2021.10.019.

Hessvik, Nina Pettersen, and Alicia Llorente. 2018. "Current Knowledge on Exosome Biogenesis and Release." *Cellular and Molecular Life Sciences*. Birkhauser Verlag AG. https://doi.org/10.1007/s00018-017-2595-9.

Hoy, Anna M., Rachel J. Lundie, Alasdair Ivens, Juan F. Quintana, Norman Nausch, Thorsten Forster, Frances Jones, et al. 2014. "Parasite-Derived MicroRNAs in Host Serum as Novel Biomarkers of Helminth Infection." *PLoS Neglected Tropical Diseases* 8 (2). https://doi.org/10.1371/journal.pntd.0002701.

Hu, Che Ming J., Ronnie H. Fang, Jonathan Copp, Brian T. Luk, and Liangfang Zhang. 2013. "A Biomimetic Nanosponge That Absorbs Pore-Forming Toxins." *Nature Nanotechnology* 8 (5): 336–40. https://doi.org/10.1038/nnano.2013.54.

Inal, Jameel M. 2020. "Decoy ACE2-Expressing Extracellular Vesicles That Competitively Bind SARS-CoV-2 as a Possible COVID-19 Therapy." *Clinical Science*. Portland Press Ltd. https://doi.org/10.1042/CS20200623.

Jahromi, L. P., & Fuhrmann, G. (2021). Bacterial extracellular vesicles: Understanding biology promotes applications as nanopharmaceuticals. *Advanced drug delivery reviews*, *173*, 125–140. https://doi.org/10.1016/j.addr.2021.03.012

Jnana, A, Sadiya S. S, Satyamoorthy K, and Murali TS 2023. "Extracellular Vesicles in Bacterial and Fungal Diseases–Pathogenesis to Diagnostic Biomarkers." *Virulence*. Taylor and Francis Ltd. https://doi.org/10.1080/21505594.2023.2180934.

Jong, Olivier G. De, Sander A. A. Kooijmans, Daniel E. Murphy, Linglei Jiang, Martijn J. W. Evers, Joost P. G. Sluijter, Pieter Vader, and Raymond M. Schiffelers. 2019. "Drug Delivery with Extracellular Vesicles: From Imagination to Innovation." *Accounts of Chemical Research* 52 (7): 1761–70. https://doi.org/10.1021/acs.accounts.9b00109.

Kadurugamuwa, Jagath L, and Terry J Beveridge. 1996. "Bacteriolytic Effect of Membrane Vesicles from Pseudomonas Aeruginosa on Other Bacteria Including Pathogens: Conceptually New Antibiotics." *JOURNAL OF BACTERIOLOGY*. Vol. 178.

Kalamvoki, Maria, Te Du, and Bernard Roizman. 2014. "Cells Infected with Herpes Simplex Virus 1 Export to Uninfected Cells Exosomes Containing STING, Viral MRNAs, and MicroRNAs." *Proceedings of the National Academy of Sciences of the United States of America* 111 (46): E4991–96. https://doi.org/10.1073/pnas.1419338111.

Keshtkar, Somayeh, Maryam Kaviani, Saeede Soleimanian, Negar Azarpira, Zahra Asvar, and Sara Pakbaz. 2022. "Stem Cell-Derived Exosome as Potential Therapeutics for Microbial Diseases." *Frontiers in Microbiology*. Frontiers Media SA https://doi.org/10.3389/fmicb.2021.786111.

Kesimer, Mehmet, Margaret Scull, Brian Brighton, Genevieve DeMaria, Kimberlie Burns, Wanda O'Neal, Raymond J. Pickles, and John K. Sheehan. 2009. "Characterization of Exosome-like Vesicles Released from Human Tracheobronchial Ciliated Epithelium: A Possible Role in Innate Defense." *The FASEB Journal* 23 (6): 1858–68. https://doi.org/10.1096/fj.08-119131.

Kojima, Ryosuke, Daniel Bojar, Giorgio Rizzi, Ghislaine Charpin El Hamri, Marie Daoud El-Baba, Pratik Saxena, Simon Ausländer, Kelly R. Tan, and Martin Fussenegger. 2018. "Designer Exosomes Produced by Implanted Cells Intracerebrally Deliver Therapeutic Cargo for Parkinson's Disease Treatment." *Nature Communications* 9 (1). https://doi.org/10.1038/s41467-018-03733-8.

Kosgodage, Uchini S., Paul Matewele, Giulia Mastroianni, Igor Kraev, Dominik Brotherton, Brigitte Awamaria, Anthony P. Nicholas, Sigrun Lange, and Jameel M. Inal. 2019. "Peptidylarginine Deiminase Inhibitors Reduce Bacterial Membrane Vesicle Release and Sensitize Bacteria to Antibiotic Treatment." *Frontiers in Cellular and Infection Microbiology* 9 (JUN). https://doi.org/10.3389/fcimb.2019.00227.

Kouwaki, Takahisa, Yoshimi Fukushima, Takuji Daito, Takahiro Sanada, Naoki Yamamoto, Edin J. Mifsud, Chean Ring Leong, et al. 2016. "Extracellular Vesicles Including Exosomes Regulate Innate Immune Responses to Hepatitis B Virus Infection." *Frontiers in Immunology* 7 (AUG). https://doi.org/10.3389/fimmu.2016.00335.

Kulp, Adam, and Meta J. Kuehn. 2010. "Biological Functions and Biogenesis of Secreted Bacterial Outer Membrane Vesicles." *Annual Review of Microbiology*. https://doi.org/10.1146/annurev.micro.091208.073413.

Kumagai, T., Shimogawara, R., Ichimura, K., & Iwanaga, S. (2022). Calpain inhibitor suppresses both extracellular vesicle-mediated secretion of miRNAs and egg production from paired adults of Schistosoma japonicum. *Parasitology international*, *87*, 102540. https://doi.org/10.1016/j.parint.2022.102540.

Lei, Chao, Jingyao Mu, Yun Teng, Liqing He, Fangyi Xu, Xiangcheng Zhang, Kumaran Sundaram, et al. n.d. "Lemon Exosome-like Nanoparticles-Manipulated Probiotics Protect Mice from C. Diff Infection." https://doi.org/10.1016/j.isci.

Li, Song Pei, Zhong Xiao Lin, Xue Yan Jiang, and Xi Yong Yu. 2018. "Exosomal Cargo-Loading and Synthetic Exosome-Mimics as Potential Therapeutic Tools." *Acta Pharmacologica Sinica*. Nature Publishing Group. https://doi.org/10.1038/aps.2017.178.

Li, Zusheng, Anthony J Clarke, and Terry J Beveridge. 1998. "Gram-Negative Bacteria Produce Membrane Vesicles Which Are Capable of Killing Other Bacteria." *JOURNAL OF BACTERIOLOGY*. Vol. 180. https://journals.asm.org/journal/jb.

Loghry, Hannah J., Wang Yuan, Mostafa Zamanian, Nicolas J. Wheeler, Timothy A. Day, and Michael J. Kimber. 2020. "Ivermectin Inhibits Extracellular Vesicle Secretion from Parasitic Nematodes." *Journal of Extracellular Vesicles* 10 (2). https://doi.org/10.1002/jev2.12036.

Luan, Xin, Kanokwan Sansanaphongpricha, Ila Myers, Hongwei Chen, Hebao Yuan, and Duxin Sun. 2017. "Engineering Exosomes as Refined Biological Nanoplatforms for Drug Delivery." *Acta Pharmacologica Sinica*. Nature Publishing Group. https://doi.org/10.1038/aps.2017.12.

Lv, Xi, Ran Chen, Taizhen Liang, Haojie Peng, Qiannan Fang, Shiqi Xiao, Sen Liu, et al. 2024. "NSP6 Inhibits the Production of ACE2-Containing Exosomes to Promote SARS-CoV-2 Infectivity." *MBio* 15 (3). https://doi.org/10.1128/mbio.03358-23.

Mashburn-Warren, Lauren, Jörg Howe, Klaus Brandenburg, and Marvin Whiteley. 2009. "Structural Requirements of the Pseudomonas Quinolone Signal for Membrane Vesicle Stimulation." *Journal of Bacteriology* 191 (10): 3411–14. https://doi.org/10.1128/JB.00052-09.

Mashburn-Warren, Lauren, Jörg Howe, Patrick Garidel, Walter Richter, Frank Steiniger, Manfred Roessle, Klaus Brandenburg, and Marvin Whiteley. 2008. "Interaction of Quorum Signals with Outer Membrane Lipids: Insights into Prokaryotic Membrane Vesicle Formation." *Molecular Microbiology* 69 (2): 491–502. https://doi.org/10.1111/j.1365-2958.2008.06302.x.

McNamara, Ryan P., and Dirk P. Dittmer. 2020. "Extracellular Vesicles in Virus Infection and Pathogenesis." *Current Opinion in Virology*. Elsevier B.V. https://doi.org/10.1016/j.coviro.2020.07.014.

Meng, Bo, Natasha C.Y. Ip, Liam J. Prestwood, Truus E. M. Abbink, and Andrew M. L. Lever. 2015. "Evidence That the Endosomal Sorting Complex Required for Transport-II (ESCRT-II) Is Required for Efficient Human Immunodeficiency Virus-1 (HIV-1) Production." *Retrovirology* 12 (1). https://doi.org/10.1186/s12977-015-0197-x.

Mori, Yasuko, Masato Koike, Eiko Moriishi, Akiko Kawabata, Huamin Tang, Hiroko Oyaizu, Yasuo Uchiyama, and Koichi Yamanishi. 2008. "Human Herpesvirus-6 Induces MVB Formation, and Virus Egress Occurs by an Exosomal Release Pathway." *Traffic* 9 (10): 1728–42. https://doi.org/10.1111/j.1600-0854.2008.00796.x.

Ñahui Palomino, Rogers A., Christophe Vanpouille, Luca Laghi, Carola Parolin, Kamran Melikov, Peter Backlund, Beatrice Vitali, and Leonid Margolis. 2019. "Extracellular Vesicles from Symbiotic Vaginal Lactobacilli Inhibit HIV-1 Infection of Human Tissues." *Nature Communications* 10 (1). https://doi.org/10.1038/s41467-019-13468-9.

Nanbo, Asuka, Eri Kawanishi, Ryuji Yoshida, and Hironori Yoshiyama. 2013. "Exosomes Derived from Epstein-Barr Virus-Infected Cells Are Internalized via Caveola-Dependent Endocytosis and Promote Phenotypic Modulation in Target Cells." *Journal of Virology* 87 (18): 10334–47. https://doi.org/10.1128/jvi.01310-13.

Nasiri Kenari, A., Cheng, L., & Hill, A. F. (2020). Methods for loading therapeutics into extracellular vesicles and generating extracellular vesicles mimetic-nanovesicles. *Methods (San Diego, Calif.)*, *177*, 103–113. https://doi.org/10.1016/j.ymeth.2020.01.001

Nasiri, Zeynab, Hoorieh Soleimanjahi, Nafiseh Baheiraei, Seyed Mahmoud Hashemi, and Mahmoud Reza Pourkarim. 2024. "The Impact Understanding of Exosome Therapy in COVID-19 and Preparations for the Future Approaches in Dealing with Infectious Diseases and Inflammation." *Scientific Reports* 14 (1). https://doi.org/10.1038/s41598-024-56334-5.

Nehls, Julia, Ramona Businger, Markus Hoffmann, Constantin Brinkmann, Birgit Fehrenbacher, Martin Schaller, Brigitte Maurer, et al. 2019. "Release of Immunomodulatory Ebola Virus Glycoprotein-Containing Microvesicles Is Suppressed by Tetherin in a Species-Specific Manner." *Cell Reports* 26 (7): 1841-1853.e6. https://doi.org/10.1016/j.celrep.2019.01.065.

Nunez, E. A., Wallis, J., & Gershon, M. D. 1974. "Secretory Processes in Follicular Cells of the Bat Thyroid. 3. The Occurrence of Extracellular Vesicles and Colloid Droplets during Arousal from Hibernation." *The American Journal of Anatomy* 141 (2): 179–201.

Osterloh, Anke. 2022. "Vaccination against Bacterial Infections: Challenges, Progress, and New Approaches with a Focus on Intracellular Bacteria." *Vaccines*. MDPI. https://doi.org/10.3390/vaccines10050751.

Ostrowski, Matias, Nuno B. Carmo, Sophie Krumeich, Isabelle Fanget, Graça Raposo, Ariel Savina, Catarina F. Moita, et al. 2010. "Rab27a and Rab27b Control Different Steps of the Exosome Secretion Pathway." *Nature Cell Biology* 12 (1): 19–30. https://doi.org/10.1038/ncb2000.

Parada, Nicol, Alfonso Romero-Trujillo, Nicolás Georges, and Francisca Alcayaga-Miranda. 2021. "Camouflage Strategies for Therapeutic Exosomes Evasion from Phagocytosis." *Journal of Advanced Research*. Elsevier B.V. https://doi.org/10.1016/j.jare.2021.01.001.

Perocheau, Dany, Loukia Touramanidou, Sonam Gurung, Paul Gissen, and Julien Baruteau. 2021. "Clinical Applications for Exosomes: Are We There Yet?" *British Journal of Pharmacology*. John Wiley and Sons Inc. https://doi.org/10.1111/bph.15432.

Piffoux, Max, Amanda K.A. Silva, Claire Wilhelm, Florence Gazeau, and David Tareste. 2018. "Modification of Extracellular Vesicles by Fusion with Liposomes for the Design of Personalized Biogenic Drug Delivery Systems." *ACS Nano* 12 (7): 6830–42. https://doi.org/10.1021/acsnano.8b02053.

Pleet, Michelle L., Allison Mathiesen, Catherine DeMarino, Yao A. Akpamagbo, Robert A. Barclay, Angela Schwab, Sergey Iordanskiy, et al. 2016. "Ebola VP40 in Exosomes Can Cause Immune Cell Dysfunction." *Frontiers in Microbiology* 7 (NOV). https://doi.org/10.3389/fmicb.2016.01765.

Popowski, Kristen D., Phuong Uyen C. Dinh, Arianna George, Halle Lutz, and Ke Cheng. 2021. "Exosome Therapeutics for COVID-19 and Respiratory Viruses." *VIEW*. John Wiley and Sons Inc. https://doi.org/10.1002/VIW.20200186.

Qian, Xijing, Chen Xu, Shuo Fang, Ping Zhao, Yue Wang, Houqi Liu, Wen Yuan, and Zhongtian Qi. 2016. "Exosomal MicroRNAs Derived from Umbilical Mesenchymal Stem Cells Inhibit Hepatitis C Virus Infection." *Stem Cells Translational Medicine* 5 (9): 1190–1203. https://doi.org/10.5966/sctm.2015-0348.

Qu, Shaoqi, Yiming Han, Ying Liu, Jiajia Zhu, Ulas Acaroz, Jianzhong Shen, and Kui Zhu. 2022. "Milk Exosomes Facilitate Oral Delivery of Drugs against Intestinal Bacterial Infections." *Journal of Agricultural and Food Chemistry* 70 (51): 16069–79. https://doi.org/10.1021/acs.jafc.2c04971.

Raghav, Alok, Richa Giri, Saurabh Agarwal, Sanjay Kala, and Goo Bo Jeong. 2022. "Protective Role of Engineered Extracellular Vesicles Loaded Quercetin Nanoparticles as Anti-Viral Therapy against SARS-CoV-2 Infection: A Prospective Review." *Frontiers in Immunology*. Frontiers Media SA https://doi.org/10.3389/fimmu.2022.1040027.

Ramakrishnaiah, Vedashree, Christine Thumann, Isabel Fofana, Francois Habersetzer, Qiuwei Pan, Petra E. De Ruiter, Rob Willemsen, et al. 2013. "Exosome-Mediated Transmission of Hepatitis C Virus between Human Hepatoma Huh7.5 Cells." *Proceedings of the National Academy of Sciences of the United States of America* 110 (32): 13109–13. https://doi.org/10.1073/pnas.1221899110.

Regente, Mariana, Georgina Corti-Monzón, Ana María Maldonado, Marcela Pinedo, Jesús Jorrín, and Laura de la Canal. 2009. "Vesicular Fractions of Sunflower Apoplastic Fluids Are Associated with Potential Exosome Marker Proteins." *FEBS Letters* 583 (20): 3363–66. https://doi.org/10.1016/j.febslet.2009.09.041.

Regev-Rudzki, Neta, Shulamit Michaeli, and Ana Claudia Torrecilhas. 2021. "Editorial: Extracellular Vesicles in Infectious Diseases." *Frontiers in Cellular and Infection Microbiology*. Frontiers Media SA https://doi.org/10.3389/fcimb.2021.697919.

Salas, Nehuén, Veronica M. Coceres, Tuanne dos Santos Melo, Antonio Pereira-Neves, Vanina G. Maguire, Tania M. Rodriguez, Bruna Sabatke, et al. 2022. "VPS32, a Member of the ESCRT Complex, Modulates Adherence to Host Cells in the Parasite Trichomonas Vaginalis by Affecting Biogenesis and Cargo Sorting of Released Extracellular Vesicles." *Cellular and Molecular Life Sciences* 79 (1). https://doi.org/10.1007/s00018-021-04083-3.

Schulz, Eilien, Adriely Goes, Ronald Garcia, Fabian Panter, Marcus Koch, Rolf Müller, Kathrin Fuhrmann, and Gregor Fuhrmann. 2018. "Biocompatible Bacteria-Derived Vesicles Show Inherent Antimicrobial Activity." *Journal of Controlled Release* 290 (November):46–55. https://doi.org/10.1016/j.jconrel.2018.09.030.

Sengupta, Vikram, Sascha Sengupta, Angel Lazo, Peter Woods, Anna Nolan, and Nicholas Bremer. 2020. "Exosomes Derived from Bone Marrow Mesenchymal Stem Cells as Treatment for Severe COVID-19." *Stem Cells and Development* 29 (12): 747–54. https://doi.org/10.1089/scd.2020.0080.

Stotz, Henrik U., Dominik Brotherton, and Jameel Inal. 2022. "Communication Is Key: Extracellular Vesicles as Mediators of Infection and Defence during Host-Microbe Interactions in Animals and Plants." *FEMS Microbiology Reviews*. Oxford University Press. https://doi.org/10.1093/femsre/fuab044.

Sundaram, Kumaran, Daniel P. Miller, Anil Kumar, Yun Teng, Mohammed Sayed, Jingyao Mu, Chao Lei, et al. 2019. "Plant-Derived Exosomal Nanoparticles Inhibit Pathogenicity of Porphyromonas Gingivalis." *IScience* 21 (November):308–27. https://doi.org/10.1016/j.isci.2019.10.032.

Suzuki, Yasuhiro, Xisheng Wang, Benard S. Jortner, Laura Payne, Yanyan Ni, Sara A. Michie, Baohui Xu, Tomoya Kudo, and Sara Perkins. 2010. "Removal of Toxoplasma Gondii Cysts from the Brain by

Perforin-Mediated Activity of CD8+ T Cells." *American Journal of Pathology* 176 (4): 1607–13. https://doi.org/10.2353/ajpath.2010.090825.

Teng, Yun, Fangyi Xu, Xiangcheng Zhang, Jingyao Mu, Mohammed Sayed, Xin Hu, Chao Lei, et al. 2021. "Plant-Derived Exosomal MicroRNAs Inhibit Lung Inflammation Induced by Exosomes SARS-CoV-2 Nsp12." *Molecular Therapy* 29 (8): 2424–40. https://doi.org/10.1016/j.ymthe.2021.05.005.

Tian, Tian, Yan Liang Zhu, Yue Yuan Zhou, Gao Feng Liang, Yuan Yuan Wang, Fei Hu Hu, and Zhong Dang Xiao. 2014. "Exosome Uptake through Clathrin-Mediated Endocytosis and Macropinocytosis and Mediating MiR-21 Delivery." *Journal of Biological Chemistry* 289 (32): 22258–67. https://doi.org/10.1074/jbc.M114.588046.

Timár, Csaba I., Ákos M. Lorincz, Roland Csépányi-Kömi, Anna Vályi-Nagy, György Nagy, Edit I. Buzás, Zsolt Iványi, et al. 2013. "Antibacterial Effect of Microvesicles Released from Human Neutrophilic Granulocytes." *Blood* 121 (3): 510–18. https://doi.org/10.1182/blood-2012-05-431114.

Tkach, Mercedes, and Clotilde Théry. 2016. "Communication by Extracellular Vesicles: Where We Are and Where We Need to Go." *Cell*. Cell Press. https://doi.org/10.1016/j.cell.2016.01.043.

Tlaposo, Graqa, Hans W Nijman, ~willem Stoorvogel, Richtje Leijendekker, Cliffordv Hardingfl Cornelis, J M Melief, and Hans J Geuze. n.d. "B Lymphocytes Secrete Antigen-PresentingVesicles." http://rupress.org/jem/article-pdf/183/3/1161/1678390/1161.pdf.

Tu, Zhengkun, Robert H. Pierce, Jonathan Kurtis, Yoshio Kuroki, I. Nicholas Crispe, and Mark S. Orloff. 2010. "Hepatitis C Virus Core Protein Subverts the Anti-viral Activities of Human Kupffer Cells." *Gastroenterology* 138 (1): 305–14. https://doi.org/10.1053/j.gastro.2009.09.009.

Uddin, Md Jamir, Popat Mohite, Shubham Munde, Nitin Ade, Tosin Ayodeji Oladosu, Vijay R. Chidrawar, Ravish Patel, Sankha Bhattacharya, Himanshu Paliwal, and Sudarshan Singh. 2024. "Extracellular Vesicles: The Future of Therapeutics and Drug Delivery Systems." *Intelligent Pharmacy*, June. https://doi.org/10.1016/j.ipha.2024.02.004.

Wan, Yuanda, Liren Li, Ruilin Chen, Jiajia Han, Qiyun Lei, Zhipeng Chen, Xiaodong Tang, Wenyu Wu, Shuwen Liu, and Xingang Yao. 2024. "Engineered Extracellular Vesicles Efficiently Deliver CRISPR-Cas9 Ribonucleoprotein (RNP) to Inhibit Herpes Simplex Virus1 Infection *in Vitro* and *in Vivo*." *Acta Pharmaceutica Sinica B* 14 (3): 1362–79. https://doi.org/10.1016/j.apsb.2023.10.004.

Wang, Yizhuo, Meng Zhao, Shuyun Liu, Jun Guo, Yanrong Lu, Jingqiu Cheng, and Jingping Liu. 2020. "Macrophage-Derived Extracellular Vesicles: Diverse Mediators of Pathology and Therapeutics in Multiple Diseases." *Cell Death and Disease*. Springer Nature. https://doi.org/10.1038/s41419-020-03127-z.

Wang, Yizhuo, Shuyun Liu, Lan Li, Ling Li, Xueli Zhou, Meihua Wan, Peng Lou, et al. 2022a. "Peritoneal M2 Macrophage-Derived Extracellular Vesicles as Natural Multitarget Nanotherapeutics to Attenuate Cytokine Storms after Severe Infections." *Journal of Controlled Release* 349 (September):118–32. https://doi.org/10.1016/j.jconrel.2022.06.063.

Witwer, Kenneth W., and Clotilde Théry. 2019. "Extracellular Vesicles or Exosomes? On Primacy, Precision, and Popularity Influencing a Choice of Nomenclature." *Journal of Extracellular Vesicles*. Taylor and Francis Ltd. https://doi.org/10.1080/20013078.2019.1648167.

Witwer, Kenneth W., and Joy Wolfram. 2021. "Extracellular Vesicles versus Synthetic Nanoparticles for Drug Delivery." *Nature Reviews Materials*. Nature Research. https://doi.org/10.1038/s41578-020-00277-6.

Woith, Eric, and Matthias F. Melzig. 2019. "Extracellular Vesicles from Fresh and Dried Plants—Simultaneous Purification and Visualization Using Gel Electrophoresis." *International Journal of Molecular Sciences* 20 (2). https://doi.org/10.3390/ijms20020357.

Wolf P. (1967). The nature and significance of platelet products in human plasma. *British journal of haematology*, *13*(3), 269–288. https://doi.org/10.1111/j.1365-2141.1967.tb08741.x.

Yang, Xiaohong, Gongming Shi, Jian Guo, Chenhui Wang, and Yun He. 2018. "Exosome-Encapsulated Antibiotic against Intracellular Infections of Methicillin-Resistant Staphylococcus Aureus." *International Journal of Nanomedicine* 13:8095–8104. https://doi.org/10.2147/IJN.S179380.

Yenuganti, Vengala Rao, Sumbul Afroz, Rafiq Ahmad Khan, Chandrima Bharadwaj, Deepti Kailash Nabariya, Nagaraj Nayak, Madhuri Subbiah, et al. 2022. "Milk Exosomes Elicit a Potent Anti-Viral Activity against Dengue Virus." *Journal of Nanobiotechnology* 20 (1). https://doi.org/10.1186/s12951-022-01496-5.

Zahoor, M. A., Yao, X. D., Henrick, B. M., Verschoor, C. P., Abimiku, A., Osawe, S., & Rosenthal, K. L. (2020). Expression profiling of human milk derived exosomal microRNAs and their targets in HIV-1 infected mothers. *Scientific reports*, *10*(1), 12931. https://doi.org/10.1038/s41598-020-69799-x.

Zahoor, Muhammad Atif, Xiao Dan Yao, Bethany M. Henrick, Chris P. Verschoor, Alash'le Abimiku, Sophia Osawe, and Kenneth L. Rosenthal. 2020a. "Expression Profiling of Human Milk Derived Exosomal MicroRNAs and Their Targets in HIV-1 Infected Mothers." *Scientific Reports* 10 (1). https://doi.org/10.1038/s41598-020-69799-x.

Zeng, Wanjia, Liwei Zheng, Yukun Li, Jing Yang, Tianhao Mao, Jing Zhang, Yanna Liu, et al. 2024. "Engineered Extracellular Vesicles for Delivering Functional Cas9/GRNA to Eliminate Hepatitis B Virus CccDNA and Integration." *Emerging Microbes and Infections* 13 (1). https://doi.org/10.1080/22221751.2023.2284286.

Zhang, Junsong, Feng Huang, Baijin Xia, Yaochang Yuan, Fei Yu, Guanwen Wang, Qianyu Chen, et al. 2021. "The Interferon-Stimulated Exosomal HACE2 Potently Inhibits SARS-CoV-2 Replication through Competitively Blocking the Virus Entry." *Signal Transduction and Targeted Therapy* 6 (1). https://doi.org/10.1038/s41392-021-00604-5.

Zhang, Zheng, Junliang Fu, Xiangsheng Xu, Siyu Wang, Ruonan Xu, Min Zhao, Weimin Nie, et al. 2013. "Safety and Immunological Responses to Human Mesenchymal Stem Cell Therapy in Difficult-to-Treat HIV-1-Infected Patients." *AIDS* 27 (8): 1283–93. https://doi.org/10.1097/QAD.0b013e32835fab77.

Zhao, Miao, Fan Zhang, Robert Zarnowski, Kenneth Barns, Ryley Jones, Jen Fossen, Hiram Sanchez, et al. 2021. "Turbinmicin Inhibits Candida Biofilm Growth by Disrupting Fungal Vesicle-Mediated Trafficking." *Journal of Clinical Investigation* 131 (5). https://doi.org/10.1172/JCI145123.

Zhou, Yu, Xu Wang, Li Sun, Li Zhou, Tong Cui Ma, Li Song, Jian Guo Wu, Jie Liang Li, and Wen Zhe Ho. 2016. "Toll-like Receptor 3-Activated Macrophages Confer Anti-HCV Activity to Hepatocytes through Exosomes." *FASEB Journal* 30 (12): 4132–40. https://doi.org/10.1096/fj.201600696R.

Zhu, Xiaolei, Kristen R. Hollinger, Yiyao Huang, Alejandra Borjabad, Boe Hyun Kim, Tanina Arab, Ajit G. Thomas, et al. 2022. "Neutral Sphingomyelinase 2 Inhibition Attenuates Extracellular Vesicle Release and Improves Neurobehavioral Deficits in Murine HIV." *Neurobiology of Disease* 169 (July). https://doi.org/10.1016/j.nbd.2022.105734.

Zhu, Ying Gang, Xiao Mei Feng, Jason Abbott, Xiao Hui Fang, Qi Hao, Antoine Monsel, Jie Ming Qu, and Michael A. Matthay. 2014. "Human Mesenchymal Stem Cell Microvesicles for Treatment of Escherichia Coli Endotoxin-Induced Acute Lung Injury in Mice." *Stem Cells* 32 (1): 116–25. https://doi.org/10.1002/stem.1504.

Zitvogel, L., Regnault, A., Lozier, A., Wolfers, J., Flament, C., Tenza, D., Ricciardi-Castagnoli, P., Raposo, G., & Amigorena, S. (1998). Eradication of established murine tumors using a novel cell-free vaccine: Dendritic cell-derived exosomes. *Nature Medicine*, 4(5), 594–600.

Section 3

Chapter 3.1

Recent Advances of Novel Smart Nanomaterials and Their Applications in Bio-Imaging

K. Vijayalakshmi[1,*,†]
Sesuraj Balasamy[2,*]
and Ashok K. Sundramoorthy[2,‡]
[1]Department of Chemistry (SFS), Madras Christian College, East Tambaram, Chennai, India
[2]Centre for Nanobiosensors, Saveetha Dental College and Hospitals,
Saveetha Institute of Medical and Technical Sciences, Chennai, India

Abstract

Due to their diverse organic compositions and varying metabolisms, the human body and live cells are considered the most complex systems. Therefore, recent research has demonstrated that advancements in newly developed nanomaterial-based bioimaging techniques and contrast agents significantly contribute to the early detection and diagnosis of diseases and understanding of biological processes at the molecular level. The unique adjustable biophysical properties and high surface area-to-volume ratio of nanoparticles have enhanced biomedical imaging, and recently, tissue-specific diagnosis has been achieved using molecularly tailored nanoparticles. The creation of theragnostic tools, which integrate therapeutic and diagnostic functions, has favored using nanomaterials for *in vivo* imaging. This approach aids in improving a patient's quality of life through the early diagnosis of serious diseases such as cardiovascular conditions, cancer, and renal failure. To address safety, therangnostic capabilities, efficacy, and multifunctional nanoplatforms were also investigated. This chapter discusses the latest developments in bioimaging and their use in diagnosing various diseases and drug delivery using zero-dimensional (0 D) carbon nanomaterials, quantum dots, metallic nanoparticles, and composite-based nanomaterials.

Keywords: biocompatible, nanomaterials, theragnostic, bioimaging, contrast agents

---

* These authors equally contributed.
† Corresponding Author's Email: kvijayalakshmi@mcc.edu.in
‡ Corresponding Author's Email: ashok.sundramoorthy@gmail.com

In: Emerging Nanomaterials in Biomedical Science
Editors: Eepsita Priyadarshini and Ajith Manayil Parambil
ISBN: 979-8-89530-600-0

1. Introduction

Nanomaterials are pivotal in nanotechnology, significantly influencing nanomedicine through their roles in imaging, diagnosis, and treatment (Chang et al. 2015). Defined as substances with dimensions under 100 nanometers, they exhibit unique properties like enhanced diagnostic specificity, extended blood circulation, and targeted organ delivery (Yetisgin et al. 2020; Paramasivam et al. 2024; Sridharan, Godwin, et al. 2024; Nagarajan, Kavitha, et al. 2023; Shivalingam, Mohan, and Ganapathy 2022). Functionalization with various agents further enhances these properties, boosting biological compatibility, targeting ability, and cellular uptake, particularly in cancer cells (Sanità, Carrese, and Lamberti 2020; Balu and Sundramoorthy 2024; Ashok K. Sundramoorthy et al. 2023). The emergence of carbon-based materials, such as graphene and carbon nanotubes (CNTs) (Dhanapalan et al. 2024; Magesh, Sundramoorthy, and Ganapathy 2022), has revolutionized material science due to their versatility and eco-friendliness (Nagarajan, Magesh, et al. 2023). Historically, carbon materials have progressed from traditional forms like activated carbon and carbon black to advanced types such as carbon fibers and graphite (Egbedina et al. 2022; Dutt et al. 2024; Sridharan et al. 2023). The discovery of carbon nanomaterials like graphene and CNTs has unlocked new possibilities in various fields (Qureshi et al. 2024; Dhanapalan et al. 2024; Ashok Kumar Sundramoorthy et al. 2013). Traditional macroscopic carbon forms, despite their strengths, often lack necessary properties like a band gap, limiting their use in luminescent applications. This has resulted in the creation of Carbon Dots (CDs), which have garnered significant attention due to their exceptional photoluminescence (PL), high quantum yield (QY), and a range of favorable properties suitable for various applications. (Mitchell et al. 2021; Murugan, Sridharan, et al. 2024). Originally identified during the purification of single-walled carbon nanotubes, CDs have progressed into essential nanomaterials. Their quantum confinement and edge effects enable tunable photoluminescence across a broad spectrum by modifying size, surface state, and doping (Rasheed et al. 2024). The high surface area-to-volume ratio and abundant functional groups on CDs enhance their capacity to adsorb or conjugate with biomolecules or analytes, improving their sensing capabilities. CD's high biocompatibility, abundance, and cost-effectiveness make them appealing for research and applications, bridging gaps left by earlier carbon materials. Research on CD synthesis and functionalization aims to optimize their properties and broaden their application range, reflecting the dynamic nature of material science driven by creativity and interdisciplinary collaboration (Titirici et al. 2022). CDs epitomize the shift from macroscopic carbon forms to nanoscale innovations, embodying scientific exploration driven by curiosity and real-world problem-solving (Sangeetha Dhanapalan, 2024). Their rise highlights the evolving material science landscape, where nanotechnology and environmental awareness converge to develop novel solutions (Yang et al. 2024).

Nanomaterials have transformed bioimaging, a non-invasive method for visualizing internal body structures. Advancements like multimodal imaging overcome the limitations of single-modality imaging by combining signals from different techniques. Examples include systems such as positron emission tomography (PET) scans, computerized tomography (CT) scans, and magnetic resonance imaging (MRI) scans, which provide thorough and efficient disease diagnosis (Lapusan, Borlan, and Focsan 2024; Ehman et al. 2017). Contrast agents, or imaging probes, improved diagnostic value by enhancing the visibility of specific tissues or organs (Balasamy and Sundramoorthy 2024). These agents must be stable in physiological conditions, maintain colloidal stability, have suitable retention time in the bloodstream, be excretable to minimize toxicity, and be biocompatible to reduce adverse reactions (Wallyn et al. 2019; Ashok K. Sundramoorthy 2024). Graphene quantum dots (GQDs), tiny graphene fragments, possess remarkable optical, electrical, and electrochemical properties, making them highly suitable for applications in biosensing and bioimaging. (Hyder et al. 2024; Chatterjee et al. 2024; Murugan, Magesh, et al. 2024).

Recent progress in synthetic chemistry, materials science, and nanotechnology has driven the development of smart nanomaterials with diverse properties, including fluorescence, magnetism, surface plasmon resonance, and photoacoustic properties, all useful for advanced diagnostic and imaging applications (Sun et al. 2023). Fluorescent nanomaterials, like quantum dots and CDs, enable sensitive detection of biomolecules, tracking cell migration, and real-time visualization of cellular structures (Wegner and Hildebrandt 2015; Atchudan et al. 2024). Magnetic nanoparticles can be guided by external magnetic fields for precise drug delivery, enhancement of MRI contrast, and magnetic hyperthermia in cancer treatment. Surface plasmon resonance nanoparticles, such as gold nanoparticles, show strong light scattering properties, allowing for sensitive analyte detection through optical property changes (Adam and Mertz 2023; Sridharan, Murugan, et al. 2024). Photoacoustic nanoparticles convert light into sound waves for high-resolution tissue imaging, offering advantages in penetration depth and sensitivity over conventional optical techniques (Balas 2009).

Integrating these functionalities into single nanomaterial platforms holds great potential for developing multimodal imaging techniques, allowing simultaneous detection of multiple biomarkers and offering a more comprehensive understanding of disease progression (Karuppasamy et al. 2024). This synergistic approach enhances diagnostic accuracy and enables more effective treatment strategies by allowing real-time monitoring of therapeutic responses and optimizing drug delivery (Ma et al. 2024). Despite challenges in scaling and replicating synthesis methods, the field of smart nanomaterials continues to progress, establishing these materials as pivotal elements in future nanotechnology-driven solutions (Walter et al. 2020).

2. Conventional Contrast Agents for Imaging and Their Limitations

Conventional contrast agents enhance visualization in imaging modalities like MRI, CT, and ultrasound (González-Mancebo et al. 2024). However, they often face efficacy and application limitations. In MRI, gadolinium-based chelates improve image contrast by altering water proton relaxation times. While effective, they can accumulate in the body, leading to potential toxicity, especially in patients with renal impairment. Gadolinium-based agents are linked to nephrogenic systemic fibrosis (NSF) in such patients (Coimbra et al. 2024). Their non-specific signal enhancement complicates differentiation between healthy and diseased tissues, and their low relaxivity requires higher doses, increasing toxicity risks (Gendron et al. 2024). CT imaging uses iodine-based agents to increase X-ray attenuation in specific tissues, effectively highlighting blood vessels and tumors. However, these agents can cause allergic reactions and nephrotoxicity (Bonlawar et al. 2024). Their limited spatial resolution often necessitates invasive biopsies for definitive diagnosis. Ultrasound employs microbubbles, gas-filled spheres that enhance contrast by scattering ultrasound waves. Despite effectively visualizing blood flow and microcirculation, microbubbles have limited tissue penetration and rapid clearance, restricting their use in deep-seated lesions (Mondal et al. 2024; Gattegno et al. 2024). Their lack of specificity and susceptibility to acoustic cavitation pose challenges in targeted imaging and therapy.

These limitations highlight the need for innovative imaging agents with improved sensitivity, specificity, and biocompatibility. Targeted and biocompatible agents are crucial for advancing diagnostics and therapeutics (Z. Chen et al. 2024; Paramasivam et al. 2024). Developing agents that target specific biomarkers or receptors enables early and accurate disease detection and monitoring (Jin et al. 2024). Enhanced sensitivity detects subtle cellular and molecular changes, facilitating early diagnosis and precise treatment monitoring. Improved biocompatibility minimizes adverse effects, ensuring safe use across diverse patient populations, including those with organ dysfunction (Passaro et al. 2024). Prolonged circulation times enable extended imaging studies and targeted drug delivery. Multimodal imaging agents offer comprehensive diagnostic information by integrating multiple imaging modalities (Riedl et al. 2024). Nanotechnology has brought forth promising nanomaterials that overcome the limitations of traditional agents. Superparamagnetic iron oxide nanoparticles (SPIONs) are effective MRI contrast agents due to their high magnetic susceptibility, which improves T2-weighted imaging and enhances lesion detection. Functionalizing SPIONs with antibodies or peptides allows targeted imaging, improving diagnostic accuracy and facilitating drug delivery. However, their limited tissue penetration and confinement to the reticuloendothelial system restrict their broader application. Due to their distinctive optical properties, gold nanoparticles have improved optical imaging techniques such as optical coherence tomography (OCT) and photoacoustic imaging. Their ability to absorb and emit light at specific wavelengths supports the development of fluorescent probes for bioimaging and diagnostics (Z. Chen et al. 2024).

Chemical Exchange Saturation Transfer (CEST) agents, which depend on the saturation transfer between exchangeable and water protons, provide high sensitivity and specificity. However, their lower sensitivity compared to paramagnetic agents requires higher concentrations for effective imaging, and specialized pulse sequences make CEST imaging challenging (Di Gregorio et al. 2024; Shaghaghi et al. 2024). Nanomaterial advances promise next-generation contrast agents, offering more accurate, sensitive, and targeted diagnostics and therapeutics. Continued research into these novel nanomaterials' synthesis, characterization, and *in vivo* evaluation is critical for clinical translation.

3. Contrast Imaging Analysis

In clinical studies, diagnosing various diseases involves examining cells from tissues to whole organisms using various bioimaging techniques, employing tools such as X-ray, ultrasound, magnetic resonance, fluorescence, electron microscopy, and positron imaging.

3.1. Ultrasound Imaging

Ultrasound (US) imaging is widely utilized in medical diagnostics due to its affordability, portability, high spatial resolution, and ability to provide real-time imaging. Ultrasound contrast agents are designed to enhance the differences in acoustic signals between target lesions and surrounding tissues, improving signal-to-noise ratios by reducing background scattering from excitation light. Recent advancements also include ultrasound-modulated luminescence imaging, which uses luminescent materials as key elements for imaging (Huynh et al. 2013).

3.2. Fluorescence Bioimaging

Fluorescence imaging is crucial in scientific research and medical diagnostics, offering non-invasive visualization at both cellular and molecular levels. Near-infrared fluorescence (NIRF) imaging offers superior spatial resolution and enhanced tissue penetration, although it faces challenges such as tissue scattering and autofluorescence. These challenges contribute to lower sensitivity in detecting fluorescence within target lesions and potential issues like photobleaching (Foucault-Collet et al. 2013). Nanoparticle fluorescence imaging is essential for the real-time tracking of therapeutic responses and is widely used in protein analysis, gene detection, early disease diagnosis, and cancer research (Cai et al. 2011).

3.3. MRI Imaging

Magnetic Resonance Imaging (MRI) utilizes the behavior of protons in magnetic fields to produce detailed tissue images of high resolution. MRI applications extend to utilizing nanomaterials as contrast agents, improving image clarity and interpretation by modifying relaxation parameters (T1 and T2) (Ronkainen and Okon 2014). This imaging technique is especially proficient in differentiating between healthy and damaged tissues, as demonstrated in stroke and prostate cancer staging, with improved toxicity profiles and imaging capabilities provided by nanoimaging agents.

3.4. CT Imaging

X-ray Computed Tomography (CT) offers detailed images of soft tissues and bones by exploiting differences in X-ray absorption and tissue density. Despite its higher radiation exposure, CT surpasses traditional radiography in clinical imaging due to its speed, efficiency, and superior spatial resolution. Contrast agents are utilized in CT to distinguish between tissues with similar X-ray absorption characteristics (Moyosore 2022).

3.5. Multimodality Imaging

To address the limitations of single-modality imaging, such as restricted tissue penetration and resolution, multimodality imaging that integrates various techniques has gained considerable recognition (Choi et al. 2007). Innovations include multifunctional nanoparticles that improve targeting and minimize contrast agent doses, enabling integrated imaging platforms that combine different materials and properties, such as graphene oxide. This structured overview highlights the diverse applications and ongoing advancements in contrast imaging techniques across various medical fields.

4. Role of Various Nanoparticles in Bioimaging and Their Mechanisms

Nanoparticles are pivotal in advancing biomedical imaging and diagnostic technologies. Their unique features, including adjustable size, high surface-area-to-volume ratio, and quantum effects at the nanoscale, enhanced strength, reactivity, and diverse optical, electrical, and magnetic properties, make them highly effective as contrast agents. They leverage the enhanced permeability and retention (EPR) effect to attain higher concentrations of contrast agents within tumors relative to adjacent healthy tissues (Huynh et al. 2013). Moreover, nanoparticles exhibit interactions with animal and plant cells, showcasing their potential to enhance biomedical applications through responsive behaviors to environmental stimuli. Smart nanomaterials, which can modify their properties in reaction to external stimuli, are advancing innovations in theragnostic devices and enabling the creation of multifunctional nanoplatforms for enhanced diagnostic and therapeutic results. Nanomaterials are essential in bioimaging, demonstrating flexibility and propelling biomedical research forward. Quantum dots, gold nanoparticles, magnetic nanoparticles, and carbon-based nanomaterials enhance imaging through their unique optical, electronic, and magnetic properties. Their versatility aids in accurate diagnostics, deeper disease insights, and the development of sophisticated therapeutic approaches (Harish et al. 2022).

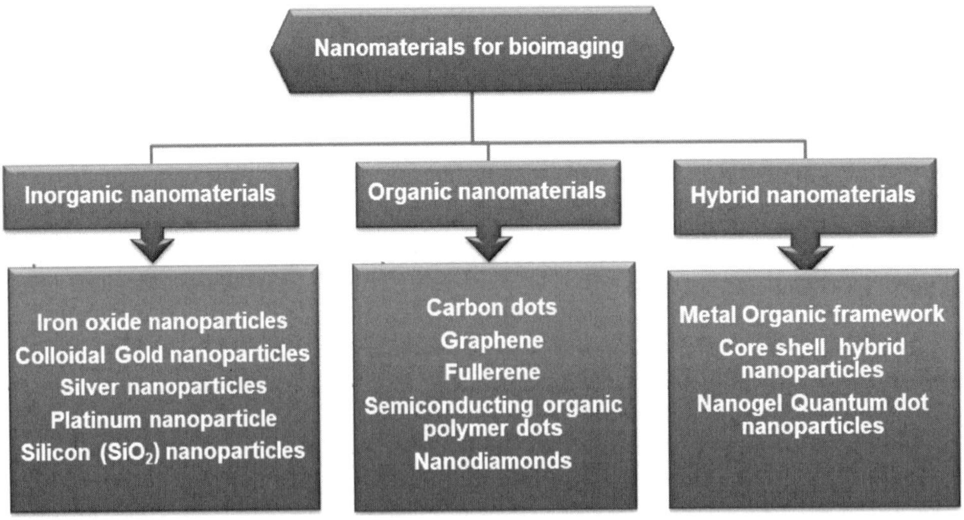

Figure 1. Various nanomaterials are utilized in the bioimaging process.

4.1. Metallic Nanoparticles in Bioimaging

Metallic nanoparticles have attracted considerable interest in biomedical research because of their highly tunable plasmonic properties, especially for diagnostic and therapeutic purposes

(L. Jing et al. 2020). In animal studies, these nanoparticles exhibit robust X-ray absorption capabilities and low toxicity profiles, making them promising candidates as short-term X-ray contrast agents. Gold nanoparticles (GNPs), known for their adjustable surface plasmon properties, are widely acknowledged as versatile nanocarriers for catalysis, sensing, diagnosis, drug delivery, and bioimaging applications.

Different metallic nanomaterials, including gold nanospheres and nanorods, offer superior contrast enhancement in dark-field optical and photothermal imaging of cellular and tissue structures. Furthermore, gold nanocages and nanoshells are well-suited for optical coherence tomography and photoacoustic imaging of deep tissues, circulatory systems, and lymph nodes thanks to their biocompatibility and non-toxic properties (Hassan et al. 2022). Additionally, metallic nanoparticles such as manganese oxide nanoparticles coated with human albumin and iron oxide-coated gelatin nanoparticles act as effective contrast agents for tumor imaging. Surface-modified nanoparticles, like Fe_3O_4 and MnO_2 hybrid nanocrystals, are employed for T1- and T2-weighted MRI, offering dual contrast functionalities (Neogi and Choudhury 2023). Figure 1 illustrates the diverse applications of metallic nanoparticles in various bioimaging contexts.

4.2. Carbon-Based Nanoparticles in Bioimaging

Highly fluorescent carbon nanoparticles (FCNs) and chemically modified nanoprobes are increasingly employed across various biological labeling and imaging applications. These FCNs, synthesized at scales ranging from milligrams to grams, exhibit adjustable emissions spanning the visible spectrum with quantum yields between 6% and 30% (Bhunia et al. 2013). Carbon-based materials, including carbon dots (CQDs), carbon nanotubes (CNTs), graphene, and carbon nanofibers (CNFs), are increasingly recognized in academic research and are expected to be pivotal in future material advancements (Ashok K. Sundramoorthy, Vignesh Kumar, and Gunasekaran 2018). CQDs are submicron-sized zero-dimensional nanomaterials with distinctive optical and chemical properties, including size-dependent color variations (Mai, Phan, and Nguyen 2020). Their versatile nature and intriguing characteristics make them suitable for applications in bioimaging, biomedicine, and biochemical analyses (Das, Mondal, and Ghosh 2023; Nagarajan and Sundramoorthy 2021).

Researchers have investigated single-walled carbon nanotubes (SWNTs) as near-infrared II (NIR-II) contrast agents in mice, using high-speed fluorescent video imaging to track their *in vivo* biodistribution (Welsher, Sherlock, and Dai 2011). This study showed the swift accumulation of SWNTs in organs like the lungs, spleen, and liver, improving organ imaging resolution through principal component analysis (PCA). Additionally, research has highlighted the effectiveness of carbon quantum dots (CQDs) in tumor diagnosis and treatment through near-infrared fluorescence (NIR FL) imaging in mouse models with H22 tumors (Bao et al. 2018). Intravenous administration of CQDs resulted in intense NIR fluorescence emissions, highlighting specific signals from tumors and indicating their potential as optical nanoprobes for advancing biological imaging. These advancements underscore the promising future of cell imaging techniques in biomedical research and clinical diagnostics, emphasizing their capability for effective *in vivo* applications.

4.3. Carbon/Metallic-Based Nanocomposites in Bio-Imaging

Carbon-based materials and their hybrid nanocomposites, incorporating metals, polymers, or ceramics, are emerging as promising materials in diverse biomedical applications (Kim et al. 2024). These materials benefit from synergistic interactions among their components, significantly enhancing overall performance. Metal-organic frameworks (MOFs) showcase this versatility by employing metal clusters and ligands for medical imaging, diagnostics, and therapy applications. MOFs also function effectively as biosensors, detecting targets such as microRNA-21, tumor biomarkers, and hydrogen peroxide with exceptional efficiency (Kong et al. 2020).

For instance, the green-emitting BUT-88 material, created from a luminescent tetracarbazolyl octa-carboxylate ligand, is used in fluorescence-based diagnostics. Similarly, Cu-metal organic frameworks (Cu-MOFs), formulated with tetradentate phosphonate ligands, function as fluorescent agents to improve cancer therapy through light radiation (Chakraborty et al. 2022).

Graphene-based carbon quantum dots (CQDs) are essential for red and near-infrared (NIR) fluorescence bioimaging, playing a critical role in monitoring therapeutics and examining their pharmacokinetics. Fullerene nanoparticles, such as Gd@C2v-C82(OH)22, are extensively investigated for their bioimaging potential and anticancer effects, particularly as MRI contrast agents noted for their natural anticancer properties and low toxicity profiles in both *in vitro* and *in vivo* research (Kim et al. 2024).

4.4. Polymeric Nanoparticles Based Bioimaging

The stability of nanoparticles against aggregation has been significantly improved by incorporating polymers, extending their circulation time by evading immune recognition (Dobrovolskaia et al. 2008). Conjugated polymer nanoparticles have recently emerged as dual-function agents, serving both as fluorescent probes and carriers for anti-tumor drugs in theranostic applications (Yao, Yang, and Duan 2014). NIR fluorescent probes have been incorporated into hyperbranched polyhydroxy polymeric nanoparticles, combined with the apoptosis-inducing protein cytochrome-c, and researchers selectively targeted them using folic acid ligands. This targeted delivery approach improves therapeutic effectiveness against different human carcinoma cells *in vitro*, employing encapsulated indocyanine green (ICG) for photon emission during imaging (Santra, Kaittanis, and Perez 2010). Semiconducting polymers (SPs) are increasingly favored over small molecule dyes in bioimaging due to their extensively delocalized p-conjugated systems and superior optical properties (Feng et al. 2013). The study achieved deep-tissue imaging of orthotopic brain tumors through an intact skull using NIR-II photoacoustic probes based on semiconductor nanoparticles (approximately 50 nm). This feat is particularly challenging due to the absence of intrinsic contrast agents within tumors, significant light absorption by the skull, and substantial background interference. The probe was formulated by encapsulating a planar SP with robust donor-acceptor interactions to optimize NIR-II absorption. *In vivo*, photoacoustic imaging demonstrated the visualization of the probe within a mouse brain tumor located approximately 3.4 mm beneath the skull upon excitation at $\lambda = 1064$ nm.

4.5. Lipid-Based Nanoparticles in Bioimaging

Nanoparticles (NPs) or nanosystems (NSs) that combine multiple functionalities on a single platform broaden the scope of biomedical applications. Lipid nanoparticles (LNPs), such as nanostructured lipid carriers, solid lipid nanoparticles, and liposomes, are notable among nanocarriers for their strong physicochemical stability, effective encapsulation, and ability to integrate delicate components into their structures, making them extensively used in imaging applications (Balasamy, Sekaran, and Eswaramoorthy 2023).

Researchers created an innovative lipid nanoparticle system loaded with the small molecular dye IR-1061, optimized for effective infrared II absorption. This system enables photoacoustic imaging and photothermal therapy for hepatocellular carcinoma (HCC)(Q. Chen et al. 2019). Extensive evaluations in nude mice with HCC demonstrated the nanoparticles' high sensitivity in detecting photoacoustic signals, exceptional photothermal conversion efficiency (45.25%), and strong absorption of laser energy at 1064 nm. *In vivo* studies demonstrated that these nanoparticles facilitated accurate intraoperative tumor removal and offered dependable, non-invasive cancer diagnosis, achieving a strong signal-to-noise ratio of 5.98 ± 0.23 at 3 hours after injection. Both *in vitro* and *in vivo* experiments validated the NIR-II photoacoustic and photothermal capabilities of IR-1061-based nanoparticles, emphasizing their photostability and biocompatibility. These attributes could potentially revolutionize the application of polymethine dyes in nanomedicine, particularly for early diagnosis and therapy of HCC (Yetisgin et al. 2020).

The biocompatible bioimaging capabilities of near-infrared probe DiR-loaded solid lipid nanoparticle suspensions (DiR-SLNS) were also investigated. These nanoparticles demonstrated outstanding physicochemical stability, retaining around 90% of their fluorescence intensity over a 10-day storage period. DiR-SLNS featured a core-shell nanostructure, with DiR primarily encapsulated within the cetyl palmitate (CP) core rather than on the SLNS surface, thereby improving their suitability for bioimaging applications. The study underscored their potential for disease diagnosis, cellular imaging, and understanding the *in vivo* behavior of nanoparticles (Shu et al. 2020).

4.6. Liposomes-Based Bioimaging

One of the leading nanomedicines in cancer treatment and bioimaging is the liposome, renowned for its versatility and multifunctional properties as a delivery system. The efficacy of liposomes as carriers for therapeutic and imaging agents is influenced by their physicochemical characteristics, particularly size, surface charge, and membrane composition. Polar lipid molecules, with a polar head group and a long hydrophobic tail, naturally assemble into liposomes, also known as vesicles. This structure allows for the incorporation of lipophilic and hydrophilic substances within the lipid bilayer and the aqueous core, facilitating the solubilization of a wide array of drugs. Examples of these molecules include PEGylated phospholipids such as 1, 2-distearoyl-sn-glycero-3-phosphoethanolamine (DSPE) and 1,2-distearoyl-sn-glycero-3-phosphocholine (DSPC) (Han et al. 2014).

Different imaging agents, including radionuclides, peptides, and nanoparticles, have been encapsulated, conjugated, or chelated onto liposomes for bioimaging applications. Magnetoliposomes, which are liposomes coated with polyethylene glycol (PEG) and filled with

PEGylated superparamagnetic iron oxide nanoparticles (SPIONs), have shown promise as contrast agents for magnetic resonance imaging (MRI) (Martins et al. 2014). Studies have demonstrated that PEGylated magneto liposomes are highly effective as negative contrast agents for MRI. Furthermore, liposomes containing hydrophobic SPIONs, called magneto liposomes, have been investigated as contrast agents, particularly for imaging the liver (Martínez-González, Estelrich, and Busquets 2016). Table 1 overviews various nanomaterials, polymers, and liposomes used in bioimaging.

5. Applications of Nanoparticles in Medical Diagnoses and Their Limitations

Nanoparticles, characterized by their size-dependent attributes and versatile functionalities, are pivotal in advancing medical diagnostics (Joseph et al. 2023). Their diminutive scale facilitates deep tissue penetration, enabling precise delivery and biomarker interaction (Cabral et al. 2023). Engineered features, such as fluorescence, magnetism, surface plasmon resonance (SPR), and photoacoustic effects render them highly suitable for advanced diagnostic platforms. Fluorescent variants like quantum dots and carbon dots emit specific wavelengths, aiding in sensitive biomolecule detection and real-time cellular visualization (Krasley et al. 2024). Magnetic counterparts respond to external magnetic fields, enhancing MRI contrast, facilitating targeted drug delivery, and supporting magnetic hyperthermia in cancer treatment. SPR nanoparticles, exemplified by gold nanoparticles, detect analytes through light scattering, essential for biosensing in intricate biological fluids (Nanda and Yi 2024). Photoacoustic nanoparticles convert light into sound waves, enabling high-resolution tissue imaging that surpasses conventional optical techniques (Anil et al. 2024; Kumar and Mahajan 2024).

Table 1. Examples of numerous nanomaterials, polymers, and liposomes utilized for bioimaging

S. No.	Imaging agent	Imaging modality	Applications	References
1	Graphene quantum dots-folate-DOX-Gd nanoparticles	MRI/Osteogenesis Imperfeta (OI)	Imaging and tumor-targeted drug delivery	(Huang et al. 2015)
2	Liposome Gd nanoparticles	MRI	Placenta imaging	(Ghaghada et al. 2017)
3	PEGylated liposomes (Lp)-Gd—positron emitting ^{89}Zr	MRI/Positron emission tomography (PET)	Cancer cells Imaging	(Abou et al. 2013)
4	Gd(III) thiolated DNA-Au nanostars (DNA Gd@stars)	MRI	Imaging pancreatic cancer cells	(Rotz et al. 2015)
5	Indocyanine Green-Loaded PLGA Nanoparticles Conjugated with Hyaluronic Acid	Near Infra-Red Fluorescence Imaging (NIR FI)	Cervical Cancer Tumours	(S. Choi et al. 2021)
6	Fullerene Nanocrystals + gold nanoparticles	Photoacoustic imaging (PI)	Imaging cancer cells	(Kawasaki et al. 2022)
7	Carbon dot incorporated mesoporous silica nanoparticles	Fluorescence imaging (FI)	Cancer therapy	(Kajani et al. 2023)
8	Ag_2S-nanoparticles encapsulated poly[di(carboxylatophenoxy) phosphazene] (AgPCPP)	Multimodality Imaging (MI)	Treatment of breast cancer	(Hsu et al. 2023)
9	Graphene quantum dots (GQDs)-Folic acid A-Evodiammine (EVO)	Fluorescence imaging (FI)	Malignant Tumour eradication	(Y. Ma et al. 2022)
10	Graphene quantum dots (GQDs)-concanavalin (ConA)@ Fe_3O_4	Fluorescence and MRI imaging	Cancer cell detection and drug delivery	(Dutta Chowdhury et al. 2018)

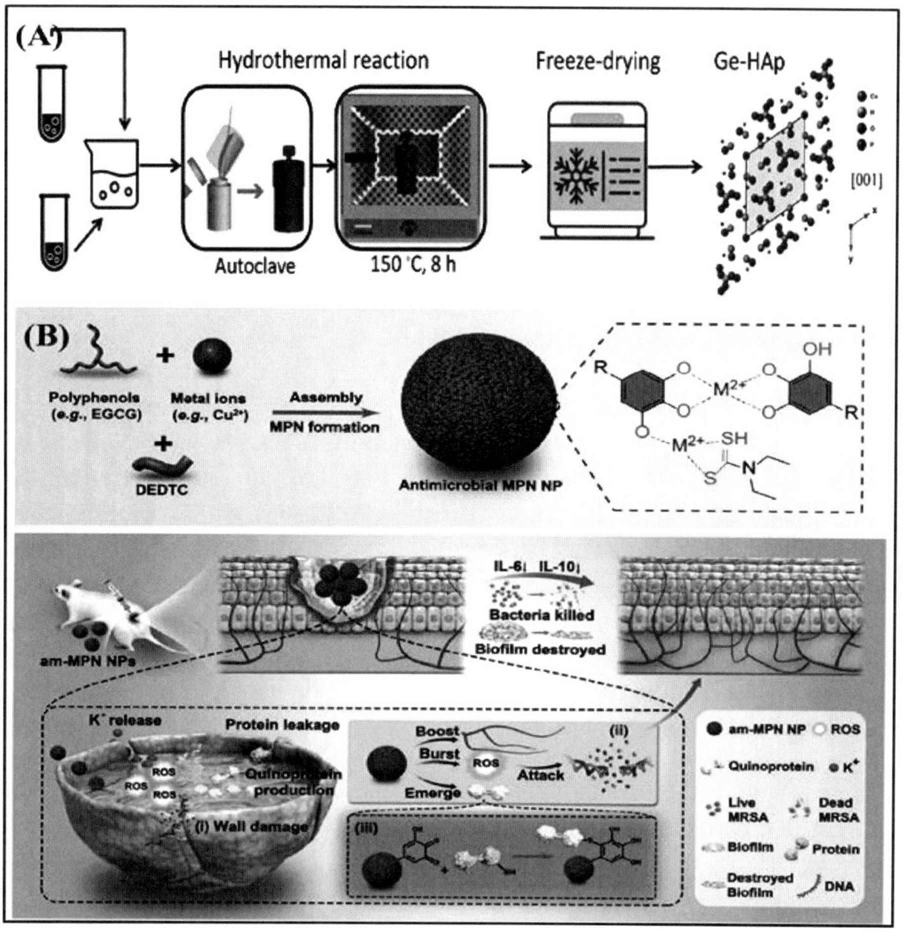

Figure 2. Schematic Illustration of Experiments: (A) Hydrothermal method employed for fabricating Germanium-doped hydroxyapatite nanoparticles (Lazarević et al. 2024), Reproduced with permission from Lazarević et al. Copyright 2024, ACS. (B) Preparation of antimicrobial metal-phenolic network nanoparticles (Yu et al. 2024), Reproduced with permission from Yu et al. Copyright 2023, Wiley. (C) Development of SMP nanocontainers, SMP–CHI MPC, NIR-activated drug release, and enhancement of osteogenic differentiation (B. Wang et al. 2024), Reproduced with permission from B. Wang et al. Copyright 2024, Royal Society of Chemistry.

Despite their potential, the clinical application of nanoparticles encounters challenges, particularly concerning their biological interactions and potential toxicity. While some exhibit biocompatibility, others may provoke cellular stress or inflammation. Evaluating the long-term behavior of nanoparticles in terms of biodistribution and clearance is crucial to ensuring safety (Sharma et al. 2024). Robust synthesis, characterization, and quality control methodologies are indispensable for achieving consistent and reliable diagnostic outcomes (Ansari et al. 2024). The synthesis of germanium-doped hydroxyapatite nanoparticles via hydrothermal methods involves coordinating germanium ions with calcium and phosphate precursors under elevated temperature and pressure, enhancing biodegradation and drug release. Incorporating germanium enhances biocompatibility, promotes protein adhesion, and supports the growth of bone marrow mesenchymal stem cells (BM-MSCs) (Figure 2A) (Lazarević et al. 2024).

Effective strategies are necessary for targeted delivery while minimizing unintended off-target effects (Chaudhari, Patel, and Kumar 2024).

Ongoing research aims to overcome these challenges by developing biocompatible and biodegradable nanoparticles, optimizing synthesis and characterization methods, and innovating targeting strategies. The future of nanoparticle-based diagnostics holds immense promise for improving disease detection, early diagnosis, and personalized medicine, ultimately leading to better patient outcomes.

5.1. Biocompatible Properties of Nanoparticles

Choosing the right materials is crucial for ensuring nanoparticle biocompatibility, such as selecting between natural biomaterials (proteins and polysaccharides) and synthetic polymers (PEG and PLGA) significantly impacts biodegradability and immunogenicity (Gil-Cabrerizo et al. 2024). Equally essential are surface modification strategies, which optimize nanoparticle interactions within biological environments. Coating nanoparticles with biocompatible polymers like PEG reduces surface energy, prevents aggregation, and enhances dispersibility in biological fluids (Arcos Rosero et al. 2024; Rezaei et al. 2024). Modifying nanoparticle surfaces with ligands, antibodies, or peptides allows for accurately targeting specific cells or tissues, enhancing therapeutic effectiveness and reducing off-target interactions (Puzzo et al. 2024). The size and shape of nanoparticles play a crucial role in determining their biocompatibility. Smaller nanoparticles typically exhibit superior tissue penetration and intracellular uptake efficiency, whereas larger particles may trigger immune responses or aggregate (J. Wang et al. 2024). For instance, elongated nanoparticles can more effectively penetrate cellular membranes than their spherical counterparts. Figure 2B illustrates the synthesis of antimicrobial metal-phenolic network nanoparticles (am-MPN NPs), demonstrating their creation by coordinating metal ions with polyphenolic compounds. These nanoparticles have antimicrobial properties by breaking down bacterial cell walls, producing reactive oxygen species (ROS) that damage DNA, and triggering oxidative stress responses via Quino proteins (Yu et al. 2024; Li et al. 2022). Assessing nanoparticle biocompatibility requires thorough *in vitro* and *in vivo* studies. Techniques such as MTT, SRB, and LDH assays evaluate cytotoxicity, while cellular uptake studies using flow cytometry and microscopy reveal uptake mechanisms and the intracellular behavior of the nanoparticles (Zawidlak-Węgrzyńska et al. 2024; Behzadi et al. 2017). Surface modifications also mitigate biofouling, influencing biodistribution and cellular interactions critical for minimizing inflammation and adverse immune reactions (Kyriakides et al. 2021; Crawford et al. 2021).

Understanding nanoparticle biodistribution and elimination pathways is essential for safe deployment. Techniques like imaging and biodistribution analyses track nanoparticle movement within the body, ensuring effective localization at target sites while avoiding potential toxicity from accumulation (Kakinen et al. 2024). Tailoring nanoparticles for specific applications (e.g., drug delivery, gene therapy, imaging) further enhances their biocompatibility and efficacy (Qiu et al. 2019; G. Liu et al. 2010). Strategic approaches such as surface modification and using biodegradable materials enable customization to maximize therapeutic benefits while minimizing adverse effects (H. H. Jing et al. 2024).

5.2. Bioimaging for Drug Delivery

Bioimaging is crucial for optimizing drug delivery through real-time visualization and tracking capabilities, essential for enhancing therapeutic outcomes (Murar, Albertazzi, and Pujals 2022; T. Wang et al. 2020). Techniques like MRI, CT, optical imaging, and ultrasound provide insights into drug carrier distribution, targeting efficiency, and therapeutic efficacy (Bai, Qiu, and Zhang 2023). MRI utilizes magnetic nanoparticles for high-resolution visualization of carrier distribution, while CT imaging employs gold nanoparticles to track carrier movement through vasculature and tissues (Molkenova et al. 2022). Optical imaging with fluorescent nanoparticles monitors real-time drug uptake, and ultrasound with microbubbles enables non-invasive monitoring of carrier delivery, particularly within blood vessels (Cooley, Wegierak, and Exner 2024). Personalized medicine benefits from bioimaging by tailoring drug delivery strategies to individual patients based on unique physiological characteristics and treatment responses, optimizing outcomes while minimizing adverse effects. Biocompatible nanoparticles are crucial in biomedical applications, ensuring safe interactions through surface chemistry modifications like PEGylation to enhance stability and prevent aggregation (Forgham et al. 2024). Figure 2C shows the experimental workflow involves three stages fabricating shape-memory polymer (SMP) nanocontainers, integrating SMP with chitosan (CHI) to form hybrid polymer chains, and initiating NIR-induced mechanisms for drug release and osteogenic differentiation. NIR light in biomedical applications offers lower phototoxicity, deeper tissue penetration, and enhanced precision in drug release measurements. Its non-invasive characteristics and adjustable intensity allow for precise regulation of drug delivery kinetics, enhancing therapeutic effectiveness with excellent spatiotemporal resolution and reducing off-target effects (B. Wang et al. 2024).

Understanding nanoparticle size, shape, degradation, and clearance mechanisms is critical for optimizing nanoparticle-cell interactions and ensuring efficacy across diverse biomedical applications (Forgham et al. 2024). Ongoing developments in biocompatible nanoparticle design propel improvements in drug delivery, imaging, diagnostics, and tissue engineering, enhancing precision medicine and personalized healthcare.

6. Perspectives

Nanomaterials are pivotal in advancing multimodal imaging techniques, providing integrated solutions that deepen our comprehension of biological systems. These materials are crucial in combining diverse imaging modalities within a single nanoparticle platform. For instance, nanoparticles can integrate fluorescent quantum dots for optical imaging with magnetic iron oxide nanoparticles for MRI enhancement (Z. Liu, Kiessling, and Gätjens 2010). They facilitate complementary imaging techniques by utilizing the inherent properties of various nanomaterials, such as their optical, electronic, and magnetic characteristics. Gold nanoparticles, for instance, serve dual roles in both photoacoustic and photothermal imaging, while carbon-based nanomaterials excel in fluorescence imaging due to their outstanding optical properties. Furthermore, nanomaterials can be customized with targeting ligands, ensuring precise accumulation in diseased tissues or cells. This capability enhances the accuracy of diagnostic and monitoring processes using multimodal imaging approaches like

PET/MRI or optical/ultrasound imaging (Yuan et al. 2021). Furthermore, nanomaterials' small size and large surface area improves imaging sensitivity and resolution, surpassing traditional contrast agents. This progress facilitates earlier disease detection and provides intricate visualizations of biological processes, thereby fostering advancements in diagnostic precision and therapeutic effectiveness (Burke, Cawthorne, and Archibald 2017).

The future of smart nanomaterials in diagnostics and bioimaging is advancing rapidly due to nanotechnology, synthetic chemistry, and materials science progress. Multimodal nanomaterials integrate multiple imaging modalities like fluorescence, magnetic resonance, surface plasmon resonance, and photoacoustic (Darwish et al. 2024). These enable sensitive detection of disease biomarkers. Customizing nanoparticles for targeted delivery to specific cells minimizes off-target effects and enhances diagnostic precision (Ghafari and Asefnejad 2024). Surface modifications with ligands or antibodies facilitate targeted drug delivery, enhancing therapeutic efficacy (Cao, Liu, and Yang 2024). Responsive nanomaterials, triggered by pH changes or enzymatic activity, promise personalized medicine by releasing therapies selectively at diseased sites (Peng et al. 2024). Biocompatible and biodegradable designs mitigate toxicity concerns, crucial for safe clinical use (Bushra et al. 2024).

Integrating artificial intelligence (AI) and machine learning (ML) with nanomaterial diagnostics enhances disease diagnosis and personalized treatment (Sher et al. 2024). Challenges include nanoparticle toxicity, biocompatibility, and standardizing synthesis and characterization methodologies for clinical reproducibility (Pandey et al. 2024). Despite these hurdles, smart nanomaterials offer the potential for more sensitive, accurate, and personalized healthcare solutions globally (Singh and Negi 2024).

Conclusion

In this chapter, we have emphasized the critical importance of smart nanomaterials in the progression of biomedical applications, especially in drug delivery and bioimaging. Nanoparticles' non-invasive nature and customizable properties permit precise control over drug delivery kinetics, thereby boosting therapeutic effectiveness and reducing off-target effects. The significance of nanoparticle size, shape, degradation, and clearance mechanisms is clear, as they are vital for optimizing interactions with biological systems and ensuring effectiveness across various biomedical applications. Additionally, integrating multimodal nanomaterials offers significant promise by combining various imaging modalities into a single platform. This integration enables highly sensitive detection of multiple disease biomarkers, thus enhancing diagnostic precision and therapeutic outcomes. Optimizing nanoparticles for targeted delivery to cells or tissues remains a key research priority to improve therapeutic effectiveness and minimize systemic side effects.

Advances in biocompatible and biodegradable nanomaterials are essential for addressing toxicity concerns, facilitating safe clinical translation, and driving progress in precision medicine. Integrating artificial intelligence and machine learning with nanomaterial-based diagnostics could improve disease detection and tailor treatment suggestions. However, nanoparticle toxicity, biocompatibility, and long-term effects must be tackled with standardized synthesis, characterization, and quality control procedures to ensure reliability and clinical effectiveness. The future of smart nanomaterials in diagnostics and bioimaging holds immense potential, promising more sensitive, accurate, and personalized healthcare solutions. Continued advancements in this field are expected to revolutionize precision medicine, significantly improving disease detection, early diagnosis, and therapeutic interventions, ultimately leading to better patient outcomes worldwide.

Acknowledgments

AKS appreciates financial support from the Science and Engineering Research Board (SERB), India (File No.: CRG/2021/001517).

References

Abou, Diane S., Daniel L. J. Thorek, Nicholas N. Ramos, Martijn W. H. Pinkse, Hubert T. Wolterbeek, Sean D. Carlin, Bradley J. Beattie, and Jason S. Lewis. 2013. "(89)Zr-Labeled Paramagnetic Octreotide-Liposomes for PET-MR Imaging of Cancer." *Pharmaceutical Research* 30 (3): 878–88. https://doi.org/10.1007/s11095-012-0929-8.

Adam, Alexandre, and Damien Mertz. 2023. "Iron Oxide@Mesoporous Silica Core-Shell Nanoparticles as Multimodal Platforms for Magnetic Resonance Imaging, Magnetic Hyperthermia, Near-Infrared Light Photothermia, and Drug Delivery." *Nanomaterials* 13 (8): 1342. https://doi.org/10.3390/nano13081342.

Anil, Anusri, Jyotsna Chaskar, Avinash B. Pawar, Abhishekh Tiwari, and Atul Changdev Chaskar. 2024. "Recent Advances in DNA-Based Probes for Photoacoustic Imaging." *Journal of Biotechnology* 382 (February): 8–20. https://doi.org/10.1016/j.jbiotec.2023.12.019.

Ansari, Shaquib Rahman, Yael Del Carmen Suárez-López, Thomas Thersleff, Lennart Häggström, Tore Ericsson, Ioannis Katsaros, Michelle Åhlén, et al. 2024. "Pharmaceutical Quality by Design Approach to Develop High-Performance Nanoparticles for Magnetic Hyperthermia." *ACS Nano*, May. https://doi.org/10.1021/acsnano.4c04685.

Arcos Rosero, Wilmmer Alexander, Angelica Bueno Barbezan, Carla Daruich de Souza, and Maria Elisa Chuery Martins Rostelato. 2024. "Review of Advances in Coating and Functionalization of Gold Nanoparticles: From Theory to Biomedical Application." *Pharmaceutics* 16 (2). https://doi.org/10.3390/pharmaceutics16020255.

Atchudan, Raji, Suguna Perumal, Ashok K. Sundramoorthy, Devaraj Manoj, Raju Suresh Kumar, Abdulrahman I. Almansour, Sambasivam Sangaraju, Wonmok Lee, and Yong Rok Lee. 2024. "Eco-Friendly Chebulic Myrobalan-Derived Porous Carbon Employed as an Electrocatalyst for the Production of Hydrogen." *The Korean Journal of Chemical Engineering* 41 (6): 1691–1702. https://doi.org/10.1007/s11814-024-00119-z.

Bai, Jing-Wen, Si-Qi Qiu, and Guo-Jun Zhang. 2023. "Molecular and Functional Imaging in Cancer-Targeted Therapy: Current Applications and Future Directions." *Signal Transduction and Targeted Therapy* 8 (1): 89. https://doi.org/10.1038/s41392-023-01366-y.

Balas, Costas. 2009. "Review of Biomedical Optical Imaging—a Powerful, Non-Invasive, Non-Ionizing Technology for Improving *in Vivo* Diagnosis." *Measurement Science & Technology* 20 (10): 104020. https://doi.org/10.1088/0957-0233/20/10/104020.

Balasamy, Sesuraj, Surya Sekaran, and Rajalakshmanan Eswaramoorthy. 2023. "Supporting Nanotechnology Safety through Nanoinformatics." In *Green Nanoremediation*, 313–32. Cham: Springer International Publishing. https://doi.org/10.1007/978-3-031-30558-0_14.

Balasamy, Sesuraj, and Ashok K. Sundramoorthy. 2024. "EGFR-Targeted Fluorescence Imaging for Precision Margin Assessment in Oral Cancer Surgery." *Oral Oncology* 150 (106712): 106712. https://doi.org/10.1016/j.oraloncology.2024.106712.

Balu, Surendar, and Ashok K. Sundramoorthy. 2024. "Label-Free Immunosensors for the Ultrasensitive Detection of Cancer Biomarkers." *Oral Oncology Reports* 10 (100273): 100273. https://doi.org/10.1016/j.oor.2024.100273.

Bao, Xin, Ye Yuan, Jingqin Chen, Bohan Zhang, Di Li, Ding Zhou, Pengtao Jing, et al. 2018. "*In Vivo* Theranostics with Near-Infrared-Emitting Carbon Dots—Highly Efficient Photothermal Therapy Based on Passive Targeting after Intravenous Administration." *Light: Science & Applications* 7 (1): 1–11. https://doi.org/10.1038/s41377-018-0090-1.

Behzadi, Shahed, Vahid Serpooshan, Wei Tao, Majd A. Hamaly, Mahmoud Y. Alkawareek, Erik C. Dreaden, Dennis Brown, Alaaldin M. Alkilany, Omid C. Farokhzad, and Morteza Mahmoudi. 2017. "Cellular Uptake of Nanoparticles: Journey inside the Cell." *Chemical Society Reviews* 46 (14): 4218–44. https://doi.org/10.1039/c6cs00636a.

Bhunia, Susanta Kumar, Arindam Saha, Amit Ranjan Maity, Sekhar C. Ray, and Nikhil R. Jana. 2013. "Carbon Nanoparticle-Based Fluorescent Bioimaging Probes." *Scientific Reports* 3: 1473. https://doi.org/10.1038/srep01473.

Bonlawar, Jyoti, Aseem Setia, Ranadheer Reddy Challa, Bhaskar Vallamkonda, Abhishesh Kumar Mehata, Vaishali, Matte Kasi Viswanadh, and Madaswamy S. Muthu. 2024. "Targeted Nanotheransotics: Integration of Preclinical MRI and CT in the Molecular Imaging and Therapy of Advanced Diseases." *Nanotheranostics* 8 (3): 401–26. https://doi.org/10.7150/ntno.95791.

Burke, Benjamin P., Christopher Cawthorne, and Stephen J. Archibald. 2017. "Multimodal Nanoparticle Imaging Agents: Design and Applications." *Philosophical Transactions. Series A, Mathematical, Physical, and Engineering Sciences* 375 (2107). https://doi.org/10.1098/rsta.2017.0261.

Bushra, Rani, Mehraj Ahmad, Kehkashan Alam, Farzad Seidi, Qurtulen, Sadaf Shakeel, Junlong Song, Yongcan Jin, and Huining Xiao. 2024. "Recent Advances in Magnetic Nanoparticles: Key Applications, Environmental Insights, and Future Strategies." *Sustainable Materials and Technologies* 40 (July): e00985. https://doi.org/10.1016/j.susmat.2024.e00985.

Cabral, Horacio, Junjie Li, Kanjiro Miyata, and Kazunori Kataoka. 2023. "Controlling the Biodistribution and Clearance of Nanomedicines." *Nature Reviews Bioengineering* 2 (3): 214–32. https://doi.org/10.1038/s44222-023-00138-1.

Cai, Zhengwei, Zhangmei Ye, Xiaowei Yang, Yanli Chang, Haifang Wang, Yuanfang Liu, and Aoneng Cao. 2011. "Encapsulated Enhanced Green Fluorescence Protein in Silica Nanoparticle for Cellular Imaging." *Nanoscale* 3 (5): 1974–76. https://doi.org/10.1039/c0nr00956c.

Cao, Ziyang, Jing Liu, and Xianzhu Yang. 2024. "Deformable Nanocarriers for Enhanced Drug Delivery and Cancer Therapy." *Exploration (Beijing, China)*, March. https://doi.org/10.1002/exp.20230037.

Chakraborty, D., D. Musib, R. Saha, A. Das, M. K. Raza, V. Ramu, S. Chongdar, K. Sarkar, and A. Bhaumik. 2022. "Highly Stable Tetradentate Phosphonate-Based Green Fluorescent Cu-MOF for Anticancer Therapy and Antibacterial Activity." *Materials Today Chemistry* 24 (June): 100882. https://doi.org/10.1016/j.mtchem.2022.100882.

Chang, Esther H., Joe B. Harford, Michael A. W. Eaton, Patrick M. Boisseau, Admire Dube, Rose Hayeshi, Hulda Swai, and Dong Soo Lee. 2015. "Nanomedicine: Past, Present and Future - A Global Perspective." *Biochemical and Biophysical Research Communications* 468 (3): 511–17. https://doi.org/10.1016/j.bbrc.2015.10.136.

Chatterjee, Shubhangini, Kavitha Kamalasekaran, Sankari Malaiappan, and Ashok K. Sundramoorthy. 2024. "Electrodeposition of Nickel Nanoparticles on Graphene/Iron (III) Phthalocyanine-4,4,'4,,"4'"-Tetrasulfonic Acid Coated Electrode for Glucose Sensing." *Inorganic Chemistry Communications* 161 (112140): 112140. https://doi.org/10.1016/j.inoche.2024.112140.

Chaudhari, Ramesh, Vishva Patel, and Ashutosh Kumar. 2024. "Cutting-Edge Approaches for Targeted Drug Delivery in Breast Cancer: Beyond Conventional Therapies." *Nanoscale Advances* 6 (9): 2270–86. https://doi.org/10.1039/d4na00086b.

Chen, Qingshan, Jingqin Chen, Mu He, Yuanyuan Bai, Huixiang Yan, Ning Zeng, Fangyan Liu, et al. 2019. "Novel Small Molecular Dye-Loaded Lipid Nanoparticles with Efficient near-Infrared-II Absorption for Photoacoustic Imaging and Photothermal Therapy of Hepatocellular Carcinoma." *Biomaterials Science* 7 (8): 3165–77. https://doi.org/10.1039/c9bm00528e.

Chen, Zhenyue, Irmak Gezginer, Quanyu Zhou, Lin Tang, Xosé Luís Deán-Ben, and Daniel Razansky. 2024. "Multimodal Optoacoustic Imaging: Methods and Contrast Materials." *Chemical Society Reviews*, May. https://doi.org/10.1039/d3cs00565h.

Choi, Jong Hyun, Freddy T. Nguyen, Paul W. Barone, Daniel A. Heller, Anthonie E. Moll, Dhaval Patel, Stephen A. Boppart, and Michael S. Strano. 2007. "Multimodal Biomedical Imaging with Asymmetric Single-Walled Carbon Nanotube/Iron Oxide Nanoparticle Complexes." *Nano Letters* 7 (4): 861–67. https://doi.org/10.1021/nl062306v.

Choi, Seonmin, San-Hui Lee, Sanghyo Park, Sun Hwa Park, Chaewon Park, and Jaehong Key. 2021. "Indocyanine Green-Loaded PLGA Nanoparticles Conjugated with Hyaluronic Acid Improve Target Specificity in Cervical Cancer Tumors." *Yonsei Medical Journal* 62 (11): 1042–51. https://doi.org/10.3349/ymj.2021.62.11.1042.

Coimbra, Susana, Susana Rocha, Nícia Reis Sousa, Cristina Catarino, Luís Belo, Elsa Bronze-da-Rocha, Maria João Valente, and Alice Santos-Silva. 2024. "Toxicity Mechanisms of Gadolinium and Gadolinium-Based Contrast Agents—A Review." *International Journal of Molecular Sciences* 25 (7): 4071. https://doi.org/10.3390/ijms25074071.

Cooley, Michaela B., Dana Wegierak, and Agata A. Exner. 2024. "Using Imaging Modalities to Predict Nanoparticle Distribution and Treatment Efficacy in Solid Tumors: The Growing Role of Ultrasound." *Wiley Interdisciplinary Reviews. Nanomedicine and Nanobiotechnology* 16 (2): e1957. https://doi.org/10.1002/wnan.1957.

Crawford, Lars, Meghan Wyatt, James Bryers, and Buddy Ratner. 2021. "Biocompatibility Evolves: Phenomenology to Toxicology to Regeneration." *Advanced Healthcare Materials* 10 (11): e2002153. https://doi.org/10.1002/adhm.202002153.

Darwish, Moustafa A., Walaa Abd-Elaziem, Ammar Elsheikh, and Abdelhameed A. Zayed. 2024. "Advancements in Nanomaterials for Nanosensors: A Comprehensive Review." *Nanoscale Advances*. https://doi.org/10.1039/D4NA00214H.

Das, Surya, Somnath Mondal, and Dhiman Ghosh. 2023. "Carbon Quantum Dots in Bioimaging and Biomedicines." *Frontiers in Bioengineering and Biotechnology* 11: 1333752. https://doi.org/10.3389/fbioe.2023.1333752.

Dhanapalan, Sangeetha, Vasanth Magesh, Raji Atchudan, Sandeep Arya, Dhanraj Ganapathy, Deepak Nallaswamy, and Ashok Sundramoorthy. 2024. "Poly-Thionine/ SWCNT Nanocomposite Coated Electrochemical Sensor for Determination of Vitamin C." *Current Nanoscience* 20 (February). https://doi.org/10.2174/0115734137289697240216070503.

Di Gregorio, Enza, Chiara Papi, Laura Conti, Antonino Di Lorenzo, Eleonora Cavallari, Marco Salvatore, Carlo Cavaliere, Giuseppe Ferrauto, and Silvio Aime. 2024. "A Magnetic Resonance Imaging-Chemical Exchange Saturation Transfer (MRI-CEST) Method for the Detection of Water Cycling across Cellular Membranes." *Angewandte Chemie* 136 (6). https://doi.org/10.1002/ange.202313485.

Dobrovolskaia, Marina A., Parag Aggarwal, Jennifer B. Hall, and Scott E. McNeil. 2008. "Preclinical Studies to Understand Nanoparticle Interaction with the Immune System and Its Potential Effects on Nanoparticle Biodistribution." *Molecular Pharmaceutics* 5 (4): 487–95. https://doi.org/10.1021/mp800032f.

Dutt, Shradha, Anoop Singh, Rajesh Mahadeva, Ashok K. Sundramoorthy, Vinay Gupta, and Sandeep Arya. 2024. "A Reduced Graphene Oxide-Based Electrochemical Sensing and Eco-Friendly 4-Nitrophenol Degradation." *Diamond and Related Materials* 141 (110554): 110554. https://doi.org/10.1016/j.diamond.2023.110554.

Dutta Chowdhury, Ankan, Akhilesh Babu Ganganboina, Yuan-Chung Tsai, Hsin-Cheng Chiu, and Ruey-An Doong. 2018. "Multifunctional GQDs-Concanavalin A@Fe3O4 Nanocomposites for Cancer Cells Detection and Targeted Drug Delivery." *Analytica Chimica Acta* 1027 (October): 109–20. https://doi.org/10.1016/j.aca.2018.04.029.

Egbedina, Abisola Opeyemi, Oladotun Paul Bolade, Ugochukwu Ewuzie, and Eder C. Lima. 2022. "Emerging Trends in the Application of Carbon-Based Materials: A Review." *Journal of Environmental Chemical Engineering* 10 (2): 107260. https://doi.org/10.1016/j.jece.2022.107260.

Ehman, Eric C., Geoffrey B. Johnson, Javier E. Villanueva-Meyer, Soonmee Cha, Andrew Palmera Leynes, Peder Eric Zufall Larson, and Thomas A. Hope. 2017. "PET/MRI: Where Might It Replace PET/CT?" *Journal of Magnetic Resonance Imaging: JMRI* 46 (5): 1247–62. https://doi.org/10.1002/jmri.25711.

Feng, Liheng, Chunlei Zhu, Huanxiang Yuan, Libing Liu, Fengting Lv, and Shu Wang. 2013. "Conjugated Polymer Nanoparticles: Preparation, Properties, Functionalization and Biological Applications." *Chemical Society Reviews* 42 (16): 6620–33. https://doi.org/10.1039/c3cs60036j.

Forgham, Helen, Jiayuan Zhu, Taoran Zhang, Xumin Huang, Xiangke Li, Ao Shen, Heather Biggs, et al. 2024. "Fluorine-Modified Polymers Reduce the Adsorption of Immune-Reactive Proteins to PEGylated Gold Nanoparticles." *Nanomedicine,* April. https://doi.org/10.2217/nnm-2023-0357.

Foucault-Collet, Alexandra, Kristy A. Gogick, Kiley A. White, Sandrine Villette, Agnès Pallier, Guillaume Collet, Claudine Kieda, et al. 2013. "Lanthanide near Infrared Imaging in Living Cells with Yb3+ Nano Metal Organic Frameworks." *Proceedings of the National Academy of Sciences of the United States of America* 110 (43): 17199–204. https://doi.org/10.1073/pnas.1305910110.

Gattegno, Roni, Lilach Arbel, Noa Riess, Hila Shinar, Sharon Katz, and Tali Ilovitsh. 2024. "Enhanced Capillary Delivery with Nanobubble-Mediated Blood-Brain Barrier Opening and Advanced High Resolution Vascular Segmentation." *Journal of Controlled Release: Official Journal of the Controlled Release Society* 369 (May): 506–16. https://doi.org/10.1016/j.jconrel.2024.04.001.

Gendron, Célia, Philippe Bourrinet, Anne Dencausse, and Nathalie Fretellier. 2024. "Preclinical Safety Assessment of Gadopiclenol: A High-Relaxivity Macrocyclic Gadolinium-Based MRI Contrast Agent." *Investigative Radiology* 59 (2): 108–23. https://doi.org/10.1097/RLI.0000000000001038.

Ghafari, Y., and A. Asefnejad. 2024. "Gold Nanoparticles in Biomedicine: Advancements in Cancer Therapy, Drug Delivery, Diagnostics, and Tissue Regeneration." *Scientific.* https://scientifichypotheses.org/index.php/scihypo/article/view/6.

Ghaghada, Ketan B., Zbigniew A. Starosolski, Saakshi Bhayana, Igor Stupin, Chandreshkumar V. Patel, Rohan C. Bhavane, Haijun Gao, et al. 2017. "Pre-Clinical Evaluation of a Nanoparticle-Based Blood-Pool Contrast Agent for MR Imaging of the Placenta." *Placenta* 57 (September): 60–70. https://doi.org/10.1016/j.placenta.2017.06.008.

Gil-Cabrerizo, Paula, Teresa Simon-Yarza, Elisa Garbayo, and María J. Blanco-Prieto. 2024. "Navigating the Landscape of RNA Delivery Systems in Cardiovascular Disease Therapeutics." *Advanced Drug Delivery Reviews* 208 (May): 115302. https://doi.org/10.1016/j.addr.2024.115302.

González-Mancebo, Daniel, Ana Isabel Becerro, Carlos Caro, Elisabet Gómez-González, María Luisa García-Martín, and Manuel Ocaña. 2024. "Nanoparticulated Bimodal Contrast Agent for Ultra-High-Field Magnetic Resonance Imaging and Spectral X-Ray Computed Tomography." *Inorganic Chemistry*, May. https://doi.org/10.1021/acs.inorgchem.4c01114.

Han, Xiao-Jian, Yong-Fang Wei, Yu-Ying Wan, Li-Ping Jiang, Jian-Feng Zhang, and Hong-Bo Xin. 2014. "Development of a Novel Liposomal Nanodelivery System for Bioluminescence Imaging and Targeted Drug Delivery in ErbB2-Overexpressing Metastatic Ovarian Carcinoma." *International Journal of Molecular Medicine* 34 (5): 1225–32. https://doi.org/10.3892/ijmm.2014.1922.

Harish, Vancha, Devesh Tewari, Manish Gaur, Awadh Bihari Yadav, Shiv Swaroop, Mikhael Bechelany, and Ahmed Barhoum. 2022. "Review on Nanoparticles and Nanostructured Materials: Bioimaging, Biosensing, Drug Delivery, Tissue Engineering, Antimicrobial, and Agro-Food Applications." *Nanomaterials (Basel, Switzerland)* 12 (3). https://doi.org/10.3390/nano12030457.

Hassan, Homa, Pradakshina Sharma, Mohd Rahil Hasan, Shiwani Singh, Deepanshi Thakur, and Jagriti Narang. 2022. "Gold Nanomaterials – The Golden Approach from Synthesis to Applications." *Materials Science for Energy Technologies* 5 (January): 375–90. https://doi.org/10.1016/j.mset.2022.09.004.

Hsu, Jessica C., Diego Barragan, Alexander E. Tward, Maryam Hajfathalian, Ahmad Amirshaghaghi, Katherine J. Mossburg, Derick N. Rosario-Berríos, et al. 2023. "A Biodegradable 'One-for-All' Nanoparticle for Multimodality Imaging and Enhanced Photothermal Treatment of Breast Cancer." *BioRxiv*. https://doi.org/10.1101/2023.11.28.568885.

Huang, Chun-Lin, Chih-Ching Huang, Fu-Der Mai, Chia-Liang Yen, Shin-Hwa Tzing, Hsiao-Ting Hsieh, Yong-Chien Ling, and Jia-Yaw Chang. 2015. "Application of Paramagnetic Graphene Quantum Dots as a Platform for Simultaneous Dual-Modality Bioimaging and Tumor-Targeted Drug Delivery." *Journal of Materials Chemistry. B, Materials for Biology and Medicine* 3 (4): 651–64. https://doi.org/10.1039/c4tb01650e.

Huynh, Nam T., Barrie R. Hayes-Gill, Fan Zhang, and Stephen P. Morgan. 2013. "Ultrasound Modulated Imaging of Luminescence Generated within a Scattering Medium." *Journal of Biomedical Optics* 18 (2): 20505. https://doi.org/10.1117/1.JBO.18.2.020505.

Hyder, Ali, Akbar Ali, Jamil A. Buledi, Ayaz Ali Memon, Muzaffar Iqbal, Talib Hussain Bangalni, Amber R. Solangi, Khalid Hussain Thebo, and Javeed Akhtar. 2024. "Nanodiamonds: A Cutting-Edge Approach to

Enhancing Biomedical Therapies and Diagnostics in Biosensing." *Chemical Record* 24 (4): e202400006. https://doi.org/10.1002/tcr.202400006.

Jin, Xuru, Golnaz Heidari, Shefa Mirani Nezhad, Minmin Shao, Zhidan Hua, Ying Lei, Ehsan Nazarzadeh Zare, et al. 2024. "Progress in Translating Bioinorganic Nanoplatform Discoveries into Clinical Lung Cancer Care: Overcoming Limitations, Targeted Drug Delivery and Imaging." *Alexandria Engineering Journal* 99 (July): 271–81. https://doi.org/10.1016/j.aej.2024.04.074.

Jing, Hong Hui, Ali A. Shati, Mohammad Y. Alfaifi, Serag Eldin I. Elbehairi, and Sreenivasan Sasidharan. 2024. "The Future of Plant Based Green Carbon Dots as Cancer Nanomedicine: From Current Progress to Future Perspectives and Beyond." *Journal of Advertising Research*, February. https://doi.org/10.1016/j.jare.2024.01.034.

Jing, Lihong, Chen Yang, Peisen Zhang, Jianfeng Zeng, Zhen Li, and Mingyuan Gao. 2020. "Nanoparticles Weaponized with Built-in Functions for Imaging-guided Cancer Therapy." *View (Beijing, China)* 1 (2). https://doi.org/10.1002/viw2.19.

Joseph, Tomy Muringayil, Debarshi Kar Mahapatra, Amin Esmaeili, Łukasz Piszczyk, Mohamed S. Hasanin, Mashhoor Kattali, Józef Haponiuk, and Sabu Thomas. 2023. "Nanoparticles: Taking a Unique Position in Medicine." *Nanomaterials (Basel, Switzerland)* 13 (3). https://doi.org/10.3390/nano13030574.

Kajani, Abolghasem Abbasi, Laleh Rafiee, Shaghayegh Haghjooy Javanmard, Nasim Dana, and Setareh Jandaghian. 2023. "Carbon Dot Incorporated Mesoporous Silica Nanoparticles for Targeted Cancer Therapy and Fluorescence Imaging." *RSC Advances* 13 (14): 9491–9500. https://doi.org/10.1039/d3ra00768e.

Kakinen, Aleksandr, Yuhao Jiang, Thomas Paul Davis, Tambet Teesalu, and Mart Saarma. 2024. "Brain Targeting Nanomedicines: Pitfalls and Promise." *International Journal of Nanomedicine* 19 (May): 4857–75. https://doi.org/10.2147/IJN.S454553.

Karuppasamy, Bavya Devi, Suguna Perumal, Raji Atchudan, Ashok K. Sundramoorthy, Devaraj Manoj, Sangaraju Sambasivam, Namachivayam Karthik, et al. 2024. "Eco-Friendly Coccinia Grandis-Derived Excitation-Dependent Fluorescence of Carbon Dots for In-Vivo Bioimaging and Fluorescent Ink Applications." *Materials Science & Engineering. B, Solid-State Materials for Advanced Technology* 303 (117300): 117300. https://doi.org/10.1016/j.mseb.2024.117300.

Kawasaki, Riku, Kosuke Kondo, Risako Miura, Keita Yamana, Hinata Isozaki, Risako Shimada, Shogo Kawamura, et al. 2022. "Theranostic Agent Combining Fullerene Nanocrystals and Gold Nanoparticles for Photoacoustic Imaging and Photothermal Therapy." *International Journal of Molecular Sciences* 23 (9). https://doi.org/10.3390/ijms23094686.

Kim, Choong-Hee, Seul-Yi Lee, Kyong Yop Rhee, and Soo-Jin Park. 2024. "Carbon-Based Composites in Biomedical Applications: A Comprehensive Review of Properties, Applications, and Future Directions." *Advanced Composites and Hybrid Materials* 7 (2): 55. https://doi.org/10.1007/s42114-024-00846-1.

Kong, Xiang-Jing, Xiaoting Ji, Tao He, Lin-Hua Xie, Yong-Zheng Zhang, Haoyuan Lv, Caifeng Ding, and Jian-Rong Li. 2020. "A Green-Emission Metal–Organic Framework-Based Nanoprobe for Imaging Dual Tumor Biomarkers in Living Cells." *ACS Applied Materials & Interfaces* 12 (31): 35375–84. https://doi.org/10.1021/acsami.0c10038.

Krasley, Andrew T., Eugene Li, Jesus M. Galeana, Chandima Bulumulla, Abraham G. Beyene, and Gozde S. Demirer. 2024. "Carbon Nanomaterial Fluorescent Probes and Their Biological Applications." *Chemical Reviews* 124 (6): 3085–3185. https://doi.org/10.1021/acs.chemrev.3c00581.

Kumar, Panangattukara Prabhakaran Praveen, and Ritu Mahajan. 2024. "Gold Polymer Nanomaterials: A Promising Approach for Enhanced Biomolecular Imaging." *Nanotheranostics* 8 (1): 64–89. https://doi.org/10.7150/ntno.89087.

Kyriakides, Themis R., Arindam Raj, Tiffany H. Tseng, Hugh Xiao, Ryan Nguyen, Farrah S. Mohammed, Saiti Halder, et al. 2021. "Biocompatibility of Nanomaterials and Their Immunological Properties." *Biomedical Materials* 16 (4). https://doi.org/10.1088/1748-605X/abe5fa.

Lapusan, Radu, Raluca Borlan, and Monica Focsan. 2024. "Advancing MRI with Magnetic Nanoparticles: A Comprehensive Review of Translational Research and Clinical Trials." *Nanoscale Advances* 6 (9): 2234–59. https://doi.org/10.1039/d3na01064c.

Lazarević, Miloš M., Nenad L. Ignjatović, Qene Mahlet, Violet V. Bumah, Milena Radunović, Jelena Milašin, Dragan P. Uskoković, and Vuk Uskoković. 2024. "Biocompatible Germanium-Doped Hydroxyapatite

Nanoparticles for Promoting Osteogenic Differentiation and Antimicrobial Activity." *ACS Applied Nano Materials* 7 (8): 8580–92. https://doi.org/10.1021/acsanm.3c05974.

Li, Yan, Miaomiao Han, Yue Cai, Bing Jiang, Yuanxin Zhang, Biao Yuan, Feng Zhou, and Chongjiang Cao. 2022. "Muscle-Inspired MXene/PVA Hydrogel with High Toughness and Photothermal Therapy for Promoting Bacteria-Infected Wound Healing." *Biomaterials Science* 10 (4): 1068–82. https://doi.org/10.1039/d1bm01604k.

Liu, Gang, Magdalena Swierczewska, Seulki Lee, and Xiaoyuan Chen. 2010. "Functional Nanoparticles for Molecular Imaging Guided Gene Delivery." *Nano Today* 5 (6): 524–39. https://doi.org/10.1016/j.nantod.2010.10.005.

Liu, Zhe, Fabian Kiessling, and Jessica Gätjens. 2010. "Advanced Nanomaterials in Multimodal Imaging: Design, Functionalization, and Biomedical Applications." *Journal of Nanomaterials* 2010 (1): 1–15. https://doi.org/10.1155/2010/894303.

Ma, Weiwei, Qiufeng Zhao, Shilong Zhu, Xinyue Wang, Chuangchuang Zhang, Daming Ma, Na Li, and Yanyan Yin. 2024. "Construction of Glutathione-Responsive Paclitaxel Prodrug Nanoparticles for Image-Guided Targeted Delivery and Breast Cancer Therapy." *RSC Advances* 14 (18): 12796–806. https://doi.org/10.1039/d4ra00610k.

Ma, Yu, Yan Liu, Yurui Wang, Yanzhu Guo, Yuanyuan Li, Ruizhi Li, Xinzi Kong, Qize Han, Rong Wei, and Jing Wang. 2022. "The Nanocomposite System Comprising Folic Acid-Modified Graphene Quantum Dots Loaded with Evodiamine in the Treatment of Oral Squamous Cell Carcinoma." *Materials & Design* 220 (August): 110838. https://doi.org/10.1016/j.matdes.2022.110838.

Magesh, Vasanth, Ashok K. Sundramoorthy, and Dhanraj Ganapathy. 2022. "Recent Advances on Synthesis and Potential Applications of Carbon Quantum Dots." *Frontiers in Materials* 9. https://doi.org/10.3389/fmats.2022.906838.

Mai, Xuan-Dung, Yen Thi Hai Phan, and Van-Quang Nguyen. 2020. "Excitation-Independent Emission of Carbon Quantum Dot Solids." *Advances in Materials Science and Engineering* 2020 (December): 1–5. https://doi.org/10.1155/2020/9643168.

Martínez-González, Raquel, Joan Estelrich, and Maria Antònia Busquets. 2016. "Liposomes Loaded with Hydrophobic Iron Oxide Nanoparticles: Suitable T_2 Contrast Agents for MRI." *International Journal of Molecular Sciences* 17 (8). https://doi.org/10.3390/ijms17081209.

Martins, M. Bárbara A. F., M. Luísa Corvo, Paulo Marcelino, H. Susana Marinho, Gabriel Feio, and Alexandra Carvalho. 2014. "New Long Circulating Magnetoliposomes as Contrast Agents for Detection of Ischemia–Reperfusion Injuries by MRI." *Nanomedicine: Nanotechnology, Biology, and Medicine* 10 (1): 207–14. https://doi.org/10.1016/j.nano.2013.06.008.

Mitchell, Sharon, Ruixuan Qin, Nanfeng Zheng, and Javier Pérez-Ramírez. 2021. "Nanoscale Engineering of Catalytic Materials for Sustainable Technologies." *Nature Nanotechnology* 16 (2): 129–39. https://doi.org/10.1038/s41565-020-00799-8.

Molkenova, A., T. S. Atabaev, S. W. Hong, C. Mao, D-W Han, and K. S. Kim. 2022. "Designing Inorganic Nanoparticles into Computed Tomography and Magnetic Resonance (CT/MR) Imaging-Guidable Photomedicines." *Materials Today Nano* 18 (June): 100187. https://doi.org/10.1016/j.mtnano.2022.100187.

Mondal, Joydip, Yue Wu, Arpit Mishra, Farzan Akbaridoust, Ivan Marusic, Parthasarathi Ghosh, and Muthupandian Ashokkumar. 2024. "Bubble Oscillations at Low Frequency Ultrasound for Biological Applications." *Ultrasonics Sonochemistry* 104 (March): 106816. https://doi.org/10.1016/j.ultsonch.2024.106816.

Moyosore, Afodun Adam. 2022. "Application of Nanoparticles in Biomedical Imaging." In *Diverse Applications of Nanotechnology in the Biological Sciences*, 73–108. Apple Academic Press. https://www.taylorfrancis.com/chapters/edit/10.1201/9781003277255-3/application-nanoparticles-biomedical-imaging-afodun-adam-moyosore.

Murar, Madhura, Lorenzo Albertazzi, and Silvia Pujals. 2022. "Advanced Optical Imaging-Guided Nanotheranostics towards Personalized Cancer Drug Delivery." *Nanomaterials (Basel, Switzerland)* 12 (3). https://doi.org/10.3390/nano12030399.

Murugan, Ridhu Varshini, Vasanth Magesh, K. Vijayalakshmi, Raji Atchudan, Sandeep Arya, and Ashok K. Sundramoorthy. 2024. "Electrochemical Sensing of Vitamin C Using Graphene/ Poly-Thionine

Composite Film Modified Electrode." *Micro and Nanosystems* 16 (February). https://doi.org/10.2174/0118764029290865240209072023.

Murugan, Ridhu Varshini, Gokul Sridharan, Raji Atchudan, Sandeep Arya, Dhanraj Ganapathy, and Ashok k. Sundramoorthy. 2024. "Fluorometric Determination of Fe3+ Ions Using Green Synthesized Carbon Quantum Dots from Damask Rose Flowers." *Current Nanoscience* 20 (May). https://doi.org/10.2174/0115734137300899240509100717.

Nagarajan, Ramila D., J. Kavitha, Raji Atchudan, Sandeep Arya, and Ashok K. Sundramoorthy. 2023. "Electrochemical Analysis of Narcotic Drugs Using Nanomaterials Modified Electrodes A Review." *Current Analytical Chemistry* 19 (6): 440–47. https://doi.org/10.2174/1573411019666230622153225.

Nagarajan, Ramila D., Vasanth Magesh, Ashok K. Sundramoorthy, Preethika Murugan, Raji Atchudan, Dhanraj Ganapathy, Sandeep Arya, Abdulnasser Mahmoud Karami, and Mani Govindasamy. 2023. "Graphene Nanoribbons/Manganese Oxide Nanocomposite Modified Electrode for Detection of Antimicrobial Drug Nitrofurantoin." *Journal of Nanomaterials* 2023 (April): 1–13. https://doi.org/10.1155/2023/5484059.

Nagarajan, Ramila D., and Ashok K. Sundramoorthy. 2021. "Recent Trends in Fabrication and Applications of Wearable Bioelectronics for Early-Stage Disease Monitoring and Diagnosis." In *Macro, Micro, and Nano-Biosensors*, 357–81. Cham: Springer International Publishing. https://doi.org/10.1007/978-3-030-55490-3_18.

Nanda, Sitansu Sekhar, and Dong Kee Yi. 2024. "Recent Advances in Synergistic Effect of Nanoparticles and Its Biomedical Application." *International Journal of Molecular Sciences* 25 (6). https://doi.org/10.3390/ijms25063266.

Neogi, Newton, and Kristi Priya Choudhury. 2023. "Utilization of Metal Based Nanoparticles in Biomedical Imaging Technologies." https://catalog.lib.kyushu-u.ac.jp/ja/recordID/7157949/?repository=yes.

Pandey, Atul, Brajesh Kumar Singh, Ekambaram Gayathiri, Sundararajan Balasubramani, Saravanamoorthy Mutharasanallur Duraisamy, Ashish Kothari, and Dinesh K. Patel. 2024. "Nanoparticles in Biomedical and Clinical Research: A Current Perspective and Future Implications." In *Nanomaterials for Biomedical and Bioengineering Applications*, edited by Rabia Javed, Jen-Tsung Chen, and Ali Talha Khalil, 415–57. Singapore: Springer Nature Singapore. https://doi.org/10.1007/978-981-97-0221-3_17.

Paramasivam, Gokul, Vishnu Vardhan Palem, Simi Meenakshy, Lakshmi Krishnaa Suresh, Moumita Gangopadhyay, Santhy Antherjanam, and Ashok K. Sundramoorthy. 2024. "Advances on Carbon Nanomaterials and Their Applications in Medical Diagnosis and Drug Delivery." *Colloids and Surfaces. B, Biointerfaces*, no. 114032 (June): 114032. https://doi.org/10.1016/j.colsurfb.2024.114032.

Passaro, Antonio, Maise Al Bakir, Emily G. Hamilton, Maximilian Diehn, Fabrice André, Sinchita Roy-Chowdhuri, Giannis Mountzios, Ignacio I. Wistuba, Charles Swanton, and Solange Peters. 2024. "Cancer Biomarkers: Emerging Trends and Clinical Implications for Personalized Treatment." *Cell* 187 (7): 1617–35. https://doi.org/10.1016/j.cell.2024.02.041.

Peng, Xueqiang, Jianjun Fang, Chuyuan Lou, Liang Yang, Shaobo Shan, Zixian Wang, Yutong Chen, Hangyu Li, and Xuexin Li. 2024. "Engineered Nanoparticles for Precise Targeted Drug Delivery and Enhanced Therapeutic Efficacy in Cancer Immunotherapy." *Yao Xue Xue Bao = Acta Pharmaceutica Sinica*, May. https://doi.org/10.1016/j.apsb.2024.05.010.

Puzzo, Marianna, Marzia De Santo, Catia Morelli, Antonella Leggio, and Luigi Pasqua. 2024. "The Advent of Molecular Targeted Therapies against Cancer. Toward Multi-targeting Drugs through Materials Engineering: A Possible Future Scenario." *Small Science*, May. https://doi.org/10.1002/smsc.202400113.

Qiu, Meng, Ajay Singh, Dou Wang, Junle Qu, Mark Swihart, Han Zhang, and Paras N. Prasad. 2019. "Biocompatible and Biodegradable Inorganic Nanostructures for Nanomedicine: Silicon and Black Phosphorus." *Nano Today* 25 (April): 135–55. https://doi.org/10.1016/j.nantod.2019.02.012.

Qureshi, Zawar Alam, Hanan Dabash, Deepalekshmi Ponnamma, and M. K. G. Abbas. 2024. "Carbon Dots as Versatile Nanomaterials in Sensing and Imaging: Efficiency and Beyond." *Heliyon* 10 (11): e31634. https://doi.org/10.1016/j.heliyon.2024.e31634.

Rasheed, P. Abdul, Menon Ankitha, Vijayamohanan K. Pillai, and Subbiah Alwarappan. 2024. "Graphene Quantum Dots for Biosensing and Bioimaging." *RSC Advances* 14 (23): 16001–23. https://doi.org/10.1039/d4ra01431f.

Rezaei, Bahareh, Parsa Yari, Sean M. Sanders, Haotong Wang, Vinit Kumar Chugh, Shuang Liang, Shahriar Mostufa, et al. 2024. "Magnetic Nanoparticles: A Review on Synthesis, Characterization, Functionalization, and Biomedical Applications." *Small* 20 (5): e2304848. https://doi.org/10.1002/smll.202304848.

Riedl, Jakob M., Florian Moik, Tamara Esterl, Sarah M. Kostmann, Armin Gerger, and Philipp J. Jost. 2024. "Molecular Diagnostics Tailoring Personalized Cancer Therapy-an Oncologist's View." *Virchows Archiv: An International Journal of Pathology* 484 (2): 169–79. https://doi.org/10.1007/s00428-023-03702-7.

Ronkainen, Niina J., and Stanley L. Okon. 2014. "Nanomaterial-Based Electrochemical Immunosensors for Clinically Significant Biomarkers." *Materials* 7 (6): 4669–4709. https://doi.org/10.3390/ma7064669.

Rotz, Matthew W., Kayla S. B. Culver, Giacomo Parigi, Keith W. MacRenaris, Claudio Luchinat, Teri W. Odom, and Thomas J. Meade. 2015. "High Relaxivity Gd(III)-DNA Gold Nanostars: Investigation of Shape Effects on Proton Relaxation." *ACS Nano* 9 (3): 3385–96. https://doi.org/10.1021/nn5070953.

Sangeetha Dhanapalan, Vasanth Magesh, Raji Atchudan, Sandeep Arya, Deepak Nallaswamy and Ashok K. Sundramoorthy. 2024. "Fluorometric Determination of Fe3+ Ions Using Carbon Dots Prepared from the Extract of Singapore Cherry." *Current Nanoscience*, April. https://doi.org/10.2174/0115734137290117240404085227.

Sanità, Gennaro, Barbara Carrese, and Annalisa Lamberti. 2020. "Nanoparticle Surface Functionalization: How to Improve Biocompatibility and Cellular Internalization." *Frontiers in Molecular Biosciences* 7 (November): 587012. https://doi.org/10.3389/fmolb.2020.587012.

Santra, Santimukul, Charalambos Kaittanis, and J. Manuel Perez. 2010. "Cytochrome C Encapsulating Theranostic Nanoparticles: A Novel Bifunctional System for Targeted Delivery of Therapeutic Membrane-Impermeable Proteins to Tumors and Imaging of Cancer Therapy." *Molecular Pharmaceutics* 7 (4): 1209–22. https://doi.org/10.1021/mp100043h.

Shaghaghi, Mehran, Frederick C. Damen, Weiguo Li, Leon M. Tai, and Kejia Cai. 2024. "Induced Saturation Transfer Recovery Steady States (ISTRESS) for Proton Exchange Rate Mapping with CEST MRI, a Preliminary Study." *Magnetic Resonance Imaging* 109 (June): 264–70. https://doi.org/10.1016/j.mri.2024.03.034.

Sharma, Nitin, Balak Das Kurmi, Dilpreet Singh, Sidharth Mehan, Kushagra Khanna, Ritu Karwasra, Shobhit Kumar, et al. 2024. "Nanoparticles Toxicity: An Overview of Its Mechanism and Plausible Mitigation Strategies." *Journal of Drug Targeting*, February, 1–13. https://doi.org/10.1080/1061186X.2024.2316785.

Sher, Farooq, Imane Ziani, Mariam Hameed, Salman Ali, and Jasmina Sulejmanović. 2024. "Advanced Nanomaterials Design and Synthesis for Accelerating Sustainable Biofuels Production – A Review." *Current Opinion in Green and Sustainable Chemistry* 47 (100925): 100925. https://doi.org/10.1016/j.cogsc.2024.100925.

Shivalingam, C., L. Mohan, and D. Ganapathy. 2022. "Current Overview on the Role of Nanoparticles in Water Desalination Technology." *Current Analytical Chemistry*. https://www.ingentaconnect.com/content/ben/cac/2022/00000018/00000009/art00005.

Shu, Lei, Fangqin Fu, Zhengwei Huang, Ying Huang, Ping Hu, and Xin Pan. 2020. "Nanostructure of DiR-Loaded Solid Lipid Nanoparticles with Potential Bioimaging Functions." *AAPS PharmSciTech* 21 (8): 321. https://doi.org/10.1208/s12249-020-01847-1.

Singh, Samridhi, and Kirti Negi. 2024. "Transformation of Healthcare through Bioimaging and Bioinformatics." *International Journal of Pharma Professional's Research (IJPPR)* 15 (2): 81–103. https://doi.org/10.69580/IJPPR.15.2.2024.96-113.

Sridharan, Gokul, Raji Atchudan, Vasanth Magesh, Sandeep Arya, Dhanraj Ganapathy, Deepak Nallaswamy, and Ashok K. Sundramoorthy. 2023. "Advanced Electrocatalytic Materials Based Biosensors for Cancer Cell Detection – A Review." *Electroanalysis* 35 (9). https://doi.org/10.1002/elan.202300093.

Sridharan, Gokul, C. Joel Trinity Godwin, Raji Atchudan, Sandeep Arya, Mani Govindasamy, Sameh M. Osman, and Ashok K. Sundramoorthy. 2024. "Iron Oxide Decorated Hexagonal Boron Nitride Modified Electrochemical Sensor for the Detection of Nitrofurantoin in Human Urine Samples." *Journal of the Taiwan Institute of Chemical Engineers*, no. 105320 (January): 105320. https://doi.org/10.1016/j.jtice.2023.105320.

Sridharan, Gokul, Ridhu Varshini Murugan, Raji Atchudan, Sandeep Arya, and Ashok K. Sundramoorthy. 2024. "Electrochemical Detection of Dopamine Using Green Synthesized Gold Nanoparticles from *Strobilanthes Kunthiana's* Leaf Extract." *Nano LIFE*, July. https://doi.org/10.1142/s1793984424500156.

Sun, Leming, Hongmei Liu, Yanqi Ye, Yang Lei, Rehmat Islam, Sumin Tan, Rongsheng Tong, Yang-Bao Miao, and Lulu Cai. 2023. "Smart Nanoparticles for Cancer Therapy." *Signal Transduction and Targeted Therapy* 8 (1): 418. https://doi.org/10.1038/s41392-023-01642-x.

Sundramoorthy, Ashok K. 2024. "Selection of Best Biomarker for the Early Detection of Oral Squamous Cell Carcinoma." *Oral Oncology Reports* 9 (100197): 100197. https://doi.org/10.1016/j.oor.2024.100197.

Sundramoorthy, Ashok K., Mary Christina Senthil Kumar, Thirugnanam Ramya, Raji Atchudan, and Sandeep Arya. 2023. "Biocompatible Heteroatom Doped Carbon Dots for Cancer Cell Imaging and Analysis." *Oral Oncology Reports* 7 (100089): 100089. https://doi.org/10.1016/j.oor.2023.100089.

Sundramoorthy, Ashok K., Thomaati H. Vignesh Kumar, and Sundaram Gunasekaran. 2018. "Graphene-Based Nanosensors and Smart Food Packaging Systems for Food Safety and Quality Monitoring." In *Graphene Bioelectronics*, 267–306. Elsevier. https://doi.org/10.1016/b978-0-12-813349-1.00012-3.

Sundramoorthy, Ashok Kumar, Sara Mesgari, Jing Wang, Raj Kumar, Mahasin Alam Sk., Siew Hooi Yeap, Qing Zhang, Siu Kwan Sze, Kok Hwa Lim, and Mary B. Chan-Park. 2013. "Scalable and Effective Enrichment of Semiconducting Single-Walled Carbon Nanotubes by a Dual Selective Naphthalene-Based Azo Dispersant." *Journal of the American Chemical Society* 135 (15): 5569–81. https://doi.org/10.1021/ja312282g.

Titirici, Magda, Sterling G. Baird, Taylor D. Sparks, Shirley Min Yang, Agnieszka Brandt-Talbot, Omid Hosseinaei, David P. Harper, et al. 2022. "The Sustainable Materials Roadmap." *Journal of Physics: Materials* 5 (3): 032001. https://doi.org/10.1088/2515-7639/ac4ee5.

Wallyn, Justine, Nicolas Anton, Salman Akram, and Thierry F. Vandamme. 2019. "Biomedical Imaging: Principles, Technologies, Clinical Aspects, Contrast Agents, Limitations and Future Trends in Nanomedicines." *Pharmaceutical Research* 36 (6): 78. https://doi.org/10.1007/s11095-019-2608-5.

Walter, Andreas, Perrine Paul-Gilloteaux, Birgit Plochberger, Ludek Sefc, Paul Verkade, Julia G. Mannheim, Paul Slezak, et al. 2020. "Correlated Multimodal Imaging in Life Sciences: Expanding the Biomedical Horizon." *Frontiers in Physics* 8. https://doi.org/10.3389/fphy.2020.00047.

Wang, Bin, Yanbin Zhao, Yangyang Li, Junyan Yao, Shunjie Wu, Guoping Miu, and Chenglin Chu. 2024. "NIR-Responsive Magnesium Phosphate Cement Loaded with Simvastatin-Nanoparticles with Biocompatibility and Osteogenesis Ability." *RSC Advances* 14 (20): 13958–71. https://doi.org/10.1039/d4ra01079e.

Wang, Jie, Zongying Zhang, Rongxiang Liang, Wujun Chen, Qian Li, Jiazhen Xu, Hongmei Zhao, and Dongming Xing. 2024. "Targeting Lymph Nodes for Enhanced Cancer Vaccination: From Nanotechnology to Tissue Engineering." *Materials Today. Bio* 26 (June): 101068. https://doi.org/10.1016/j.mtbio.2024.101068.

Wang, Tingting, Di Zhang, Dong Sun, and Jingkai Gu. 2020. "Current Status of *in Vivo* Bioanalysis of Nano Drug Delivery Systems." *Journal of Pharmaceutical Analysis* 10 (3): 221–32. https://doi.org/10.1016/j.jpha.2020.05.002.

Wegner, K. David, and Niko Hildebrandt. 2015. "Quantum Dots: Bright and Versatile *in Vitro* and *in Vivo* Fluorescence Imaging Biosensors." *Chemical Society Reviews* 44 (14): 4792–4834. https://doi.org/10.1039/c4cs00532e.

Welsher, Kevin, Sarah P. Sherlock, and Hongjie Dai. 2011. "Deep-Tissue Anatomical Imaging of Mice Using Carbon Nanotube Fluorophores in the Second near-Infrared Window." *Proceedings of the National Academy of Sciences of the United States of America* 108 (22): 8943–48. https://doi.org/10.1073/pnas.1014501108.

Yang, Zhongwei, Jiawei Guo, Longwei Wang, Jian Zhang, Longhua Ding, Hong Liu, and Xin Yu. 2024. "Nanozyme-Enhanced Electrochemical Biosensors: Mechanisms and Applications." *Small* 20 (14): e2307815. https://doi.org/10.1002/smll.202307815.

Yao, Jun, Mei Yang, and Yixiang Duan. 2014. "Chemistry, Biology, and Medicine of Fluorescent Nanomaterials and Related Systems: New Insights into Biosensing, Bioimaging, Genomics, Diagnostics, and Therapy." *Chemical Reviews* 114 (12): 6130–78. https://doi.org/10.1021/cr200359p.

Yetisgin, Abuzer Alp, Sibel Cetinel, Merve Zuvin, Ali Kosar, and Ozlem Kutlu. 2020. "Therapeutic Nanoparticles and Their Targeted Delivery Applications." *Molecules* 25 (9). https://doi.org/10.3390/molecules25092193.

Yu, Rongxin, Hongping Chen, Jian He, Zhichao Zhang, Jiajing Zhou, Qinqin Zheng, Zhouping Fu, et al. 2024. "Engineering Antimicrobial Metal-Phenolic Network Nanoparticles with High Biocompatibility for Wound Healing." *Advanced Materials* 36 (6): e2307680. https://doi.org/10.1002/adma.202307680.

Yuan, Hui, Hanyu Liang, Peidong Hou, and Juan Li. 2021. "Advanced Nanomaterials for Multimodal Molecular Imaging." *Chemical Research in Chinese Universities* 37 (4): 840–45. https://doi.org/10.1007/s40242-021-1196-1.

Zawidlak-Węgrzyńska, Barbara, Joanna Rydz, Marta Musioł, and Aneta Radziwon-Balicka. 2024. "Polymer-Drug Anti-Thrombogenic and Hemocompatible Coatings as Surface Modifications." *Pharmaceutics* 16 (3). https://doi.org/10.3390/pharmaceutics16030432.

Chapter 3.2

Nanomaterials in Regenerative Medicine: Recent Developments in Smart Biomimetic Scaffolds

Soniya Rani
and Prem Shankar Gupta*

Department of Pharmacology, GITAM School of Pharmacy, GITAM (Deemed to be University), Hyderabad, Telangana, India
Department of Pharmaceutics, Teerthanker Mahaveer College of Pharmacy, Teerthanker Mahaveer University, Moradabad, Uttar Pradesh, India

Abstract

Regenerative Medicine (RM) combinestissue engineering (TE), cell transplantation, nanotechnology, and biochemistry to restore the normal function of human cells, tissues, or organs compromised by congenital defects, disease, or senescence. RM has made significant progress in addressing ethical issues related to embryonic stem cells and obstacles such as organ shortages using biomaterials, scaffolds, and gels. These biomaterials play a crucial role in TE within RM because they are biocompatible and non-toxic, making them ideal for tissue regeneration. Recent advancements in nanotechnology and biomaterials have shown promise/applications in several medical fields, such as cardiovascular, orthopedic, and dental. TE uses 3D porous scaffolds to promote cell proliferation and differentiation, which are essential for developing functional tissue. Synthetic and natural polymers offer mechanical stability and biocompatibility which are crucial for effective TE. Nanoscaffolds (NS), mimicking the extracellular matrix at the nanoscale, enhance cellular behavior and enable the controlled release of bioactive molecules. Progress in biomimetic nanomaterials, such as biodegradable polymer nanofibrous scaffolds and hydrogel-based nanoparticles, provide mechanical durability, biological activity, and regulated therapeutic release. Additionally, stimuli-responsive nanomaterials react to changes in pH, temperature, light, or magnetic fields, thereby improving the accuracy of treatments. Techniques such as electrospinning and 3D printing are used to enhance the design of nanomaterials by imitating the architectures seen in natural tissues. This chapter elucidates the significance of scaffolds fabricated from polymers, metals, and ceramics for facilitating tissue regeneration. In addition, the chapter discusses biomimetic nanomaterials and stimuli-sensitive nanomaterials (scaffold) in the field of RM. Ultimately, advancements in nanotechnology and its potential to deliver

---

* Corresponding Author's Email: prem.shankar@mmumullana.org; premsgupta.rs.bme17@iitbhu.ac.in

In: Emerging Nanomaterials in Biomedical Science
Editors: Eepsita Priyadarshini and Ajith Manayil Parambil
ISBN: 979-8-89530-600-0

effective and precise treatment for tissue damage and degenerative illnesses has also been highlighted.

Keywords: nano scaffolds, nano biomaterials, polymers, regenerative nanomedicine, tissue regeneration

1. Introduction

The concept of "Regenerative Medicine" (RM) was first coined by Leland Kaiser in 1992 through his publication. Currently, "RM" refers to a burgeoning field of medicine dedicated to the replacement, manipulating, repairing, engineering, or restoration of human cells, tissues, or organs that have been compromised, damaged, or lost due to congenital abnormalities, diseases, or aging process in order to reinstate normal functionality (Ip 2012). According to several RM experts, the word "RM" was introduced by William Haseltine in 1999 at a symposium held in Lake Como. Haseltine used this phrase to describe the fields of tissue engineering (TE), cell transplantation, nanotechnology, and biochemistry (Sampogna, Guraya, and Forgione 2015). RM technology has significant potential to improve the quality of life of humans by treating diseases (Santa Cruz-Pavlovich et al. 2024). There is an ongoing debate regarding the application of RM, questioning whether it should be limited to disease treatment or extended to pursue indefinite human lifespan extension, potentially venturing into transhuman or posthuman realms (Ip 2012). Additionally, the deficit and increasing need for organs has spurred progress in molecular and cellular biology, leading to novel opportunities in RM. Potential strategies to address organ and tissue shortages encompass cell transplantation and TE. These methodologies focus on the development of functional tissues and organs *in vitro* for transplant purposes and the application of cells for therapeutic interventions.

Approximately seventy years ago, a team led by Dr. Joseph Murray, John Merrill, and J. Hartwell Harrison performed the first successful organ transplant in Boston (Tan and Merchant 2019). This occurrence signified the inception of organ transplantation as a specialized field and represented a ground-breaking accomplishment that initiated a novel area of organ replacement therapy. This enabled for the first time the comprehensive treatment of a patient with a critical organ-related illness. During the late 1950s and 1960s, pioneering advancements in cellular therapies, including bone marrow transplants, were realized. (Watson and Dark 2012). For the first time, a patient with leukemia was treated with the implantation of allogeneic marrow in Seattle, by Dr. Don Thomas (Thomas et al. 1957). In 1968, Dr. Robert Good, who led the technique, successfully cured an immunodeficient patient at the University of Minnesota by performing an allogeneic bone marrow transplant from the patient's sister (Gatti et al. 1968). Since then, numerous attempts have been made to achieve successful organ transplantation, cell treatment, and gene therapy. However, these efforts have been unsuccessful over several decades. Nevertheless, the significant scientific and clinical interest in these fields has paved the way for the initial wave of accomplishments in RM, which are currently being implemented in clinical settings (Morris 2004; Daar 1998; Nielsen 1998).

RM frequently employs hyperosmotic glucose solutions, platelet-enriched plasma, patient-derived mesenchymal stem cells, and various allogenic biological products in its treatment strategies. RM therapies are clearly beneficial for treating common musculoskeletal ailments at level one, and there is increasing evidence supporting their effectiveness for spinal disorders

(Chang Chien and Stogicza 2021). Presently, human stem cells are highly regarded for their regenerative capabilities and their potential application in treating various diseases and disabilities. However, the use of embryonic stem cells (ESCs) in research and therapy is a topic of significant ethical debate, given the moral considerations associated with the status of embryos. The process of isolating ESCs necessitates the destruction of embryos, which many equate to taking a human life. Therefore, the ethical and philosophical debate around the RM is very tough (Ip 2012). Further, Leland Kaiser suggested using biomaterials to accelerate the healing process or repair missing or damaged body tissue (Ip 2012). Because of limitations in traditional approaches and breakthroughs in the field of biomaterials (Figure 1), they have revolutionized and propelled RM and TE to new heights (Stupp 2005). The chemical nature of biomaterials make it an excellent substrate for use in RM. Every biomaterial category has distinct characteristics that make it appropriate for specific applications. The primary characteristic of biomaterials is their behavior when implanted in living organisms, namely their ability to avoid triggering immune rejection reactions or generating harmful by-products in other animal tissues (Troy et al. 2021). In addition, they possess complementary advantages, such as the capacity to be included into a 3D framework that can be used to accommodate the initial growth and multiplication of a specific cell category (Loh and Choong 2013). In addition to traditional scaffold fabrication methods, contemporary technologies like hydrogels and cryogels offer interconnected porous networks that enhance cell migration and growth. These gel matrices, made from either natural or synthetic biomaterials, have demonstrated advantages in numerous tissue engineering applications investigated across various studies (Carriero et al. 2023). Moreover, biomaterials have shown their use in the treatment of several diseases such as cardiovascular, orthopedic, and dental implants (Abraham and Venkatesan 2023). Additionally, recent advancements in RM have demonstrated that, in addition to using stem cells from one's own body and from other individuals, stem cells (as mentioned before) might be a feasible alternative. The interactions between stem cells and biomaterials have demonstrated importance in RM, thereby offering substantial potential for future applications. This chapter elucidates the use of nanostructured biomaterials for tissue regeneration and TE.

Nano-biomaterials are the substances that are deliberately engineered into nanoscale structures and serve as a basis for composite fabrication due to their advantageous properties, such as biocompatibility, non-toxicity, and ease of degradation/elimination (Pandya et al. 2022). Given that cellular interactions occur at the nanoscale level in the biological milieu, the use of nano biomaterials provides an advantage in altering cellular and matrix functions to better imitate native tissues and organs (Chaudhury et al. 2014). Nanomaterials employed in RM are predominantly produced using one or more techniques reviewed elsewhere (Baig, Kammakakam, and Falath 2021). The application of nanotechnology enables the bottom-up synthesis of new nano materials that display unique characteristics at the nanoscale level compared to their larger-scale bulk forms. Instances of such nanomaterials utilized in RM encompass self-assembled structures, polymeric and inorganic NPs, dendritic polymers, quantum dots (QDs), carbon nanotubes (CNTs), and layer-by-layer (LbL) assemblies (Khademhosseini et al. 2010). Additionally, the materials employed in RM and TE can be classified into three primary groups: polymers, metallic substances, and ceramic materials (Figure 1).

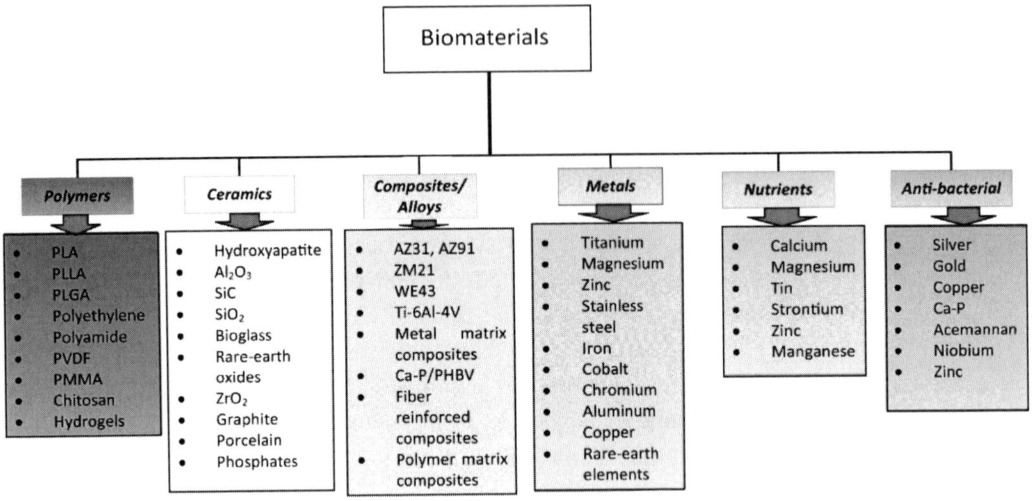

Figure 1. Biomaterials used in regenerative medicine (RM).

2. Importance of Scaffolds in Guiding Tissue Regeneration

To achieve fully functional tissues, it is essential to ensure adequate mechanical strength and structural characteristics. TE aims to enable distinct biochemical activities by utilizing cells embedded in artificial support structures. Researchers have recognized that generating cell-laden tissue-engineered constructs can maintain cell proliferation and differentiation. Consequently, a key objective in TE is the designing of such constructs that promote cellular activities (Ai et al. 2012).

An adequate supply of nutrients and oxygen is necessary for the survival and proper upkeep of transplanted cells. Insufficient supply results in the persistence of only a restricted number of pre-seeded cells or those that migrate into the scaffold from neighbouring tissues. In conventional 2D cell culture, the growth of cells is initially fast but stops once they reach confluency. To address the constraints of supply, researchers have fabricated 3D porous materials (Milián et al. 2024; Qian et al. 2018). These materials provide a greater surface area for cell attachment and proliferation compared to 2D structures. They are also suitable for creating 3D cell constructions that mimic the shapes and functions of real tissues. The 3D structures enable the effective spread of nutrients, oxygen, and bioactive substances, promoting the survival of groups of cells for long durations (Kim et al. 2013). Furthermore, the incorporation of extracellular matrix (ECM) components and soluble factors within these constructs enhances angiogenesis and the differentiation of encapsulated stem cells, thus addressing certain limitations of traditional TE approaches. Neovascularization, which is the formation of a rapid vascular network at the site of tissue transplantation, is vital for supplying stem cells with necessary nutrients, like the processes seen in normal physiological growth and wound healing. Growth factors like basic fibroblast growth factor (bFGF) have been shown to promote angiogenesis (Polverini 1995); however, owing to its short half-life and instability in the biological milieu, it must be encapsulated in carriers like alginate, gelatin, or poly (ethylene-co-vinyl acetate) for controlled release and prolonged effectiveness. Ideally, a bioresorbable carrier that integrates with tissue development would function both as a framework for tissue regeneration and as a medium for sustained release (Edelman et al. 1991).

Engineered materials offer benefits for TE due to their adjustable chemical and physical characteristics, including porosity and mechanical strength. Polymeric scaffolds offer a complex internal architecture that supports cell attachment and maintains differentiation without hindering proliferation (Gunatillake, Adhikari, and Gadegaard 2003). An optimal scaffold should exhibit the following attributes (Soman et al. 2012):

1. Surface characteristics that facilitate cell attachment, proliferation, growth, and specialization.
2. Biocompatibility
3. Sufficient mechanical characteristics to withstand stressful environment under *in vivo* conditions.
4. Elevated porosity and a substantially large surface-area-to-volume ratio with interlinked pores to enhance the transport of nutrients and waste products.

Research has shown that cell growth in 3D scaffolds outperforms 2D scaffolds, with sustained proliferation over longer periods (Xie, Yang, and Kniss 2001). Developing materials that control cellular behavior is essential for fabricating effective tissue engineering scaffolds. Implanted biomaterials should replicate the structure and biological functions of the native ECM to induce a positive immune response (Rosenberg 1963). ECM proteins such as collagen and elastin, which are composed of nanoscale fibers (60-200 nm), significantly influence cell behavior. Nanoscale features enhance cell adhesion and other cell activities, making artificial nanoscale fibers promising in biomaterials and TE. The nanoscale size of cell membrane receptors, including integrins, is essential for cell interactions with scaffold materials (Webster et al. 2001).

Successful cell proliferation in 3D scaffolds requires an environment rich in oxygen and nutrients. Self-assembling molecules forming artificial scaffolds provide suitable support for cell growth and facilitate the diffusion of soluble factors and cell migration. These extensively hydrated matrices guarantee the viability of substantial cell populations for prolonged durations (Edelman et al. 1991). 3D scaffolds can differentiate progenitor cells into specific lineages by incorporating soluble factors or delivering chemicals, hormones, and growth factors. Designing patterned scaffolds capable of developing multiple lineages and hybrid organ structures offers a promising alternative for TE (Rosenberg 1963). Metal scaffolds made of titanium, are biocompatible and ideal for hard-tissue applications, including the differentiation of dental pulp progenitor cells. Coating these metal scaffolds with biological compounds can improve their effectiveness in promoting osteogenic differentiation (Edelman et al. 1991; Mehrabani et al. 2013).

Organic material-based scaffolds create a biomimetic environment conducive to stem cell activity, fostering the regeneration of bone, cartilage, and adipose tissue (Rosenberg 1963). These scaffolds can be optimized with bioactive agents to direct the fate of stem cell. Synthetic polymers provide enhanced control over mechanical properties and degradation rates, supporting the differentiation of progenitor cells into functional tissues. Additionally, the controlled release of regulatory molecules from synthetic scaffolds can further refine spatial organization and stem cell differentiation (Hosseinkhani et al. 2006; Iwakura et al. 2001).

Consequently, the integration of scaffold materials with stem cell technology presents significant potential for tissue regeneration advancements.

Organic material-based scaffolds provide a biomimetic environment for stem cells, enabling regeneration of bone, cartilage, and adipose tissue. These scaffolds can be functionalized with biological agents to influence the fate of stem cell. Synthetic polymers offer greater control over mechanical degradation, facilitating the differentiation of progenitor cells into functional tissues. The controlled release of regulatory molecules from synthetic scaffolds can further improve spatial organization and stem cell differentiation (Hosseinkhani et al. 2006; Iwakura et al. 2001). Therefore, the integration of scaffolding materials with stem cell technologies holds immense potential for advancing tissue regeneration.

2.1. Polymers

Both natural and synthetic polymers are essential in RM, having a multitude of microscopic repetitive units. These biomaterials offer various applications that extend beyond the fields of tissue regeneration and RM. For instance, polymethyl methacrylate (PMMA) is utilized in orthopedic cements (Arora et al. 2013), polyglycolic acid (PGA) in bioresorbable surgical sutures (Manoukian et al. 2019), poly(glycolide-co-lactide) (PGLA) in osteosynthetic screws (Hedelin et al. 2020), and polyvinyl siloxane in dental prosthetics (Ghahremanloo et al. 2017). In addition, poly (hydroxyethyl methacrylate) (PHEMA) is also used in the production of soft contact lenses (Zare et al. 2021).

The synthetic polymers commonly employed in RM encompass silicone elastomer (SE), polyethene (PE), polypropene (PP), poly (ethylene terephthalate) (PET), polytetrafluoroethylene (PTFE), poly (methyl methacrylate) (PMMA), polyvinyl chloride (PVC), poly (2-hydroxyethyl methacrylate) (PHEMA), and poly (ethylene glycol) (PEG). Poly (methyl methacrylate) (PMMA), a hydrophobic polymer, is utilized as bone cement in orthopedic procedures and, due to its superior optical property, is also used as a material for intraocular lenses (IOLs) and rigid contact lenses manufacturing (Ramanathan et al. 2024; Samir et al. 2019). Poly (2-hydroxyethyl methacrylate) (PHEMA) is cross-linked with ethylene glycol dimethacrylate (EGDMA) to ensure dimensional integrity in soft contact lenses. Incorporating minute quantities of poly (methacrylic acid) enhances the wettability of the contact lenses (Arora et al. 2013).

Polymers, including polyacrylic acids, are employed in dental cement and serve as mucoadhesive agents in mucosal drug delivery systems. Polypropylene (PP) is a durable and rigid isotactic crystalline polymer utilized in surgical sutures and hernia repair (Saha et al. 2022). Polyvinyl alcohol (PVA) is used in biomedical settings for various purposes, such as manufacturing tubing and blood storage bags (Abbondandolo and Brewer 2024). PLGA (poly lactic-co-glycolic acid), a bioresorbable polymer, is employed in absorbable surgical sutures, pharmaceutical delivery systems, and orthopedic fixation devices (Kuperkar et al. 2024). During degradation, it generates harmless metabolites like glycolic acid and lactic acid. Tetrafluoroethylene and hexafluoropropylene (referred to as FEP Teflon) are some of the co-polymers synthesized from various monomeric units and are crucial in biomedical research. They are specifically used to manufacture tubing connections and catheters (Cardoso et al. 2018).

Biomedical research and RM are also significantly influenced by natural polymers. Alginate, collagen, hyaluronic acid, and fibrin gels are some of the most highly investigated biomaterials. Alginate, composed of L-guluronic acid and D-mannuronic acid residues, undergoes gelation when exposed to calcium ions, rendering it appropriate for encapsulating growth factors or specific cell types (Mahheidari et al. 2024). Alginate is efficacious in the process of cartilage repair and regeneration (Milián et al. 2024), as well as in the preservation of the chondrocyte phenotype and long-term storage of chondrocyte for a duration of up to 8 months (Häuselmann et al. 1994).

Collagen has significant potential for skin regeneration because of its inherent presence in the skin. The use of bovine collagen, in conjunction with human keratinocytes and fibroblasts, has shown favorable outcomes in the treatment of severe burns (Kondej et al. 2024). Collagen-based bi-layered artificial skin, Integra, which consists of collagen and glycosaminoglycans, is used for the purpose of skin regeneration (Taupin, Gandhi, and Saini 2023). Another example is Matriderm, which employs bovine dermal collagen types I, III, and elastin (Dickson et al. 2023).

Hyaluronan can undergo chemical modifications to enhance its stability, despite its drawbacks, such as high-water solubility and rapid removal from host tissue. HYAFF 11 is a degradable scaffold used for cartilage healing (Costa et al. 2023). It was created by esterifying free carboxylic groups with benzyl alcohol. Fibrin, a crucial component in the wound healing process, may serve as a framework for tissue regeneration. Tissucol and Tisseel, a type of modified fibrin gel, has enhanced resistance to biodegradation and promotes the growth of ECM by chondrocytes (Wang et al. 2022). Cellulose, agarose, chitosan, and gelatin, among other natural polymers, show potential as scaffolding materials in RM (Samrot et al. 2023).

2.2. Metals

Metals are integral to RM and exert a significant economic influence on the biomaterial industry, specifically within RM and TE. Frequently utilized metals in these contexts encompass stainless steel (316, 316L), silver, tantalum, cobalt, F-75, and assorted alloys of titanium (Ti), chromium (Cr), cobalt (Co), and molybdenum (Mo). These metals are extensively utilized to support and repair structural tissues, particularly bone (Muster 1992). However, metallic biomaterials present certain challenges, such as biocompatibility issues, susceptibility to corrosion, potential physiological environment changes at the implant site, and mechanical strength discrepancies compared to natural tissue. In RM, metallic implants are used in various forms, including staples, plates, and wires, as well as in specialized implants for dental, penile, and facial reconstruction applications (Bazaka et al. 2021).

2.2.1. Ceramics

Ceramics are distinguished by excellent biocompatibility, resistance to corrosion and low electrical and thermal conductivity, rendering them ideal candidates for implants in RM. (Vaiani et al. 2023). Ceramics, vitreous materials, and glass-ceramics are classified under the extensive group of inorganic, non-metallic compounds. Frequently utilized ceramic biomaterials in RM and biomedical applications encompass alumina, calcium aluminate, titanium dioxide, calcium phosphates, carbon-based materials, and bio glass. Ceramics are extensively utilized in biomedical applications, including the production of eyewear, diagnostic devices, laboratory equipment, and temperature measurement instruments (Sáenz et al. 1999). In bone regeneration, hydroxyapatite (HAp) materials, both alone and in combination with other ceramics, have shown excellent performance. However, ceramics also have disadvantages, such as complexity in processing and fabrication along with low impact resistance. Ceramic biocompatible materials are predominantly utilized for dental components, surface coatings, bone grafts, otologic implants, medical instruments, and various prosthetic devices. Moreover, ceramic-based composites are employed in heart valve prostheses, knee replacements, and hip arthroplasties. (Nandhini, Sudhakar, and Premkumar 2021).

2.4. Nano Scaffolds (NS's)

Nano scaffolds represent a substantial advancement in the areas of RM and TE, providing unparalleled promise for the restoration and renewal of damaged tissues and organs. These constructs replicate the nanoscale dimensions and characteristics of the ECM, as exemplified by the PLGA50-64K nano-construct, which promotes osteogenesis (Wei et al. 2007). They are specifically intended to promote cell development and specialization. Their nanometric architecture closely replicates the complex arrangement of actual cells, which is crucial in promoting cellular adhesion, growth, and motility (Smith et al. 2009). Cellulose, agarose, chitosan, and gelatin, among other natural polymers, show potential as scaffolding materials in RM (Perez-Puyana et al. 2020).

A significant advantage of nanoscale systems is their ability to enhance the structural properties and bioactivity of the scaffold substances. Nanofibers can be fabricated utilizing electrospinning, a technique that generates fibers with desired diameters spanning from several tens to several hundreds of nanometers. These nanofibers can mimic the collagen fibrils found in native extracellular matrices, establishing an optimal milieu for tissue regeneration (Zulkifli et al. 2023). The nano-systems possess an elevated surface-area-to-volume ratio, facilitating a higher concentration of bioactive compounds, including growth factors and peptides. These compounds can be incorporated into the matrix to enhance their tissue repair potential (Hasan et al. 2018).

The efficacy of NS in clinical settings is contingent upon their biocompatibility and biodegradability, which are the essential parameters. Polycaprolactone (PCL), polylactic acid (PLA), and their copolymers are frequently utilized in the fabrication of nanocarriers due to their favorable degradation properties and minimal cytotoxicity (Castro et al. 2023). Natural polymers such as collagen, gelatin, and chitosan are often used because they are naturally compatible with living organisms and have the capacity to facilitate the adherence and development of cells (Reddy et al. 2021). In addition, the use of nanomaterials such as graphene oxide and carbon nanotubes in scaffolds may highly improve their mechanical strength and electrical conductivity. This is especially advantageous for the development of brain and cardiac tissues (Kandhola et al. 2023).

NS has utility that goes beyond providing plain structural support. These scaffolds can be modified with certain ligands to guide cell activity and stimulate specific biological reactions (Alkhursani et al. 2022). An instance of enhancing nanoparticle efficacy involves modifying their surface with RGD (arginine-glycine-aspartic acid) peptides, which can augment cellular adhesion and proliferation (Qian et al. 2018). In addition, NS can be designed to provide medicinal substances in a precise and regulated way. Nanoparticles integrated into the scaffold framework facilitate targeted and sustained release of therapeutics, growth factors, or genetic material, thereby creating a favorable environment for tissue repair and regeneration (Habibzadeh et al. 2022; Perán et al. 2012).

Progress in nanotechnology has enabled the development of sophisticated NS that can respond to environmental clues (Zhang et al. 2024). These scaffolds can modify their physical or chemical characteristics in the presence of changes in pH, temperature, or certain biomolecules. This enables them to deliver tailored assistance that is appropriate for tissue repair. In addition, the incorporation of biosensing components into NS enables continuous monitoring of tissue regeneration processes, providing vital information that can be used to enhance treatment approaches (Qu et al. 2024).

In summary, NS has great potential for advancing the science of RM and TE. Owing to their ability to precisely replicate the native extracellular matrix and their versatility in material composition and functionalization (Qian et al. 2018), they are an optimal platform for facilitating tissue repair and regeneration. Continued exploration and progress in this domain are anticipated to yield more sophisticated nanostructured systems, which have the potential to revolutionize the methodology for addressing diverse tissue injuries and degenerative conditions.

3. Biomimetic Nanomaterials in RM

Restoration of function in injured tissues and organs has been a key focus of biomedical research, with particular emphasis on tissue regeneration. A multitude of literature sources have documented the latest progress in biomimetic nanomaterials (Figure 2) (Gallo and Villasante 2023; Ciulla et al. 2023; Gong et al. 2022), demonstrating substantial potential in improving biological applications ranging from diagnostics to cancer therapy. Biomimetic nanomaterials are specifically engineered to imitate the natural ECM to achieve tissue repair and regeneration. These materials offer a milieu that fosters cellular attachment, proliferation, and differentiation, which are crucial processes for regenerative applications (Liu et al. 2023).

A very innovative method entails using nanofibrous scaffolds made from biodegradable polymers such as PLGA and PCL (Bazgir et al. 2021; Zhang et al. 2023). When biologically active molecules like trophic factors or polypeptides are incorporated into these matrices, they can replicate the mechanical and biochemical cues of the native extracellular matrix. This, in turn, stimulates cellular processes that contribute to tissue regeneration (Igwe Idumah 2022; Guo et al. 2015). A novel approach to creating biomimetic biomaterials involves the use of hydrogel-based nanoparticles (Lin et al. 2019). Hydrogels offer a moist three-dimensional milieu that enhances cellular viability and functionality, attributable to their elevated water content and modifiable mechanical properties. Researchers have improved the mechanical strength and bioactivity of hydrogels by adding nanoparticles like silica or gold. Gold nanostructures can be utilized to augment the electrical conductivity of hydrogels, thereby promoting the regeneration of electrically responsive tissues such as myocardial and neural tissues (Esmaeili et al. 2024). Furthermore, these nanocomposite hydrogels can be optimized or tailored to deliver therapeutic agents in a precisely regulated manner, thereby augmenting the regenerative process (Chen et al. 2023).

Nanomaterials play an essential role in delivering genes that promote tissue regeneration (Hosseinkhani et al. 2023). Lipids, polymers, or inorganic substances in the form of nanocarriers can be used to deliver nucleic acids (DNA, RNA) to specific cells, thereby regulating gene expression to promote tissue healing (Mendes et al. 2022). siRNA-loaded nanoparticles have been used to suppress genes that impede tissue regeneration, thus facilitating the healing process (Mottaghitalab et al. 2017). The combination of gene therapy with biomimetic scaffolds has a synergistic effect that greatly enhances the results of tissue regeneration (Esmaeili et al. 2023).

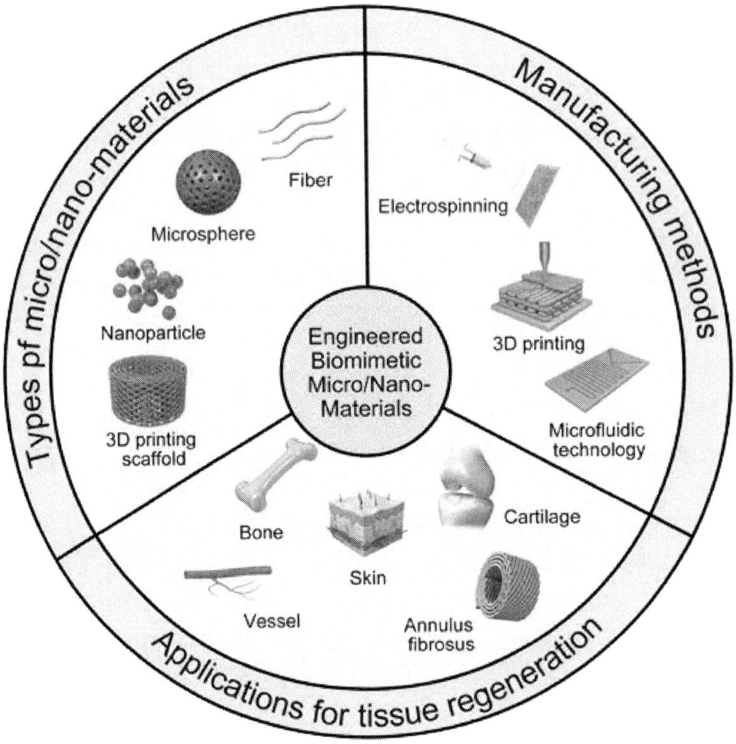

Figure 2. Schematic demonstrating the development of bioengineered biomimetic micro- and nanomaterials tailored for tissue regeneration. It highlights various micro and nanostructures such as fibers, nanoparticles, microspheres, and 3D-printed scaffolds. The biomedical applications of these materials are exemplified through their use in bone repair, annulus fibrosus (AF) repair, and vascular grafts. Additionally, manufacturing techniques for these diverse micro and nanomaterials are presented (Han, Meng, et al. 2023) (CC BY 4.0 license).

Moreover, the surface properties of biomimetic nanoparticles can be engineered to modulate cellular interactions. Various surface alteration methods, such as plasma treatment and chemical grafting, can be employed to introduce functional groups that enhance cellular behaviors such as adhesion and proliferation (Bhatt et al. 2023; Wieszczycka et al. 2021). For instance, the addition of RGD peptides, which are recognized by integrins available on cell surface, to nanomaterial surfaces has been shown to greatly assist in augmenting cell attachment and spreading, resulting in enhanced tissue regeneration (Kumar et al. 2023).The incorporation of biomimetic nanoparticles into the domains of regenerative nanomedicine and TE is further enhanced by advances in nanotechnology and material science. Electrospinning, 3D printing, and self-assembly are advanced methods that provide meticulous control over the structure and composition of nanomaterials. This allows for the development of highly ordered structures that precisely resemble the hierarchical organization seen in natural tissues (Chen et al. 2020). The use of these advanced fabrication techniques improves the capabilities and functionalities of nanomaterials, enhancing their efficacy in directing cellular activity and facilitating tissue regeneration (Bertsch et al. 2023). Therefore, the incorporation of biomimetic nanomaterials into tissue regeneration strategies provides a versatile and novel method for improving restoration and regrowth of injured tissues. These nanomaterials can greatly transform RM by imitating the natural ECM, offering mechanical and bioactive assistance, enabling gene delivery, and allowing precise control over material characteristics. Further investigations are necessary to examine the relationship between the characteristics of nanomaterials and the reactions of cells to them. This will help improve and advance these novel methods, making them more suitable for use in therapeutic settings.

4. Stimuli Responsive Nanomaterials for RM

The advent of stimuli-reactive nanomaterials has profoundly revolutionized the domain of tissue regeneration by introducing innovative methods for the restoration and revitalization of damaged tissues. Advanced materials capable of responding to specific biological or local environmental cues such as pH, temperature, light, or magnetic fields enable precise modulation of therapeutic delivery and cellular activities (Municoy et al. 2020). One promising approach is the use of pH-responsive nanoparticles that take advantage of the acidic milieu in injured or diseased tissues to selectively release therapeutic drugs at the desired location (Han, Yuan, et al. 2023; Haidari et al. 2021). For instance, a pH-responsive intelligent hydrogel (Figure 3) embedded with silver nanoparticles has demonstrated effectiveness in managing infected lesions, minimizing systemic toxicity, and enhancing the therapeutic efficacy (Kumar et al. 2023).

The fabrication of thermo-responsive nanomaterials is an advanced technique used in RM and TE (Sponchioni, Capasso Palmiero, and Moscatelli 2019). Poly(N-isopropylacrylamide) (PNIPAAm) and similar materials exhibit phase transitions at certain temperatures, enabling precise release of growth factors or other bioactive compounds in response to local temperature variations (Nagase 2021). This characteristic is especially advantageous in hyperthermia therapies because it allows targeted heating to induce drug release. This promotes tissue regeneration while minimizing collateral damage to nearby healthy tissues (Gu et al. 2023).

Photo-responsive nanomaterials have received considerable interest in the fields of RM and TE (Municoy et al. 2020). These substances, such as gold nanoparticles (Mendes et al. 2020) and up-conversion nanoparticles (Gültekin et al. 2024), have the ability to transform light into heat or reactive oxygen species when exposed to certain wavelengths. This feature enables precise removal of unhealthy cells or deliberate discharge of therapeutic substances, thereby facilitating restoration and rejuvenation of tissues. Gold nanorods have been used in photothermal treatment to specifically eradicate cancer cells and subsequently induce an immunological response that facilitates tissue regeneration (Nanda et al. 2022).

Magnetically sensitive nanoparticles provide a new and innovative approach for improving tissue regeneration (Inam et al. 2024). Superparamagnetic iron oxide nanomaterials (SPIONs) can be directed to specific tissue sites within the body through the application of an external magnetic field (Friedrich, Cicha, and Alexiou 2021). Subsequently, these nanoparticles could be used for the purpose of delivering active constituents such as growth factors, and genes to specific regions, hence enabling precise and targeted treatment. In addition, magnetic fields can provide mechanical stimulation to cells, a phenomenon that has been shown to improve the process of cellular differentiation and proliferation. These processes are essential for tissue regeneration (Xie et al. 2024). Combining these nanoparticles with a scaffold offers a mutually beneficial method for TE. Scaffolds based on stimuli-responsive nanoparticles may provide a dynamic milieu that emulates physiological tissue processes (Municoy et al. 2020). An example of this is a scaffold that includes pH-responsive nanoparticles, which may continuously release growth factors when exposed to the acidic conditions of a wound. This promotes more effective healing, as reported by Zarur et al. in 2023 (Zarur et al. 2023).

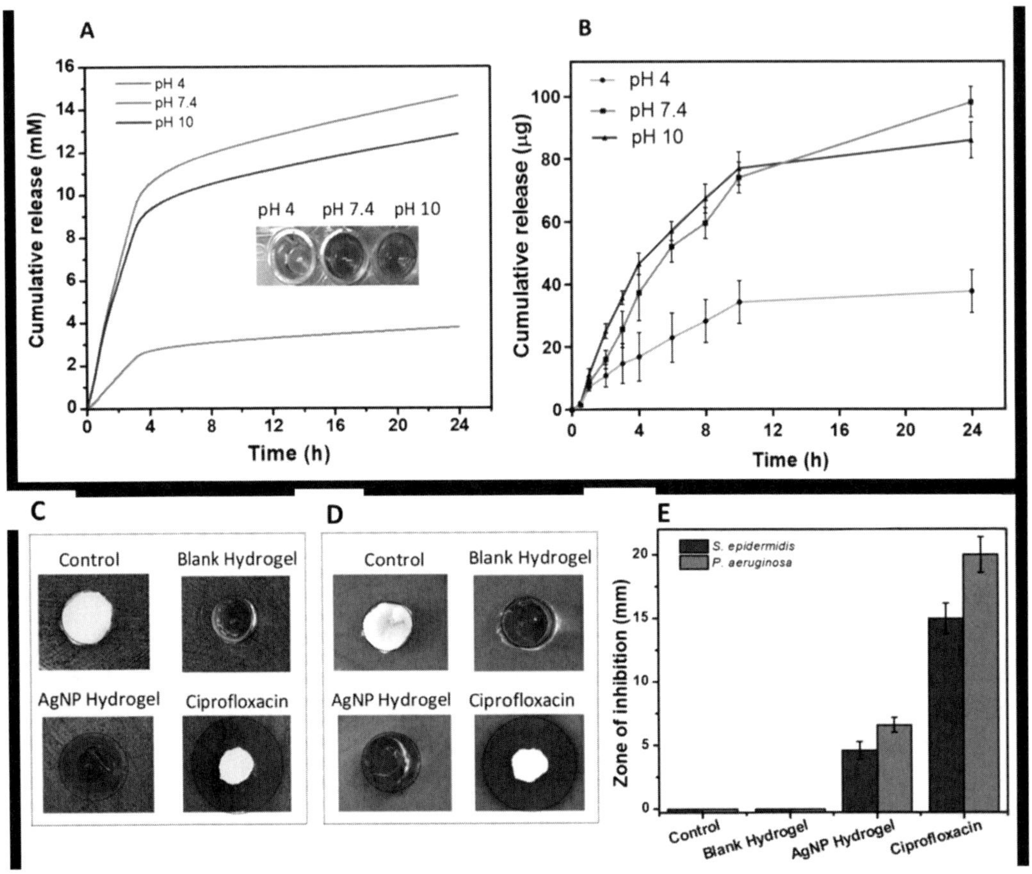

Figure 3. Cumulative release and zone of inhibition of crystal violet and AgNPs under influence of pH. Release of crystal violet (A) and Ag NPs (B) from smart hydrogel (B); Zone of inhibition images (C, D) in S. epidermidis (C) and P. aeruginosa (D). (E) Assessment of zones of inhibition (Haidari et al. 2021). (CC BY 4.0 license).

Furthermore, progress in nanotechnology has facilitated the development of multi-stimuli responsive materials capable of responding to several types of stimuli at the same time (Serna et al. 2021). These intricate systems may provide enhanced accuracy and adaptability for tissue regeneration applications. An instance of this is a hydrogel that can respond to changes in both pH and temperature. This hydrogel can offer a meticulously controlled milieu for the sustained delivery of medicinal agents, adapting to the dynamic conditions of the tissue microenvironment (Knipe and Peppas 2014). Stimuli-sensitive nanomaterials possess the capability to transform tissue repair by providing enhanced and accurate regulation of therapeutic delivery and cellular interactions. The incorporation of these materials into scaffold technologies and their capacity to react to various stimuli provide novel opportunities in the field of RM, offering more efficient and focused therapies for various tissue injuries and disorders.

5. Challenges and Prospects

RM and TE may encounter numerous obstacles similar to other cutting-edge transdisciplinary domains, but the amalgamation of nanotechnology-based biomimetic NS might soon be able to propel the development of cell- and stem-cell-based treatments. The primary concerns are the safety and cytotoxicity profile of NS, as well as their unclear effect on organ or tissue regeneration. It is difficult to track the real-time effects of NS in cell-based therapies. In this context, comprehensive and deep research is required to understand how cells interact with NS, how they are digested inside cells, and how they alter cell function. Other issues include correctly handling, characterizing, and individualizing 3D nanostructures for RM and TE. Future research is needed with smart tracking technology that can theranostically report the meticulous effects of biomimetic NS. Currently, this is an area of active exploration for further research and study.

Conclusion

In conclusion, the recent advances in nanomaterials and smart biomimetic scaffolds represent a transformative leap in RM and TE. Since its inception in the early 1990s, RM has aimed to replace or restore damaged tissues and organs, offering the potential for true cures and enhanced quality of life. Despite initial challenges, such as organ shortages and ethical concerns surrounding certain stem cell uses, the field has rapidly progressed, due to innovations in molecular and cell biology. The development of biomaterials has been pivotal, with polymers, metals, and ceramics providing suitable environments for cell growth and tissue development. Advancements in nanotechnology have improved these materials, facilitating the development of nanoscale architectures that more closely emulate native tissues. These nanostructures, synthesized from biopolymers such as cellulose, chitosan, and gelatin, simulate the nanoscale characteristics of the extracellular matrix, thereby augmenting cellular adhesion, proliferation, and migration. The increased surface-area-to-volume ratio of these scaffolds permits the incorporation of substantial quantities of bioactive molecules, thereby promoting efficient tissue regeneration. Techniques like electrospinning produce nanofibers that simulate natural collagen, creating an ideal environment for tissue regeneration. Additionally, biomimetic nanomaterials enhance regenerative processes by imitating ECM, resulting in improved mechanical strength and bioactivity. Innovations in gene delivery, surface modification, and advanced fabrication methods like 3D printing contribute to the precision and efficacy of these materials. Stimuli-responsive nanomaterials, which respond to environmental cues such as pH and thermal changes, enable regulated therapeutic delivery, thereby augmenting the healing process. The incorporation of these sophisticated scaffolds and nanomaterials into regenerative medicine holds significant potential, providing more effective, precise, and adaptable treatments for various tissue injuries and degenerative disorders. As research progresses, these innovations are anticipated to transform tissue regeneration, expanding the limits of what is achievable in RM and ultimately enhancing patient outcomes.

Acknowledgments

The authors would like to acknowledge the Maharishi Markandeshwar (Deemed to be University), Mullana, Ambala, Haryana for providing the necessary facilities.

References

Abbondandolo, Antonio G., and Erik C. Brewer. 2024. "Assessing the degradation profiles of a thermoresponsive polyvinyl alcohol (PVA)-based hydrogel for biomedical applications." *Polymers for Advanced Technologies* 35 (2):e6305. doi: https://doi.org/10.1002/pat.6305.

Abraham, Adarsh Mathew, and Subramani Venkatesan. 2023. "A review on application of biomaterials for medical and dental implants." *Proceedings of the Institution of Mechanical Engineers, Part L: Journal of Materials: Design and Applications* 237 (2):249-273. doi: https://doi.org/10.1177/14644207221121981.

Ai, J., S. Ebrahimi, A. Khoshzaban, T. S. Jafarzadeh Kashi, and D. Mehrabani. 2012. "Tissue engineering using human mineralized bone xenograft and bone marrow mesenchymal stem cells allograft in healing of tibial fracture of experimental rabbit model." *Iran Red Crescent Med J* 14 (2):96-103.

Alkhursani, S. A., M. M. Ghobashy, S. A. Al-Gahtany, A. S. Meganid, S. M. Abd El-Halim, Z. Ahmad, F. S. Khan, G. A. N. Atia, and S. Cavalu. 2022. "Application of Nano-Inspired Scaffolds-Based Biopolymer Hydrogel for Bone and Periodontal Tissue Regeneration." *Polymers (Basel)* 14 (18). doi: https://doi.org/10.3390/polym14183791.

Arora, M., E. K. Chan, S. Gupta, and A. D. Diwan. 2013. "Polymethylmethacrylate bone cements and additives: A review of the literature." *World J Orthop* 4 (2):67-74. doi: https://doi.org/10.5312/wjo.v4.i2.67.

Baig, Nadeem, Irshad Kammakakam, and Wail Falath. 2021. "Nanomaterials: a review of synthesis methods, properties, recent progress, and challenges." *Materials Advances* 2 (6):1821-1871. doi: https://doi.org/10.1039/D0MA00807A.

Bazaka, O., K. Bazaka, P. Kingshott, R. J. Crawford, and E. P. Ivanova. 2021. "Metallic Implants for Biomedical Applications." In *The Chemistry of Inorganic Biomaterials*, edited by Christopher Spicer, 0. The Royal Society of Chemistry.

Bazgir, M., W. Zhang, X. Zhang, J. Elies, M. Saeinasab, P. Coates, M. Youseffi, and F. Sefat. 2021. "Fabrication and Characterization of PCL/PLGA Coaxial and Bilayer Fibrous Scaffolds for Tissue Engineering." *Materials (Basel)* 14 (21). doi: https://doi.org/10.3390/ma14216295.

Bertsch, Christelle, Hélène Maréchal, Varvara Gribova, Benjamin Lévy, Christian Debry, Philippe Lavalle, and Léa Fath. 2023. "Biomimetic Bilayered Scaffolds for Tissue Engineering: From Current Design Strategies to Medical Applications." *Advanced Healthcare Materials* 12 (17):2203115. doi: https://doi.org/10.1002/adhm.202203115.

Bhatt, P., V. Kumar, V. Subramaniyan, K. Nagarajan, M. Sekar, S. V. Chinni, and G. Ramachawolran. 2023. "Plasma Modification Techniques for Natural Polymer-Based Drug Delivery Systems." *Pharmaceutics* 15 (8). doi: https://doi.org/10.3390/pharmaceutics15082066.

Cardoso, Vanessa F., Daniela M. Correia, Clarisse Ribeiro, Margarida M. Fernandes, and Senentxu Lanceros-Méndez. 2018. "Fluorinated Polymers as Smart Materials for Advanced Biomedical Applications." *Polymers* 10 (2):161. doi: https://doi.org/10.3390/polym10020161.

Carriero, V. C., L. Di Muzio, S. Petralito, M. A. Casadei, and P. Paolicelli. 2023. "Cryogel Scaffolds for Tissue-Engineering: Advances and Challenges for Effective Bone and Cartilage Regeneration." *Gels* 9 (12). doi: https://doi.org/10.3390/gels9120979.

Castro, J. I., D. G. Araujo-Rodríguez, C. H. Valencia-Llano, D. López Tenorio, M. Saavedra, P. A. Zapata, and C. D. Grande-Tovar. 2023. "Biocompatibility Assessment of Polycaprolactone/Polylactic Acid/Zinc Oxide Nanoparticle Composites under *In Vivo* Conditions for Biomedical Applications." *Pharmaceutics* 15 (9). doi: https://doi.org/10.3390/pharmaceutics15092196.

Chang Chien, George C., and Agnes Stogicza. 2021. "Regenerative Medicine." In *Pain Care Essentials and Innovations*, edited by Sanjong Pangarkar, Quynh G. Pham and Blessen C. Eapen, 245-253. Elsevier.

Chaudhury, K., V. Kumar, J. Kandasamy, and S. RoyChoudhury. 2014. "Regenerative nanomedicine: current perspectives and future directions." *Int J Nanomedicine* 9:4153-67. doi: https://doi.org/10.2147/ijn.S45332.

Chen, Weixing, Yue Ming, Min Wang, Mingshu Huang, Hongyu Liu, Yisheng Huang, Zhijie Huang, Ling Qing, Qin Wang, and Bo Jia. 2023. "Nanocomposite Hydrogels in Regenerative Medicine: Applications and Challenges." *Macromolecular Rapid Communications* 44 (15):2300128. doi: https://doi.org/10.1002/marc.202300128.

Chen, Yujie, Muhammad Shafiq, Mingyue Liu, Yosry Morsi, and Xiumei Mo. 2020. "Advanced fabrication for electrospun three-dimensional nanofiber aerogels and scaffolds." *Bioactive Materials* 5 (4):963-979. doi: https://doi.org/10.1016/j.bioactmat.2020.06.023.

Ciulla, Maria G., Alessio Massironi, Michela Sugni, Matthew A. Ensign, Stefania Marzorati, and Mahdi Forouharshad. 2023. "Recent Advances in the Development of Biomimetic Materials." *Gels* 9 (10):833. doi: https://doi.org/10.3390/gels9100833.

Costa, F. R., M. D. S. Santos, R. A. Martins, C. B. Costa, P. C. Hamdan, M. B. Da Silva, G. O. M. Azzini, L. Pires, Z. Menegassi, G. S. Santos, and J. F. Lana. 2023. "The Synergistic Effects of Hyaluronic Acid and Platelet-Rich Plasma for Patellar Chondropathy." *Biomedicines* 12 (1). doi: https://doi.org/10.3390/biomedicines12010006.

Daar, A. S. 1998. "Xenotransplants: proceed with caution." *Nature* 392 (6671):12-12. doi: https://doi.org/10.1038/32023.

Dickson, Kathryn, Kwang Chear Lee, Abdulrazak Abdulsalam, Ezekwe Amirize, Hadyn K N Kankam, Britt ter Horst, Fay Gardiner, Amy Bamford, Rahul K Hejmadi, and Naiem Moiemen. 2023. "A Histological and Clinical Study of MatriDerm® Use in Burn Reconstruction." *Journal of Burn Care & Research* 44 (5):1100-1109. doi: https://doi.org/10.1093/jbcr/irad024.

Edelman, Elazer R, Edith Mathiowitz, Robert Langer, and Michael Klagsbrun. 1991. "Controlled and modulated release of basic fibroblast growth factor." *Biomaterials* 12 (7):619-626.

Esmaeili, H., A. Patino-Guerrero, R. A. Nelson, Jr., N. Karamanova, M. Fisher T, W. Zhu, F. Perreault, R. Q. Migrino, and M. Nikkhah. 2024. "Engineered Gold and Silica Nanoparticle-Incorporated Hydrogel Scaffolds for Human Stem Cell-Derived Cardiac Tissue Engineering." *ACS Biomater Sci Eng* 10 (4):2351-2366. doi: https://doi.org/10.1021/acsbiomaterials.3c01256.

Esmaeili, Yasaman, Elham Bidram, Ashkan Bigham, Mehdi Atari, Reyhaneh Nasr Azadani, Mohamadreza Tavakoli, Saeideh Salehi, Marjan Mirhaj, Arefeh Basiri, Zahra Mirzavandi, Maryam Boshtam, Mohammad Rafienia, Anousheh Zargar Kharazi, Saeid Karbasi, Laleh Shariati, and Ali Zarrabi. 2023. "Exploring the evolution of tissue engineering strategies over the past decade: From cell-based strategies to gene-activated matrix." *Alexandria Engineering Journal* 81:137-169. doi: https://doi.org/10.1016/j.aej.2023.08.080.

Friedrich, R. P., I. Cicha, and C. Alexiou. 2021. "Iron Oxide Nanoparticles in Regenerative Medicine and Tissue Engineering." *Nanomaterials (Basel)* 11 (9). doi: https://doi.org/10.3390/nano11092337.

Gallo, Juan, and Aranzazu Villasante. 2023. "Recent Advances in Biomimetic Nanocarrier-Based Photothermal Therapy for Cancer Treatment." *International Journal of Molecular Sciences* 24 (20):15484. doi: https://doi.org/10.3390/ijms242015484.

Gatti, R. A., H. J. Meuwissen, H. D. Allen, R. Hong, and R. A. Good. 1968. "Immunological reconstitution of sex-linked lymphopenic immunological deficiency." *Lancet* 2 (7583):1366-9. doi: https://doi.org/10.1016/s0140-6736(68)92673-1.

Ghahremanloo, A., M. Seifi, J. Ghanbarzade, S. M. Abrisham, and R. A. Javan. 2017. "Effect of Polyvinyl Siloxane Viscosity on Accuracy of Dental Implant Impressions." *J Dent (Tehran)* 14 (1):40-47.

Gong, Y., H. Liu, S. Ke, L. Zhuo, and H. Wang. 2022. "Latest advances in biomimetic nanomaterials for diagnosis and treatment of cardiovascular disease." *Front Cardiovasc Med* 9:1037741. doi: https://doi.org/10.3389/fcvm.2022.1037741.

Gu, R., H. Zhou, Z. Zhang, Y. Lv, Y. Pan, Q. Li, C. Shi, Y. Wang, and L. Wei. 2023. "Research progress related to thermosensitive hydrogel dressings in wound healing: a review." *Nanoscale Adv* 5 (22):6017-6037. doi: https://doi.org/10.1039/d3na00407d.

Gültekin, H. E., G. Yaşayan, A. Bal-Öztürk, A. Bigham, A. A. Simchi, A. Zarepour, S. Iravani, and A. Zarrabi. 2024. "Advancements and applications of upconversion nanoparticles in wound dressings." *Mater Horiz* 11 (2):363-387. doi: https://doi.org/10.1039/d3mh01330h.

Gunatillake, Pathiraja A, Raju Adhikari, and N Gadegaard. 2003. "Biodegradable synthetic polymers for tissue engineering." *Eur Cell Mater* 5 (1):1-16. doi: https://doi.org/10.22203/eCM.v005a01

Guo, B., B. Lei, P. Li, and P. X. Ma. 2015. "Functionalized scaffolds to enhance tissue regeneration." *Regen Biomater* 2 (1):47-57. doi: https://doi.org/10.1093/rb/rbu016.

Habibzadeh, Faezeh, Seyed Mahdi Sadraei, Roghayeh Mansoori, Narendra Pal Singh Chauhan, and Ghasem Sargazi. 2022. "Nanomaterials supported by polymers for tissue engineering applications: A review." *Heliyon* 8 (12):e12193. doi: https://doi.org/10.1016/j.heliyon.2022.e12193.

Haidari, H., Z. Kopecki, A. T. Sutton, S. Garg, A. J. Cowin, and K. Vasilev. 2021. "pH-Responsive "Smart" Hydrogel for Controlled Delivery of Silver Nanoparticles to Infected Wounds." *Antibiotics (Basel)* 10 (1). doi: https://doi.org/10.3390/antibiotics10010049.

Han, Feng, Qingchen Meng, En Xie, Kexin Li, Jie Hu, Qianglong Chen, Jiaying Li, and Fengxuan Han. 2023. "Engineered biomimetic micro/nano-materials for tissue regeneration." *Frontiers in Bioengineering and Biotechnology* 11. doi: https://doi.org/10.3389/fbioe.2023.1205792.

Han, Zeyu, Mujie Yuan, Lubin Liu, Kaiyue Zhang, Baodong Zhao, Bin He, Yan Liang, and Fan Li. 2023. "pH-Responsive wound dressings: advances and prospects." *Nanoscale Horizons* 8 (4):422-440. doi: https://doi.org/10.1039/D2NH00574C.

Hasan, A., M. Morshed, A. Memic, S. Hassan, T. J. Webster, and H. E. Marei. 2018. "Nanoparticles in tissue engineering: applications, challenges and prospects." *Int J Nanomedicine* 13:5637-5655. doi: https://doi.org/10.2147/ijn.S153758.

Häuselmann, H. J., R. J. Fernandes, S. S. Mok, T. M. Schmid, J. A. Block, M. B. Aydelotte, K. E. Kuettner, and E. J. Thonar. 1994. "Phenotypic stability of bovine articular chondrocytes after long-term culture in alginate beads." *J Cell Sci* 107 (Pt 1):17-27. doi: https://doi.org/10.1242/jcs.107.1.17.

Hedelin, Henrik, Hanna Hebelka, Helena Brisby, and Tero Laine. 2020. "MRI evaluation of resorbable poly lactic-co-glycolic acid (PLGA) screws used in pelvic osteotomies in children—a retrospective case series." *Journal of Orthopaedic Surgery and Research* 15 (1):329. doi: https://doi.org/10.1186/s13018-020-01858-5.

Hosseinkhani, H., A. J. Domb, G. Sharifzadeh, and V. Nahum. 2023. "Gene Therapy for Regenerative Medicine." *Pharmaceutics* 15 (3). doi: https://doi.org/10.3390/pharmaceutics15030856.

Hosseinkhani, Hossein, Mohsen Hosseinkhani, Ali Khademhosseini, Hisatoshi Kobayashi, and Yasuhiko Tabata. 2006. "Enhanced angiogenesis through controlled release of basic fibroblast growth factor from peptide amphiphile for tissue regeneration." *Biomaterials* 27 (34):5836-5844. doi: https://doi.org/10.1016/j.biomaterials.2006.08.003.

Igwe Idumah, Christopher. 2022. "Emerging trends in Poly(lactic-co-glycolic) acid bionanoarchitectures and applications." *Cleaner Materials* 5:100102. doi: https://doi.org/10.1016/j.clema.2022.100102.

Inam, Hina, Simone Sprio, Marta Tavoni, Zahid Abbas, Federico Pupilli, and Anna Tampieri. 2024. "Magnetic Hydroxyapatite Nanoparticles in Regenerative Medicine and Nanomedicine." *International Journal of Molecular Sciences* 25 (5):2809. doi: https://doi.org/10.3390/ijms25052809.

Ip, K. T. 2012. Regenerative Medicine. In *Encyclopedia of Applied Ethics*, edited by Ruth Chadwick. San Diego: Academic Press.

Iwakura, Atsushi, Yasuhiko Tabata, Nobushige Tamura, Kazuhiko Doi, Kazunobu Nishimura, Tatsuo Nakamura, Yasuhiko Shimizu, Masatoshi Fujita, and Masashi Komeda. 2001. "Myocardial Protection and Vascular Biology-Gelatin Sheet Incorporating Basic Fibroblast Growth Factor Enhances Healing of Devascularized Sternum in Diabetic Rats." *Circulation-Hagertown* 104 (12):I325. doi: https://doi.org/10.1161/hc37t1.094544.

Kandhola, G., S. Park, J. W. Lim, C. Chivers, Y. H. Song, J. H. Chung, J. Kim, and J. W. Kim. 2023. "Nanomaterial-Based Scaffolds for Tissue Engineering Applications: A Review on Graphene, Carbon Nanotubes and Nanocellulose." *Tissue Eng Regen Med* 20 (3):411-433. doi: https://doi.org/10.1007/s13770-023-00530-3.

Khademhosseini, Ali, Bimal Rajalingam, Satoshi Jinno, and Robert Langer. 2010. "Nanoengineered Systems for Tissue Engineering and Regeneration." In *Nanotechnology*, 361-384.

Kim, J. A., J. H. Choi, M. Kim, W. J. Rhee, B. Son, H. K. Jung, and T. H. Park. 2013. "High-throughput generation of spheroids using magnetic nanoparticles for three-dimensional cell culture." *Biomaterials* 34 (34):8555-63. doi: https://doi.org/10.1016/j.biomaterials.2013.07.056.

Knipe, J. M., and N. A. Peppas. 2014. "Multi-responsive hydrogels for drug delivery and tissue engineering applications." *Regen Biomater* 1 (1):57-65. doi: https://doi.org/10.1093/rb/rbu006.

Kondej, K., M. Zawrzykraj, K. Czerwiec, M. Deptuła, A. Tymińska, and M. Pikuła. 2024. "Bioengineering Skin Substitutes for Wound Management-Perspectives and Challenges." *Int J Mol Sci* 25 (7). doi: https://doi.org/10.3390/ijms25073702.

Kumar, V. B., O. S. Tiwari, G. Finkelstein-Zuta, S. Rencus-Lazar, and E. Gazit. 2023. "Design of Functional RGD Peptide-Based Biomaterials for Tissue Engineering." *Pharmaceutics* 15 (2). doi: https://doi.org/10.3390/pharmaceutics15020345.

Kuperkar, K., L. I. Atanase, A. Bahadur, I. C. Crivei, and P. Bahadur. 2024. "Degradable Polymeric Bio(nano)materials and Their Biomedical Applications: A Comprehensive Overview and Recent Updates." *Polymers (Basel)* 16 (2). doi: https://doi.org/10.3390/polym16020206.

Lin, Fengcai, Zi Wang, Yanping Shen, Lirong Tang, Pinle Zhang, Yuefang Wang, Yandan Chen, Biao Huang, and Beili Lu. 2019. "Natural skin-inspired versatile cellulose biomimetic hydrogels." *Journal of Materials Chemistry A* 7 (46):26442-26455. doi: https://doi.org/10.1039/C9TA10502F.

Liu, Shuai, Jiang-Ming Yu, Yan-Chang Gan, Xiao-Zhong Qiu, Zhe-Chen Gao, Huan Wang, Shi-Xuan Chen, Yuan Xiong, Guo-Hui Liu, Si-En Lin, Alec McCarthy, Johnson V. John, Dai-Xu Wei, and Hong-Hao Hou. 2023. "Biomimetic natural biomaterials for tissue engineering and regenerative medicine: new biosynthesis methods, recent advances, and emerging applications." *Military Medical Research* 10 (1):16. doi: https://doi.org/10.1186/s40779-023-00448-w.

Loh, Q. L., and C. Choong. 2013. "Three-dimensional scaffolds for tissue engineering applications: role of porosity and pore size." *Tissue Eng Part B Rev* 19 (6):485-502. doi: https://doi.org/10.1089/ten.TEB.2012.0437.

Mahheidari, Naimeh, Mohammad Kamalabadi-Farahani, Mohammad Reza Nourani, Amir Atashi, Morteza Alizadeh, Niloofar Aldaghi, and Majid Salehi. 2024. "Biological study of skin wound treated with Alginate/Carboxymethyl cellulose/chorion membrane, diopside nanoparticles, and Botox A." *npj Regenerative Medicine* 9 (1):9. doi: https://doi.org/10.1038/s41536-024-00354-2.

Manoukian, Ohan S., Naseem Sardashti, Teagen Stedman, Katie Gailiunas, Anurag Ojha, Aura Penalosa, Christopher Mancuso, Michelle Hobert, and Sangamesh G. Kumbar. 2019. "Biomaterials for Tissue Engineering and Regenerative Medicine." In *Encyclopedia of Biomedical Engineering*, edited by Roger Narayan, 462-482. Oxford: Elsevier.

Mehrabani, Davood, Golshid Mehrabani, Shahrokh Zare, and Ali Manafi. 2013. "Adipose-derived stem cells (ADSC) and aesthetic surgery: a mini review." *World Journal of Plastic Surgery* 2 (2):65.

Mendes, Bárbara B., João Conniot, Aviram Avital, Dongbao Yao, Xingya Jiang, Xiang Zhou, Noga Sharf-Pauker, Yuling Xiao, Omer Adir, Haojun Liang, Jinjun Shi, Avi Schroeder, and João Conde. 2022. "Nanodelivery of nucleic acids." *Nature Reviews Methods Primers* 2 (1):24. doi: https://doi.org/10.1038/s43586-022-00104-y.

Mendes, Carolini, Daniela Pacheco dos Santos Haupenthal, Rubya Pereira Zaccaron, Gustavo de Bem Silveira, Maria Eduarda Anastácio Borges Corrêa, Laura de Roch Casagrande, Samara de Sousa Mariano, Jennyffer Ione de Souza Silva, Thiago Antônio Moretti de Andrade, Paulo Emilio Feuser, Ricardo Andrez Machado-de-Ávila, and Paulo Cesar Lock Silveira. 2020. "Effects of the Association between Photobiomodulation and Hyaluronic Acid Linked Gold Nanoparticles in Wound Healing." *ACS Biomaterials Science & Engineering* 6 (9):5132-5144. doi: https://doi.org/10.1021/acsbiomaterials.0c00294.

Milián, L., M. Oliver-Ferrándiz, I. Peregrín, M. Sancho-Tello, J. J. Martín-de-Llano, C. Martínez-Ramos, C. Carda, and M. Mata. 2024. "Alginate Improves the Chondrogenic Capacity of 3D PCL Scaffolds *In Vitro*: A Histological Approach." *Curr Issues Mol Biol* 46 (4):3563-3578. doi: https://doi.org/10.3390/cimb46040223.

Morris, Peter J. 2004. "Transplantation — A Medical Miracle of the 20th Century." *New England Journal of Medicine* 351 (26):2678-2680. doi: https://doi.org/10.1056/NEJMp048256.

Mottaghitalab, Fatemeh, Ali Rastegari, Mehdi Farokhi, Rassoul Dinarvand, Hossein Hosseinkhani, Keng-Liang Ou, Daniel W. Pack, Chuanbin Mao, Meshkat Dinarvand, Yousef Fatahi, and Fatemeh Atyabi. 2017. "Prospects of siRNA applications in regenerative medicine." *International Journal of Pharmaceutics* 524 (1):312-329. doi: https://doi.org/10.1016/j.ijpharm.2017.03.092.

Municoy, S., M. I. Álvarez Echazú, P. E. Antezana, J. M. Galdopórpora, C. Olivetti, A. M. Mebert, M. L. Foglia, M. V. Tuttolomondo, G. S. Alvarez, J. G. Hardy, and M. F. Desimone. 2020. "Stimuli-Responsive Materials for Tissue Engineering and Drug Delivery." *Int J Mol Sci* 21 (13). doi: https://doi.org/10.3390/ijms21134724.

Muster, Dominique. 1992. "Biomaterials: Hard Tissue Repair and Replacement."

Nagase, Kenichi. 2021. "Thermoresponsive interfaces obtained using poly(N-isopropylacrylamide)-based copolymer for bioseparation and tissue engineering applications." *Advances in Colloid and Interface Science* 295:102487. doi: https://doi.org/10.1016/j.cis.2021.102487.

Nanda, Sitansu Sekhar, Tuntun Wang, Md Imran Hossain, Hong Yeol Yoon, Subramanian Tamil Selvan, Kwangmeyung Kim, and Dong Kee Yi. 2022. "Gold-Nanorod-Based Scaffolds for Wound-Healing Applications." *ACS Applied Nano Materials* 5 (6):8640-8648. doi: https://doi.org/10.1021/acsanm.2c02230.

Nandhini, A., T. Sudhakar, and J. Premkumar. 2021. "Ceramics and Nanoceramics in Biomedical Applications." In *Handbook of Polymer and Ceramic Nanotechnology*, edited by Chaudhery Mustansar Hussain and Sabu Thomas, 763-779. Cham: Springer International Publishing.

Nielsen, Claus. 1998. "Sequences lead to tree of worms." *Nature* 392 (6671):25-25. doi: https://doi.org/10.1038/32058.

Pandya, Anjali, Prashant Upadhaya, Shruti Lohakare, Tishya Srivastava, Susmit Mhatre, Sreeranjini Pulakkat, and Vandana B. Patravale. 2022. "Nanobiomaterials for regenerative medicine." In *Nanotechnology in Medicine and Biology*, edited by Huinan H. Liu, Tolou Shokuhfar and Sougata Ghosh, 141-187. Elsevier.

Perán, M., M. A. García, E. López-Ruiz, M. Bustamante, G. Jiménez, R. Madeddu, and J. A. Marchal. 2012. "Functionalized nanostructures with application in regenerative medicine." *Int J Mol Sci* 13 (3):3847-3886. doi: https://doi.org/10.3390/ijms13033847.

Perez-Puyana, V., M. Jiménez-Rosado, A. Romero, and A. Guerrero. 2020. "Polymer-Based Scaffolds for Soft-Tissue Engineering." *Polymers (Basel)* 12 (7). doi: https://doi.org/10.3390/polym12071566.

Polverini, P. J. 1995. "The pathophysiology of angiogenesis." *Crit Rev Oral Biol Med* 6 (3):230-47. doi: https://doi.org/10.1177/10454411950060030501.

Qian, Yun, Xiaotian Zhao, Qixin Han, Wei Chen, Hui Li, and Weien Yuan. 2018. "An integrated multi-layer 3D-fabrication of PDA/RGD coated graphene loaded PCL nanoscaffold for peripheral nerve restoration." *Nature Communications* 9 (1):323. doi: https://doi.org/10.1038/s41467-017-02598-7.

Qu, Zhi, Yang Wang, Yanhong Dong, Xinmeng Li, Lingwan Hao, Liwei Sun, Lu Zhou, Rujian Jiang, and Weihua Liu. 2024. "Intelligent Electrospinning Nanofibrous Membranes for Monitoring and Promotion of Wound Healing." *Materials Today Bio* 26:101093. doi: https://doi.org/10.1016/j.mtbio.2024.101093.

Ramanathan, S., Y. C. Lin, S. Thirumurugan, C. C. Hu, Y. F. Duann, and R. J. Chung. 2024. "Poly(methyl methacrylate) in Orthopedics: Strategies, Challenges, and Prospects in Bone Tissue Engineering." *Polymers (Basel)* 16 (3). doi: https://doi.org/10.3390/polym16030367.

Reddy, M. Sai Bhargava, Deepalekshmi Ponnamma, Rajan Choudhary, and Kishor Kumar Sadasivuni. 2021. "A Comparative Review of Natural and Synthetic Biopolymer Composite Scaffolds." *Polymers* 13 (7):1105. doi: https://doi.org/10.3390/polym13071105.

Rosenberg, Murray D. 1963. "Cell Guidance by Alterations in Monomolecular Films." *Science* 139 (3553):411-412. doi: https://doi.org/10.1126/science.139.3553.411.

Sáenz, Alejandro, E Rivera, Witold Brostow, and Victor M Castaño. 1999. "Ceramic biomaterials: an introductory overview." *Journal of Materials Education* 21 (5/6):267-276.

Saha, Tanushree, Xin Wang, Rajiv Padhye, and Shadi Houshyar. 2022. "A review of recent developments of polypropylene surgical mesh for hernia repair." *OpenNano* 7:100046. doi: https://doi.org/10.1016/j.onano.2022.100046.

Samir, Čanović, Konjevoda Suzana, Pavičić Ana Didović, and Stanić Robert. 2019. "Intraocular Lens (IOL) Materials." In *Intraocular Lens*, edited by Wang Xiaogang and M. Ferreri Felicia, Ch. 2. Rijeka: IntechOpen.

Sampogna, G., S. Y. Guraya, and A. Forgione. 2015. "Regenerative medicine: Historical roots and potential strategies in modern medicine." *J Microsc Ultrastruct* 3 (3):101-107. doi: https://doi.org/10.1016/j.jmau.2015.05.002.

Samrot, A. V., M. Sathiyasree, S. B. A. Rahim, R. E. Renitta, K. Kasipandian, S. Krithika Shree, D. Rajalakshmi, N. Shobana, S. Dhiva, S. Abirami, S. Visvanathan, B. K. Mohanty, G. S. Sabesan, and S. V. Chinni. 2023. "Scaffold Using Chitosan, Agarose, Cellulose, Dextran and Protein for Tissue Engineering-A Review." *Polymers (Basel)* 15 (6). doi: https://doi.org/10.3390/polym15061525.

Santa Cruz-Pavlovich, F. J., A. J. Bolaños-Chang, X. I. Del Rio-Murillo, G. A. Aranda-Preciado, E. M. Razura-Ruiz, A. Santos, and J. Navarro-Partida. 2024. "Beyond Vision: An Overview of Regenerative Medicine and Its Current Applications in Ophthalmological Care." *Cells* 13 (2). doi: https://doi.org/10.3390/cells13020179.

Serna, Julian A., Laura Rueda-Gensini, Daniela N. Céspedes-Valenzuela, Javier Cifuentes, Juan C. Cruz, and Carolina Muñoz-Camargo. 2021. "Recent Advances on Stimuli-Responsive Hydrogels Based on Tissue-Derived ECMs and Their Components: Towards Improving Functionality for Tissue Engineering and Controlled Drug Delivery." *Polymers* 13 (19):3263. doi: https://doi.org/10.3390/polym13193263.

Smith, I. O., X. H. Liu, L. A. Smith, and P. X. Ma. 2009. "Nanostructured polymer scaffolds for tissue engineering and regenerative medicine." *Wiley Interdiscip Rev Nanomed Nanobiotechnol* 1 (2):226-36. doi: https://doi.org/10.1002/wnan.26.

Soman, Pranav, Brian TD Tobe, Jin Woo Lee, Alicia AM Winquist, Ilyas Singec, Kenneth S Vecchio, Evan Y Snyder, and Shaochen Chen. 2012. "Three-dimensional scaffolding to investigate neuronal derivatives of human embryonic stem cells." *Biomedical microdevices* 14:829-838. doi: https://doi.org/10.1007/s10544-012-9662-7.

Sponchioni, Mattia, Umberto Capasso Palmiero, and Davide Moscatelli. 2019. "Thermo-responsive polymers: Applications of smart materials in drug delivery and tissue engineering." *Materials Science and Engineering: C* 102:589-605. doi: https://doi.org/10.1016/j.msec.2019.04.069.

Stupp, Samuel I. 2005. "Biomaterials for regenerative medicine." *Mrs Bulletin* 30 (7):546-553. doi: https://doi.org/10.5005/jp-journals-10028-1018.

Tan, S. Y., and J. Merchant. 2019. "Joseph Murray (1919-2012): First transplant surgeon." *Singapore Med J* 60 (4):162-163. doi: https://doi.org/10.11622/smedj.2019032.

Taupin, P., A. Gandhi, and S. Saini. 2023. "Integra® Dermal Regeneration Template: From Design to Clinical Use." *Cureus* 15 (5):e38608. doi: https://doi.org/10.7759/cureus.38608.

Thomas, E. Donnall, Harry L. Lochte, Wan Ching Lu, and Joseph W. Ferrebee. 1957. "Intravenous Infusion of Bone Marrow in Patients Receiving Radiation and Chemotherapy." *New England Journal of Medicine* 257 (11):491-496. doi: https://doi.org/doi:10.1056/NEJM195709122571102.

Troy, E., M. A. Tilbury, A. M. Power, and J. G. Wall. 2021. "Nature-Based Biomaterials and Their Application in Biomedicine." *Polymers (Basel)* 13 (19). doi: https://doi.org/10.3390/polym13193321.

Vaiani, L., A. Boccaccio, A. E. Uva, G. Palumbo, A. Piccininni, P. Guglielmi, S. Cantore, L. Santacroce, I. A. Charitos, and A. Ballini. 2023. "Ceramic Materials for Biomedical Applications: An Overview on Properties and Fabrication Processes." *J Funct Biomater* 14 (3). doi: https://doi.org/10.3390/jfb14030146.

Wang, Li, Xiaolei Guo, Jiaqing Chen, Zhen Zhen, Bin Cao, Wenqian Wan, Yuandong Dou, Haobo Pan, Feng Xu, Zepu Zhang, Jianmei Wang, Daisong Li, Quanyi Guo, Qing Jiang, Yanan Du, Jiakuo Yu, Boon Chin Heng, Qianqian Han, and Zigang Ge. 2022. "Key considerations on the development of biodegradable biomaterials for clinical translation of medical devices: With cartilage repair products as an example." *Bioactive Materials* 9:332-342. doi: https://doi.org/10.1016/j.bioactmat.2021.07.031.

Watson, C. J. E., and J. H. Dark. 2012. "Organ transplantation: historical perspective and current practice." *British Journal of Anaesthesia* 108:i29-i42. doi: https://doi.org/10.1093/bja/aer384.

Webster, Thomas J, Linda S Schadler, Richard W Siegel, and Rena Bizios. 2001. "Mechanisms of enhanced osteoblast adhesion on nanophase alumina involve vitronectin." *Tissue engineering* 7 (3):291-301. doi: https://doi.org/10.1089/10763270152044152.

Wei, Guobao, Qiming Jin, William V. Giannobile, and Peter X. Ma. 2007. "The enhancement of osteogenesis by nano-fibrous scaffolds incorporating rhBMP-7 nanospheres." *Biomaterials* 28 (12):2087-2096. doi: https://doi.org/10.1016/j.biomaterials.2006.12.028.

Wieszczycka, Karolina, Katarzyna Staszak, Marta J. Woźniak-Budych, Jagoda Litowczenko, Barbara M. Maciejewska, and Stefan Jurga. 2021. "Surface functionalization – The way for advanced applications of smart materials." *Coordination Chemistry Reviews* 436:213846. doi: https://doi.org/10.1016/j.ccr.2021.213846.

Xie, W., C. Song, R. Guo, and X. Zhang. 2024. "Static magnetic fields in regenerative medicine." *APL Bioeng* 8 (1):011503. doi: https://doi.org/10.1063/5.0191803.

Xie, Yubing, ST Yang, and Douglas A Kniss. 2001. "Three-dimensional cell-scaffold constructs promote efficient gene transfection: implications for cell-based gene therapy." *Tissue engineering* 7 (5):585-598. doi: https://doi.org/10.1089/107632701753213200.

Zare, M., A. Bigham, M. Zare, H. Luo, E. Rezvani Ghomi, and S. Ramakrishna. 2021. "pHEMA: An Overview for Biomedical Applications." *Int J Mol Sci* 22 (12). doi: https://doi.org/10.3390/ijms22126376.

Zarur, Mariana, Alejandro Seijo-Rabina, Alvaro Goyanes, Angel Concheiro, and Carmen Alvarez-Lorenzo. 2023. "pH-responsive scaffolds for tissue regeneration: *In vivo* performance." *Acta Biomaterialia* 168:22-41. doi: https://doi.org/10.1016/j.actbio.2023.07.025.

Zhang, T., K. Chen, X. Wu, and X. Xiao. 2023. "Preparation of nanofibrous poly (L-lactic acid) scaffolds using the thermally induced phase separation technique in dioxane/polyethylene glycol solution." *Des Monomers Polym* 26 (1):77-89. doi: https://doi.org/10.1080/15685551.2023.2194175.

Zhang, T., X. Sun, X. Chen, W. Chen, H. Tang, and C. Y. Li. 2024. "Intelligent near-infrared light-activatable DNA machine with DNA wire nano-scaffold-integrated fast domino-like driving amplification for high-performance imaging in live biological samples." *Biosens Bioelectron* 259:116412. doi: https://doi.org/10.1016/j.bios.2024.116412.

Zulkifli, Muhammad Zikri Aiman, Darman Nordin, Norazuwana Shaari, and Siti Kartom Kamarudin. 2023. "Overview of Electrospinning for Tissue Engineering Applications." *Polymers* 15 (11):2418. doi: https://doi.org/10.3390/polym15112418.

Chapter 3.3

Graphene-Based Magnetic Nanoparticles for Enhanced Disease Diagnosis

Pralaya Kumar Sahoo[1]
Sangeeta Chhotaray[1]
Samukarani Swain[2]
and Soumya Jal[1,*]

[1]School of Paramedics and Allied Health Sciences, Centurion University of Technology and Management, Bhubaneswar, Odisha, India
[2]Centre for Industrial Biotechnology and Research, SOA University, Bhubaneswar, Odisha, India

Abstract

This chapter provides an in-depth examination of the latest advancements, challenges, and possible applications of graphene-based nanomaterials for disease diagnosis. It discusses the synthesis and scalability of graphene-based nanomaterials, including Graphene Oxide (GO) and reduced Graphene Oxide (rGO), which offer enhanced electrical conductivity and biocompatibility for diagnostic applications. Next, functionalization options to improve the specificity and targeting of these nanomaterials toward disease biomarkers are discussed. It then explores the various uses of graphene-based nanomaterials in diagnostic modalities, such as medication delivery systems, imaging agents, and biosensors. Accurate illness localization and monitoring are made possible by their exceptional electrical, optical, and magnetic capabilities, which allow for the accurate detection of biomolecules specific to a given disease. The chapter also discusses important issues related to the clinical translation of graphene-based nanomaterials, including biocompatibility, long-term stability, and regulatory approval. It suggests methods to address these issues by surface functionalization and biocompatible encapsulation. The chapter concludes by highlighting the enormous potential of graphene-based nanomaterials to transform illness detection by offering incredible sensitive, precise, and affordable solutions. The significance of interdisciplinary research endeavors in surmounting residual obstacles and proficiently integrating these advancements into medical practice is underscored. Additionally, it offers insights into the changing landscape of illness diagnosis by discussing new developments and directions in integrating graphene-based nanomaterials

* Corresponding Author's Email: soumya.jal@cutm.ac.in

In: Emerging Nanomaterials in Biomedical Science
Editors: Eepsita Priyadarshini and Ajith Manayil Parambil
ISBN: 979-8-89530-600-0

with cutting-edge technologies like microfluidics and Artificial Intelligence (AI) for development of point-of-care diagnostics.

Keywords: graphene-based nanomaterials, biocompatibility, disease diagnosis, biomarker, point-of-care diagnosis

1. Introduction

Many established and cutting-edge techniques are available to diagnose illnesses and their biomarkers with considerable precision and sensitivity (Haque et al. 1993; Lazcka, et al. 2007). The typical methods include the polymerase chain reaction, electrochemical techniques, lateral flow immunoassay, DNA sequencing, microarrays, fluorescent microarrays, and Enzyme-Linked Immunosorbent Assay (ELISA) protocols (Josephson et al. 1993; Wilson et al. 2002). To obtain sensitive and accurate detection, these methods, however, necessitate using expensive reagents, exact instruments, laborious quantification procedures, and complex sample preparation steps (Lindsley et al. 2011; Schulze et al. 2015).

Furthermore, the efficacy of these methods for real-time disease detection is limited. The employment of sensors, which are relatively cheap, straightforward, and highly specific for detecting biomolecules, is one example of a novel approach. These sensors have numerous clinical uses since they can be employed in real-time for monitoring and diagnosing illnesses (Wilson & Charles, 1999; Sapsford et al. 2008; Patolsky et al. 2006; Zhu et al. 2010). The ability to detect diseases early and the need for less invasive treatments are two additional benefits of using sensors. Nanomaterials are used to fabricate these sensors to enhance their electrical and chemical characteristics and, consequently, their sensitivity (Morales-Narváez et al. 201). Many nanomaterials, including graphene and graphene-based nanomaterials, are employed to make these sensors, which have shown impressive features such as better signal detection (Chauhan et al. 2017; Janegitz et al. 2017; Wang et al. 2011). Because of its remarkable biocompatibility with many biological molecules, such as proteins, enzymes, antibodies, DNA, and cells, graphene-based nanomaterials are being greatly extensively investigated (Liu et al. 2012). Biosensors refers to sensors immobilized with biological molecules and are used to detect various chemicals, biomolecules and cells (Shams et al. 2016; Singh et al. 2016). Graphene was developed in 2004 and is a carbon-based nanomaterial composed of a single layer of sp^2 carbon sheet with atoms organized in a honeycomb lattice (Geim & Novoselov, 2010; Park et al. 2009). Among its many distinctive qualities is its tremendous inherent mobility. Graphene is a highly selective and sensitive choice for biosensing applications because of its ability to be functionalized with various linker molecules due to its two-dimensional structure (Stoller et al. 2008; Park et al. 2010; Krajewska et al. 2017). There are several types of graphene, each with unique qualities and uses. Graphene oxide types, including GO and rGO, are also used in biosensing applications. Other forms of graphene include few-layer, multilayer, and Graphene Nano-Platelets (GNP) (Nieto et al. 2012; Pei et al. 2012; Alam et al. 2017; Compton et al. 2010; Eda et al. 2010; Wang et al. 2011; Chen et al. 2011).

The ultrasensitive detection of biomolecules is made possible by the facile molecular interaction and electron transfer from the adsorbed molecules through graphene, which is possible due to the presence of graphene atoms throughout the surface (Yang et al. 2018;

Silicene et al. 2012). Real-time health monitoring and the early diagnosis of several serious diseases and life-threatening conditions have become possible because of graphene-based biosensors. For example, graphene biosensors demonstrated successful diagnosis by identifying the biomarkers of various cancer types and cardiovascular disorders. Patient's quality of life might be improved by real-time disease monitoring. In addition, graphene-based electrochemical biosensors have demonstrated great promise for implantable devices that could identify biomarkers in real time and gauge the severity of a disease.

2. Structure and Properties of Graphene

Graphene's discovery has broadened the carbon materials family by enabling its transformation into 0D fullerenes, stacking into 3D graphite, and rolling into 1D carbon nanotubes. Consequently, graphene can be considered the fundamental building block of the family of carbon-based compounds. Layers that refer to the number of functional groups that include oxygen and other chemical components, are used to categorize graphene-based nanomaterials (GBNs). According to one study, the structural variations of GBNs' determine their unique physical and chemical properties (Fasolino et al. 2007). Research on graphene's structure and characteristics has been largely ignored, leading to its mockery as an "impractical material (Zhao et al. 2017)." The planar 2D structure contradicts Landau's predictions, Mermin-Wagner's theory (Zhao et al. 2017), and atomic force microscopy, revealing that graphene is not a perfect 2D crystal (Figure 1) (Dral et al. 2018).

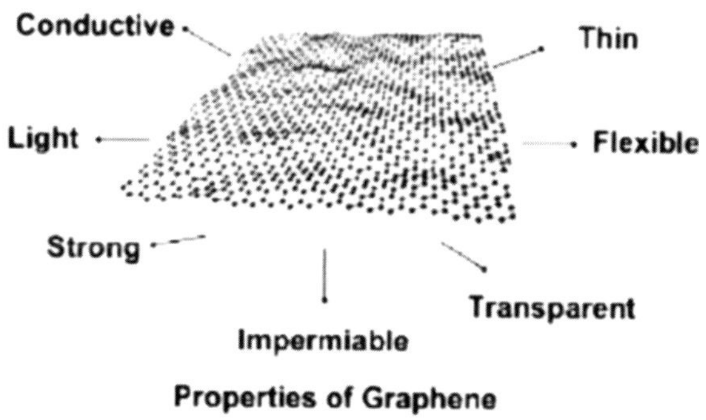

Figure 1. Properties of Graphene.

Planar hexagons make up the monolayer carbon of standard graphene's graphite structure. Graphene, with a thickness of only 0.335 nm, is the thinnest 2D substance ever found. Originally, graphite was exfoliated mechanically or chemically using CVD (Chen et al. 2013). Three additional identical carbons form a connection with one carbon atom. The σ-bond generated by sp^2 hybrid orbitals gives rise to the C–C bond, distinguished by its bond angle of 120° and bond length of 1.42 Å. Each atom contributes an electron in a π–π system formed by the p orbitals of neighboring carbon atoms. Hence, the structure of graphene consists of a honeycomb lattice of pairs of electron clouds and σ-bonds. Large specific areas, significant intrinsic mobility, and advantageous thermal properties are only a few of graphene's excellent mechanical, thermal, and electrical properties. Because graphene has no active groups on its surface, it is extremely stable, inert, and insoluble in water. Graphene's biocompatibility, biochemical functionalization, and cost-effective manufacturing technology make it a desirable nanomaterial for various applications in biotechnology, including medicine delivery, diagnostics, photothermal therapy, biological sensing, tissue engineering, and antimicrobial activity.

3. Synthesis of Graphene

There are multiple ways to synthesize graphene, each with benefits and drawbacks. Mechanical exfoliation, which uses sticky tape or other mechanical tools, is one of the most used techniques for removing graphene layers from graphite. While this process yields high-quality graphene with few flaws and great purity, it takes a long time and is challenging to scale up for mass production. Chemical exfoliation is another technique that uses chemicals or solvents to remove graphene layers from graphite. This process is more scalable than mechanical exfoliation and can yield graphene with certain features. However, it frequently calls for unfavorable circumstances and can produce flawed graphene.

Chemical synthesis refers to the use of chemical processes to generate graphene from smaller carbon units. This technique is more scalable than mechanical exfoliation and can yield high-quality graphene with specific features. However, managing it can be challenging and frequently necessitates intricate chemical processes. Chemical Vapor Deposition (CVD) is a thermal synthesis method wherein carbon atoms are deposited onto a substrate utilizing heat energy. This technique is more scalable than mechanical exfoliation and can yield high-quality graphene with certain features. However, it can be challenging to regulate and frequently demands high temperatures.

On the other hand, bottom-up flash graphene synthesis creates graphene by quickly heating carbon sources in a single step. Compared to other processes, this one is more scalable and can yield high-quality graphene with low concentrations of defects (Table 1).

Table 1. Synthesis of graphene

Graphene-Nanomaterial	Synthesis technique	Forerunner	Reducing/plugging/ stabilizing agent	Temperature	References
Nanoplatelet	Microwave-assisted exfoliation method	Graphite powder	H_2SO_4 and HNO_3	60°C	(Anzar et al. 2020)
Nanoparticle	Hummer system	Graphite powder	H_2SO_4 $KMnO_4$ H_2O_2 $NaNo_3$	80°C	(Cataldi et al. 2018)
Nanoribbon	CVD method	–		–	(Fu et al. 2018)
	Unzipping of CNT	SWCNT	H_2SO_4/HNO_3	100°C	(Ou et al. 2016)
Nanosheet	Green method	Graphene oxide	Glucose (reducing sugar)	95°C	
Nanoflake	Ball milling method	Natural graphite	NaCl	85°C	(Han et al. 2020)

4. Application of Graphene in Disease Diagnosis

GBNs are classified as "nova materials" due to their simple surface modification, rich oxygen-containing functional groups, 2D planar reticular structure, high hydrophilicity, and biocompatibility. They also offer outstanding mechanical, electrical, and photothermal qualities. Consequently, GBNs are widely used in electrochemistry, solar cells, catalysis, new energy sources, novel transistors, and molecular sensors. In recent years, substantial developments have been witnessed in the use of GBNs in the biomedical sciences, including biosensors, biological imaging, antineoplastic drug delivery, and tumor PTT (Figure 2).

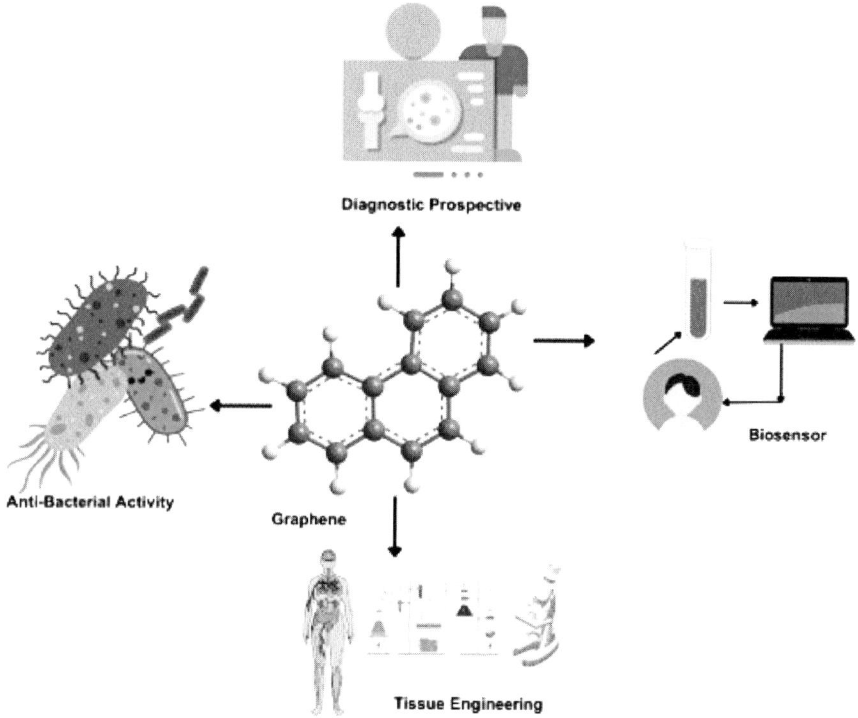

Figure 2. Applications of Graphene.

4.1. Drug/Gene Delivery

GBNs are deemed superior drug/gene carriers due to their intriguing characteristics:

- The 2D monolayer structure with a large specific surface area (2600 $m^{2}\cdot g^{-1}$) enables drug loading with a rate of up to 200-400%.
- GBNs are highly durable and chemically stable, making them suitable for various physical contexts.
- While GO and pure graphene have some cytotoxicity, functionalizing GBNs is simple. By controlling drug release and preventing the build-up of pharmaceuticals or genes, GBN modifications can decrease toxicity, enable targeting, enhance physiological stability, and increase biocompatibility. Currently, two main types of modified materials are employed to enhance the drug delivery capabilities of GBNs: inorganic and organic (Song et al. 2018).

4.2. Biosensor

Biosensors are vital for disease detection, diagnosis, and treatment, utilizing biological and electrical sensing components with excellent electrochemical and optical properties and providing active sites for biological receptors. Because GBNs increase sensitivity, they are an ideal substrate for producing biosensors. Currently, GBN-based biosensors are classified into three groups based on the conduction mechanism:

- Field-effect transistors, or FETs, operate using electrical signals.
- GBN-modified electrodes for electrochemical biosensors.
- Optical biosensors like Foster Resonance Energy Transfer (FRET), Surface-Enhanced Raman Scattering (SERS), Chemiluminescence Resonance Energy Transfer (CRET), Electrochemiluminescence (ECL), Surface-Plasma Resonance (SPR) detect RNA/DNA, protein, and biomarkers like glucose, ATP, and hydrogen peroxide, nicotinamide dinucleotide adenine is commonly detected with GBN biosensors.

Most reactions in traditional sensing methods cannot be quantified in real time, requiring expensive chemicals and high-precision equipment. Graphene-based sensors have several biomedical uses and are currently used as a substitute for identifying proteins linked to specific diseases. Excellent sensing capabilities, affordability, nontoxicity, and ease of use characterize graphene biosensors (Gurunathan et al. 2018). A sensor typically consists of two parts: a transducer, which transforms chemical information into signals, and a receptor, which is connected to the target molecule. Graphene or GO-based biosensors function as transducers, converting the interaction between a target chemical and a receptor into a detectable signal. Receptors like enzymes, antibodies, or nucleic acids are frequently affixed to transducers to facilitate the interaction with target molecules (Tweedie et al. 2020; Liu et al. 2012).

4.3. Microbes Detection

Graphene biosensors are capable of detecting microbes via spectroscopy (Jin et al. 2019), wherein they are attached with antibodies, nucleic acids, proteins, and enzymes. Such graphene biosensors have been used to detect infectious diseases like the *Ebola* virus (Maity et al. 2018; Thakur et al. 2018), *Escherichia coli* (Afsahi et al. 2018) and Zika virus (Sign et al. 2016) and *Salmonella typhimurium* (Fan et al. 2019), HCV (Li et al. 2020; Veerapandian et al. 2016), and avian influenza virus H7 (Yang et al. 2018) using silver and gold nanoparticles.

4.4. Detection of Nucleic Acid and Gene

The DNA-based graphene biosensor employs electrochemical and fluorescent detection methods to identify various biomarkers, including viruses, genes, and tiny molecules and proteins. The study uses electrochemical techniques to immobilize DNA on graphene biosensor surfaces, detecting voltage and current shifts caused by conductivity changes and electron depletion. CV, differential pulse voltammetry, and electrochemical impedance spectroscopy were utilized to measure voltage and current shifts influenced by conductivity changes and electron depletion due to oxidation or hybridization.

4.5. Uncovering of Enzymes and Other Molecules

Enzyme-immobilized graphene biosensors detect various enzymes and related biomolecules through covalent bonding, physical entrapment, EDC/NHS chemistry, and adsorption methods, allowing the detection of substances like 2,6-dimethozyphenol, hydrogen peroxide, caffeic acid, glucose, bilirubin, and 17β-estradiol.

4.6. Bioimaging

In research and clinical practice, bioimaging is essential, especially when it comes to identifying the aberrant processes associated with disease and comprehending how cells, tissues, and organs function. Notably, inadequate sensitivity, specificity, and targeting are linked to the restricted use of bioimaging technology in medicine. As a result, developing various imaging techniques for such purposes is essential. GBNs, with GQDs as the main material, are widely used in bioimaging systems for Fluorescence Imaging (FRI), Magnetic Resonance Imaging (MRI), Raman Spectroscopy Imaging (RSI), and Photo-Acoustic Imaging (PAI).

4.7. Fluorescence Imaging

GBNs have strong fluorescence properties for fluorescence *in vivo* imaging, but their low quantum yield limits their use. Fluorescent dyes can functionalize GBNs to increase FRI signals (Song et al. 2019) and enable real-time medication delivery and distribution *in vivo* monitoring. Highly visible/NIR-II luminescent hydrophilic lanthanum-doped nanocrystals (Ln^{3+}-NCs) were synthesized (Yan et al. 2019).

4.8. Magnetic Resonance Imaging

With superior spatial resolution, depth of tissue penetration, and non-invasiveness, Magnetic Resonance Imaging (MRI) is the preferred method in bioimaging. Manganese (Mn) and Gadolinium (Gd) are commonly used contrast agents in MRI (Cai et al. 2016; Li et al. 2019), but they are known for their nonspecific toxicity *in vivo*. Different functional groups on GBNs easily bond with dangerous paramagnetic metal ions, lowering their toxicity. Mn-NGQDs and Gd-NGQDs are novel nitrogen-containing GQDs that have been doped with Mn^{2+} or Gd^{3+}, according to research conducted by Lee and colleagues (Lee et al. 2020). Chelation of GQDs enhances the biocompatibility of Mn^{2+} and Gd^{3+} *in vitro*, enabling dual imaging modes of MRI and FRI.

4.9. Tissue Engineering

The tissues and organs within the human body have different biological and physical properties. Therefore, organ damage caused by serious or chronic disorders might not be resolved by a single substance. A recent development in biomedicine is tissue engineering, a field of study that combines expertise in biology, medicine, engineering, and other relevant fields. Its goal is to create biodegradable replacements that can alter an organ's function or restore and preserve its features (Bakhshandeh et al. 2017; Vega et al. 2017). Currently, the following are the key parameters that determine the materials or processes used to heal injured tissues or organs: The repair of tissue damage resulting from wounds, infections, cancers, and malformation is greatly limited by

- Conventional transplantation
- Compatibility, toxicity, and anticoagulant properties
- Materials lacking tissue-induced activity, such as calcium silicate and calcium phosphate, which delay wound healing
- Materials lacking extensibility, aggregation, and mobility, such as tio2 and hap
- Other materials, such as hydrogels, lack the mechanical strength for cell adhesion and diffusion.

The extensive application of Graphene-Based Nanomaterials (GBNs) in tissue engineering can be attributed to their distinct physical, chemical, and biological characteristics, including high strength, flexibility, and significant surface adaptation. Furthermore, they can control how cells interact with one another and stimulate stem cells to develop into tissue cells with certain roles. GBNs have the potential to be both neurogenic and osteogenic. Furthermore, GBNs' surface oxygen-containing groups can be easily modulated, facilitating their interaction with various biomolecules, such as peptides, nucleic acids, and proteins (Ghosh et al. 2020). GBNs are beneficial for tissue engineering due to their antibacterial properties and manufacturing flexibility, but their widespread use in various tissue cells hinders regeneration.

4.10. Antibacterial

The antibacterial community has been seriously threatened in recent decades by bacterial and fungal resistance brought about by antibiotic abuse. Scientists are investigating novel strategies to counteract multidrug resistance, such as creating and combining novel medications. The morphological, physical, and chemical characteristics of GBNs—such as their degree of oxidation, hydrophilicity, charge, lateral size, purity, and structural defects—significantly impact their antibacterial activity. GBNs' antibacterial activity is influenced by various components (Xia et al. 2019). Modifying GBN surfaces can create low-toxicity, high-antibacterial nanocomposites and can be used in biomedical applications like medicine delivery, food packaging, and microbial infection control (Yan et al. 2019).

4.11. Biocompatibility Evaluation

Biocompatibility is the compatibility of inert materials with live bodily tissues, as defined by the International Standards Organization (ISO). When biomaterials are put into the human body, they affect the biological tissues surrounding them, and vice versa. The cyclic effect continues until balance is achieved or the implant is removed. One of the most common issues in biomaterials research is biocompatibility.

The material's chemical nature and intended application have a big impact on biocompatibility. In brief, material's shape, size, charge, and other physical and chemical characteristics—as well as any hazardous materials left over from production and breakdown products within the body—all impact how biocompatible the material is. The material is poisonous, irritating, teratogenic, and produces localized inflammation in both cells and the body as a whole when it comes into short-term contact with the body. In addition to causing aberrant blood coagulation and hemolysis, prolonged contact with blood can have mutagenic, teratogenic, and carcinogenic effects. Therefore, biocompatibility is a crucial criterion that needs to be looked at and evaluated when assessing the use of materials in the biomedical industry.

Size, shape, and surface functionalization all affect GBNs' physical, chemical, and biological characteristics. To properly utilize GBNs in biomedical applications, an assessment of their *in vivo* and *in vitro* biocompatibility is required; this is also necessary for more systematic research into GBNs.

4.11. *In Vitro* Cytocompatibility

The foundation of GBNs' use in biomedicine is their cyto-compatibility, which is primarily determined by their physical and chemical characteristics, such as their solubility in water, diameter, concentration, number of layers, action time, and surface modification (Geng et al. 2019). Similar to GO and GQDs, GBNs are typically linked to enhanced cytocompatibility and reduced toxicity. On the other hand, GO shows dose-time-dependent substantial cytotoxicity at greater doses. Furthermore, hydrophilic GBNs, such as those found in dental pulp stem cells, potently stimulate cell differentiation and adhesion (Rosa et al. 2016). Large GO may be intercepted in the membrane cytoskeleton during aggregation. As a result, GO aggregation causes cell death, distortion, or fracture.

Therefore, by appropriately modifying the surface of GBNs, researchers can make them more soluble in water and stop them from building up. In the end, minimal cytotoxicity and long-term stability are attained, which raises cytocompatibility. It was previously discovered by Bhusari et al. that functionalized GO promotes myoblast development (Bhusari et al. 2019). However, no discernible difference in the number of cells was reported after three days in culture. Furthermore, there was a strong correlation between cytocompatibility and GBN size. The experiment showed that nano-sized graphene oxide (GO) produced a less inflammatory response than micron-sized GO, indicating better cytocompatibility than micron GO.

4.12. *In Vivo* Cytocompatibility

Concentration, physical and chemical properties, functional alteration, and action duration all affect GBNs' biocompatibility *in vivo*. This effect's primary mechanism is comparable to *in vitro* cytocompatibility (Li et al. 2021). Numerous investigations have demonstrated that altering co-polymers can enhance GBNs' *in vivo* biocompatibility. The toxicity *in vivo* is even lower than GO when the unoxidized graphene surface mosaic is pluronic (Shamsi et al. 2020). Furthermore, co-polymer modification provides a dependable and secure solution for GBNs in the biomedical industry.

The biodegradation rate and delivery mode impact GBNs' *in vivo* biocompatibility. An *in vivo* GO toxicity investigation showed that following an intravenous injection, the lungs received the initial concentration of GO through blood flow. PEG-GO then accumulated primarily in the liver and spleen and was eliminated by the kidney and feces over time. Following a 3-month dose of 20 mg kg^{-1}, the experimental group showed no discernible signs of PEG-GO toxicity (Li et al. 2014). Furthermore, graphene broke down in plasma into aromatic hydrocarbons and porosity sheets. Active plasma molecules like OH^- and O_2^- break down in graphene's sp^2 plane, where carbon atoms connected to oxygen-containing groups are present. Furthermore, oxidation has a direct impact on the pace of deterioration. Significant connections exist between immune cell infiltration, absorption, and clearance and reduced GBN oxidation (Li et al. 2019).

The preparation method of GBNs impacts their biocompatibility, with liquid-phase exfoliation reducing toxicity and high levels of Reactive Oxygen Species (ROS) in graphene neurons, reducing the apoptotic number.

5. Timeline of Graphene-Based Disease Diagnosis Research

5.1. Discovery of Graphene: 2004

The discovery of graphene in 2004 by Andre Geim and Konstantin Novoselov at the University of Manchester opened new applications for sensing and diagnostics applications. Graphene's unique properties, such as high surface area, excellent conductivity, and biocompatibility, made it an attractive material for biosensor development.

5.2. Exploration of Graphene-Based Biosensors: 2010

During the 2010s, extensive research was conducted on graphene and graphene-based materials for the fabrication of biosensors. Researchers explored various methods to synthesize and functionalize graphene to enhance its sensitivity, selectivity, and stability. These efforts led to the development of graphene-based biosensors with improved performance compared to traditional sensing platforms.

5.3. Focus on Cancer Biomarker Detection: 2015-2018

From 2015 to 2018, research in graphene-based disease diagnosis focused primarily on developing electrochemical biosensors for the ultrasensitive detection of cancer biomarkers and risk factors. Graphene's high surface area and excellent conductivity enabled the fabrication of highly sensitive sensors capable of detecting cancer-related molecules at low concentrations.

5.4. Graphene-Based Sensors for COVID-19 Detection: 2020

In 2020, the COVID-19 pandemic highlighted the urgent need for rapid and accurate diagnostic tools. Researchers leveraged the advantages of graphene-based sensors to develop rapid tests for detecting SARS-CoV-2, the virus causing COVID-19. These sensors demonstrated the ability to detect the virus in under 5 minutes, showcasing the potential of graphene-based diagnostics in addressing global health challenges.

6. Current Trends in Graphene-Based Disease Diagnosis

6.1. Graphene-Based Biosensors for Early Disease Detection

One of the current trends in graphene-based disease diagnosis is the development of biosensors for the early detection of life-threatening diseases such as cancer, Alzheimer's, and COVID-19. Early diagnosis is crucial for improving treatment outcomes and patient prognosis. Graphene-based sensors offer high sensitivity and selectivity, enabling the detection of disease biomarkers at low concentrations, even in complex biological samples.

6.2. Graphene Oxide in Disease Diagnosis and Treatment

Graphene oxide, a graphene derivative, has shown potential as a valuable tool for diagnosing and treating diseases due to its extraordinary mechanical properties and ability to interact with various biomolecules. Researchers are exploring the use of graphene oxide in targeted drug delivery, gene therapy, and theranostics applications, combining diagnostic and therapeutic capabilities.

6.3. Graphene-Based Wearable Sensors

The development of graphene-based wearable sensors is another current trend in disease diagnosis. These sensors can be integrated into clothing, accessories, or directly onto the skin, enabling continuous monitoring of physiological parameters and early detection of health issues. Graphene's flexibility, durability, and conductivity make it an ideal material for wearable sensor applications.

6.4. Commercialization of Graphene-Based Biosensors

The commercialization of graphene-based biosensors is a race among researchers and companies worldwide. The goal is to develop cost-effective, user-friendly diagnostic devices that can be easily deployed in clinical settings, point-of-care facilities, and even at home. Successful commercialization will depend on overcoming challenges related to consistent quality, reliability, and scalability of graphene production.

7. Challenges and Future Prospects

Despite the significant progress made in graphene-based disease diagnosis, several challenges remain. Consistent quality, reliability, and cost-effectiveness of bulk graphene production are crucial for widely adopting graphene-based diagnostic tools. Additionally, concerns about graphene-based materials' potential toxicity and long-term biocompatibility must be addressed through extensive safety studies and regulatory approvals.

Prospects for graphene-based disease diagnosis include the development of multifunctional sensors capable of detecting multiple biomarkers simultaneously, enabling more comprehensive and accurate diagnosis. Integrating graphene-based sensors with advanced technologies such as artificial intelligence and machine learning can enhance data analysis and decision-making capabilities. Furthermore, combining graphene-based sensors with microfluidic devices and lab-on-a-chip platforms can lead to the development of portable, point-of-care diagnostic systems.

Conclusion

Graphene-based biosensors are developed using nanomaterials like virgin graphene, GO, rGO, and graphene quantum dots. These biosensors showed promise for target molecule detection applications in biomedical and non-biomedical industries. Graphene's unique properties, including mobility, large surface area, transparency, and nontoxicity, enable the creation of sensitive biosensors for early disease detection, enhancing biosensing platforms. Advancements in research on graphene-based nanomaterials reveal their exceptional potential for specific applications. Cost-effectiveness, miniaturization, precision and reproducibility are further advantages of graphene-based biosensors. However, it's necessary to pay attention to the serious drawbacks of graphene-based technologies. Graphene's ability to adsorb nontarget molecules may result in false-positive detection, particularly when employing noncovalent functionalization techniques. Conversely, covalent functionalization methods require fine conditional control and are challenging. External environmental factors and intrinsic attributes of graphene, including type, molecular weight, and surface properties, influence biomolecule loading onto graphene surfaces. Further research and improved production methods are necessary for design and development of biosensing devices.

Acknowledgments

The authors thank the Senior Management Team of Centurion University of Technology and Management for their kind support.

References

Afsahi, Savannah, Mitchell B. Lerner, Jason M. Goldstein, Joo Lee, Xiaoling Tang, Dennis A. Bagarozzi Jr, Deng Pan et al. "Novel graphene-based biosensor for early detection of Zika virus infection." *Biosensors and Bioelectronics* 100 (2018): 85-88.

Alam, Syed Nasimul, Nidhi Sharma, and Lailesh Kumar. "Synthesis of graphene oxide (GO) by modified hummers method and its thermal reduction to obtain reduced graphene oxide (rGO)." *Graphene* 6, no. 1 (2017): 1-18.

Anzar, Nigar, Rahil Hasan, Manshi Tyagi, Neelam Yadav, and Jagriti Narang. "Carbon nanotube-A review on Synthesis, Properties and plethora of applications in the field of biomedical science." *Sensors International* 1 (2020): 100003.

Bakhshandeh, Behnaz, Payam Zarrintaj, Mohammad Omid Oftadeh, Farid Keramati, Hamideh Fouladiha, Salma Sohrabi-Jahromi, and Zarrintaj Ziraksaz. "Tissue engineering; strategies, tissues, and biomaterials." *Biotechnology and genetic engineering reviews* 33, no. 2 (2017): 144-172.

Bhusari, Shardul Atul, Vidushi Sharma, Suryasarathi Bose, and Bikramjit Basu. "HDPE/UHMWPE hybrid nanocomposites with surface functionalized graphene oxide towards improved strength and cytocompatibility." *Journal of the Royal Society Interface* 16, no. 150 (2019): 20180273.

Cai, Xiaoxia, Qingxia Zhu, Yun Zeng, Qi Zeng, Xueli Chen, and Yonghua Zhan. "Manganese oxide nanoparticles as MRI contrast agents in tumor multimodal imaging and therapy." *International journal of nanomedicine* (2019): 8321-8344.

Cataldi, Pietro, Athanassia Athanassiou, and Ilker S. Bayer. "Graphene nanoplatelets-based advanced materials and recent progress in sustainable applications." *Applied Sciences* 8, no. 9 (2018): 1438.

Chauhan, Neha, Toru Maekawa, and Dasappan Nair Sakthi Kumar. "Graphene-based biosensors—Accelerating medical diagnostics to new dimensions." *Journal of Materials Research* 32, no. 15 (2017): 2860-2882.

Chen, Ji, Bowen Yao, Chun Li, and Gaoquan Shi. "An improved Hummers method for eco-friendly synthesis of graphene oxide." *Carbon* 64 (2013): 225-229.

Chen, Liuyun, Yanhong Tang, Ke Wang, Chengbin Liu, and Shenglian Luo. "Direct electrodeposition of reduced graphene oxide on glassy carbon electrode and its electrochemical application." *Electrochemistry communications* 13, no. 2 (2011): 133-137.

Compton, Owen C., and SonBinh T. Nguyen. "Graphene oxide, highly reduced graphene oxide, and graphene: versatile building blocks for carbon-based materials." *small* 6, no. 6 (2010): 711-723.

Dral, A. Petra, and Johan E. ten Elshof. "2D metal oxide nanoflakes for sensing applications: Review and perspective." *Sensors and Actuators B: Chemical* 272 (2018): 369-392.

Eda, Goki, and Manish Chhowalla. "Chemically derived graphene oxide: towards large-area thin-film electronics and optoelectronics." *Advanced materials* 22, no. 22 (2010): 2392-2415.

Fan, Jialong, Liqun Yuan, Qingxin Liu, Chunyi Tong, Wenmiao Wang, Feng Xiao, Bin Liu, and Xuanming Liu. "An ultrasensitive and simple assay for the Hepatitis C virus using a reduced graphene oxide-assisted hybridization chain reaction." *Analyst* 144, no. 13 (2019): 3972-3979.

Fasolino, Annalisa, Los, and Mikhail Katsnelson. "Intrinsic ripples in graphene." *Nature materials* 6, no. 11 (2007): 858-861.

Fu, Liangjie, Zhaoli Yan, Qihang Zhao, and Huaming Yang. "Novel 2D nanosheets with potential applications in heavy metal purification: A review." *Advanced Materials Interfaces* 5, no. 23 (2018): 1801094.

Geim, and Novoselov. "The rise of graphene, nanoscience and technology: a collection of reviews from nature journals." *World Sci,* 11–19. (2010).

Geng, Hao, Tianbo Wang, Huiliang Cao, Hongqin Zhu, Zengfeng Di, and Xuanyong Liu. "Antibacterial ability, cytocompatibility and hemocompatibility of fluorinated graphene." *Colloids and Surfaces B: Biointerfaces* 173 (2019): 681-688.

Ghosh, Santanu, and Kaushik Chatterjee. "Poly (Ethylene glycol) functionalized graphene oxide in tissue engineering: A review on recent advances." *International Journal of Nanomedicine* (2020): 5991-6006.

Gurunathan, Sangiliyandi, and Jin-Hoi Kim. "Synthesis, toxicity, biocompatibility, and biomedical applications of graphene and graphene-related materials." *International journal of nanomedicine* (2016): 1927-1945.

Han, Xiao-Min, Ke-Wen Zheng, Rui-Long Wang, Shu-Fen Yue, Jing Chen, Zi-Wei Zhao, Fang Song, Yan Su, and Qiang Ma. "Functionalization and optimization-strategy of graphene oxide-based nanomaterials for gene and drug delivery." *American Journal of Translational Research* 12, no. 5 (2020): 1515.

Haque, Rashidul, Kendra Kress, Sheila Wood, Terry FHG Jackson, David Lyerly, Tracy Wilkins, and William A. Petri Jr. "Diagnosis of pathogenic Entamoeba histolytica infection using a stool ELISA based on monoclonal antibodies to the galactose-specific adhesin." *Journal of Infectious Diseases* 167, no. 1 (1993): 247-249.

Janegitz, Bruno Campos, Tiago Almeida Silva, Ademar Wong, Laís Ribovski, Fernando Campanhã Vicentini, Maria del Pilar Taboada Sotomayor, and Orlando Fatibello-Filho. "The application of graphene for *in vitro* and *in vivo* electrochemical biosensing." *Biosensors and Bioelectronics* 89 (2017): 224-233.

Jin, Xin, Hong Zhang, Yu-Tao Li, Meng-Meng Xiao, Zhi-Ling Zhang, Dai-Wen Pang, Gary Wong, Zhi-Yong Zhang, and Guo-Jun Zhang. "A field effect transistor modified with reduced graphene oxide for immunodetection of Ebola virus." *Microchimica Acta* 186 (2019): 1-9.

Josephson, Charles Gerba, and IL182487 Pepper. "Polymerase chain reaction detection of nonviable bacterial pathogens." *Applied and Environmental Microbiology* 59, no. 10 (1993): 3513-3515.

Krajewska, Aleksandra, Iwona Pasternak, Grzegorz Sobon, Jaroslaw Sotor, Aleksandra Przewloka, Tymoteusz Ciuk, Jan Sobieski, Justyna Grzonka, Krzysztof M. Abramski, and Wlodek Strupinski. "Fabrication and applications of multi-layer graphene stack on transparent polymer." *Applied Physics Letters* 110, no. 4 (2017).

Lazcka, Olivier, Javier Del Campo, and Xavier Munoz. "Pathogen detection: A perspective of traditional methods and biosensors." *Biosensors and bioelectronics* 22, no. 7 (2007): 1205-1217. 17.

Lee, Bong Han, Md Tanvir Hasan, Denise Lichthardt, Roberto Gonzalez-Rodriguez, and Anton V. Naumov. "Manganese–nitrogen and gadolinium–nitrogen Co-doped graphene quantum dots as bimodal magnetic resonance and fluorescence imaging nanoprobes." *Nanotechnology* 32, no. 9 (2020): 095103.

Li, Bo, Xiao-Yong Zhang, Jian-Zhong Yang, Yu-Jie Zhang, Wen-Xin Li, Chun-Hai Fan, and Qing Huang. "Influence of polyethylene glycol coating on biodistribution and toxicity of nanoscale graphene oxide in mice after intravenous injection." *International journal of nanomedicine* (2014): 4697-4707.

Li, Dandan, Xiangang Hu, and Suyan Zhang. "Biodegradation of graphene-based nanomaterials in blood plasma affects their biocompatibility, drug delivery, targeted organs and antitumor ability." *Biomaterials* 202 (2019): 12-25.

Li, Hao, and Thomas J. Meade. "Molecular magnetic resonance imaging with Gd (III)-based contrast agents: challenges and key advances." *Journal of the American Chemical Society* 141, no. 43 (2019): 17025-17041.

Li, Jingwen, Yifei Li, Xingwei Zhai, Ya Cao, Jing Zhao, Yuguo Tang, and Kun Han. "Sensitive electrochemical detection of hepatitis C virus subtype based on nucleotides assisted magnetic reduced graphene oxide-copper nanocomposite." *Electrochemistry Communications* 110 (2020): 106601.

Li, Zhentao, Xin Yan, Keke Wu, Yanpeng Jiao, Changren Zhou, and Jingxin Yang. "Surface modification of reduced graphene oxide beads: integrating efficient endotoxin adsorption and improved blood compatibility." *ACS Applied Biomaterials* 4, no. 6 (2021): 4896-4906.

Lindsley, Mark, Nanthawan Mekha, Henry Baggett, Yupha Surinthong, Rinrapas Autthateinchai, Pongpun Sawatwong, Julie Harris, et al. "Evaluation of a newly developed lateral flow immunoassay for the diagnosis of cryptococcosis." *Clinical infectious diseases* 53, no. 4 (2011): 321-325.

Liu, Jingquan, Jianguo Tang, and J. Justin Gooding. "Strategies for chemical modification of graphene and applications of chemically modified graphene." *Journal of materials chemistry* 22, no. 25 (2012): 12435-12452.

Liu, Jingquan, Jianguo Tang, and J. Justin Gooding. "Strategies for chemical modification of graphene and applications of chemically modified graphene." *Journal of materials chemistry* 22, no. 25 (2012): 12435-12452.

Maity, Arnab, Xiaoyu Sui, Bing Jin, Haihui Pu, Kai J. Bottum, Xingkang Huang, Jingbo Chang, Guihua Zhou, Ganhua Lu, and Junhong Chen. "Resonance-frequency modulation for rapid, point-of-care Ebola-Glycoprotein diagnosis with a graphene-based field-effect biotransistor." *Analytical chemistry* 90, no. 24 (2018): 14230-14238.

Morales-Narváez, Eden, Luis Baptista-Pires, Alejandro Zamora-Gálvez, and Arben Merkoçi. "Graphene-based biosensors: going simple." *Advanced Materials* 29, no. 7 (2017): 1604905.

Nieto, Andy, Debrupa Lahiri, and Arvind Agarwal. "Synthesis and properties of bulk graphene nanoplatelets consolidated by spark plasma sintering." *Carbon* 50, no. 11 (2012): 4068-4077.

Ou, Lingling, Bin Song, Huimin Liang, Jia Liu, Xiaoli Feng, Bin Deng, Ting Sun, and Longquan Shao. "Toxicity of graphene-family nanoparticles: a general review of the origins and mechanisms." *Particle and fibre toxicology* 13 (2016): 1-24.

Park, Hye Jin, Jannik Meyer, Siegmar Roth, and Viera Skákalová. "Growth and properties of few-layer graphene prepared by chemical vapor deposition." *Carbon* 48, no. 4 (2010): 1088-1094.

Park, Sungjin, and Rodney Ruoff. "Chemical methods for the production of graphenes." *Nature nanotechnology* 4, no. 4 (2009): 217-224.

Patolsky, Fernando, Gengfeng Zheng, and Charles Lieber. "Nanowire sensors for medicine and the life sciences." (2006): 51-65.

Pei, Songfeng, and Hui-Ming Cheng. "The reduction of graphene oxide." *Carbon* 50, no. 9 (2012): 3210-3228.

Rosa, Vinicius, Han Xie, Nileshkumar Dubey, Thulasi T. Madanagopal, Sneha S. Rajan, Julien Luc Paul Morin, Intekhab Islam, and Antônio Hélio Castro Neto. "Graphene oxide-based substrate: physical and surface characterization, cytocompatibility and differentiation potential of dental pulp stem cells." *Dental Materials* 32, no. 8 (2016): 1019-1025.

Sapsford, Kim, Christopher Bradburne, James Delehanty, and Igor Medintz. "Sensors for detecting biological agents." *Materials Today* 11, no. 3 (2008): 38-49.

Schulze, Holger, Maya Rubtsova, and Till T. Bachmann. "DNA microarrays for pathogen detection." *Modern Techniques for Pathogen Detection* (2015): 113-220.

Shams, Nafiseh, Hong Ngee Lim, Reza Hajian, Nor Azah Yusof, Jaafar Abdullah, Yusran Sulaiman, Izwaharyanie Ibrahim, and Nay Ming Huang. "Electrochemical sensor based on gold nanoparticles/ethylenediamine-reduced graphene oxide for trace determination of fenitrothion in water." *RSC advances* 6, no. 92 (2016): 89430-89439.

Shamsi, Suhaili, Addison Alvin Alagan, Seri Narti Edayu Sarchio, and Faizah Md Yasin. "Synthesis, characterization, and toxicity assessment of Pluronic F127-functionalized graphene oxide on the embryonic development of Zebrafish (Danio Rerio)." *International Journal of Nanomedicine* (2020): 8311-8329.

Sign, Chandan, and Gajjala Sumana. "Antibody conjugated graphene nanocomposites for pathogen detection." *In Journal of Physics: Conference Series* 704, no. 1, (2016): 012014.

Silicene, Germanene Be Minigap Semiconductors. "A Stable "Flat "Form of Two-Dimensional Crystals: Could Graphene." (2012).

Singh, Rajesh Kumar, Rajesh Kumar, and Dinesh Pratap Singh. "Graphene oxide: strategies for synthesis, reduction and frontier applications." *RSC Advances* 6, no. 69 (2016): 64993-65011.

Song, Xiaorong, Shihua Li, Hanhan Guo, Wenwu You, Xiaoying Shang, Renfu Li, Datao Tu et al. "Graphene-oxide-modified lanthanide nanoprobes for tumor-targeted visible/NIR-II luminescence imaging." *Angewandte Chemie International Edition* 58, no. 52 (2019): 18981-18986.

Song, Yuan-Yang, Cheng Li, Xiao-Quan Yang, Jie An, Kai Cheng, Yang Xuan, Xue-Meng Shi et al. "Graphene oxide coating core–shell silver sulfide@ mesoporous silica for active targeted dual-mode imaging and chemo-photothermal synergistic therapy against tumors." *Journal of Materials Chemistry* B 6, no. 29 (2018): 4808-4820.

Stoller, Meryl, Sungjin Park, Yanwu Zhu, Jinho An, and Rodney Ruoff. "Graphene-based ultracapacitors." *Nano letters* 8, no. 10 (2008): 3498-3502.

Thakur, Bhawana, Guihua Zhou, Jingbo Chang, Haihui Pu, Bing Jin, Xiaoyu Sui, Xiaochen Yuan, Ching-Hong Yang, Matthew Magruder, and Junhong Chen. "Rapid detection of single E. coli bacteria using a graphene-based field-effect transistor device." *Biosensors and Bioelectronics* 110 (2018): 16-22.

Tweedie, Martin EP, Veerle Kersemans, Stuart Gilchrist, Sean Smart, and Jamie H. Warner. "Electromagnetically transparent graphene respiratory sensors for multimodal small animal imaging." *Advanced Healthcare Materials* 9, no. 21 (2020): 2001222.

Veerapandian, Murugan, Robert Hunter, and Suresh Neethirajan. "Dual immunosensor based on methylene blue-electro adsorbed graphene oxide for rapid detection of the influenza A virus antigen." *Talanta* 155 (2016): 250-257.

Vega, Sebastián L., Mi Y. Kwon, and Jason A. Burdick. "Recent advances in hydrogels for cartilage tissue engineering." European cells & materials 33 (2017): 59.

Wang, Xin, and Sheng Chen. "Graphene-based nanocomposites." *Physics and Applications of Graphene-Experiments* (2011): 135-168.

Wang, Ying, Zhaohui Li, Jun Wang, Jinghong Li, and Yuehe Lin. "Graphene and graphene oxide: biofunctionalization and applications in biotechnology." *Trends in biotechnology* 29, no. 5 (2011): 205-212.

Wilson, Charles B. "Sensors in medicine." *BMJ* 319, no. 7220 (1999): 1288.

Wilson, Strout, DeSantis, Stilwell, Carrano, and Andersen. "Sequence-specific identification of 18 pathogenic microorganisms using microarray technology." *Molecular and cellular probes* 16, no. 2 (2002): 119-127.

Xia, Meng-Ying, Yu Xie, Chen-Hao Yu, Ge-Yun Chen, Yuan-Hong Li, Ting Zhang, and Qiang Peng. "Graphene-based nanomaterials: the promising active agents for antibiotics-independent antibacterial applications." *Journal of Controlled Release* 307 (2019): 16-31.

Yan, Yayuan, Chunya Li, Haoping Wu, Jie Du, Jing Feng, Jingxian Zhang, Langhuan Huang, Shaozao Tan, and Qingshan Shi. "Montmorillonite-modified reduced graphene oxide stabilizes copper nanoparticles and enhances bacterial adsorption and antibacterial activity." *ACS Applied Biomaterials* 2, no. 5 (2019): 1842-1849.

Yan, Yibo, Jun Gong, Jie Chen, Zhiping Zeng, Wei Huang, Kanyi Pu, Jiyang Liu, and Peng Chen. "Recent advances on graphene quantum dots: from chemistry and physics to applications." *Advanced materials* 31, no. 21 (2019): 1808283.

Yang, Gao, Lihua Li, Wing Bun Lee, and Man Cheung Ng. "Structure of graphene and its disorders: a review." *Science and technology of advanced materials* 19, no. 1 (2018): 613-648.

Yang, Limin, Bo Liu, Meimei Wang, Jia Li, Wei Pan, Xiaonan Gao, Na Li, and Bo Tang. "A highly sensitive strategy for fluorescence imaging of microRNA in living cells and *in vivo* based on graphene oxide-enhanced signal molecules quenching of molecular beacon." *ACS applied materials & interfaces* 10, no. 8 (2018): 6982-6990.

Zhao, Guoke, Xinming Li, Meirong Huang, Zhen Zhen, Yujia Zhong, Qiao Chen, Xuanliang Zhao et al. "The physics and chemistry of graphene-on-surfaces." *Chemical Society Reviews* 46, no. 15 (2017): 4417-4449.58. J.

Zhao, Hong, Ruihua Ding, Xin Zhao, Yiwei Li, Liangliang Qu, Hao Pei, Lara Yildirimer, Zhengwei Wu, and Weixia Zhang. "Graphene-based nanomaterials for drug and/or gene delivery, bioimaging, and tissue engineering." *Drug Discovery Today* 22, no. 9 (2017): 1302-1317.

Zhu, Yanwu, Shanthi Murali, Weiwei Cai, Xuesong Li, Ji Won Suk, Jeffrey Potts, and Rodney Ruoff. "Graphene and graphene oxide: synthesis, properties, and applications." *Advanced Materials* 22, no. 35 (2010): 3906-3924.

Chapter 3.4

Nano-Encapsulated Marine Fish Oil and Its Potential Effect on Rheumatoid Arthritis Therapy

Sushree Swati Rout[1,*]
and Bhaskar Behera[1,2]
[1]P. G. Department of Zoology, Fakir Mohan University, Odisha, India
[2]P. G. Department of Bioscience and Biotechnology, Fakir Mohan University, Odisha, India

Abstract

Metal-based nanoparticles are composed of metals like gold, silver, iron, and others. These nanoparticles have exclusive activities due to their insignificant dimension and high external area, making it useful in many biomedical applications. Marine fish oil, particularly eicosa-pentaenoic acids, and docosahexaenoic acids in omega-3 fatty acids, are recommended for joint pain and arthritis. These omega-3s have anti-inflammatory activities, which can help decrease irritation in the joints, thereby alleviating pain and stiffness associated with arthritis. Common sources of marine fish oil include salmon, mackerel, sardines, and anchovies. However, the oil is typically extracted from these fishes and processed into medicinal supplements. It is important to discuss with doctors and medical professionals before initiating a new fish oil medicinal supplement if patients have different health disorders or are taking other medications. Nanoparticles can enhance the bioavailability and efficacy of marine fish oil compounds, particularly omega-3 fatty acids like eicosa-pentaenoic acids associated with docosa-hexaenoic acids, which contain anti-inflammatory effects. Nanoparticles can protect the fatty acids with omega-3 from degradation in the digestive system and improve their absorption in the body, leading to higher concentrations reaching the target tissues, including the joints. By delivering a higher concentration of fatty acids with omega-3 to the joints, marine fish fat or oil nanoparticles may exert more potent anti-inflammatory effects, potentially reducing pain and stiffness associated with arthritis., In this chapter, we have also discussed the importance and efficacy of marine fish oil nanoparticles for joint pain and arthritis.

Keywords: cardinal fish, *Apogonichthyoides sialis*, fish oil, omega-3, knee pain

---

* Corresponding Author's Email: swatirout32@gmail.com

In: Emerging Nanomaterials in Biomedical Science
Editors: Eepsita Priyadarshini and Ajith Manayil Parambil
ISBN: 979-8-89530-600-0

1. Introduction

Metal-based nanoparticles (MNPs) are a class of nanomaterials that have a crucial role because of their exclusive pharmacological activities and potential effects across different health care fields. In this study a detailed overview of metal-based nanoparticles, covering their types, synthesis methods, properties, applications, and challenges, (Alhalili 2023) has been discussed. Metal nanoparticles like gold are known for their stability, quick functionalization and optical properties. Silver nanoparticles are notable for their antimicrobial properties. Platinum nano is valued for their catalytic activities. Palladium nanomaterials are used in catalysis and hydrogen storage while copper nanoparticles exhibit antimicrobial and catalytic properties. Some metal oxide nanoparticles like TiO_2 are widely used in sunscreens and photocatalysis (Naseem and Tayyiba 2021). ZnO is known for UV protection and antimicrobial effects while iron oxide nanoparticles possess important magnetic properties, essential in biomedical imaging. Cerium oxide is utilized in catalysis and antioxidant applications. When combinations of two different metals are used, they enhance properties associated with monometallic counterparts (Atanasoy et al. 2023). Metal-based nanomaterials (MNPs) are significant in bone fracture treatment and biomedical applications because of their unique effects like biocompatibility, mechanical strength, antimicrobial effect, and enhanced imaging capabilities (Mohanty et al. 2014). In bone regeneration and healing, some nanoparticles mimic the mineral component of bone, promoting osteointegration and bone regeneration. Nanoparticles are often combined with polymers to create scaffolds for bone tissue engineering. Different bioactive nanoparticles stimulate bone cell growth and improve the mechanical properties of bone grafts (Garner et al. 2014). Calcium phosphate nanoparticles enhance bone repair by providing a calcium phosphate source, critical for bone mineralization (Tanaka 2021). In targeted drug delivery systems, metal nanoparticles (e.g., gold and silver) can deliver antibiotics, resolve bone regeneration issues, and provide anti-inflammatory medications directly at the fracture site. These medications also improve remedial and reduce infection risks. Nanomaterials can be concocted to release medications precisely, confirming sustained healing consequences at the fracture site (Alhalili 2023). Metal nanoparticles can reinforce bone graft materials, enhancing their mechanical properties, similar to that of natural bone (Table 1). Iron oxide and gold nanoparticles are used as divergence mediators in imaging techniques like CT scans, MRI scans, and X-rays, providing clearer images and aiding in diagnosing bone fractures and other conditions (Guo et al. 2018). Quantum dots and other metal-based nanoparticles can be used for fluorescent labeling, enabling precise imaging at the cellular level. Gold nanoparticles are used in photothermal therapy to target and destroy cancer cells with minimal damage to surrounding tissues (He et al. 2020). Magnetic nanoparticles (e.g., iron oxide) can be used in hyperthermia-based treatment of cancer cells. Coating medical devices and implants with silver or copper nanoparticles can prevent biofilm formation and reduce the infection rate. By delivering a higher concentration of omega-3 to the joints, marine fish oil nanoparticles may exert more potent anti-inflammatory effects, potentially reducing pain and stiffness associated with arthritis (Kovalchuk and Auernhammer 2023; Yang et al. 2020; Keeler and Sprittles 2023).

2. Metal-Based Nanoparticles

MNPs are tiny particles composed of metals or metal oxides that range in size from 1-100 nanometers. Because of micro shape and high superficial areas, these nanoparticles show exclusive chemical, physical, and biotic activities that fluctuate expressively from their majority complements (Trejo et al. 2021). These properties make MNPs highly useful in various bio-applications across multiple fields such as medicines, electronic particles, and life science. MNPs often have enhanced catalytic properties due to their high surface area and reactivity, making them valuable in chemical reactions and industrial processes. MNPs are also used for targeted drug delivery, imaging, and diagnostics (Cao et al. 2019). For example, gold NPs transport painkillers straight to cancerous tissues, dropping adjacent properties on healthy cells.

2.1. Mechanisms of Metal-Based Nanoparticles

Bone fracture healing and joint pain is a complex physiological process involving the repair and regeneration of bone tissue. Recent advancements in nanotechnology have introduced MNPs as a promising approach to enhance bone healing (Singh 2005; Trejo et al. 2021). These nanoparticles can be incorporated into various delivery systems to provide localized treatment and improve the healing process.

2.2. Interaction with Bone Cells

The interaction of MNPs with bone cells, specifically osteoblasts and osteoclasts, is critical in bone fracture healing. These interactions can significantly influence bone remodeling process, which involves balanced activities of bone development by osteoblasts as well as bone resorption with osteoclasts (Wan et al. 2020). Osteoblasts are responsible for bone formation, and their interaction with metal-based nanoparticles can enhance their function and promote bone regeneration. Metal nanoparticles such as gold (AuNPs) and zinc oxide (ZnO NPs) can activate signaling pathways like Wnt/β-catenin, which is crucial for osteoblast differentiation and proliferation. Nanoparticles can facilitate the deposition of calcium phosphate minerals, essential for forming the bone matrix (Ebadi and Hikmat 2017). For instance, hydroxyapatite nanoparticles (a form of calcium phosphate) closely mimic natural bone minerals, promoting mineralization. Nanoparticles can be functionalized to release development features like bone-morphogenetic proteins (BMPs), further stimulating osteoblast activity and bone formation (Gehrke et al. 2009). Exposure to nanoparticles can upregulate genes associated with osteogenesis, with osteocalcin, alkaline phosphatase (ALP), and collagen I-type, enhancing the bone-forming capabilities of osteoblasts. MNPs incorporated into scaffolds can improve their mechanical properties, providing a stable environment for osteoblast attachment, proliferation, and differentiation (Garner et al. 2014; Tanaka 20121). Osteoclasts are responsible for bone resorption, and their interaction with metal-based nanoparticles can help balance the bone remodelling process by modulating their activity (Wan et al. 2020). Nanoparticles can inhibit the variation of osteoclasts' predecessors into mature osteoclasts by downregulating markers like cathepsin K and tartrate-resistant acid phosphatase (TRAP). The receptor activators of the nuclear factor kappa-B ligand (RANKL) cycle is vital for osteoclast variations and activations (Mahmoud et al. 2020; Arkusz et al. 2020). Some nanoparticles can interfere with this pathway, reducing osteoclast formation and activity. MNPs, such as silver NPs and gold NPs, can reduce the bone-resorbing activity of osteoclasts, leading to a net gain in bone mass (Song et al. 2023). Certain nanoparticles can induce apoptosis in osteoclasts, reducing their number and bone resorption capacity.

Table 1. Some metal-based nanoparticles and their applications

Metals	Applications	References
Al	Coating additives, fuel additives	Connell et al. 2019
Cu	Antibiotic treatment, catalyst, enhanced conductivity, wear resistance	Sun et al. 2022; Abu-zied et al. 2018; Vorotilo et al. 2020
Ce	Polishing, modulating the oxidative stress	Song et al. 2023; Hoseinifar et al. 2020; Zengin 2021
Au	Photodynamic therapy and cellular imaging promote osteogenic differentiation	Mahmoud et al. 2020; Wang et al. 2016; Arkusz et al. 2020
Mn	Catalyst	Guo et al. 2018; Hoseinifar et al. 2020
Fe	Magnetic imaging, bone healing	Arkusz et al. 2020; Wan et al. 2020; Zhang et al. 2023
Si	Drug carrier, gene delivery, catalyst	Inglis et al. 2022; Peng et al. 2019; Cao et al. 2019; Mora-Raimundo et al. 2021
Ag	Bio-imaging prevent infection	Shen et al. 2022
Ni	Magnetic properties	Alexis et al. 201
Ti	Sterilization, coating, photocatalyst	Wang et al. 2020
Zn	Skin protection, antibacterial activity	Cole et al. 2009; Liu et al. 2019

2.3. Challenges and Limitations

While metal-based nanoparticles offer promising advantages for bone fracture healing, their toxicity and biocompatibility are major concerns that must be addressed to ensure safe and effective clinical use. High concentrations of metal nanoparticles can be toxic to cells (Wang et al. 2021). For example, silver nanoparticles (AgNPs) are well-known for their antimicrobial properties, but at higher doses, they can be toxic to human cells, including osteoblasts and osteoclasts (Mahmoud et al. 2020; Wang et al. 2016). Some metal nanoparticles, such as zinc oxide (ZnO) and silver (Ag), can generate ROS, leading to oxidative stress, cellular damage, and apoptosis. Direct interaction of nanoparticles with cell membranes can disrupt membrane integrity, leading to tissue lysis and cell death (Wang et al. 2021; Zhou et al. 2021). Nanoparticles can accumulate in various organs, such as the liver, spleen, kidneys, and lungs, potentially leading to organ-specific toxicity. This accumulation can result from poor clearance mechanisms (Zhao et al. 2021; Feghhi et al. 2021). The prolonged presence of nanoparticles in the kidneys and liver can impair their function, leading to nephrotoxicity and hepatotoxicity. The immune system recognizes nanoparticles as foreign, leading to an inflammatory response (Gama et al. 2024). This can result in chronic inflammation, detrimental to tissue healing and overall health. In some cases, nanoparticles can provoke allergic reactions, leading to hypersensitivity and adverse immune responses.

There is potential for some metal nanoparticles to cause DNA damage, leading to mutations and possibly carcinogenesis (Wang et al. 2016). Persistent inflammation due to nanoparticle exposure can contribute to chronic diseases and hinder long-term health outcomes. Coating nanomaterials with biocompatible particles such as polyethylene-glycol (PEG), proteins, or polysaccharides can decrease their immunomodulation and increase biocompatibility (Fan et al. 2019). Functionalizing the surface of nanoparticles with targeting ligands can direct them specifically to bone tissue, minimizing off-target effects and reducing systemic exposure. Encapsulating nanoparticles in biodegradable matrices can provide a controlled release of therapeutic agents, reducing peak concentrations and minimizing toxicity (Zhai et al. 2021). Designing nanoparticles that release their payload in response to the acidic environment of bone fractures can enhance local efficacy while minimizing systemic exposure. Choosing metals known for their biocompatibility, such as gold (Au) and titanium (Ti), can reduce the risk of adverse reactions (Odintsov et al. 2023). Using biodegradable nanoparticles that break down into non-toxic by-products can enhance biocompatibility. For instance, calcium phosphate nanoparticles degrade into calcium and phosphate ions, natural bone constituents.

2.4. Ethical Considerations

Regulatory and ethical considerations are paramount when developing and implementing MNPs for bone fracture healing. These considerations ensure that such advanced materials are safe, effective, and aligned with ethical standards in medical practice (Andre et al. 2023). Before clinical use, nanoparticles must undergo rigorous preclinical check-ups, with *in vitro* and *in vivo* experiments, to estimate their protection, toxicity, biocompatibility, and efficiency. These studies are essential to gathering sufficient data on potential adverse effects and therapeutic benefits (Junaid and Jun 2022). Once preclinical data supports the safety and efficacy of the nanoparticles, they can proceed to clinical trials. Clinical trials are conducted in phases (I, II, III) to systematically assess safety, dosage, efficacy, and potential side effects in humans (Wang et al. 2013). Phase one focuses on small-scale trials, while phase two focuses on scale trials to evaluate the efficacy and side effects. In phase three trials, large-scale efficacy, monitoring of side effects, and standard treatments are compared (Shen et al. 2024).

3. Novelties in Fish Oil Nanoparticle Synthesis and Characterization

Nano-encapsulation of marine fish oil is a promising approach to improve its beneficial potential for various medications, including rheumatoid arthritis (RA). Marine fish oil, rich in omega-3 fatty acids like eicosa-pentaenoic acid (EPA) and docosa-hexaenoic acid (DHA), has recognized anti-inflammatory and immunomodulatory properties (Arunachalam et al. 2024). Nano-encapsulation can improve these advantageous components' stability, bioavailability, and targeted transfer. Consistent consumption of appropriate amounts of polyunsaturated fatty acid (PUFA) are necessary to decrease the frequencies of innumerable types of disorders and long-lasting sicknesses, including psoriasis, bowel diseases, psychological illness, cancer, bone rheumatoid arthritis, cardiovascular disease, diabetes, respiratory disorder, coordination disorder, movement illness, obesity-related issues and weak bone regeneration (Rizwana et al. 2021). Fish oil derived from marine resources is one of the major nutritional bases of PUFA, especially eicosa-pentaenoic acid and docosa-hexaenoic acid, which are supposed to have effective biological actions and health benefits (Moosavv et al. 2023; Yaghoubi et al. 2015).

Table 2. Different fish species and their oil percentage (Adapted from Ayeloja et al. 2024)

Fish Species	Scientific name	Oil (%)
Tuna fish	*Allothunnus fallai*	17.7
Bight redfish	*Centroberyx gerrardi*	1.6
Spotted mackerel	*Scomberomorus munroi*	2.5
Morwong	*Cheilodactylus spectabilis*	3.5
Redfish	*Beryx splendens*	5.9
School mackerel	*Scomberomorus queenslandicus*	4.9
Whitebait	*Lovettia sealii*	1.6
Australian Hering	*Arripus georgianus*	2.3
Tailor fish	*Pomatomus saltatrix*	2.6
Trevally	*Caranx sexfasciatus*	5.8
Whitebait	*Galaxias maculatus*	4.9
Mackerel marine fish	*Scomberomorus astralasicus*	2.8
Bonito Australian	*Sarda australis*	3.8
Marine Gem fish	*Rexea solandri*	3.8
Rudder fish	*Centrolophus niger*	10.5
Spanish mackerel fish	*Scomberomorus commerson*	4.5
Sweep	*Scorpis lineolatus*	2.7
Western blue grouper	*Achoerodus gouldii*	4.7
Boarfish	*Pentaceros decacanthus*	2.7
Australian salmon fish	*Arripis trutta*	2.3
Threadfin emperor	*Lethrinus genivittatus*	1.7
Grey mackerel	*Scomberomorus semifasciatus*	2.5
Australian sardine	*Sardinops neopilchardus*	3.4
Blue-eye trevella	*Schedophilus labyrinthica*	2.3
Swordfish	*Xiphias gladius*	8.8
Twinbar cardinal fish	*Apogonichthyoides sialis*	5.8

3.1. Twinbar Fish Medicinal Effect

Apogonichthyoides sialis, also recognized as the "cardinal fish," is a species within the Apogonidae family. These fish contain more omega-3, primarily EPA and DHA (Abd et al. 2022). Different fish have different oil percentages (Table 2). The concentration of all these fatty acids has significant anti-inflammatory effects, which can be valuable for conditions like rheumatoid arthritis, cardiovascular diseases, and inflammatory bowel disease. Omega-3 support decreases the risk of heart arrhythmias with lower blood pressure, contributing to overall cardiovascular health (Ayeloja et al. 2024; Moosavv et al. 2023). The high-quality protein in marine fish provides essential amino acids crucial for muscle growth and recovery, benefiting athletes and individuals recovering from injuries or surgery. Amino acids in these fish are vital for various cellular functions, including hormone production, immune response, and enzyme activity. It is an excellent source of essential vitamins D, Selenium, Iodine, and Zinc (Odoli et al. 2019). Vitamin D is vibrant for bone regeneration and immune functions. Selenium and Zinc are important for defense and immune support. Twin bar fish oil contains antioxidants like Selenium and Vitamin E. These components help condense oxidative stress and decrease the risk of long-lasting disease, cancer, and neurodegenerative conditions (Ramos et al. 2010). Peptides derived from marine fish have been found to exhibit antihypertensive effects by inhibiting angiotensin components, which help regulate blood pressure. Certain peptides have antimicrobial properties, making them potential candidates for developing new antibiotics (Moniruzzaman et al. 2021). Collagen supplements can help improve joint health and reduce symptoms of osteoarthritis.

3.2. Extraction of Fish Oil

In traditional fish oils (FO), extraction is accomplished from entire fish fat, exact lipid-rich tissues, and fish surplus. Conventionally, FO extraction is supported by utilizing conservative solvent procedures with a mixture of ethanol, methanol, and chloroform as solvents (Ayeloja et al. 2022). Equally, bipolar and nonpolar phospholipids can be secluded by using these procedures. Firstly, the ratio of fish materials with a combination of chloroform and methanol will be 2:1. Distilled water can be added for layer separation. The bottom layer comprises nonpolar and polar lipids. The top methanolic layer with water comprises all additional materials (Moniruzzaman et al. 2021). After that mixture of chloroform and different solvent layer can be melted, leaving the fats. These procedures have been functional for fish wax and liver to extract fatty acids (Matos et al. 2019). The fish oil can be detached from other fatty fish using manual procedures, like hydraulic pressure and warmth extraction methods. After that, the residue of the fish oil can be detached utilizing biological solvent extractions procedure (Rafat et al. 2023). In the green method, supercritical enzymes, fluids, and microwaves are used for the extraction of omega-3 due to the high temperature of the organic solvent.

3.3. Characterization Methods

FO's nourishment, physiochemical, and safety activities depend on their biochemical configuration. So, various advanced scientific instruments are used to evaluate their compound structure, quantities, quality, and composition (Figure 1). Gas chromatography-mass spectrometry (GCMS) is used to quantify the chemical composition of nonpolar lipids in FO samples. Firstly, the oil is frozen with different solvents. Therefore, glycerols are transformed into unstable, unrestricted fatty acids. GC columns can easily separate it (Hantoush et al. 2015). This provides a sequence of chromatogram peaks corresponding to Omega 3 fatty acids of dissimilar chain lengths. Conventionally, the compounds are enumerated by determining the area percentage and recognized by comparison with recognized standards of the NIST library. A comprehensive study can be conducted by coupling the GC instrument with a mass spectrometer (Zainol et al. 2020). Raw data evidence about different lipids available in fish oil, like monoacylglycerols, di-triacylglycerols, fatty acids, and different phospholipids, can be evaluated by utilizing (TLC) chromatography analysis. The microstructure of the TLC peak can be estimated using different ranges of optical microscopy, (SEM) and (TEM) microscopy (Matos et al. 2019). The physical stage of wall powder materials, which are either amorphous or crystalline, can be recognized by XRD and (DSC) calorimetry technology. Evidence about biochemical configuration and molecular exchanges in powder particles can be obtained by analyzing Fourier transform infrared (FTIR) and Raman spectroscopy (Saji et al. 2024). Figure 1 illustrates that metal-based NPs are functionalized with fish oil therapeutic agents. The functionalized NPs circulate through the bloodstream and target tissues like inflamed joint areas.

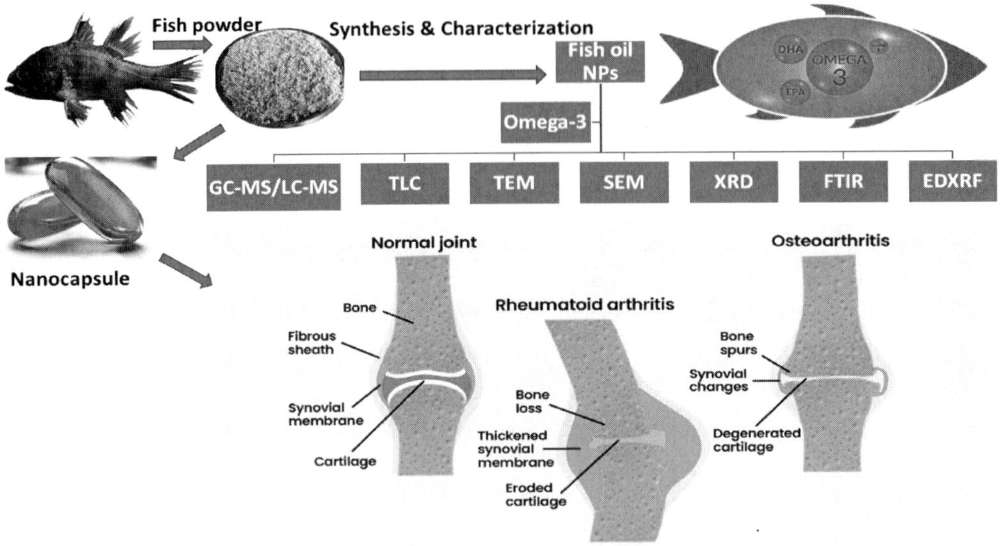

Figure 1. Graphical representation of fish oil NPs and their effect in Rheumatoid arthritis.

4. Rheumatoid Arthritis Mechanism

Rheumatoid arthritis (RA) is a chronic autoimmune disease categorized by swelling of the joints, which can lead to joint damage and deformity. Arthritis is characterized as the inflammation of single or additional joints. There are numerous symptoms of sickness, including inflammation of the joints, swelling of joints, joint distress, and joint problems (Ayeloja et al. 2024). Rheumatoid arthritis (RA) and osteoarthritis (OA) are the frequent forms of arthritis. Inflammatory medications are one of the most expensive treatment groups for health care. Specific HLA-DRB1 alleles, known as the "shared epitope," increase the risk of developing RA by influencing the immune system's response to self-antigens (Bajpai et al. 2022). Certain infections, such as those caused by Epstein-Barr virus (EBV) and *Porphyromonas gingivalis* (a bacterium associated with periodontal disease), have been implicated in the onset of RA (Acharya et al. 2022). These pathogens may contribute to molecular mimicry, where the immune system confuses self-antigens with foreign ones. In RA, proteins in the joints are often citrullinated, a process where the amino acid arginine is converted to citrulline (Hu et al. 2022). This modification can make proteins appear foreign to the immune system. The immune system produces ACPAs that target these citrullinated proteins, leading to an autoimmune attack on the joint tissues. Rheumatoid factor (RF) is an autoantibody directed against the Fc region of IgG antibodies (Algabbani 2022). RF can form immune complexes that contribute to RA inflammation and tissue damage. In immune system dysregulation T-cell activation plays an effective role. CD4+ T cells play a critical role in the pathogenesis of RA (Oktariani et al. 2023). These cells become activated in response to self-antigens presented by antigen-presenting cells (APCs) in the synovial membrane. Th1 and Th17 cells are particularly important in RA. Th1 cells produce interferon-gamma (IFN-γ), and Th17 cells produce interleukin-17 (IL-17), both of which contribute to inflammation (Chigurupati et al. 2022). B cells are responsible for producing autoantibodies, including ACPAs and RF. They are contemporary antigens to T cells and conceal pro-inflammatory cytokines. In RA, there is an overproduction of pro-inflammatory cytokines, such as tumor necrosis factor-alpha (TNF-α), interleukin-1 (IL-1), IL-6, and IL-17. These cytokines drive the inflammatory process and contribute to joint damage (Zafar et al. 2022).

4.1. Inflammation and Joint Damage

The synovial membrane becomes inflamed and hyperplastic, forming a pannus. The pannus is an abnormal layer of fibrovascular tissue that invades and destroys cartilage and bone (Dutta et al. 2022). Infiltration of the synovial tissue by immune cells, including T cells, B cells, macrophages, and neutrophils, contributes to chronic inflammation (Alhalili 2023). Matrix metalloproteinases are enzymes produced by synovial fibroblasts and other cells that degrade extracellular matrix components, leading to cartilage destruction. Osteoclasts, responsible for bone resorption, become overactive in RA due to the influence of pro-inflammatory cytokines like RANKL (Receptor Activator of Nuclear factor Kappa-B Ligand). This results in bone erosion and joint deformities (Atanasoy et al. 2023).

4.2. Osteogenesis Stimulation

Metal-based nanoparticles can be functionalized to release osteoinductive growth factors, like bone morphogenetic protein (BMP), which endorse the differentiation of progenitor tissues into osteoblasts, the bone-forming cells (Guo et al. 2018). Nanoparticles can enhance the osteogenic variation of mesenchymal stem cells (MSC) by providing a conducive microenvironment and delivering osteogenic signals (Tanaka 2021).

4.3. Angiogenesis Promotion

Some metal nanoparticles can be loaded with other angiogenic factors, which promote the formation of new blood vessels, ensuring adequate blood supply to the healing bone (Peng et al. 2019). Certain metal-based nanoparticles, like those containing gold NPs or silver NPs, have inherent pro-angiogenic properties that stimulate endothelial cell proliferation and migration. Metal nanoparticles, particularly silver and zinc oxide nanoparticles, have strong antibacterial properties that prevent infection at the fracture site, reducing complications and promoting faster healing (Shipman et al. 2020). These nanoparticles can disrupt bacterial biofilms, which are often resistant to conventional antibiotics, thereby enhancing the effectiveness of infection control (Wang et al. 2013).

4.4. Inflammation Modulation

Metal nanoparticles such as gold and cerium oxide have anti-inflammatory effects that support and modulate the inflammatory responses, ensuring they are balanced to facilitate healing without excessive inflammation that could impair bone repair (Pelegri et al. 2008). They can influence the proclamation of pro-inflammatory and anti-inflammatory cytokines, creating an optimal environment for bone regeneration (Gama et al. 2024). Metal nanoparticles can be incorporated into bone scaffolds, enhancing their mechanical properties and mimicking the natural bone matrix more closely. They help distribute mechanical stress more evenly, reducing the risk of further fractures during the healing process (Dechet et al. 2021; Liu et al. 2021).

4.5. Cellular Effects

Nanoparticles can enhance the proliferation of osteoblasts and other bone cells, increasing the cellular activity required for bone formation. They can activate specific signaling cycles, like Wnt/β-catenin pathways, crucial for joint growth and development (Kumar et al. 2010; Marriott et al. 2020).

5. Emerging Applications in Orthopedics Medicine

Another potential challenge for some fish oil formulations is their highly variable oral bioavailability, which depends on the molecular form of the omega-3 fatty acids and the nature of the surrounding food matrix (Islam et al. 2023). The main cause of this effect is the relatively low solubility of omega-3 oils in the aqueous gastrointestinal fluids inside the human gut, as well as their susceptibility to chemical degradation during passage through the gut (Dutta et al. 2022; Periyasamy et al. 2018). After ingestion, fish oils pass through the mouth, esophagus, stomach, and small intestine, where they may be absorbed. They are, therefore, exposed to various digestive enzymes, pH conditions, mechanical forces, flow profiles, mineral ions, bile salts, and other gastrointestinal constituents (Fouda et al. 2022). Bulk fish oils are converted into coarse oil-in-water emulsions in the mouth and stomach, whereas emulsified oils are colloidal before ingestion (Table 3). Gastric and pancreatic lipases then adsorb to the surfaces of the lipid droplets and initiate lipid digestion, which involves converting triacylglycerols (TAGs) into free fatty acids (FFAs) and monoacylglycerols (MAGs) (Garner et al. 2014; Tanaka 2021). The resulting digestion products (FFAs and MAGs) then interact with endogenous bile salts and phospholipids to form mixed micelles that carry the lipids through the mucus layer to the surfaces of the epithelium cells, where they can be absorbed. Studies have shown that the bioavailability of omega-3 oils is higher when they are ingested in an emulsified form than in a bulk form. This can be attributed to the greater surface area of lipases that adsorb in emulsions (Bilal et al. 2020). Consequently, controlling the initial colloidal state of fish oils, such as the composition, size, and surface characteristics of the particles they are encapsulated within, can have a major impact on their bioavailability profiles. In addition, the bioavailability of omega-3 oils can be enhanced by incorporating absorption enhancers into supplement formations, which increase the permeability of the epithelium cells (Wang et al. 2019).

Table 3. Different delivery systems used to encapsulate omega-3 fish oil

Delivery system	Details	Reference
Nano-lipid carriers	Little emulsifier-covered oil beads scattered inside 100 nm water droplet.	Venugopalan et al. 2021
Microgels	Fatty acid oils combined with nanomaterials would be integrated into the microgel capsules.	Zhang et al. 2014
Nanofiber materials	Nanofibers comprise extensive slender stringy compounds regularly gathered from biopolymers, similar to enzymes and polysaccharides derived from food sources.	Huynh et al. 2009
Inclusion complexes	Fish oil could be embodied inside cyclodextrin consideration buildings with high exemplification proficiency.	Wang et al. 2021
Microencapsulation	The fish oil fatty acid is blended with an aqueous substance contaning wall material. The powders coming about because of these cycles comprise micro-sized capsules encompassing various oil drops installed inside a wall substance.	Manoharan et al. 2019
Spray drying	Splash drying embodies fish oil utilizing wall substances like lactose, Maltodextrin, casein, dextrose, altered starch, and Arabic gum.	Mohanty et al. 2016
Freeze dry	Freeze drying is also utilized to change fluid fish oil into a powder structure.	Abd et al. 2022; Manoharan et al. 2019)
Electro spraying and spinning	Fish oil is transformed into a blended form assorted with different biopolymers before spinning.	Ayeloja et al. 2024

Conclusion

Inflammatory medications are the most exclusive treatment types for pharmacological medical systems. Nonsteroidal inflammatory painkillers, glucocorticoids, modified antirheumatic drugs, and microbial mediators are generally prescribed to treat RA (Moosayy et al. 2023; Ayeloja et al. 2024). Modified drug-induced nephrotoxicity, liver injury, high blood pressure, and heart attack are probable side effects of these advanced drugs. Marine fish are easily available worldwide. So, smart fish oil nanoparticulate systems have been designed to release the therapeutic agent for RA therapy (Ramos et al. 2010). Compressing omega-3 fatty acids associated with nano-encapsulate particles gives a possible approach for refining the energetic nutrients' bioavailability, strength, and embattled management. Researchers have efficiently moderated corrosion, unwanted sensitivity, and limited absorption related to omega-3, consuming many compressing approaches like nano emulsions to solid lipid and polymeric nanoparticles (Mohanty et al. 2016). Nanoparticle inhibition contains numerous health benefits and immunity-boosting effects. Nanomaterials also have some disadvantages, addressed to improve acceptability (Moniruzzaman et al. 2021). Additionally, the nano-scale measurement and distribution structures simplify cellular fascination, transit genetic blocks, and recover therapeutic properties (Arunachalam et al. 2024). Further research is required to enhance the design of fish oil nanoparticles, like their molecular structures, dimensions, surface characterization, and encapsulation, to exploit the effectiveness and bioavailability of omega-3 (Matos et al. 2019). Moreover, in various scientific contexts, lifelong constancy clinical trials and *in vivo* investigations are essential to estimate the therapeutic kinetics, lifelong security, and healing potential of nanoparticle-associated omega-3 fish oil fatty acid (Odoli et al. 2019). *Apogonichthyoides sialis*, like other marine fish, holds potential medicinal benefits due to the major concentration of omega-3, superior quality enzymes, essential vitamin D, and many minerals as well as unique bioactive compounds. While specific research on this species is limited, the general health benefits of marine fish provide a strong foundation for exploring its potential therapeutic effects (Bajpai et al. 2022). Further research, including biochemical analyses, preclinical studies, and clinical trials, is required to realize and validate the medicinal properties of *Apogonichthyoides sialis*.

Acknowledgments

The authors are greatly obliged to Fakir Mohan University, Odisha, for providing web access, software's and manuscript services.

References

Abd Elmeged, Lobna Saad Mohammed, and Abdullah AA Alghamdi. "The effect of consuming different proportions of hummer fish on biochemical and histopathological changes of hyperglycemic rats." *Saudi Journal of Biological Sciences* 29, no. 1 (2022): 140-147.

Abu-Zied, Bahaa M., Mahmoud A. Hussein, Anish Khan, and Abdullah M. Asiri. "Cu-Cu2O@ graphene nanoplatelets nanocomposites: Facile synthesis, characterization, and electrical conductivity properties." *Materials Chemistry and Physics* 213 (2018): 168-176.

Acharya, Prakruti P., Manjushree H. Kupendra, Aneesa Fasim, Sunil Shivajirao More, and V. Krishna Murthy. "A comparative assessment of collagen type 1 from silver carp (fresh water) and milk shark (marine) fish waste." 3 *Biotech* 12, no. 3 (2022): 82.

Alexis, Joël, Bernard Etcheverry, Jean-Denis Béguin, and Jean-Pierre Bonino. "Structure, morphology and mechanical properties of electrodeposited composite coatings Ni–P/SiC." *Materials Chemistry and Physics* 120, no. 2-3 (2010): 244-250.

AlGabbani, Qwait. "Mutations in TP53 and PIK3CA genes in hepatocellular carcinoma patients are associated with chronic Schistosomiasis." *Saudi Journal of Biological Sciences* 29, no. 2 (2022): 848-853.

Alhalili, Zahrah. "Metal oxides nanoparticles: general structural description, chemical, physical, and biological synthesis methods, role in pesticides and heavy metal removal through wastewater treatment." *Molecules* 28, no. 7 (2023): 3086.

André, Chantale, Samuel Turgeon, Caroline Peyrot, Kevin James Wilkinson, Joëlle Auclair, Nadia Ménard, and François Gagné. "Comparative toxicity of micro and nanopolystyrene particles in Mya arenaria clams." *Marine Pollution Bulletin*, 192 (2023): 115052.

Arkusz, Katarzyna, Ewa Paradowska, Marta Nycz, Justyna Mazurek-Popczyk, and Katarzyna Baldy-Chudzik. "Evaluation of the antibacterial activity of Ag-and Au-nanoparticles loaded TiO2 nanotubes." *Journal of Biomedical Nanotechnology* 16, no. 9 (2020): 1416-1425.

Arunachalam, Selva Sudha, Vaisali Chandrasekar, and Prasanna D. Belur. "Synthesis and characterization of 3, 4-dihydroxyphenyl acetic acid esters and study of their efficacy in bulk fish oil." *Food Chemistry*, 441 (2024): 138380.

Atanasov, Nikola, Yana Evstatieva, and Dilyana Nikolova. "Probiotic Potential of Lactic Acid Bacterial Strains Isolated from Human Oral Microbiome" *Microbiology Research* 14, no. 1(2023): 262-278. https://doi.org/10.3390/microbiolres14010021

Ayeloja, Ayodeji Ahmed, Wasiu Adeyemi Jimoh, and Taiwo Victoria Oyewole. "Nutritional quality of fish oil extracted from selected marine fish species." *Food and Humanity* 2, (2024): 100212.

Bajpai, Vertika, Divya Pragyan, Kirty Suman, Jyotirmaya Mohanty, and Pramoda Kumar Sahoo. "Viral diseases in Indian freshwater and marine water pisciculture." *Current Science*, (2022): 267-280.

Bilal, Muhammad, Maimoona Qindeel, Leonardo Vieira Nunes, Marco Thúlio Saviatto Duarte, Luiz Fernando Romanholo Ferreira, Renato Nery Soriano, and Hafiz MN Iqbal. "Marine-derived biologically active compounds for the potential treatment of rheumatoid arthritis." *Marine drugs* 19, no. 1 (2020): 10.

Cao, Ye, Yang Fei Tan, Yee Shan Wong, Melvin Wen Jie Liew, and Subbu Venkatraman. "Recent advances in chitosan-based carriers for gene delivery." *Marine drugs* 17, no. 6 (2019): 381.

Chigurupati, Sridevi, Atheer Al-Murikhy, Suliman A. Almahmoud, Yosif Almoshari, Amira Saber Ahmed, Shantini Vijayabalan, Shatha Ghazi Felemban, and Vasanth Raj Palanimuthu. "Molecular docking of phenolic compounds and screening of antioxidant and antidiabetic potential of Moringa oleifera ethanolic leaves extract from Qassim region, Saudi Arabia." *Saudi Journal of Biological Sciences* 29, no. 2 (2022): 854-859.

Cole, D. P., H. A. Bruck, and A. L. Roytburd. "Nanomechanical characterisation of graded NiTi films fabricated through diffusion modification." *Strain* 45, no. 3 (2009): 232-237.

Connell Jr, Terrence L., Richard A. Yetter, Grant A. Risha, Zachary J. Huba, Albert Epshteyn, and Brian T. Fisher. "Enhancement of solid fuel combustion in a hybrid rocket motor using amorphous Ti–Al–B nanopowder additives." *Journal of Propulsion and Power* 35, no. 3 (2019): 662-665.

Dechet, Maximilian A., Juan S. Gomez Bonilla, Moritz Grünewald, Kevin Popp, Johannes Rudloff, Marieluise Lang, and Jochen Schmidt. "A novel, precipitated polybutylene terephthalate feedstock material for powder bed fusion of polymers (PBF): Material development and initial PBF processability." *Materials & Design* 197 (2021): 109265.

Dutta, Joystu, Sufia Zaman, Tarun Kumar Thakur, Sandeep Kaushik, Abhijit Mitra, Poonam Singh, Ravinder Kumar et al. "Assessment of the bioaccumulation pattern of Pb, Cd, Cr and Hg in edible fishes of East kolkata Wetlands, India." *Saudi Journal of Biological Sciences* 29, no. 2 (2022): 758-766.

Ebadi, Abdol Ghaffar, and Hikmat Hisoriev. "The prevalence of heavy metals in Cladophora glomerata L. from Farahabad Region of Caspian Sea–Iran." *Toxicological & Environmental Chemistry* 99, no. 5-6 (2017): 883-891.

Fan, Xinyi, Xiaopeng Li, Ningping Tao, Jing Zhang, Mingfu Wang, Xueli Qian, Hong Su, and Jian Zhong. "Effect of salt addition time on the nutritional profile of thunnus obesus head soup and the formation of micro/nano-sized particle structure." *Molecules*, 24, no. 24 (2019): 4447.

Feghhi, Maryam, Jafar Rezaie, Ali Akbari, Nasrollah Jabbari, Hessam Jafari, Farzad Seidi, and Sławomir Szafert. "Effect of multi-functional polyhydroxylated polyhedral oligomeric silsesquioxane (POSS) nanoparticles on the angiogenesis and exosome biogenesis in human umbilical vein endothelial cells (HUVECs)." *Materials & Design*, 197 (2021): 109227.

Fouda, Sarah EE, Fathy MA El-Saadony, Ahmed M. Saad, Samy M. Sayed, Mohamed El-Sharnouby, Amira M. El-Tahan, and Mohamed T. El-Saadony. "Improving growth and productivity of faba bean (Vicia faba L.) using chitosan, tryptophan, and potassium silicate anti-transpirants under different irrigation regimes." *Saudi Journal of Biological Sciences* 29, no. 2 (2022): 955-962.

Gama, Tsvakai, Michael B. Farrar, Mahshid Tootoonchy, Helen M. Wallace, Stephen J. Trueman, Iman Tahmasbian, and Shahla Hosseini Bai. "Hyperspectral imaging predicts free fatty acid levels, peroxide values, and linoleic acid and oleic acid concentrations in tree nut kernels." *LWT* 199 (2024): 116068.

Garner, Rozeena, Tina Ding, and Chris Deighton. "Management of rheumatoid arthritis." *Medicine* 42, no. 5 (2014): 237-242.

Gehrke, Gretchen E., Joel D. Blum, and Philip A. Meyers. "The geochemical behavior and isotopic composition of Hg in a mid-Pleistocene western Mediterranean sapropel." *Geochimica et Cosmochimica Acta* 73, no. 6 (2009): 1651-1665.

Guo, Qiang, Yuxiang Wang, Dan Xu, Johannes Nossent, Nathan J. Pavlos, and Jiake Xu. "Rheumatoid arthritis: pathological mechanisms and modern pharmacologic therapies." *Bone research* 6, no. 1 (2018): 15.

Guo, Yu, Entezar Mehrabi Nasab, Fatemeh Hassanpour, and Seyyed Shamsadin Athari. "Linsitinib and aspirin as the IGF1-R antagonists, inhibit regorafenib-resistant chemotherapy in colon cancer." *Saudi Journal of Biological Sciences* 29, no. 2 (2022): 872-877.

Hantoush, A. A., Q. H. Al-Hamadany, A. S. Al-Hassoon, and H. J. Al-Ibadi. "Nutritional value of important commercial fish from Iraqi waters." *International Journal of Marine Science* 5, no. 11 (2015): 1-5.

He, Shan, Nikita Joseph, Marzieh Mirzamani, Scott J. Pye, Ahmed Hussein Mohammed Al-Anataki, Andrew E. Whitten, Yaonan Chen, Harshita Kumari, and Colin L. Raston. "Vortex fluidic mediated encapsulation of functional fish oil featuring in situ probed small angle neutron scattering." *npj Science of Food* 4, no. 1 (2020): 12.

Hoseinifar, Seyed Hossein, Samira Yousefi, Hien Van Doan, Ghasem Ashouri, Giorgia Gioacchini, Francesca Maradonna, and Oliana Carnevali. "Oxidative stress and antioxidant defense in fish: The implications of probiotics, prebiotics and synbiotics." *Reviews in Fisheries Science & Aquaculture* 29, no. 2, (2020): 198-217.

Hu, Qiong, Zhenfu Li, Xiaoyu Zhai, and Hao Zheng. "Development of hydraulic lifting system of Deep-Sea mineral resources." *Minerals* 12, no. 10 (2022): 1319.

Huynh, Minh Dieu, and David D. Kitts. "Evaluating nutritional quality of pacific fish species from fatty acid signatures." *Food Chemistry* 114, no. 3 (2009): 912-918.

Inglis, Brendan, Peter Schwarzenberg, Karina Klein, Brigitte von Rechenberg, Salim Darwiche, and Hannah L. Dailey. "Biomechanical duality of fracture healing captured using virtual mechanical testing and validated in ovine bones." *Scientific Reports* 12, no. 1 (2022): 2492.

Islam, Md Rakibul, Momota Yeasmin, Sultana Sadia, Md Sadek Ali, Ahmed Redwan Haque, and Vikash Chandra Roy. "Small indigenous fish: A potential source of valuable nutrients in the context of Bangladesh." *Hydrobiology* 2, no. 1 (2023): 212-234.

Junaid, Muhammad, and Jun Wang. "Interaction of micro (nano) plastics with extracellular and intracellular biomolecules in the freshwater environment." *Critical Reviews in Environmental Science and Technology*, 52, no. 23 (2022): 4241-4265.

Keeler, J. S., and J. E. Sprittles. "Finding the point of no return: Dynamical systems theory applied to the moving contact-line instability." *Current Opinion in Colloid & Interface Science* (2023): 101724.

Kovalchuk, V. I., and G. K. Auernhammer. "Adsorption Layer and Flow within Liquid Meniscus in Forced Dewetting." *Current Opinion in Colloid & Interface Science* (2023): 101723.

Kumar, Bhupander, Kurunthachalam Senthil Kumar, Mahalakshmi Priya, Debapriyo Mukhopadhyay, and Rita Shah. "Distribution, partitioning, bioaccumulation of trace elements in water, sediment and fish from sewage fed fish ponds in eastern Kolkata, India." *Toxicological & Environ Chemistry* 92, no. 2 (2010): 243-260.

Liu, Jingqi, Ning Hu, Xuyang Liu, Yaolu Liu, Xuewei Lv, Liangxiao Wei, and Shoutao Zheng. "Microstructure and mechanical properties of graphene oxide-reinforced titanium matrix composites synthesized by hot-pressed sintering." *Nanoscale Research Letters* 14, (2019): 1-12.

Liu, Lu, Guobing Ying, Dong Wen, Kaicheng Zhang, Cong Hu, Yongting Zheng, Chen Zhang, Xiang Wang, and Cheng Wang. "Aqueous solution-processed MXene (Ti3C2Tx) for non-hydrophilic epoxy resin-based composites with enhanced mechanical and physical properties." *Materials & Design* 197, (2021): 109276.

Mæhre, H. K., K. Hamre, and E. O. Elvevoll. "Nutrient evaluation of rotifers and zooplankton: feed for marine fish larvae." *Aquaculture Nutrition* 19, no. 3 (2013): 301-311.

Mahmoud, Nadia S., Hanaa H. Ahmed, Mohamed R. Mohamed, Khalda S. Amr, Hadeer A. Aglan, Mohamed AM Ali, and Mohamed A. Tantawy. "Role of nanoparticles in osteogenic differentiation of bone marrow mesenchymal stem cells." *Cytotechnology* 72, no. 1 (2020): 1-22.

Manoharan, Shobana, Raghavan Kuppu, and Ramesh Uthandakalaipandian. "Comparative assessment of nutritional composition in raw and cooked Indian freshwater fish Lepidocephalus thermalis (V.)." *In Proceedings of the Zoological Society*, vol. 72, pp. 154-159. Springer India, 2019.

Marriott, A. L., T. J. Kelly, S. K. Sarkar, S. R. N. Chenery, D. Rakshit, B. D. Bhattacharya, and M. J. Watts. "Elemental composition of aquaculture fish from West Bengal, India: nutrition versus food safety." *Environmental geochemistry and health* 42, (2020): 1211-1228.

Matos, Ângelo Paggi, Anastácio Castelo Matos, and Elisa Helena Siegel Moecke. "Polyunsaturated fatty acids and nutritional quality of five freshwater fish species cultivated in the western region of Santa Catarina, Brazil." *Brazilian Journal of Food Technology* 22, (2019): e2018193.

Mohanty, Bimal P., T. V. Sankar, Satabdi Ganguly, Arabinda Mahanty, R. Anandan, Kajal Chakraborty, B. N. Paul et al. "Micronutrient composition of 35 food fishes from India and their significance in human nutrition." *Biological Trace Element Research* 174, (2016): 448-458.

Mohanty, Bimal, Arabinda Mahanty, Satabdi Ganguly, T. V. Sankar, Kajal Chakraborty, Anandan Rangasamy, Baidyanath Paul et al. "Amino acid compositions of 27 food fishes and their importance in clinical nutrition." *Journal of amino acids* 2014, no. 1 (2014): 269797.

Moniruzzaman, Mohammad, Sonia Sku, Parvez Chowdhury, Mohosena Begum Tanu, Selina Yeasmine, Md Nazmul Hossen, Taesun Min, Sungchul C. Bai, and Yahia Mahmud. "Nutritional evaluation of some economically important marine and freshwater mollusc species of Bangladesh." *Heliyon* 7, no. 5 (2021).

Moosavy, Mir-Hassan, Miguel de la Guardia, Ahad Mokhtarzadeh, Seyed Amin Khatibi, Neda Hosseinzadeh, and Nasser Hajipour. "Green synthesis, characterization, and biological evaluation of gold and silver nanoparticles using Mentha spicata essential oil." *Scientific Reports* 13, no. 1 (2023): 7230.

Mora-Raimundo, Patricia, Daniel Lozano, Manuel Benito, Francisca Mulero, Miguel Manzano, and María Vallet-Regí. "Osteoporosis remission and new bone formation with mesoporous silica nanoparticles." *Advanced Science* 8, no. 16 (2021): 2101107.

Naseem, Taiba, and Tayyiba Durrani. "The role of some important metal oxide nanoparticles for wastewater and antibacterial applications: A review." *Environmental Chemistry and Ecotoxicology* 3 (2021): 59-75.

Odintsov, Vyacheslav S., Alexander A. Karpenko, and Maxim A. Karpenko. "Degradation of micro-nano-sized polytetrafluoroethylene and acrylic fluorinated copolymer particles in the periwinkle digestive tract." *Environmental Science and Pollution Research* 30, no. 10 (2023): 25972-25980.

Odoli, Cyprian O., Horace Owiti, Nyakeya Kobingi, Maurice Obiero, Zachary Ogari, James Mugo, Chrisphine Nyamweya, and Christopher M. Aura. "Post-harvest interventions in small-scale fisheries: a boon or bane to food and nutritional security in Kenya?." *Food security* 11, (2019): 855-868.

Oktariani, Adnorita Fandah, Putu Eka Sudaryatma, Yan Ramona, I. Made Gelgel Wirasuta, Ida Bagus Gede Darmayasa, Putu Angga Wiradana, and Tamaki Okabayashi. "Heavy metals content in fresh tuna and swordfish caught from Hindian and Pacific Oceans: Health risk assessment of dietary exposure." *Veterinary World* 16, no. 4 (2023): 858.

Pelegri, Assimina A., and Xiaoqin Huang. "Nanoindentation on soft film/hard substrate and hard film/soft substrate material systems with finite element analysis." *Composites Science and Technology* 68, no. 1 (2008): 147-155.

Peng, Jinfeng, Jiajia Zhao, Yanlin Long, Yanling Xie, Jiaming Nie, and Lili Chen. "Magnetic materials in promoting bone regeneration." *Frontiers in Materials* 6, (2019): 268.

Periyasamy, Thirunavukkarasu, and Ming-Wei Lu. "Molecular characterization and expression of malabar grouper (Epinephelus malabaricus) IRF8 gene response to immuno-stimulants and nervous necrosis virus." *Pathways* 8, (2018): 16.

Ramos Filho, Manoel Mendes, Maria Isabel Lima Ramos, Priscila Aiko Hiane, and Elizabeth Maria Talá de Souza. "Nutritional value of seven freshwater fish species from the Brazilian Pantanal." *Journal of the American Oil Chemists' Society* 87, no. 12 (2010): 1461-1467.

Rifat, M. A., Md Abdul Wahab, Muhammad Arifur Rahman, Md Nahiduzzaman, and Abdullah-Al Mamun. "Nutritional value of the marine fish in Bangladesh and their potential to address malnutrition: A review." *Heliyon* 9, no. 2 (2023).

Rizwana, Humaira, Mona S. Alwhibi, Hadeel A. Aldarsone, Manal Ahmed Awad, Dina A. Soliman, and Ramesa Shafi Bhat. "Green synthesis, characterization, and antimicrobial activity of silver nanoparticles prepared using Trigonella foenum-graecum L. leaves grown in Saudi Arabia." *Green Processing and Synthesis*, 10, no. 1 (2021): 421-429.

Saji, Rakendhu, Akshay Ramani, Kamal Gandhi, Raman Seth, and Rajan Sharma. "Application of FTIR spectroscopy in dairy products: a systematic review." *Food and Humanity*, (2024): 100239.

Shen, Haijun, Yane Ma, Yi Qiao, Chun Zhang, Jialing Chen, and Ran Zhang. "Application of Deferoxamine in Tissue Regeneration Attributed to Promoted Angiogenesis." *Molecules*, 29, no. 9 (2024): 2050.

Shen, Xingyu, Chao Han, Xinwei Tao, Mingting Zhang, Dandan Meng, Ling Weng, Junwang Liu, and Lizhu Guan. "Preparation and dielectric properties of multilayer Ag@ FeNi-MOF/PVDF composites." *Journal of Materials Science: Materials in Electronics* 33, no. 8 (2022): 5311-5324.

Shipman, Joshua T., Hanna T. Nguyen, and Heather Desaire. "So, you discovered a potential glycan-based biomarker; now what? we developed a high-throughput method for quantitative clinical glycan biomarker validation." *ACS omega* 5, no. 12 (2020): 6270-6276.

Singh, N. B. "The activation effect of K2SO4 on the hydration of gypsum anhydrite, CaSO4 (II)." *Journal of the american ceramic society* 88, no. 1 (2005): 196-201.

Song, Changyou, Cunxin Sun, Bo Liu, and Pao Xu. "Oxidative Stress in Aquatic Organisms" *Antioxidants* 12, no. 6 (2023): 1223.

Sun, Wenquan, Zhiqiang Xiao, Yongjun Sun, Lei Ding, and Jun Zhou. "Preparation of Cu-Ce@ γ-Al2O3 and study on catalytic ozone oxidation for the treatment of RO concentrate water." *Water* 14, no. 18 (2022): 2881.

Tanaka, Yoshiya. "Genome-oriented treatment strategies for autoimmune diseases." *Inflammation and Regeneration* 41, no. 1 (2021): 30.

Trejo, Jhonattan, Ramón Tolosa, Néstor Ruiz, Pablo Ninin, Carmen Fuenmayor, Marcos Zambrano, Oriana Palma, and Yamid Nuñez. "Comparative study of the main properties associated with thin layers of coatings with the cobalt-chromium-tungsten alloy (stellite) and hard chromium plating used as reinforcements for wood sawing." *Mechanics of Materials* 152, (2021): 103637.

Venugopalan, Vishnu Kalladathvalappil, Lekshmi Ramadevi Gopakumar, Ajeeshkumar Kizhakkeppurath Kumaran, Niladri Sekhar Chatterjee, Vishnuja Soman, Shaheer Peeralil, Suseela Mathew, David Julian McClements, and Ravishankar Chandragiri Nagarajarao. "Encapsulation and protection of omega-3-rich fish oils using food-grade delivery systems." *Foods* 10, no. 7 (2021): 1566.

Vorotilo, Stepan, Pavel Alexandrovich Loginov, Alexandr Yuryevich Churyumov, Alexey Sergeevich Prosviryakov, Marina Yakovlevna Bychkova, Sergey Ivanovich Rupasov, Anton Sergeevich Orekhov, Philipp Vladimirovich Kiryukhantsev-Korneev, and Evgeny Alexandrovich Levashov. "Manufacturing of conductive, wear-resistant nanoreinforced Cu-Ti alloys using partially oxidized electrolytic copper powder." *Nanomaterials* 10, no. 7 (2020): 1261.

Wan, Zhuqing, Ping Zhang, Longwei Lv, and Yongsheng Zhou. "NIR light-assisted phototherapies for bone-related diseases and bone tissue regeneration: A systematic review." *Theranostics* 10, no. 25 (2020): 11837.

Wang, Dengdong, Xiyang Chen, Xinhui Zhang, Jia Li, Yunhai Yi, Chao Bian, Qiong Shi et al. "Whole genome sequencing of the giant grouper (Epinephelus lanceolatus) and high-throughput screening of putative antimicrobial peptide genes." *Marine drugs* 17, no. 9 (2019): 503.

Wang, Jingya, Guanglong Xu, Xiaoqin Zeng, Javier Llorca, and Yuwen Cui. "Exploring the correlation between solvent diffusion and creep resistance of Mg-Ga HCP alloys from high throughput liquid-solid diffusion couple." *Materials & Design*, 197 (2021): 109243.

Wang, Qiwei, Bo Chen, Meng Cao, Jianfei Sun, Hao Wu, Peng Zhao, Jing Xing et al. "Response of MAPK pathway to iron oxide nanoparticles *in vitro* treatment promotes osteogenic differentiation of hBMSCs." *Biomaterials* 86, (2016): 11-20.

Wang, Shuai, LuJun Huang, Lin Geng, Yuan Sun, Hua-Xin Peng, and ShaoXing Qu. "Microstructure evolution and damage mechanism of layered titanium matrix composites under tensile loading." *Materials Science and Engineering*, A 777 (2020): 139067.

Wang, Xiaomeng, Sabu Abraham, Jenny AG McKenzie, Natasha Jeffs, Matthew Swire, Vineeta B. Tripathi, Ulrich FO Luhmann et al. "LRG1 promotes angiogenesis by modulating endothelial TGF-β signalling." *Nature* 499, no. 7458 (2013): 306-311.

Yaghoubi, Alireza, Morteza Ghojazadeh, Sakhavat Abolhasani, Hossein Alikhah, and Fatemeh Khaki-Khatibi. "Correlation of serum levels of vitronectin, malondialdehyde and Hs-CRP with disease severity in coronary artery disease." *Journal of cardiovascular and thoracic research* 7, no. 3 (2015): 113.

Yang, Junsi, and Ozan N. Ciftci. "*In vitro* bioaccessibility of fish oil-loaded hollow solid lipid micro-and nanoparticles." *Food & function* 11, no. 10 (2020): 8637-8647.

Zafar, Samavia, Huma Ajab, Sofia Baig, Ayesha Baig, Zeshan Habib, Farrukh Jamil, Muhammad Ibrahim, Sumaira Kanwal, and Muhammad Asif Rasheed. "Prediction and evaluation of multi epitope based sub-unit vaccine against Salmonella typhimurium." *Saudi journal of biological sciences* 29, no. 2 (2022): 1092-1099.

Zainol Abidin, Intan Zarina, Shazrul Fazry, Nur Hidayah Jamar, Herryawan Ryadi Ediwar Dyari, Zaidah Zainal Ariffin, Anis Nabilah Johari, Nur Suhanawati Ashaari, Nor Azfa Johari, Rohaya Megat Abdul Wahab, and Shahrul Hisham Zainal Ariffin. "The effects of Piper sarmentosum aqueous extracts on zebrafish (Danio rerio) embryos and caudal fin tissue regeneration." *Scientific reports* 10, no. 1 (2020): 14165.

Zengin, Hatayi. "The effects of feeding and starvation on antioxidant defence, fatty acid composition and lipid peroxidation in reared Oncorhynchus mykiss fry." *Scientific Reports* 11, no. 1 (2021): 16716.

Zhai, Zheng-yan, Xin-yi Fan, and Ning-ping Tao. "Effect of salt concentration on the formation of micro/nano particles in tuna head soup." *Food Science*, 42, no. 3 (2021): 21-29.

Zhang, Zhenhe, Rong Wang, Hang Xue, Samuel Knoedler, Yongtao Geng, Yuheng Liao, Michael Alfertshofer et al. "Phototherapy techniques for the management of musculoskeletal disorders: strategies and recent advances." *Biomaterials Research* 27, no. 1 (2023): 123.
Zhang, Zhimin, Lianhua Liu, Congxin Xie, Dapeng Li, Jun Xu, Meng Zhang, and Min Zhang. "Lipid contents, fatty acid profiles and nutritional quality of nine wild caught freshwater fish species of the Yangtze Basin, China." *Journal of Food and Nutrition Research* 2, no. 7 (2014): 388-394.
Zhao, Liang, Junjie Zhang, Janine Pfetzing, Masud Alam, and Alexander Hartmaier. "Depth-sensing ductile and brittle deformation in 3C-SiC under Berkovich nanoindentation." *Materials & Design*, 197 (2021): 109223.
Zhou, Xia, Yuting Zhao, Xingchi Chen, Zifan Liu, Junlei Li, and Yafu Fan. "Fabrication and mechanical properties of novel CFRP/Mg alloy hybrid laminates with enhanced interface adhesion." *Materials & Design*, 197 (2021): 109251.

Chapter 3.5

Nucleic Acid Nano-Architectonics Assisted Heavy Metal Biosensors

Mohamed Nabeel Mattath[1,*]
Preeti Prajapati[2]
and Jouharsha Afthab[1]
[1]School of Chemical Science and Engineering, Tongji University, Shanghai, PRC
[2]UNEP-Tongji Institute of Environment for Sustainable Development,
College of Environmental Science and Engineering, Tongji University, Shanghai, PRC

Abstract

Heavy metal contamination poses significant threats to human health and safety. Metals such as arsenic, cadmium, lead, and mercury are persistent pollutants. They can accumulate in ecosystems and enter the food chain, leading to severe health issues. Nucleic acid biosensors (NABs) leverage the unique ability of nucleic acids (DNA or RNA) to detect and quantify specific substances because of their inherent specificity, sensitivity, and versatility. They employ various tools such as DNA/RNA probes, DNAzymes, aptamers, and DNA hybrid architectures specifically designed for heavy metal recognition. NABs can detect metals through responses to temperature, pH, light, and small ions or molecules, making nucleic acids valuable for biological and nonbiological applications. Integrating nucleic acid recognition elements with DNA nano biosensor-mediated transduction methods enhances performance through label-free and labeled detection approaches. Here, we explored the distinct binding mechanisms of NABs and examined the factors influencing their specificity and sensitivity. Various readout techniques, including fluorescence, colorimetric, and electrochemical methods, are discussed in the context of their suitability for different heavy metal detection scenarios. Additionally, this study addresses the prospects and challenges in the field of NABs, providing insights into potential avenues for further research and development. Overall, this chapter underscores the tremendous potential of NABs in addressing critical public health challenges posed by toxic heavy metal contamination.

Keywords: nucleic acids, heavy metal toxicity, aptamers, DNAzymes, nanobiosensors

---

* Corresponding Author's Email: bilusilu@gmail.com

In: Emerging Nanomaterials in Biomedical Science
Editors: Eepsita Priyadarshini and Ajith Manayil Parambil
ISBN: 979-8-89530-600-0

1. Introduction

Nucleic acid biosensors (NABs) are employed to sense metal toxicity during food ingestion and to ensure human health. These compact analytical devices feature a biological sensing element, typically an oligonucleotide, a short sequence of deoxyribonucleic acid (DNA) or ribonucleic acid (RNA), either integrated with a physicochemical transducer or linked within the device (Palchetti et al. 2008; Zhou et al. 2017; J. Xu et al. 2022). The increase in toxic heavy metals such as arsenic, cadmium, lead, and mercury in food, industrial products, and the environment poses a significant public health concern (G. Zhu et al. 2014; Balali-Mood et al. 2021). Even at low concentrations, these metals can cause severe health issues, including cancer, organ damage, neurological disorders, and developmental problems (Balali-Mood et al. 2021; Mitra et al. 2022). Advanced detection technologies, especially NABs, have gained attention for their ability to selectively and sensitively detect heavy metals.

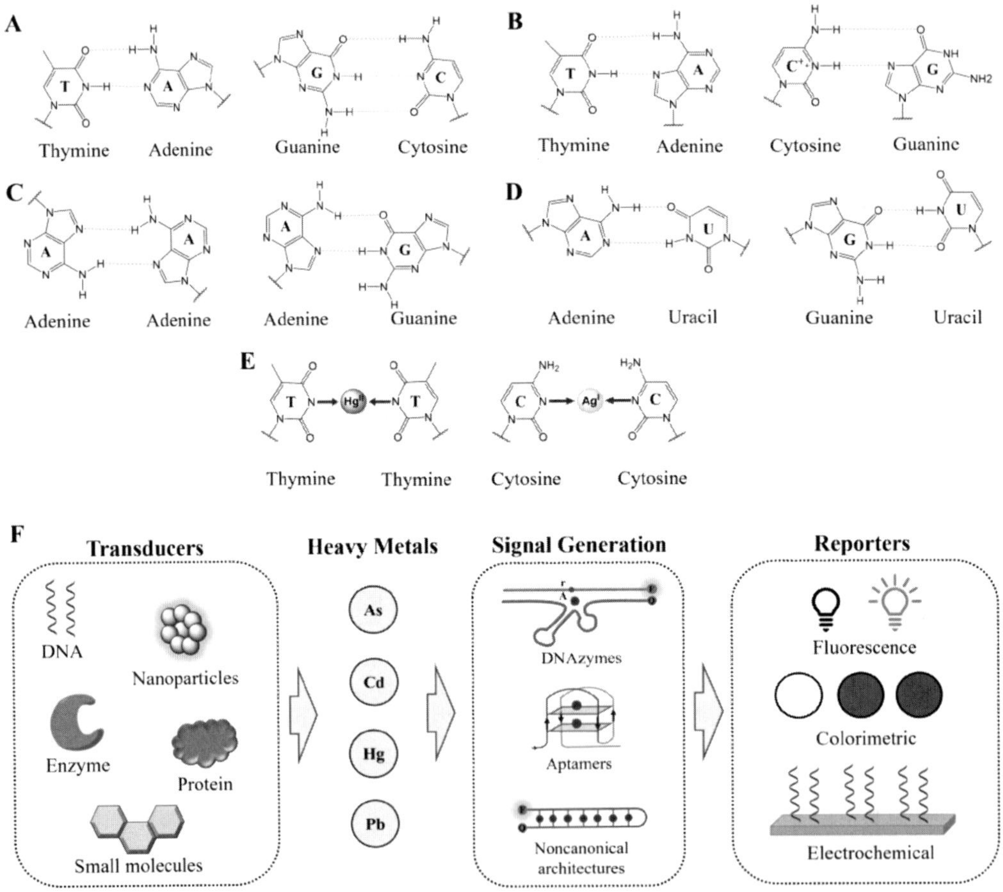

Figure 1. Schematic structure for (A) adenine-thymine (A: T) and guanine-cytosine (G:C) base pairs. (B) A:T and G:C base pairs, (C) A:A and A:G noncanonical hydrogen bonds (D) G:U and A:U wobble base pairs. (E) Noncanonical T-Hg-T and C-Ag-C DNA metal complexation. (F) The basic sensing components are detection of analytes, the target recognition platform, and the signal reporter.

1.1. Nucleic Acid Architectures

Nucleic acids exhibit diverse molecular nanoarchitectures essential for storing, processing, and transmitting genetic information. Researchers have applied this molecular organization of nucleic acids in material science applications (Govindaraju 2019). The adenine (A), guanine (G), cytosine (C), and thymine (T) nucleobases form A:T and G:C base pairings via Watson-Crick (WC) hydrogen bonding to form a DNA double helical structure. These nucleobases also engage in canonical (WC) and noncanonical (non-WC) hydrogen bonding in the form of A:T, G:C, A:A, and A:G interactions (Figure 1A-C) (Roy et al. 2019). The ability of A to form A:T, A:A, and A:G pairs has encouraged the development of hybrid DNA ensembles (Pandeeswar et al. 2016; Narayanaswamy et al. 2016; D. Ghosh et al. 2020). Due to the influence of RNA architecture, guanine-uracil (G:U) base pairs (wobble base pairs) are frequently used to replace G:C or A:U base pairs, resulting in RNA secondary structures such as helices, stem-loops, and pseudoknots (Figure 1D) (Govindaraju 2019). The stability and cost-effectiveness of DNA make it preferable for use as a nanomaterial, while RNA offers structural diversity. Ono and Togashi first showed that a mercury ion (Hg^{2+}) has a strong affinity for T and selectively forms a T–Hg^{2+}–T metal complex (Miyake et al. 2006). The metallo-duplex formed T-Hg^{2+}-T mismatched bases were found to be distinguishable from those of T:A or A:T (Torigoe et al. 2010). The target sequence specifically binds to the complementary probe through complementary base pairing. On the other hand, silver ions (Ag^{+}) have a strong affinity for C and selectively form C-Ag^{+}-C metal complexes (Figure 1E) (Miyake et al. 2006; Tong et al. 2014). These versatile NAB structures enable tailored designs in molecular and material sciences (Avinash et al. 2018; H. Moorthy et al. 2021).

This chapter highlights the growing importance of NABs in detecting toxic heavy metals, driven by advancements in biosensor types such as DNA, RNA, DNAzyme-based, aptamer-based, and DNA hybridization-based systems. Each biosensor type's design, sensing mechanism, and sensitivity to specific heavy metal ions were examined. Key components of NABs include the recognition element (probe), which is designed to interact specifically with the target analyte, and the transduction element, which converts the binding event into a measurable signal (Zhou et al. 2017; J. Xu et al. 2022). These biosensors utilize diverse recognition elements, such as nucleic acids, proteins, enzymes, nanoparticles, or small organic molecules, coupled with transduction methods such as fluorescence, colorimetry, and electrochemistry to achieve quantitative analysis (Figure 1F) (Zhou et al. 2017; Guo et al. 2021; J. Xu et al. 2022). NABs leverage nucleic acids' specificity and versatility to overcome traditional detection methods' limitations, offering promising solutions for accurate heavy metal detection.

2. Toxic Heavy Metals

Toxic heavy metals are high molecular weight elements that can have severe side effects on human health even at low concentrations. Among the heavy metals, arsenic, cadmium, chromium, lead, and mercury are the most well-known toxic pollutants (Mitra et al. 2022; S. Moorthy et al. 2023). The following discussion focuses on their impact on human health and food safety.

2.1. Arsenic Toxicity

Arsenic is a widespread hazardous element found in various industries and agricultural practices. Inorganic forms, such as arsenic trioxide and pentoxide, are more toxic (Punshon et al. 2017; Kuivenhoven et al. 2019). As^{3+} (arsenite) and As^{5+} (arsenate) are the two oxidation states of arsenic, in which As^{3+} is more toxic due to its slow excretion rate, easy cell entry, and allosteric inhibition of pyruvate dehydrogenase (Mochizuki 2019). Arsenic-induced toxicity involves epigenetic modifications, DNA damage, carcinogenesis, and hormonal regulation changes (J. Ghosh et al. 2023).

2.2. Cadmium Toxicity

Cadmium is a metal that is often used in industrial processes, batteries, and pigments. The World Health Organization (WHO) has set an approx. permissible limit of 0.005 mg/L for cadmium in drinking water, although industrial sources may lead to higher limits in some countries (Tchounwou et al. 2012). Long-term exposure to cadmium can lead to kidney damage, bone disorders, and respiratory problems. At the cellular level, cadmium interferes with calcium channels, impairs vitamin D metabolism, and disrupts estrogenic signaling (Bishak et al. 2015). It also induces oxidative stress, damages mitochondria, and inhibits DNA repair mechanisms. These mechanisms contribute to cadmium-induced toxicity and its association with genomic instability and increased expression of proto-oncogenes (Gobrecht et al. 2017).

2.3. Lead Toxicity

Lead, a nonessential heavy metal, poses a significant health risk and is found in various natural resources. Lead exposure, mainly through inhalation and ingestion, can lead to neurological defects, behavioral issues in children, neurodegeneration, elevated blood pressure, and bone deformities in adults (Wani et al. 2015). From the detection perspective, monitoring lead levels in drinking water and the environment is crucial for public health, with the WHO setting a 0.01 mg/L value in the provisional guidelines for lead content in drinking water (Organization 2017).

2.4. Mercury Toxicity

Mercury exists in both organic and inorganic forms, and methylmercury is the most toxic (Rana 2008). Industrial processes and coal-fired power plants release mercury into the air and can end up in water bodies, transforming it into methylmercury and bioaccumulates in fish and seafood. Mercury acts as an enzyme poison by binding to sulfhydryl groups, rendering enzymes inactive and contributing to liver toxicity (Ung et al. 2010). Mercury also triggers metabolic pathways that lead to programmed cell death (apoptosis) in liver cells (Ung et al. 2010). Monitoring and detecting these heavy metals in various samples, such as blood, serum, urine, and biological tissues, are essential for assessing human exposure and contamination levels.

3. Nucleic Acid Biosensors

NABs are innovative and versatile analytical devices that influence the specific interactions between nucleic acids (DNA or RNA) and target molecules to detect and quantify a broad spectrum of analytes (O'Sullivan 2002; Bunka et al. 2006). Since the beginning of the 1990s, many DNA molecules that can bind various compounds with great affinity and specificity have been identified (Lu et al. 2006). When DNA was initially demonstrated to function as a catalyst in 1994, it represented a significant advancement in the creation of functional nucleic acids. Combining the two fascinating fields of DNA nanotechnology and the study of functional nucleic acids has led to creation of a new, multidisciplinary field that uses NABs to control and fine-tune the structure and dynamics of DNA nanostructures and materials, given the tremendous progress made in both areas. Comparing a genosensor to a conventional DNA hybridization experiment, it is possible to collect sequence-specific information more quickly, easily, and affordably. Numerous NABs for identifying interactions and damage to DNA have also been identified (Palchetti et al. 2008). The use of nanomaterials, such as metal oxide nanostructures, carbon nanotubes, and nanoparticles, will undoubtedly significantly influence the advancement of biosensors based on NABs for biomedical surveillance (Palchetti et al. 2008). Future directions for all biosensor research undoubtedly include the combination of nanotechnology, microfluidics, and bioanalytical systems. Due to their exceptional qualities, which include easy modification, low cost, and good specificity, aptamers have been utilized extensively in the past three decades in a variety of sectors, including targeted drug administration and disease diagnostics (Guo et al. 2021; Khoshbin et al. 2023). The working principle involves recognition and binding between the target analyte and the probe, triggering a change in the transduction element and generating a detectable signal. The signal intensity correlates with the concentration of the target analyte in the sample. They provide insight into the use of innovative elements such as DNAzymes, aptamers, hybrid DNA architectures, and label-free/labeled DNA nanobiosensors.

3.1. Advantages of Nucleic Acid Biosensors

NABs are excellent molecular recognition tools for bacteria and viruses and promising therapeutic medications for infectious disorders. Their most popular classes are aptamers and DNAzymes, which have the added advantages of high stability and an adaptable design (L. Zhu et al. 2021). NABs offer a number of benefits over antibodies. First, due to their reduced molecular weight, NABs exhibit superior penetration for *in vivo* applications. Second, because NABs are nucleic acid molecules, they can be readily modified chemically and are inexpensive to prepare in large quantities using solid-phase synthesis, eliminating batch effects. Third, NABs have lower immunogenicity than antibodies, which is important for testing the immune system. Fourth, different NABs versus targets can be enhanced with distinct functions because the library has many random sequences, and the selection of NABs is routine (L. Zhu et al. 2021). Organic molecule-based sensors, which have the advantages of simplicity and low detection limits, have been developed to overcome the limitations of traditional analytical methods. However, most of these sensors inherently involve an organic solvent and suffer from quenched emissions and poor selectivity in aqueous media (G. Zhu et al. 2014). Organic molecule-based sensors, while simpler and offering low detection limits, often suffer from solvent dependence and lack selectivity in aqueous media. In contrast, NAB-based fluorescent metal-ion sensors are cost-effective and programmable, ensuring accurate metal ion detection without the complexities of traditional methods. Colorimetric tests are inexpensive, instrument-free, and easily observable. Electrochemical sensors provide high sensitivity with nanomolar detection limits by detecting changes in electrode-surface distances (J. Xu et al. 2022; Khoshbin et al. 2023).

3.2. DNAzyme-Based Biosensors

DNAzyme-based biosensors utilize synthetic single-stranded DNA (ssDNA) sequences called DNAzymes to catalyze specific chemical reactions upon binding to their target molecules. These biosensors control the catalytic properties of DNAzymes to detect and quantify target molecules effectively (Liu et al. 2009). DNAzyme-based sensors use target molecules as cofactors for specific catalytic reactions, with enhanced target specificity achieved through selection processes (Bruesehoff et al. 2002). The key features of DNAzyme-based biosensors center around the design of the DNAzyme itself, which possesses a specific catalytic core, typically an RNA-like structure, that efficiently cleaves a nucleic acid substrate upon interaction with the target analyte (Liu et al. 2009). A DNAzyme probe complements the target analyte sequence and contains the catalytic core. The DNAzyme probe forms a specific complex and undergoes a conformational change that activates its catalytic core in the presence of the target analyte. The activated DNAzyme then binds to a nucleic acid substrate carrying a specific recognition sequence, leading to cleavage of the substrate and generation of a detectable signal. DNAzymes are particularly effective for metal detection, require metal ions as auxiliary factors, and offer advantages over protein enzymes, including cost-effectiveness, stability, and resistance to hydrolysis (Niazov et al. 2004; Liu et al. 2005, 2006; Shen et al. 2007; Cheng et al. 2014; Y. Zhang et al. 2014).

The advanced approaches for detecting Pb^{2+} exemplified by leadzyme and the 8-17 DNAzyme are illustrated in Figure 2.A(a), respectively (Pan et al. 1992a, 1992b; Liu et al. 2009). The Leadzyme, discovered by Pan and Uhlenbeck, is an RNA ribozyme that requires Pb2+ for activity, with cleavage and ligation sites highlighted in red on the green and black substrate strands. The 8-17 DNAzyme, extensively studied by Lee and colleagues, exhibits robust catalytic efficiency and cleavage capabilities in the presence of Pb^{2+} (Cruz et al. 2004). Figure 2.A(b) shows the Pb^{2+}-dependent 8-17 DNA enzyme. The substrate cleavage site can be found either at a single RNA linkage site (rA) or on an RNA strand. The DNAzyme exhibits substantial activity in the presence of Pb^{2+} (Brown et al. 2003). Only Pb^{2+} can cause the enzyme to become active at low metal concentrations in the nanomolar to micromolar range (Liu et al. 2002; Cruz et al. 2004; H.-K. Kim et al. 2007; Ma et al. 2013). Figure 2B(a) shows various configurations of catalytic beacons for metal sensing, molecular beacons with fluorophore and quencher labels (Figure 2B(b)), fluorophore-quencher pairs at substrate ends (Figure 2B(c)), fluorophore placement on the substrate for reduced background fluorescence (Figure 2B(d)), dual quenching strategies (Figure 2B(e)), quencher placement adjacent to the cleavage site (Figure 2B(g)), and immobilized DNAzymes eliminating the need for quenchers (Figure 2B(h)). These configurations have specific advantages and limitations regarding sensitivity and quenching efficiency for detecting metal ions.

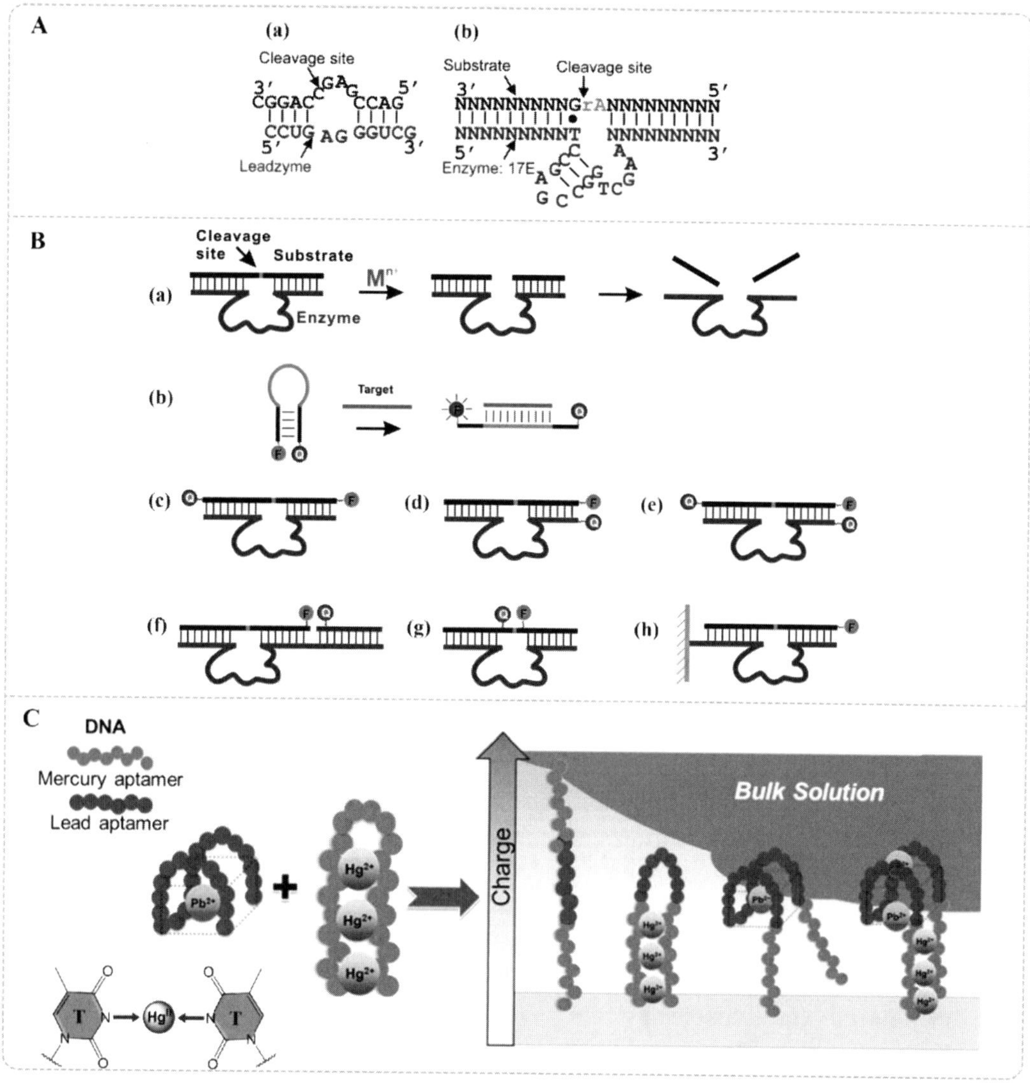

Figure 2. A) DNAzyme structures for detecting Pb^{2+} ions. B) Principal for nucleic acid enzyme-based fluorescence detection. Reproduced with permission from Liu et al. Copyright 2009, American Chemical Society. C) Schematic representation of a multi-target aptamer sensor with double ligand binding targets Reproduced with permission from Mayne et al. Copyright 2018, American Chemical Society.

3.3. Aptamer-Based Biosensors

Aptamer-based biosensors use recognition elements with short ssDNA or ssRNA sequences to detect and quantify specific target molecules (Guo et al. 2021). The selection of aptamers through the systematic evolution of ligands by exponential enrichment (SELEX) enables their selectivity and high affinity for a wide range of analytes, including proteins, toxins, small molecules, and even whole cells (McConnell et al. 2020; Khoshbin et al. 2023; Pratihar et al. 2023). The design involves using aptamers with specific binding sites, immobilizing aptamers on sensor surfaces or signal transducers, and detecting conformational changes induced by target binding through methods such as fluorescence, colorimetry, or electrochemistry. For example, a fluorescent biosensor developed by Li et al. detects lead ions indirectly using a G-rich aptamer in the presence of lead ions that form a G-quadruplex, which interacts with the malachite green fluorophore to produce a significant fluorescence signal (S. Li et al. 2018). This biosensor demonstrates high sensitivity and a broad dynamic range for lead detection, outperforming traditional methods.

3.4. DNA Hybridization-Based Biosensors

DNA hybridization-based biosensors detect and quantify specific target sequences through complementary base pairing between DNA strands (Mattath, Ghosh, Pratihar, et al. 2022). The design involves a DNA probe complementary to the target sequence, ensuring specific recognition and binding according to WC base pairing rules (Govindaraju 2019). The nucleic acid nanoarchitecture of these materials involves replacing traditional WC base-pairing interactions with metal complexation. Highly specific aptamers based on mismatched base pairs have been successfully used for detecting Hg^{2+}, forming a T–Hg^{2+}–T conformation that has proven effective in assays (Ono et al. 2004; Miyake et al. 2006). Mayne et al. utilized aptamer recognition and resistive pulse sensor technology to develop an aptamer sensor for monitoring Pb^{2+} and Hg^{2+} and characterizing samples by single particle resolution (Mayne et al. 2018). By modifying nanomaterials with aptamers and passing them through the sensor, information such as the density and charge state can be experimentally determined and correlated with the target concentration. The sensor detected various targets using various aptamers (Figure 2C). DNA hybridization-based biosensors provide advantages such as high specificity through complementary base pairing, versatility in detecting a range of target sizes, simple assay procedures, and potential for multiplexing.

3.5. DNA-Based Nanobiosensors

DNA-based nanobiosensors are advanced technologies integrating DNA molecules and nanomaterials to create innovative nanoscale devices. DNA nanomaterials can be functionalized and enhanced with additional molecules to improve functionality and biosensor performance. Nanotechnology-based biosensors typically employ metal nanoparticles, nanoclusters, quantum dots, or magnetic nanoparticles as transduction elements (Liu et al. 2009; Mattath, Ghosh, Dong, et al. 2022; Mattath et al. 2023; Parambil et al. 2023; Rajendran et al. 2024). By combining 8-17 DNAzymes and AuNPs, a fluorescent sensor for detecting Pb^{2+} ions was developed (J. H. Kim et al. 2011). Initially, in this system, the substrate strand binds to the AuNPs via hybridization with the enzyme strand, which quenches the fluorophore. Adding Pb^{2+} causes the substrate strand to be cleaved, detaching the fluorophore from the AuNPs and significantly increasing the fluorescence intensity. The detection limit was 5 nM, and rapid analysis was performed for 20 minutes. Another similar sensor with a lower detection limit of 0.25 nM was developed utilizing the GR-5 DNAzyme and AuNPs (H.-B. Wang et al. 2013). Fluorescent sensors utilizing AuNPs for the detection of Hg^{2+} have been developed, but they often suffer from low sensitivity (H. Wang, Wang, et al. 2008; Tan et al. 2013). To enhance sensitivity, nanometal surface energy transfer between quantum dots and AuNPs has been employed in constructing fluorescent sensors (Figure 3A) (M. Li et al. 2011). The DNA-conjugated quantum dots hybridize with DNA-conjugated AuNPs in the presence of Hg^{2+}, leading to quenching of the quantum dot fluorescence. The limit of detection of this sensor was 0.4 ppb, and the sensor demonstrated high selectivity for Hg^{2+} over other metal ions. Using quantum dots as fluorophores provides high sensitivity and addresses the photobleaching issues common with organic dyes.

3.6. Label-Free and Labeled Biosensors

Label-free and labeled biosensors are two distinct biosensing approaches for detecting and quantifying target analytes. Label-free biosensors measure direct interactions between the sensing element and the target analyte by observing changes in physical properties such as refractive index, mass, electrical impedance, or surface charge. Label-free platforms such as nanoclusters, nanoparticles, carbon dots, and small molecules utilize the unique affinity and selectivity of DNA for specific metal ions through sequence design and structural modifications (Pandeeswar et al. 2016; Zhou et al. 2017; Mattath, Ghosh, Pratihar, et al. 2022; Mattath, Ghosh, Dong, et al. 2022). Label-free DNA detection methods are simple, cost-effective, and capable of miniaturization, making them ideal for field applications and point-of-care testing.

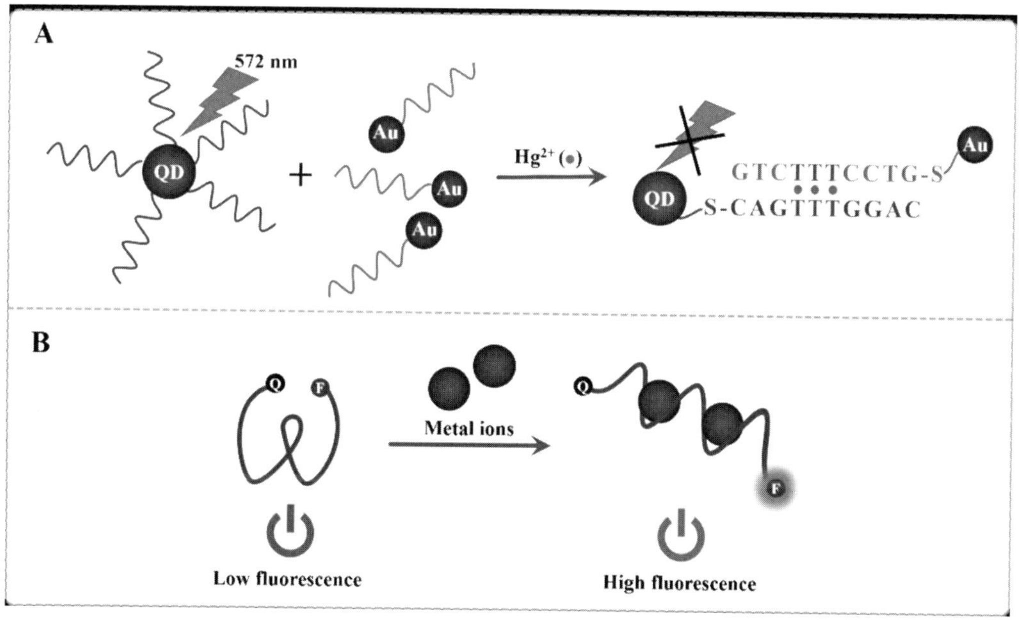

Figure 3. A) Fluorescent sensors for Hg^{2+} detection utilize nanometal surface energy transfer between AuNPs and quantum dots. Reproduced with permission from M. Li et al. Copyright 2011, American Chemical Society. B) Fluorescent sensors using aptamers for the detection of Pb^{2+}. Reproduced with permission from Mathivanan et al. Copyright 2022, American Chemical Society.
C) Illustration of the assay principle for target-triggered detection of As^{3+} based on a successive signal amplification strategy.

Labeled detection methods attach tags or labels to the target molecule or sensing element to facilitate detection, generating measurable signals upon binding. This approach offers higher sensitivity through signal amplification and enables multiplexing to detect multiple targets. A fluorescent sensor detects T-Hg^{2+}-T mismatches, causing hairpin structure formation in DNA and quenching fluorescence via fluorescence resonance energy transfer (FRET). This sensor has a simple detection limit of 40 nM but may produce a turn-off signal due to false positive results (Miyake et al. 2006). To improve the design, a fluorescent sensor was developed based on the structure switching model (Z. Wang, Heon Lee, et al. 2008). The quencher-labeled strand is hybridized to the fluorophore-labeled strand without Hg^{2+}, resulting in fluorescence quenching. When Hg^{2+} is present, a hairpin structure is formed by noncomplementary segments, releasing the quencher-labeled strand and enhancing fluorescence. Furthermore, it has been demonstrated that a molecular beacon-based sensor can detect Hg^{2+} (H. Xu et al. 2011). The molecular beacon opens in the presence of Hg^{2+}, releasing the fluorophore from the quencher and restoring the fluorescence signal. Recently, a fluorophore-labeled Pb^{2+}-binding aptamer was developed (Figure 3B) (Mathivanan et al. 2022). This aptamer shows increased fluorescence upon binding Pb^{2+}, allowing for quantitative measurement. This strategy exhibits strong specificity for Pb^{2+} over other metals and functions effectively in complex biological fluids, highlighting its potential for practical metal ion detection applications. Fluorescent labels such as dyes or quantum dots are commonly used in fluorescence-based detection assays, where heavy metal ions induce conformational changes or binding events in the labeled DNA probe (Guo et al. 2021; J. Xu et al. 2022).

4. Biosensing Readout Methods

Biosensing readout methods vary widely based on analyte and biosensor design. Optical methods, such as fluorescence and luminescence, are used to measure the influence of analyte concentration on light emission. Colorimetric detection observes color changes upon analyte binding. Electrochemical biosensors detect analytes through changes in electrical properties.

4.1. Fluorescence Sensing

Fluorescence detection techniques offer high sensitivity, detecting concentrations below 1 nM, and even single-molecule fluorescence with advanced instruments. Natural nucleic acid sensors need external fluorescent groups since they are nonfluorescent. Using a target-triggered signal amplification strategy with an aptamer sequence and exonuclease III (Exo III)-mediated DNA recycling, a highly sensitive biosensor was developed for the detection of trace amounts of As^{3+}, with a detection limit of 2 pM (Figure 3C) (Zeng, Zhou, et al. 2019). Using the FRET process and the fluorescence quenching of graphene oxide, Koshbin et al. created a paper-based apta-sensor for ultrasensitive Pb^{2+} detection (Khoshbin et al. 2019). To achieve a detection limit of 0.5 pM, the conformation of the aptamer changed from a random coil to a G-quadruplex structure. This sensor effectively detects Pb^{2+} in water, milk, and human blood serum. Zhu et al. used a Cd-4 aptamer modified with 6-carboxyfluorescein (6-FAM) and GGGG sequences, which quench fluorescence through photoinduced electron transfer, to create a fluorescence sensor for Cd^{2+} detection with a limit of 2.15 nM (Y.-F. Zhu et al. 2017). Recovery tests confirmed its potential for accurate and rapid heavy metal ion detection in water and serum samples.

4.2. Colorimetric Sensing

Colorimetric methods are simple and affordable, often allowing detection by the naked eye without expensive instruments. AuNPs are frequently used due to their optical properties and high extinction coefficients. Liu et al. developed a colorimetric sensor for Pb^{2+} detection that modifies color according to the state of aggregation of DNAzyme-functionalized AuNPs, as illustrated in Figure 4A(a) (Liu et al. 2003). The AuNPs functionalized with the DNAzyme substrate aggregate and appear blue when Pb^{2+} is absent. When Pb^{2+} is present, substrate cleavage occurs, and the AuNPs disaggregate, changing the color from blue to red. The sensor design involves a "tail" structure to manage steric effects and improve AuNP aggregation at room temperature (Figure 4A(b)) (Liu et al. 2004).

Lee et al. developed a colorimetric sensor for detecting Hg^{2+} using either labeled or label-free approaches. In their labeling approach, AuNPs functionalized with T-T base pair-containing DNA aggregate when exposed to Hg^{2+}, forming Hg^{2+}-mediated T-T base pairs and resulting in detectable color change at concentrations as low as 0.1 μM (Lee et al. 2007). This method requires precise temperature control, limiting its field applicability. To overcome this, a new sensor design incorporating oligonucleotide connections and controlled mismatches was created to function efficiently at room temperature (D. Li et al. 2008). In a label-free approach, ssDNA stops AuNP aggregation from occurring without Hg^{2+} (Xia et al. 2010). When Hg^{2+} is present, it induces DNA hairpin structure formation, causing the AuNPs to aggregate and change color from red to blue, with detection limits as low as 10 nM. This assay is selective and effective even in complex biological media such as target-doped blood serum. A thymine-rich DNA functionalized with polyacrylamide hydrogel was used to detect Hg^{2+} through green fluorescence, a specific interaction between Hg^{2+} and thymine bases (Figure 4B) (Dave et al. 2010). Without Hg^{2+}, the DNA adopts a random coil structure, resulting in weak fluorescence when introducing SYBR Green I. Upon binding with Hg^{2+}, the DNA undergoes a conformational change to form a hairpin structure that enhances SYBR Green I binding, leading to an approximately 9-fold increase in fluorescence emission. This color change is due to the formation of hairpin structures in DNA upon Hg^{2+} binding, resulting in a visible green fluorescence signal for detecting Hg^{2+}.

A sensitive and simple colorimetric method for rapidly detecting Cd^{2+} utilizes aptamer-functionalized AuNPs (Gan et al. 2020). In high-salt solutions, AuNPs typically aggregate, but aptamers enhance their stability and prevent aggregation. When Cd^{2+} ions are present, they bind to the aptamers, reducing free aptamers and causing the AuNPs to aggregate. This results in a color change that can be analyzed with a smartphone-based system for quantitative detection within 10 minutes. A personal glucose meter (PGM) is used for point-of-use Cd^{2+} detection. This method uses aptamer-Cd^{2+} recognition and Exo III for amplification (Zeng, Gong, et al. 2019). Magnetic beads release single-stranded DNA through cleavage and hybridization reactions. Invertase-conjugated DNA complementary to the released DNA hydrolyzes sucrose into blood glucose, which is then quantified by the PGM. The low detection limit of 5 pM achieved by this method, along with its portability, affordability, and user-friendliness, make it appropriate for routine heavy metal detection by the public.

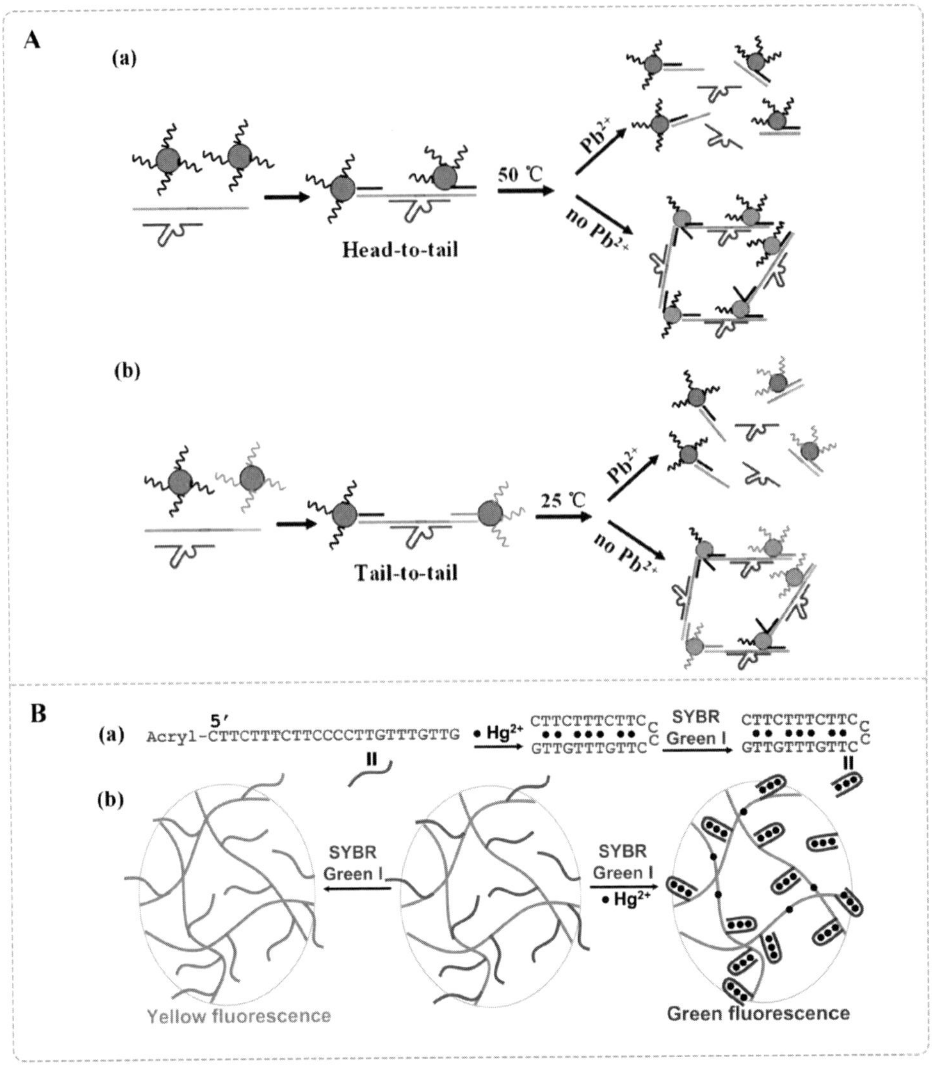

Figure 4. A) Colorimetric Pb^{2+} detection by aligned AuNPs (a) head-to-tail and (b) tail-to-tail. Figure 4A(a) is reproduced with permission from Liu et al. Copyright 2003, American Chemical Society. Figure 4A(b) is reproduced with permission from Liu et al. Copyright 2004, American Chemical Society. B) A polyacrylamide hydrogel functionalized with thymine-rich DNA was utilized for detecting Hg^{2+}. (a) The DNA sequence of acrydite-Hg-DNA shows a modification at the 5′-end with an acrydite group for attachment to the hydrogel, facilitating fluorescence signal generation for Hg^{2+} detection. (b) Covalent immobilization of DNA within the polyacrylamide hydrogel enables interactions with Hg^{2+} and SYBR Green I, resulting in a detectable fluorescence signal visible to the naked eye. Reproduced with permission from Dave et al. Copyright 2010, American Chemical Society.

4.3. Electrochemical Sensing

Electrochemical sensors offer better sensitivity and designs and use electrochemical labels to identify changes in the distance between double-stranded DNA chains. On gold electrode surfaces, they achieve nanomolar detection limits (Wu et al. 2010; Ziping Zhang et al. 2011; Ziping Zhang et al. 2013). Tang et al. developed a Hg^{2+} sensor with a three-dimensional network of AuNPs and DNA. This sensor achieved a remarkable detection limit of 7.38 pM (Tang et al. 2012). Through DNA hybridization, T-Hg^{2+}-T interactions produced an internal binding layer. Subsequently, DNA1-AuNP complexes were bound to DNA2-AuNPs, forming an outer layer and supramolecular structures.

Zhang et al. synthesized a nanostructured composite, Fe-MOF@mFe_3O_4@mC, for detecting trace heavy metal ions (Pb^{2+} and As^{3+}) (Zhihong Zhang et al. 2017). Strong binding between aptamer strands targeted by heavy metal ions was facilitated by the nanocomposite. Hairpin aptamers release their stem-loop structure and form a stabilized G-quadruplex specific to Pb^{2+} or As^{3+} when these metals are present in an aqueous solution. This conformational change in aptamers can be detected using electrochemical techniques. The aptasensor shows specific selectivity for detecting Pb^{2+} and As^{3+} with 2.27 and 6.73 pM detection limits, respectively, in water and human blood serum. It demonstrates excellent regenerability, selectivity, and reproducibility, highlighting its potential for biomedical applications.

5. Challenges and Prospects

In medical diagnostics, NABs are crucial for detecting and monitoring conditions caused by heavy metal exposure, including lead poisoning and mercury toxicity (Mitra et al. 2022). In the food industry, they help identify harmful heavy metal contaminants, enhancing food safety and quality control. A significant challenge in metal ion detection is achieving high sensitivity and specificity due to low concentrations in complex samples. Innovative sensor materials and advanced surface functionalization techniques are in greater demand to improve the selectivity of these biosensors for specific metal ions. Biological samples often contain interfering substances that can affect the response of nucleic acid biosensors, leading to false readings. Therefore, careful sample preparation is essential to minimize interference and improve detection reliability. Few studies have focused on heavy metal detection in biological samples such as blood, serum, and urine. Emerging studies should concentrate on developing durable sensor materials and manufacturing processes to improve their stability and reproducibility. Lowering the production costs of biosensors will increase accessibility, especially in resource-limited areas affected by heavy metal pollution. The future promises significant potential for biosensors to transform healthcare, industry, food safety, sustainability, disease management, and personalized medicine. By addressing these challenges, metal ion-detecting biosensors could significantly impact food safety and human health.

Conclusion

Biosensors are powerful analytical tools that combine chemistry, biology, nanotechnology, and engineering for precise analyte detection. They provide real-time insights using innovative elements such as DNAzymes, aptamers, hybrid DNA architectures, and label-free/labeled DNA nanobiosensors. In heavy metal detection, NAB techniques such as DNAzymes and aptamers are superior to conventional antibody approaches. Fluorescence methods offer sensitive and selective detection using labeled or label-free DNA probes. Colorimetric and electrochemical sensors exhibit enhanced sensitivity and detection ability. There is a growing demand for DNA biosensors to detect toxic metals such as chromium. These methods show promise for biomedical applications, clinical diagnostics, and personalized treatments related to metal-induced health effects.

References

Avinash, MB, and Thimmaiah Govindaraju. 2018. "Architectonics: Design of Molecular Architecture for Functional Applications." *Accounts of Chemical Research* 51 (2): 414-426. https://doi.org/https://doi.org/10.1021/acs.accounts.7b00434.

Balali-Mood, Mahdi, Kobra Naseri, Zoya Tahergorabi, Mohammad Reza Khazdair, and Mahmood Sadeghi. 2021. "Toxic Mechanisms of Five Heavy Metals: Mercury, Lead, Chromium, Cadmium, and Arsenic." *Frontiers in Pharmacology* 12. https://doi.org/https://doi.org/10.3762/bjnano.11.11.

Bishak, Yaser Khaje, Laleh Payahoo, Alireza Osatdrahimi, and Alireza Nourazarian. 2015. "Mechanisms of Cadmium Carcinogenicity in the Gastrointestinal Tract." *Asian Pacific Journal of Cancer Prevention* 16 (1): 9-21. https://doi.org/https://doi.org/10.7314/apjcp.2015.16.1.9.

Brown, Andrea K, Jing Li, Caroline M-B Pavot, and Yi Lu. 2003. "A Lead-Dependent DNAzyme with a Two-Step Mechanism." *Biochemistry* 42 (23): 7152-7161. https://doi.org/https://doi.org/10.1021/bi027332w.

Bruesehoff, Peter J, Jing Li, Anthony J Augustine III, and Yi Lu. 2002. "Improving Metal Ion Specificity During *in Vitro* Selection of Catalytic DNA." *Combinatorial Chemistry & High Throughput Screening* 5 (4): 327-335. https://doi.org/https://doi.org/10.2174/1386207023330264.

Bunka, David HJ, and Peter G Stockley. 2006. "Aptamers Come of Age–at Last." *Nature Reviews Microbiology* 4 (8): 588-596. https://doi.org/https://doi.org/10.1038/nrmicro1458.

Cheng, Yan, Yin Huang, Jianping Lei, Lei Zhang, and Huangxian Ju. 2014. "Design and Biosensing of Mg^{2+}-Dependent DNAzyme-Triggered Ratiometric Electrochemiluminescence." *Analytical Chemistry* 86 (10): 5158-5163. https://doi.org/https://doi.org/10.1021/ac500965p.

Cruz, Rani PG, Johanna B Withers, and Yingfu Li. 2004. "Dinucleotide Junction Cleavage Versatility of 8-17 Deoxyribozyme." *Chemistry & Biology* 11 (1): 57-67. https://doi.org/https://doi.org/10.1016/j.chembiol.2003.12.012.

Dave, Neeshma, Michelle Y. Chan, Po-Jung Jimmy Huang, Brendan D. Smith, and Juewen Liu. 2010. "Regenerable DNA-Functionalized Hydrogels for Ultrasensitive, Instrument-Free Mercury(II) Detection and Removal in Water." *Journal of the American Chemical Society* 132 (36): 12668-12673. https://doi.org/https://doi.org/10.1021/ja106098j.

Gan, Ying, Tao Liang, Qiongwen Hu, Longjie Zhong, Xinyi Wang, Hao Wan, and Ping Wang. 2020. "In-Situ Detection of Cadmium with Aptamer Functionalized Gold Nanoparticles Based on Smartphone-Based Colorimetric System." *Talanta* 208: 120231. https://doi.org/https://doi.org/10.1016/j.talanta.2019.120231.

Ghosh, Debasis, Lakshmi P Datta, and Thimmaiah Govindaraju. 2020. "Molecular Architectonics of DNA for Functional Nanoarchitectures." *Beilstein Journal of Nanotechnology* 11 (1): 124-140. https://doi.org/https://doi.org/10.3762/bjnano.11.11.

Ghosh, Jyotirmoy, and Parames C Sil. 2023. "Mechanism for Arsenic-Induced Toxic Effects." In *Handbook of Arsenic Toxicology*, 223-252. Elsevier.

Gobrecht, Jillian, Claire McDyre, James Comotto, and Mindy Reynolds. 2017. "Induction of Cytotoxic and Genotoxic Damage Following Exposure of V79 Cells to Cadmium Chloride." *Mutation Research/Genetic Toxicology and Environmental Mutagenesis* 816: 12-17. https://doi.org/https://doi.org/10.1016/j.mrgentox.2017.03.001.

Govindaraju, Thimmaiah. 2019. *Templated DNA nanotechnology: Functional DNA Nanoarchitectonics*. edited by Thimmaiah Govindaraju. New York City, USA: Jenny Stanford Publishing.

Guo, Wenfei, Chuanxiang Zhang, Tingting Ma, Xueying Liu, Zhu Chen, Song Li, and Yan Deng. 2021. "Advances in Aptamer Screening and Aptasensors' Detection of Heavy Metal Ions." *Journal of Nanobiotechnology* 19 (1): 1-19. https://doi.org/https://doi.org/10.1186/s12951-021-00914-4.

Khoshbin, Zahra, Mohammad Reza Housaindokht, Mohammad Izadyar, Asma Verdian, and Mohammad Reza Bozorgmehr. 2019. "A Simple Paper-Based Aptasensor for Ultrasensitive Detection of Lead (II) Ion." *Analytica Chimica Acta* 1071: 70-77. https://doi.org/https://doi.org/10.1016/j.aca.2019.04.049.

Khoshbin, Zahra, Nadiyeh Rouhi, Khalil Abnous, Seyed Mohammad Taghdisi, and Asma Verdian. 2023. "Aptamer-Based Analysis of Heavy Metal Ions." In *Aptamers for Food Applications*, 127-137. Elsevier.

Kim, Hee-Kyung, Juewen Liu, Jing Li, Nandini Nagraj, Mingxi Li, Caroline M-B Pavot, and Yi Lu. 2007. "Metal-Dependent Global Folding and Activity of the 8-17 DNAzyme Studied by Fluorescence Resonance Energy Transfer." *Journal of the American Chemical Society* 129 (21): 6896-6902. https://doi.org/https://doi.org/10.1021/ja0712625.

Kim, Joong Hyun, Sang Ho Han, and Bong Hyun Chung. 2011. "Improving Pb^{2+} Detection Using DNAzyme-Based Fluorescence Sensors by Pairing Fluorescence Donors with Gold Nanoparticles." *Biosensors and Bioelectronics* 26 (5): 2125-2129. https://doi.org/https://doi.org/10.1016/j.bios.2010.09.018.

Kuivenhoven, M, and K Mason. 2019. *Arsenic (Arsine) Toxicity.StatPearls Publishing*.

Lee, Jae-Seung, Min Su Han, and Chad A Mirkin. 2007. "Colorimetric Detection of Mercuric Ion (Hg^{2+}) in Aqueous Media Using DNA-Functionalized Gold Nanoparticles." *Angewandte Chemie International Edition* 46 (22): 4093-4096. https://doi.org/https://doi.org/10.1002/anie.200700269.

Li, Di, Agnieszka Wieckowska, and Itamar Willner. 2008. "Optical Analysis of Hg^{2+} Ions by Oligonucleotide–Gold-Nanoparticle Hybrids and DNA-Based Machines." *Angewandte Chemie International Edition* 47 (21): 3927-3931. https://doi.org/https://doi.org/10.1002/anie.200705991.

Li, Ming, Qiaoyi Wang, Xiaodong Shi, Lawrence A Hornak, and Nianqiang Wu. 2011. "Detection of Mercury (II) by Quantum Dot/DNA/Gold Nanoparticle Ensemble Based Nanosensor via Nanometal Surface Energy Transfer." *Analytical Chemistry* 83 (18): 7061-7065. https://doi.org/https://doi.org/10.1021/ac2019014.

Li, Shiya, Hongwen Liu, Guiying Yang, Shimeng Liu, Ran Liu, and Changyin Lv. 2018. "Detection of Radon with Biosensors Based on the Lead(II)-Induced Conformational Change of Aptamer HTG and Malachite Green Fluorescence Probe." *Journal of Environmental Radioactivity* 195: 60-66. https://doi.org/https://doi.org/10.1016/j.jenvrad.2018.09.021.

Liu, Juewen, Zehui Cao, and Yi Lu. 2009. "Functional Nucleic Acid Sensors." *Chemical Reviews* 109 (5): 1948-1998. https://doi.org/https://doi.org/10.1021/cr030183i.

Liu, Juewen, and Yi Lu. 2002. "FRET Study of a Trifluorophore-Labeled DNAzyme." *Journal of the American Chemical Society* 124 (51): 15208-15216. https://doi.org/https://doi.org/10.1021/ja027647z.

Liu, Juewen, and Yi Lu. 2003. "A Colorimetric Lead Biosensor Using DNAzyme-Directed Assembly of Gold Nanoparticles." *Journal of the American Chemical Society* 125 (22): 6642-6643. https://doi.org/https://doi.org/10.1021/ja034775u.

Liu, Juewen, and Yi Lu. 2004. "Accelerated Color Change of Gold Nanoparticles Assembled by DNAzymes for Simple and Fast Colorimetric Pb^{2+} Detection." *Journal of the American Chemical Society* 126 (39): 12298-12305. https://doi.org/https://doi.org/10.1021/ja046628h.

Liu, Juewen, and Yi Lu. 2005. "Stimuli-Responsive Disassembly of Nanoparticle Aggregates for Light-Up Colorimetric Sensing." *Journal of the American Chemical Society* 127 (36): 12677-12683. https://doi.org/https://doi.org/10.1021/ja053567u.

Liu, Juewen, and Yi Lu. 2006. "Fluorescent DNAzyme Biosensors For Metal Ions Based on Catalytic Molecular Beacons." *Fluorescent Energy Transfer Nucleic Acid Probes: Designs and Protocols*: 275-288. https://doi.org/https://doi.org/10.1385/1-59745-069-3:275.

Lu, Yi, and Juewen Liu. 2006. "Functional DNA Nanotechnology: Emerging Applications of DNAzymes and Aptamers." *Current Opinion in Biotechnology* 17 (6): 580-588. https://doi.org/https://doi.org/10.1016/j.copbio.2006.10.004.

Ma, Dik-Lung, Hong-Zhang He, Daniel Shiu-Hin Chan, and Chung-Hang Leung. 2013. "Simple DNA-Based Logic Gates Responding to Biomolecules and Metal Ions." *Chemical Science* 4 (9): 3366-3380. https://doi.org/https://doi.org/10.1039/C3SC50924A.

Mathivanan, Johnsi, Hehua Liu, Jianhua Gan, Arun Richard Chandrasekaran, and Jia Sheng. 2022. "Fluorescent Aptaswitch for Detection of Lead Ions." *ACS Applied Bio Materials* 5 (11): 5089-5093. https://doi.org/10.1021/acsabm.2c00368. https://doi.org/10.1021/acsabm.2c00368.

Mattath, Mohamed Nabeel, Debasis Ghosh, Chunyan Dong, Thimmaiah Govindaraju, and Shuo Shi. 2022. "Mercury Mediated DNA–Au/Ag Nanocluster Ensembles to Generate a Gray Code Encoder for Biocomputing." *Materials Horizons* 9 (8): 2109-2114. https://doi.org/https://doi.org/10.1039/D2MH00598K.

Mattath, Mohamed Nabeel, Debasis Ghosh, Sumon Pratihar, Shuo Shi, and Thimmaiah Govindaraju. 2022. "Nucleic Acid Architectonics for pH-Responsive DNA Systems and Devices." *ACS Omega* 7 (4): 3167-3176. https://doi.org/https://doi.org/10.1021/acsomega.1c06464.

Mattath, Mohamed Nabeel, Haibin Zhang, Debasis Ghosh, Thimmaiah Govindaraju, and Shuo Shi. 2023. "Nanoclusters with Specific DNA Overhangs: Modifying Configurability, Engineering Contrary Logic Pairs and the Parity Generator/Checker for Error Detection." *Nanoscale* 15 (43): 17386-17397. https://doi.org/https://doi.org/10.1039/D3NR04167K.

Mayne, Laura, Chih-Yuan Lin, Steven DR Christie, Zuzanna S Siwy, and Mark Platt. 2018. "The Design and Characterization of Multifunctional Aptamer Nanopore Sensors." *ACS Nano* 12 (5): 4844-4852. https://doi.org/https://doi.org/10.1021/acsnano.8b01583.

McConnell, Erin M, Julie Nguyen, and Yingfu Li. 2020. "Aptamer-Based Biosensors for Environmental Monitoring." *Frontiers in Chemistry* 8: 434. https://doi.org/https://doi.org/10.3389/fchem.2020.00434.

Mitra, Saikat, Arka Jyoti Chakraborty, Abu Montakim Tareq, Talha Bin Emran, Firzan Nainu, Ameer Khusro, Abubakr M. Idris, Mayeen Uddin Khandaker, Hamid Osman, Fahad A. Alhumaydhi, and Jesus Simal-Gandara. 2022. "Impact of Heavy Metals on the Environment and Human Health: Novel Therapeutic Insights to Counter the Toxicity." *Journal of King Saud University - Science* 34 (3): 101865. https://doi.org/https://doi.org/10.1016/j.jksus.2022.101865.

Miyake, Yoko, Humika Togashi, Mitsuru Tashiro, Hiroshi Yamaguchi, Shuji Oda, Megumi Kudo, Yoshiyuki Tanaka, Yoshinori Kondo, Ryuichi Sawa, and Takashi Fujimoto. 2006. "Mercury^{II}-Mediated Formation of Thymine– Hg^{II}– Thymine Base Pairs in DNA Duplexes." *Journal of the American Chemical Society* 128 (7): 2172-2173. https://doi.org/https://doi.org/10.1021/ja056354d.

Mochizuki, Hitoshi. 2019. "Arsenic Neurotoxicity in Humans." *International Journal of Molecular Sciences* 20 (14): 3418. https://doi.org/https://doi.org/10.3390%2Fijms20143418.

Moorthy, Hariharan, Lakshmi Priya Datta, and Thimmaiah Govindaraju. 2021. "Molecular Architectonics-Guided Design of Biomaterials." *Chemistry: An Asian Journal* 16 (5): 423-442. https://doi.org/https://doi.org/10.1002/asia.202001445.

Moorthy, Sathishkumar, K. T Mohammed Hisham, and Mohamed Nabeel Mattath. 2023. *Aqueous Metal Induced Contamination in Wastewater and its Adverse Effects*. In *EIACP: Geodiversity and Impact on Environment*: Geodiversity And Impact On Environment Jnu-Eiacp.

Narayanaswamy, Nagarjun, Raji R Nair, YV Suseela, Deepak Kumar Saini, and T Govindaraju. 2016. "A Molecular Beacon-Based DNA Switch for Reversible pH Sensing in Vesicles and Live Cells." *Chemical Communications* 52 (56): 8741-8744. https://doi.org/https://doi.org/10.1039/c6cc02705a.

Niazov, Tamara, Valeri Pavlov, Yi Xiao, Ron Gill, and Itamar Willner. 2004. "DNAzyme-Functionalized Au Nanoparticles for the Amplified Detection of DNA or Telomerase Activity." *Nano Letters* 4 (9): 1683-1687. https://doi.org/https://doi.org/10.1021/nl0491428.

O'Sullivan, Ciara K. 2002. "Aptasensors–the Future of Biosensing?" *Analytical and Bioanalytical Chemistry* 372: 44-48. https://doi.org/https://doi.org/10.1007/s00216-001-1189-3.

Ono, Akira, and Humika Togashi. 2004. "Highly Selective Oligonucleotide-Based Sensor for Mercury (II) in Aqueous Solutions." *Angewandte Chemie International Edition* 116 (33): 4400-4402. https://doi.org/https://doi.org/10.1002/anie.200454172.

Organization, World Health. 2017. "Guidelines for Drinking-Water Quality: First Addendum to the Fourth Edition."

Palchetti, Ilaria, and Marco Mascini. 2008. "Nucleic Acid Biosensors for Environmental Pollution Monitoring." *Analyst* 133 (7): 846-854. https://doi.org/https://doi.org/10.1039/b802920m.

Pan, Tao, and Olke C Uhlenbeck. 1992a. "*In Vitro* Selection of RNAs that Undergo Autolytic Cleavage with Lead (2+)." *Biochemistry* 31 (16): 3887-3895. https://doi.org/https://doi.org/10.1021/bi00131a001.

Pan, Tao, and Olke C Uhlenbeck. 1992b. "A Small Metalloribozyme with a Two-Step Mechanism." *Nature* 358 (6387): 560-563. https://doi.org/https://doi.org/10.1038/358560a0.

Pandeeswar, M, Satyaprasad P Senanayak, and T Govindaraju. 2016. "Nanoarchitectonics of Small Molecule and DNA for Ultrasensitive Detection of Mercury." *ACS Applied Materials & Interfaces* 8 (44): 30362-30371. https://doi.org/https://doi.org/10.1021/acsami.6b10527.

Parambil, Ajith Manayil, Mohamed Nabeel Mattath, Paulraj Rajamani, Phuong V Pham, Gopalakrishnan Kumar, and Vinoth Kumar Ponnusamy. 2023. "Biogenic Fluorescent Carbon Dots Modulated Fabrication of Concatenate Logic Library and Pattern-Mediated Molecular Keypad Lock for Chemical Sensing Application." *Chemical Engineering Journal* 463: 142354. https://doi.org/https://doi.org/10.1016/j.cej.2023.142354.

Pratihar, Sumon, Mohamed Nabeel Mattath, and Thimmaiah Govindaraju. 2023. "Coronavirus Genomic cDNA Derived G-Quadruplex as a Selective Target for Fluorometric Detection." *Chemical Communications* 59 (38): 5717-5720. https://doi.org/https://doi.org/10.1039/D3CC00740E.

Punshon, Tracy, Brian P Jackson, Andrew A Meharg, Todd Warczack, Kirk Scheckel, and Mary Lou Guerinot. 2017. "Understanding Arsenic Dynamics in Agronomic Systems to Predict and Prevent Uptake by Crop Plants." *Science of the Total Environment* 581: 209-220. https://doi.org/https://doi.org/10.1016/j.scitotenv.2016.12.111.

Rajendran, Kalimuthu, Mattath Mohamed Nabeel, Kannaiyan Pandian, Velu Manikandan, Sivarasan Ganesan, Huang-Mu Lo, Nagappan Rajendiran, Xinghui Liu, and Oluwatobi Samuel Oluwafemi. 2024. "Fabrication of Silicon Quantum Dots-Methyl Viologen Nanohybrids: Turn-On-Off-On Fluorescence Nanoprobe for the Detection of *d*-Penicillamine." *Microchemical Journal* 197: 109803. https://doi.org/https://doi.org/10.1016/j.microc.2023.109803.

Rana, Suresh Vir Singh. 2008. "Metals and Apoptosis: Recent Developments." *Journal of Trace Elements in Medicine and Biology* 22 (4): 262-284. https://doi.org/https://doi.org/10.1016/j.jtemb.2008.08.002.

Roy, Bappaditya, Madhu Ramesh, and Thimmaiah Govindaraju. 2019. "DNA-Based Nanoswitches and Devices." In *Templated DNA Nanotechnology*, 365-408. Jenny Stanford Publishing.

Shen, Yutu, Gillian Mackey, Nicholas Rupcich, Darin Gloster, William Chiuman, Yingfu Li, and John D Brennan. 2007. "Entrapment of Fluorescence Signaling DNA Enzymes in Sol− gel-Derived Materials for Metal Ion Sensing." *Analytical Chemistry* 79 (9): 3494-3503. https://doi.org/https://doi.org/10.1021/ac070235u.

Tan, Daidi, Yue He, Xiaojing Xing, Yu Zhao, Hongwu Tang, and Daiwen Pang. 2013. "Aptamer Functionalized Gold Nanoparticles Based Fluorescent Probe for the Detection of Mercury (II) Ion in Aqueous Solution." *Talanta* 113: 26-30. https://doi.org/https://doi.org/10.1016/j.talanta.2013.03.055.

Tang, Xuemei, Huixiang Liu, Binghua Zou, Danbi Tian, and He Huang. 2012. "A Fishnet Electrochemical Hg^{2+} Sensing Strategy Based on Gold Nanoparticle-Bioconjugate and Thymine–Hg^{2+}–Thymine Coordination Chemistry." *Analyst* 137 (2): 309-311. https://doi.org/https://doi.org/10.1039/C1AN15908A.

Tchounwou, Paul B, Clement G Yedjou, Anita K Patlolla, and Dwayne J Sutton. 2012. *Heavy Metal Toxicity and the Environment*. Vol. 3. Vol. Environmental Toxicology*Molecular, Clinical and Environmental Toxicology*.

Tong, Dingyi, Haifeng Duan, Hejing Zhuang, Jungang Cao, Zhonglin Wei, and Yingjie Lin. 2014. "Using T–Hg–T and C–Ag–T: a Four-Input Dual-Core Molecular Logic Gate and its New Application in Cryptography." *RSC Advances* 4 (11): 5363-5366. https://doi.org/https://doi.org/10.1039/C3RA44650F.

Torigoe, Hidetaka, Akira Ono, and Tetsuo Kozasa. 2010. "Hg^{II} Ion Specifically Binds with T: T Mismatched Base Pair in Duplex DNA." *Chemistry–A European Journal* 16 (44): 13218-13225. https://doi.org/https://doi.org/10.1002/chem.201001171.

Ung, Choong Yong, Siew Hong Lam, Mya Myintzu Hlaing, Cecilia Lanny Winata, Svetlana Korzh, Sinnakaruppan Mathavan, and Zhiyuan Gong. 2010. "Mercury-Induced Hepatotoxicity in Zebrafish: *In Vivo* Mechanistic Insights from Transcriptome Analysis, Phenotype Anchoring and Targeted Gene Expression Validation." *BMC Genomics* 11: 1-14. https://doi.org/https://doi.org/10.1186/1471-2164-11-212.

Wang, Hai-Bo, Lan Wang, Ke-Jing Huang, Shu-Ping Xu, Hong-Qi Wang, Ling-Ling Wang, and Yan-Ming Liu. 2013. "A Highly Sensitive and Selective Biosensing Strategy for the Detection of Pb^{2+} Ions Based on GR-5 DNAzyme Functionalized AuNPs." *New Journal Of Chemistry* 37 (8): 2557-2563. https://doi.org/https://doi.org/10.1039/C3NJ00328K.

Wang, Hao, Yongxiang Wang, Jianyu Jin, and Ronghua Yang. 2008. "Gold Nanoparticle-Based Colorimetric and "Turn-on" Fluorescent Probe for Mercury (II) Ions in Aqueous Solution." *Analytical Chemistry* 80 (23): 9021-9028. https://doi.org/https://doi.org/10.1021/ac801382k.

Wang, Zidong, Jung Heon Lee, and Yi Lu. 2008. "Highly Sensitive "Turn-on" Fluorescent Sensor for Hg^{2+} in Aqueous Solution based on Structure-Switching DNA." *Chemical Communications* (45): 6005-6007. https://doi.org/http://dx.doi.org/10.1039/B812755G.

Wani, AL, A Ara, and JA Usmani. 2015. "Lead Toxicity: A Review." *Interdisciplinary Toxicology* 8 (2): 55-64. https://doi.org/https://doi.org/10.1515/intox-2015-0009.

Wu, Danhong, Qing Zhang, Xia Chu, Haibo Wang, Guoli Shen, and Ruqin Yu. 2010. "Ultrasensitive Electrochemical Sensor for Mercury (II) Based on Target-Induced Structure-Switching DNA." *Biosensors and Bioelectronics* 25 (5): 1025-1031. https://doi.org/https://doi.org/10.1016/j.bios.2009.09.017.

Xia, Fan, Xiaolei Zuo, Renqiang Yang, Yi Xiao, Di Kang, Alexis Vallée-Bélisle, Xiong Gong, Jonathan D Yuen, Ben BY Hsu, and Alan J Heeger. 2010. "Colorimetric Detection of DNA, Small Molecules, Proteins, and Ions Using Unmodified Gold Nanoparticles and Conjugated Polyelectrolytes." *Proceedings of the National Academy of Sciences* 107 (24): 10837-10841. https://doi.org/https://doi.org/10.1073/pnas.1005632107.

Xu, Huifeng, Xi Zhu, Hongzhi Ye, Lishuang Yu, Xianxiang Liu, and Guonan Chen. 2011. "A Simple "Molecular Beacon"-Based Fluorescent Sensing Strategy for Sensitive and Selective Detection of Mercury (II)." *Chemical Communications* 47 (44): 12158-12160. https://doi.org/https://doi.org/10.1039/C1CC14265H.

Xu, Jiaqi, Rundong Jiang, Yuxin Feng, Zhuqing Liu, Jin Huang, Changbei Ma, and Kemin Wang. 2022. "Functional Nucleic Acid-Based Fluorescent Probes for Metal Ion Detection." *Coordination Chemistry Reviews* 459: 214453. https://doi.org/https://doi.org/10.1016/j.ccr.2022.214453.

Zeng, Lingwen, Junyu Gong, Peisen Rong, Chengshuai Liu, and Junhua Chen. 2019. "A Portable and Quantitative Biosensor for Cadmium Detection Using Glucometer as the Point-of-Use Device." *Talanta* 198: 412-416. https://doi.org/https://doi.org/10.1016/j.talanta.2019.02.045.

Zeng, Lingwen, Danhua Zhou, Junyu Gong, Chengshuai Liu, and Junhua Chen. 2019. "Highly Sensitive Aptasensor for Trace Arsenic (III) Detection Using DNAzyme as the Biocatalytic Amplifier." *Analytical Chemistry* 91 (3): 1724-1727. https://doi.org/https://doi.org/10.1021/acs.analchem.8b05466.

Zhang, Yan, Li-juan Wang, and Chun-yang Zhang. 2014. "Highly Sensitive Detection of Telomerase Using a Telomere-Triggered Isothermal Exponential Amplification-Based DNAzyme Biosensor." *Chemical Communications* 50 (15): 1909-1911. https://doi.org/https://doi.org/10.1039/c3cc48518h.

Zhang, Zhihong, Hongfei Ji, Yingpan Song, Shuai Zhang, Minghua Wang, Changchang Jia, Jia-Yue Tian, Linghao He, Xiaojing Zhang, and Chun-Sen Liu. 2017. "Fe (III)-Based Metal–Organic Framework-Derived Core–Shell Nanostructure: Sensitive Electrochemical Platform for High Trace Determination of Heavy Metal Ions." *Biosensors and Bioelectronics* 94: 358-364. https://doi.org/https://doi.org/10.1016/j.bios.2017.03.014.

Zhang, Ziping, Aman Tang, Shuzhen Liao, Pengfei Chen, Zhaoyang Wu, Guoli Shen, and Ruqin Yu. 2011. "Oligonucleotide Probes Applied for Sensitive Enzyme-Amplified Electrochemical Assay of Mercury (II) Ions." *Biosensors and Bioelectronics* 26 (7): 3320-3324. https://doi.org/http://dx.doi.org/10.1016/j.bios.2011.01.006.

Zhang, Ziping, Jungang Yin, Zhaoyang Wu, and Ruqin Yu. 2013. "Electrocatalytic Assay of Mercury (II) Ions Using a Bifunctional Oligonucleotide Signal Probe." *Analytica Chimica Acta* 762: 47-53. https://doi.org/https://doi.org/10.1016/j.aca.2012.12.008.

Zhou, Wenhu, Runjhun Saran, and Juewen Liu. 2017. "Metal Sensing by DNA." *Chemical Reviews* 117 (12): 8272-8325. https://doi.org/https://doi.org/10.1021/acs.chemrev.7b00063.

Zhu, Guichi, and Chun-yang Zhang. 2014. "Functional Nucleic Acid-Based Sensors for Heavy Metal Ion Assays." *Analyst* 139 (24): 6326-6342. https://doi.org/https://doi.org/10.1039/C4AN01069H.

Zhu, Lin, Jiajun Ling, Zhi Zhu, Tian Tian, Yanling Song, and Chaoyong Yang. 2021. "Selection and Applications of Functional Nucleic Acids for Infectious Disease Detection and Prevention." *Analytical and Bioanalytical Chemistry* 413: 4563-4579. https://doi.org/https://doi.org/10.1007/s00216-020-03124-3.

Zhu, Yu-Feng, Yong-Sheng Wang, Bin Zhou, Jun-Hui Yu, Li-Lan Peng, Yan-Qin Huang, Xue-Jiao Li, Si-Han Chen, Xian Tang, and Xiao-Feng Wang. 2017. "A Multifunctional Fluorescent Aptamer Probe for Highly Sensitive and Selective Detection of Cadmium (II)." *Analytical and Bioanalytical Chemistry* 409: 4951-4958. https://doi.org/https://doi.org/10.1007/s00216-017-0436-1.

Chapter 3.6

The Antimicrobial Mechanisms of Inorganic Nanoparticles

Ahmed Al-Jumaili[1,2,*]
Avishek Kumar[3]
Saif A. Mohammed[4]
Michael S. A. Kamel[1,5]
and Mohan V. Jacob[1,†]

[1]Electronics Materials Lab, College of Science and Engineering, James Cook University, Townsville, Australia
[2]Department of Medical Physics, College of Applied Science, University of Fallujah, Al-Anbar, Iraq
[3]Assistant professor, Department of Metallurgy and Materials Engg, Iiest shibpur, Howrah, West Bengal, India
[4]School of Engineering and Natural Sciences, Altinbas University, Istanbul, Turkey
[5]Physics Department, Faculty of Science, Minia University, Minia, Egypt

Abstract

The antimicrobial activity of nanomaterials, especially metallic nanoparticles, has gained noticeable global attention due to their broad-spectrum biocidal performance and compatibility with various delivery methods, besides the need for innovative antimicrobial agents that can effectively overcome drug-resistant pathogenic infections. The biological activity of these nanoparticles (NPs) is sufficiently distinct from synthetic antibiotics, providing a complementary pathway for microorganism inactivation. Even though numerous studies have investigated the bactericidal activities of NPs, the antibacterial mechanisms are not fully understood due to NPs diversity. This chapter will focus on several prominent pathways that govern the antibacterial performance of NPs, including their physico-chemical interference with cell wall functions, generation of reactive oxygen species, the release of toxic ions, disruption of cell signaling, and prevention of DNA replication. These pathways are discussed concerning key determinants of the NP bioactivity, such as their size, shape, and the properties of the medium.

* Corresponding Author's Email: ahmed.aljumaili@my.jcu.edu.au
† Corresponding Author's Email: mohan.Jacob@jcu.edu.au

In: Emerging Nanomaterials in Biomedical Science
Editors: Eepsita Priyadarshini and Ajith Manayil Parambil
ISBN: 979-8-89530-600-0

Keywords: inorganic nanoparticles, antibacterial mechanisms, cell wall rupture, reactive oxygen species generation, toxic ions releasing

1. Introduction

Engineering a material on a nanoscale can lead to major variations in its fundamental properties, as the percentage of atoms at the surface of the material becomes substantial (Holister et al. 2003). From the theoretical point of view, with uniform spherical particles, a reduction in the particle dimension from ~10 μm to 10 nm will significantly enlarge the contact surface area by 10^9 (Pal, Tak, and Song 2007); Hence, nanostructured materials have extraordinary characteristics (mechanical, electrical, biological, and optical, etc.) that have led to numerous novel applications (Siddiqui et al. 2018, Al-Jumaili et al. 2017b, Perelshtein et al. 2015, Moritz and Geszke-Moritz 2013, Abramova et al. 2013, Al-Jumaili et al. 2017a, Mu et al. 2016, Kamel et al. 2022, Kumar et al. 2017b).

In the biological field, nanoparticles are being investigated comprehensively, *in vivo* and *in vitro*, due to their efficacy and similarity in size with most molecules and cells. Important usage of these nanoparticles is as antimicrobial agents to target highly pathogenic and drug-resistant microorganisms (Gharpure, Akash, and Ankamwar 2020, A 2015, Karimi 2019, Vasantharaj et al. 2019). The most serious problem with antibiotic resistance is that some pathogenic bacteria have now become resistant to almost all standard antibiotics (Mingeot-Leclercq and Décout 2016, Woon and Fisher 2016). Important examples are methicillin-resistant *Staphylococcus aureus* (MRSA), multi-drug-resistant *Mycobacterium tuberculosis* (MDR-TB) and *Klebsiella pneumoniae* carbapenemase-producing bacteria (2015, Al-Jumaili, Bazaka, and Jacob 2017). Therefore, a large and growing body of literature have investigated the utilization/implementation of nanoparticles to kill/inhibit microorganisms and substitute traditional antibiotics. It is well established that several NPs are effective biocides, particularly those based on metallic elements, such as Ag (Martinez-Castanon et al. 2008), Cu (Kaweeteerawat et al. 2015), Zn (Tobaldi et al. 2016), Au (Troncarelli et al. 2013), MgO (Bindhu et al. 2016), Al_2O_3 (Simon-Deckers, Loo, Mayne-L'hermite, Herlin-Boime, Menguy, Reynaud, Gouget, and Carrière 2009) and TiO_2 (Ahmad et al. 2015). These agents have revealed broad-spectrum antimicrobial properties toward both gram-negative and gram-positive bacteria and can inhibit the growth (bacteriostatic), killing (bactericidal) and/or physically damaging cells (bacteriolytic) (J. Díaz-Visurragal 2011, Musee, Thwala, and Nota 2011, Pourhajibagher, Sodagar, and Bahador 2020).

The field of antimicrobial nanoparticle research has experienced significant advancements in recent years, with focus on improving the synthesis and functionalization of these materials to enhance their antimicrobial properties (Gold et al. 2018). Researchers have explored various green synthesis methods, such as the use of plant extracts, to produce nanoparticles with improved biocompatibility and reduced toxicity (Vanlalveni et al. 2021). Additionally, the development of bimetallic nanoparticles have shown promise in enhancing antimicrobial efficacy and overcoming potential resistance mechanisms (Arora, Thangavelu, and Karanikolos 2020). Overall, the current research landscape in the field of the antimicrobial mechanisms of inorganic nanoparticles is marked by a focus on understanding the relationship between the physicochemical properties of nanoparticles and their antimicrobial activity, as well as the development of novel nanoparticle-based antimicrobial strategies to address the growing threat of antibiotic resistance.

The antibacterial mechanism of NPs is principally different from synthetic antibiotics, offering an important pathway to potentially diminish microbial resistance. Although there is a significant number of papers demonstrating the advantages of antimicrobial NPs, only a limited number of reports have endeavored to clarify the mechanisms of antibacterial activity of these inorganic NPs (Wyszogrodzka et al. 2016, Marambio-Jones and Hoek 2010), with further efforts needed to systematically explore all aspects of this activity and translate the findings into optimization criteria for antibacterial applications. To guarantee the reliable and effective contribution of these particles as antimicrobial agents into synthetic biomaterials, the specific mechanisms that control the antibacterial activity should be elucidated. In this chapter, we highlight several antibacterial mechanisms of inorganic nanoparticles including reactive oxygen species (ROS) formation, ion release phenomena, physical rupture of microbial cells, modulation of bacterial signals, and damage to DNA molecules. In addition, the influences of various factors (e.g., size, shape, surface charges, coatings, and medium/corona) on the biological performance of NPs will be thoroughly discussed.

2. The Antimicrobial Mechanisms of Inorganic NPs

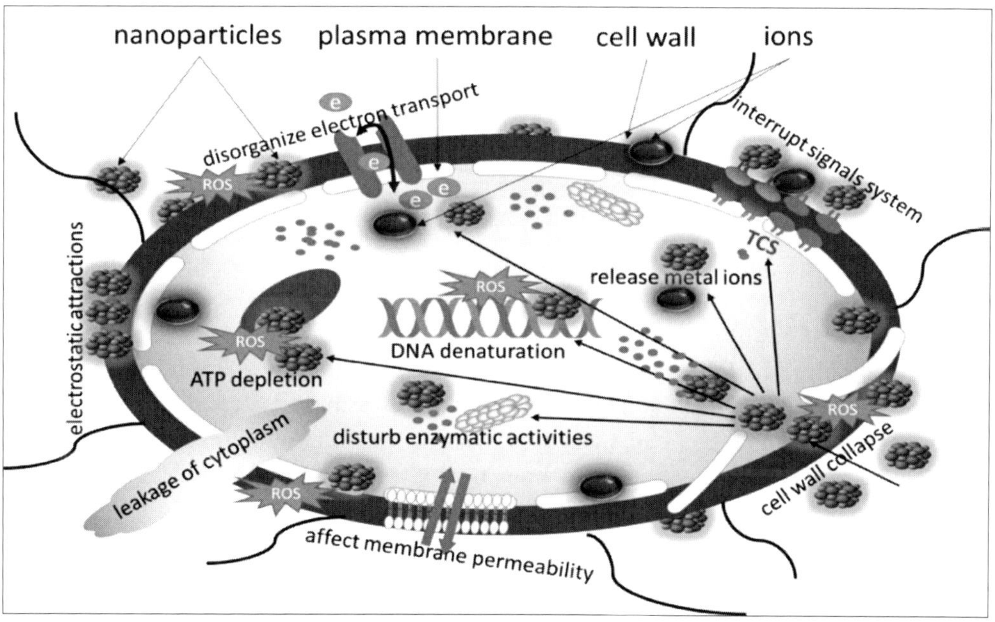

Figure 1. Schematic of the probable interactions between inorganic NPs and bacterial cell: (1) the release of toxic metal ions, capable of killing microorganisms, from NPs into bacterial microenvironment; (2) NPs and/or ions accumulate in the cell membrane, affecting its functions, such as changing membrane permeability and disorganizing electron transportation, followed by the rupture of the wall and leakage of the cytoplasmic content (3) NPs and/or ions generate highly biologically-active ROS that cause serious damages to bacterial biomolecules (e.g., depletion of intracellular ATP production and disruption of enzymatic activities); (4) NPs interfere with bacterial signaling systems (both TCS and QS), preventing bacterial communications and gene expression; (5) DNA can be highly affected by NPs, ions and ROS, preventing its replication and cell division.

Although in the past few decades, intensive efforts have been dedicated to investigating nanoparticles, the mechanisms of toxicity of NPs against microbial cells are not fully understood and are still under debate (Pleskova et al. 2016, Lopes et al. 2016, Li et al. 2008, Franco et al. 2022). The lethal action may potentially involve at least five distinct mechanisms: (I) nanoparticles may release deadly ions into the bacterial microenvironment; (II), NPs generate toxic substances (e.g., ROS that affect bacterial biological activities; (III) NPs and/or products of their degradation may interact chemically with the microbial cell wall and building molecules; (IV) NPs modulate bacterial signal systems (both systems: two-component regulatory system (TCS) and quorum sensing) that disturb cell growth and communication; (V) NPs may cause shearing/denaturation of bacterial DNA that interrupt its replication, and thus prevent cell division. The proposed mechanisms reported in the literature are summarized in Figure 1.

2.1. Interactions of NPs with Cell Membranes

Principally, a bacterial cell is a biological body consisting of a cell wall, cell membrane, and cytoplasm. The cell wall is located on the outer portion of the cell membrane and mainly consists of a homogeneous peptidoglycan layer, containing amino acids and sugars (Sirelkhatim et al. 2015, Mendes et al. 2022, Czyżowska and Barbasz 2022, Gudkov et al. 2021). Microbial cell membranes are very complex dynamic structures, containing various components such as proteins, lipopolysaccharides, phospholipid bilayers, and extracellular polymeric substances (Chen and Bothun 2013). Physicochemical changes often occur in cell membranes in response to the surrounding microenvironment and substances that come into direct contact with the microorganism. NPs are capable of anchoring to the outer surface of the microorganism's cell wall due to their extremely small size, as seen in Figure 2. This interaction may disturb the proper functioning of the microbial cell wall (Naddeo et al. 2015). For instance, large variations could occur in the permeability of the cell membrane, leading to biological death (Prabhu and Poulose 2012). Direct microscopic observations showed acutely damaged cell membranes of *E. coli* cells on exposure to truncated triangular silver nanoplates, causing unrestrained transport of fluids through the plasma membrane and eventual death of the microbial cell (Pal, Tak, and Song 2007). Similarly, plasmolysis (a process of cytoplasm separating from the microbial cell wall) in *P. aeruginosa* and the inhibition of bacterial cell wall structure in *S. aureus* were detected when treated with a NP antimicrobial agent (Song et al. 2006). Sondi et al reported that the physical contact between silver NPs and the "building elements" of the outer membrane of microorganisms led to structural changes, disintegration and, finally, microbial demise (Sondi and Salopek-Sondi 2004). More findings showed that even a brief exposure of bacterial cells to NPs resulted in an accumulation of the latter in the envelope proteins, with ensuing changes to the cell membrane structure and function, and a dissipation of the proton motive force (Lok et al. 2006, Agarwal and Shanmugam 2020). It is worth mentioning that the bacterial cell wall has suffered from serious deterioration/ deformation when exposed to nanostructures (e.g., graphene nanowalls and carbon nanotubes) (Chen et al. 2013, Al-Jumaili, Zafar, et al. 2022).

Several components in the structure of the cell wall, such as carboxyl, hydroxyl groups, amide, phosphate, and carbohydrate-linked moieties provide possible sites for interactions with the active NPs (Jiang, Mashayekhi, and Xing 2009). A highly electronegative carboxyl group function is found to serve as a preferred site of adsorption for metal cations (Lemire, Harrison, and Turner 2013). Thus, electrostatic forces may play a significant role in the attachment and penetration of the cell wall by charged NPs. According to Sirelkhatim et al, opposite charges increase the overall effect by creating electrostatic attractions, working as a powerful bond between NPs and the microbial wall (Sirelkhatim et al.). Neutral and negatively charged particles absorbed significantly less and had lower rate of internalization by the negatively charged cell membrane as compared to the positively charged NPs (Verma and Stellacci 2010). In contrast, Huang et al reported that ZnO NPs had similar biocidal action against both gram-positive *S. aureus* and gram-negative *S. agalactiae* (Huang et al. 2008).

The configuration of cell walls varies between species and strains, as well as the environment and cell developmental stage (Bazaka et al. 2015). Gram-positive microorganisms have structurally thick walls comprised of several polymeric layers of peptidoglycan, and one plasma membrane, while gram-negative bacteria have only a thin layer of peptidoglycan, and a complex cell wall consisting of two cell membranes (an outer and a plasma membrane) (Fu, Vary, and Lin 2005). Consequently, the cell membrane of gram-negative bacteria consists of a comparatively high fraction of negatively charged groups (e.g., lipopolysaccharides), which attract weakly positively charged particles (Shrivastava et al. 2007). This might explain why gram-negative microorganisms are often reported to be more resistant to various biochemical toxic agents, including NPs, than gram-positive cells (Nazzaro et al. 2013).

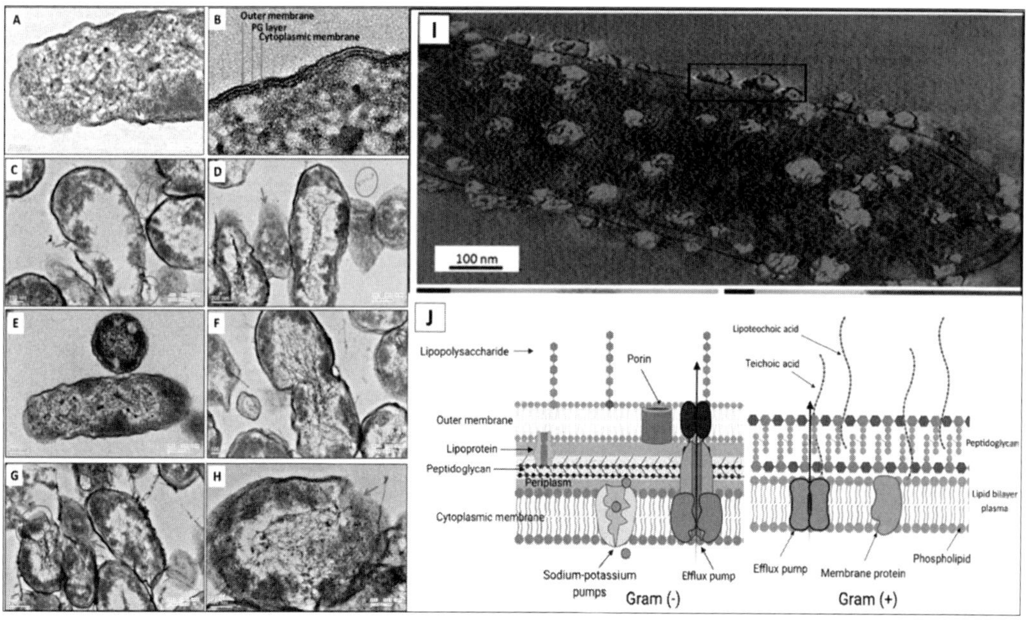

Figure 2. TEM images of untreated cells of (A) *P. mosselii*, (B) a magnified view of cell envelope, and after interaction with 500 μg mL–1 each of (C) Ag-NPs and (D) ZnO-NPs, (E) untreated cells of *A. chroococcum*, (F) a magnified view of the cell envelope, and after interaction with 500 μg mL–1 each of (G) Ag-NPs and (H) ZnO-NPs. Red circles show leakage of the cellular content, while arrows indicate cellular damage. Reproduced with permission from (Ahmed et al. 2020). (I) 3D reconstruction of a bacterium decorated with Fe nanoaggregates (orange) clearly demonstrates that NPs attach to the outer membrane of the bacterial cell wall. Reproduced with permission from (Luef et al. 2013). (J) Differences in cell wall structure in gram-negative and gram-positive bacterial cell. Reproduced with permission from (Rapacka-Zdończyk et al. 2021).

2.2. The Generation Reactive Oxygen Species

In live cells, ROS are naturally produced at low rate as a result of aerobic respiration and normal metabolic activities (Vatansever et al. 2013, Hernansanz-Agustín and Enríquez 2021, Canaparo et al. 2020, Barhoum et al. 2022). ROS are highly chemically reactive and short-lived species, with a lifetime of about 10^{-9} seconds, with typical ROS including hydrogen peroxide (H_2O_2), superoxide radicals (O^{-}_{2}), hydroxyl radicals ($OH^{\bullet}$), and singlet oxygen (O^{*}_{2}) (Naddeo et al. 2015, Yin et al. 2013, Gao et al. 2021, Sathiyaseelan et al. 2023). These free radical molecules consist of one or more unpaired electrons, excited in their highest occupied atomic or molecular level (Boonstra and Post 2004). Under normal conditions, cells can neutralize oxygen to water via their electron transport chains and using enzymes like superoxide dismutases (SOD) that convert superoxide to hydrogen peroxide, as well as use some catalases that convert hydrogen peroxide to oxygen and water (Choi and Hu 2008). For example, *E. coli* uses three forms of SOD enzymes that have different metallic cationic centers, namely MnSOD, FeSOD, and CuZnSOD. (Vatansever et al. 2013). Undoubtedly, in active bacteria, the regulation of both external ROS and ROS from normal respiration is a key survival factor (Kumar et al. 2017a).

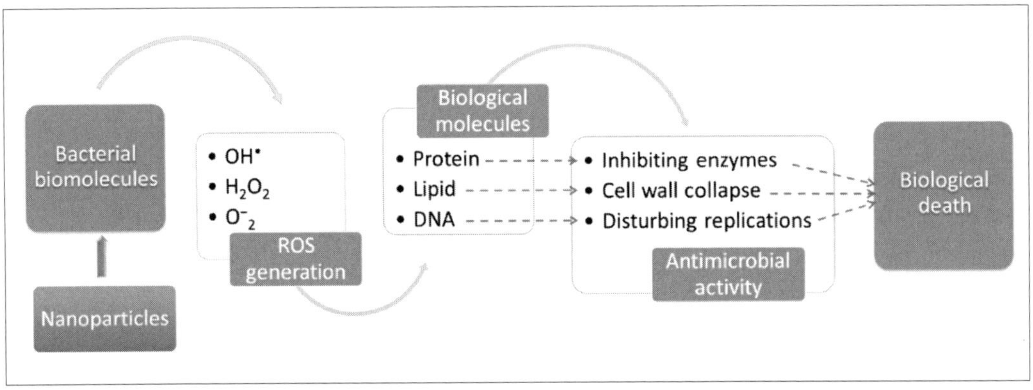

Figure 3. The pathway of ROS generation induced by nanoparticles during interaction with bacterial biomolecules.

Upon the interaction with cells, toxic NPs can dramatically increase the rate of ROS creation in the targeted microorganisms. ROS over-production can cause severe damage due to the deterioration of peptidoglycan, DNA, ribosomes, and proteins, a decrease in enzymatic activities and amino acid synthesis, these events eventually lead to cell lysis (Pelgrift and Friedman 2013, Naddeo et al. 2015), as seen in Figure 3. Hydroxyl radicals are well recognized to be a powerful oxidant with significant damage of biological molecules (Pacher, Beckman, and Liaudet 2007). It has also been hypothesized that superoxide and hydroxyl cannot enter the cell membrane first due to their negative charge, while hydrogen peroxide, which is less damaging to membranes, can penetrate the cell (Fu, Vary, and Lin 2005). Furthermore, $O_2^{-\bullet}$ may interact with other free radicals in the cell, such as nitric oxide ($NO^{\bullet}$), in a process controlled mainly by the degree of diffusion of both radicals (Turrens 2003). This interaction will generate peroxynitrite that is also a very strong oxidant and can negatively impact the microorganism (Szabo, Ischiropoulos, and Radi 2007). Importantly, ROS generation is highly dependent on NPs concentration. Figure 4 gives an example of the generation of ROS from Ag nanoparticles and their effect on the treated microorganism. There have been several studies ting the generation of ROS due to NPs and their significance in the NP antibacterial activity. For example, *E. coli* cells are subject to significant cellular damage and the inhibition of vital enzymes when exposed to extracellular hydrogen peroxide or agents that increase the formation of superoxide (Lemire, Harrison, and Turner 2013). Simon-Deckers et al reported that exposure to Al_2O_3 NPs induced an increase in intracellular ROS in *Cupriavidus metallidurans* and *E. coli* bacteria (Simon-Deckers, Loo, Mayne-L'hermite, Herlin-Boime, Menguy, Reynaud, Gouget, and Carrière 2009). In the case of NPs with high photocatalytic properties (e.g., ZnO), a photon creates electron and hole pairs that can directly oxidize organic macromolecules or oxidize water generating hydrogen peroxide and radicals (e.g., $OH^{\bullet}$ and $O_2^{-\bullet}$)(Muñoz-Bonilla and Fernández-García 2012). Under UV radiation, ZnO NPs in aqueous medium have phototoxic activity that can yield superoxide ions (O_2^-) and hydrogen peroxide (H_2O_2) (Sirelkhatim et al.). It is important to mention that several types of typical synthetic broad-spectrum antibiotics (e.g., fluoroquinolones) eradicate bacterial cells by stimulating the intracellular generation of ROS (Dwyer, Kohanski, and Collins 2009).

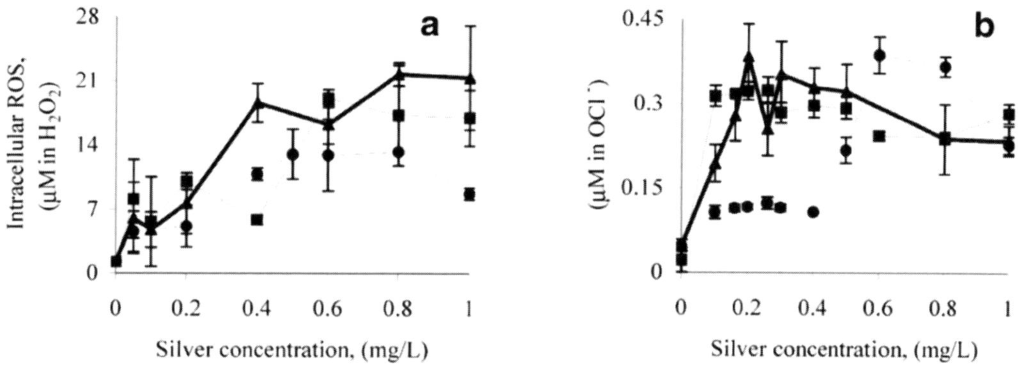

Figure 4. The increase of intracellular ROS (a) and photocatalytic ROS concentrations (b) as function of the concentration of Ag nanoparticle. Reproduced with permission from (Choi and Hu 2008).

2.3. Metal Ions

NPs can undergo decomposition in the proximity of or within the microorganism. The associated release of metallic ions may explain the toxicological profiles of these NPs (Heinlaan et al. 2008, Kasemets et al. 2009, Godoy-Gallardo et al. 2021). In the structure of bacterial biomolecules, minerals bind to some atoms of donor ligands (e.g., O, N, S) by strong and selective interactions. Hence, presence of external metallic ions (released by NPs) can substitute the original metals existing in biomolecules, affecting their cellular function (Palza 2015). For instance, in *E. coli*, (Ni (II) can affect an enzyme aldolase by substituting the non-essential (Zn (II) through binding to the carboxylate sidechains, causing changes in the enzyme structure and impacting its ability to be active (allosterically) (Macomber, Elsey, and Hausinger 2011). Ag^+ can also bind to some functional groups of microorganisms' proteins, leading to protein denaturation (Sondi and Salopek-Sondi 2004). According to Kumar et al. Ag in an atomic state can absorb surrounding oxygen. The atomic oxygen absorbed by silver will readily interact with the sulfhydryl (-S-H) components existing on the bacterial wall to remove the hydrogen atoms (as water), causing the sulfur atoms to form an R–S–S–R bond. This process leads to the blockage of bacterial respiration and cellular death (Kumar et al. 2004). Binding of Ag is often achieved through interactions with the thiol groups of the l-cysteine residue of proteins leading to the inactivation of enzymatic activities (Feng et al. 2000, You et al. 2012). In the same way, metal replacements at non-catalytic metal-binding sites have also been shown to alter enzyme activity (Lemire, Harrison, and Turner 2013). More researchers investigated the toxicity of nanosized and bulk ZnO, CuO and TiO_2 toward different microorganisms and indicated that metallic oxide nanoparticles do not necessarily have to enter the microorganisms to cause the toxicity. Indeed, the physical contact between the bacterial cell wall and nanoparticle may cause sufficient variations in the microenvironment in the proximity of the microorganism to increase the solubilization of metals or to create ROS that may harm cell membranes (Heinlaan et al. 2008).

The rate of ion release is largely dependent on the surface area of NPs and can thus be the influential factor for the antibiotic action (Knetsch and Koole 2011). It has been reported that 1.3% of Ag was released as ions from 10 nm Ag NPs, while 0.6% of Ag was dissolved from 80 nm Ag NPs (both in ultrapure water) (Ivask et al. 2014). Furthermore, the level of oxygen in the ambient environment is important for the generation of ions and as such NP biological activity. Xiu et al. found that the toxicity of Ag NPs declined significantly under anaerobic conditions, suggesting that oxidative dissolution under aerobic conditions leads to a substantial increase in the concentration of soluble metal ions (Xiu et al. 2012). However, in some cases, it was observed that nanoparticles are more toxic against bacteria than their dissolved ions (Choi and Hu 2008). In the literature, there are many contradictory results of different studies on the antibacterial ions mechanism. For instance, Kasemets et al demonstrated that the toxicity of ZnO NPs toward *S. cerevisiae* bacteria is related to the solubility of Zn^{2+} in the medium containing the bacterial cells (Kasemets et al. 2009). On the other hand, Sawai considered the role of Zn^{2+} on the bactericidal efficiency of ZnO NPs against *E. coli* and S. *aureus* to be negligible due to the low concentrations of solubilized Zn ions (Sawai 2003). Also, several microorganisms are capable of living, and sometimes grow, in the presence of considerable metal ion concentrations, due to their resistance to that metal (Prabhu and Poulose 2012). It has been shown that the sensitivity of several bacterial types toward copper-iodide nanoparticles varied dramatically in the order of DH5α> *E. coli* ≈ *S. dysentry* > *S. aureus* > *B. subtilis* (Pramanik et al. 2012). This suggests more complex mechanisms that govern cell-ions-NPs interactions, where the interplay of various factors (e.g., NP concentration, exposure to air, pH, temperature, the viability of bacteria and microbial species) contribute synergistically to the overall bactericidal outcome of NPs (Hajipour et al. 2012).

2.4. Interruption of Bacterial Signaling

A bacterial cell has a two-component system (TCS) that senses environmental conditions (e.g., stress) and transduces the signal inside the cells for further adaption (Wang et al. 2016). Briefly, a TCS system involves a histidine kinase (HK) sensor located in the inner membrane and a cognate response regulator (RR) that resides in the cytoplasm (Utsumi 2008), as seen in Figure 5. Upon sensing of a stimulus, the HK sensor is activated and auto-phosphorylation occurs at a conserved histidine, inducing gene expression by phosphorylating its cognate RR at a preserved aspartate group (Prabhakaran, Ashraf, and Aqma 2016). More than 40 dissimilar two-component systems have been recognized in *E. coli* alone, each responding to different environmental stimuli (D'Agostino and Kieber 1999). Interestingly, the accumulation of selected NPs (e.g., Ag, Si) can modulate the signal transduction in microorganisms. According to Shrivastava et al, NPs would interfere with the microbial growth signaling pathways by modulating the tyrosine phosphorylation of putative peptides necessary for cells' viability (Shrivastava et al. 2007). The tyrosine phosphorylation process is known to activate several vital protein substrates, such as RNA polymerase sigma factor and UDP glucose dehydrogenases (Singh et al. 2008). Reduction of phosphorylation impacts on the activity of enzymes essential for bacterial growth (Prabhu and Poulose 2012). Indeed, understanding and manipulating TCS pathways hold enormous potential for minimizing bacterial infections.

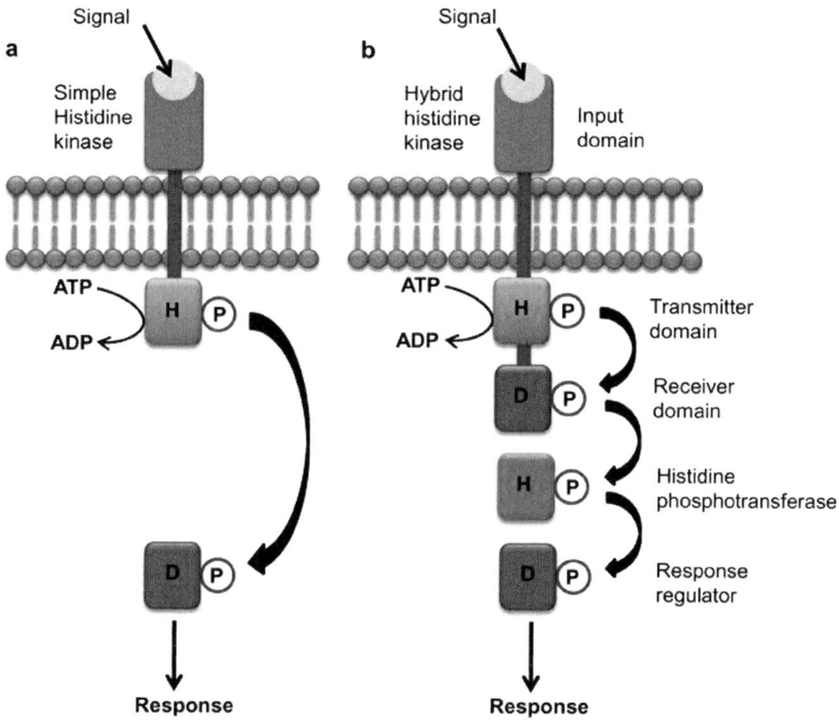

Figure 5. A typical TCS pathway in bacteria. Reproduced with permission from (Foster et al. 2004).

On the other hand, bacteria can communicate with each other through diffusible signals that may perform as 'autoinducers,' the process known as quorum sensing (QS). The QS system allows a cluster of microorganisms to transfer information (such as susceptibility to toxic substances) to synchronize the gene expression of the bacterial community, such that it behaves in unison (Prabhakaran, Ashraf, and Aqma 2016, Ng and Bassler 2009). Quorum sensing is a population density dependent phenomenon that occurs in both gram positive and gram-negative bacteria (Mukherji and Prabhune 2015). It is vital to bacterial endurance in harsh conditions and hence provides a unique target for antimicrobial agents. There are two possible pathways to target intercellular communication; one way is via quorum-sensing agonists that simulate autoinducers and encourage non-virulent microbial activities, which could be used against bacteria that naturally repress maliciousness behaviors through quorum sensing. The second way could be via quorum sensing antagonists that would perform by inhibiting/preventing cell–cell communications, which could be used against pathogens that employ quorum sensing to trigger harmful effects (Lu et al. 2015). Miller et al demonstrated that 50 nm Si-NPs (functionalized with β-cyclodextrin) were able to bind signaling molecules (e.g., acylhomoserine lactones) of *Vibrio fischeri* bacteria and down-regulate its quorum sensing genes. At high concentrations of these NPs, the chemical signals were observed to be effectively removed from the immediate bacterial micro-environment (Miller et al. 2015). Singh et al reported that modified Ag-NPs (myco-fabricated) could attenuate virulence and biofilm development of *P. aeruginosa*. The modified Ag-NPs significantly downregulate quorum sensing-regulated genes, specifically those responsible for the secretion of virulence effects (Singh et al. 2015). Clearly, interrupting the quorum sensing system could be the key for the reduction of pathogenic infections (Holban, Gestal, and Grumezescu 2016).

2.5. DNA Replication

DNA is the most important carrier of genetic information within bacteria, being a molecule that is irregularly shaped, not membrane-bound, and freely floating through the cytoplasm (Feng et al. 2000, Thanbichler, Wang, and Shapiro 2005). Various protein structures act together to fold and condense bacterial DNA, mainly at sulfur and phosphorus sites; where nanoparticles, such as silver, can interact with these soft sites and disturb the DNA. This may cause shearing or denaturation of the DNA and disrupt cell division (Hsueh et al. 2015). It has been reported that Ag^{+} is capable of binding to bacterial DNA nucleotide bases, affecting denaturation by displacing H-bonds between adjacent purines and pyrimidines, which disrupts the double helical structure of the DNA (Klueh et al. 2000). Feng et al observed that microbial DNA molecules become significantly condensed and lose their replication abilities due to the denaturation effects of metallic ions (Feng et al. 2000), preventing cell reproduction (Damm, Münstedt, and Rösch 2008). Also, a significant level of degradation was detected in the chromosomal DNA of gram-positive *B. subtilis* exposed to sufficiently high concentrations of silver nanoparticles (Hsueh et al. 2015). Other findings showed that weakly charged cationic nanoparticles (e.g., gold) are capable of causing structural changes in DNA, including local denaturing and compaction of the DNA (Railsback et al. 2012).

There is a correlation between ROS generation and DNA mutilation that signifies the role of oxidative stress as an important pathway for the induction of DNA damage in bacterial cells. Several studies reported that NP-induced oxidative stress is the main source of DNA damage (Radzig et al. 2013). ZnO and TiO_2 treatments of *E. coli* were demonstrated to initiate substantial DNA destruction through ROS formation (Kumar et al. 2011). ROS attacks DNA molecules, inducing chain breaks and causing chemical alteration of carbohydrate structures, nitration, methylation, and/or deamination reaction (Kumar et al. 2011). Interestingly, several studies on various bacterial cell types proposed that ROS formation might damage cell DNA and prompt apoptosis (programmed cell death) with no detectible membrane destruction (Choi and Hu 2008).

The damage to the DNA can be fatal if not repaired before bacterial division; thus, different bacterial species have developed a wide range of mechanisms to tolerate elevated levels of NPs and ensure this repair is carried out in a timely manner (Wigley 2013). For instance, *P. aeruginosa*, and *Sal. typhimurium* upregulate the translation and transcription of genes responsible for fixing of DNA and altering the metal homeostasis in the incidence of NPs (Hajipour et al. 2012). *E. coli* cells act to protect DNA molecules through the repair of mutator genes such as *mutH, mutL, mutS*, and *uvrD* (Qiu et al. 2015). Remarkably, some nanoparticles are able to deactivate bacterial repair systems as well as induce DNA impairment. Ag NP is a good example; since it is capable of triggering strand breaks, DNA damage, and mutations in DNA repair genes (e.g., mutY, mutS) in E.coli, producing mutant strains that are more susceptible to antibacterial agents than the parent strain (Dakal et al. 2016). It is worth mentioning that DNA structures are not completely stable since they are subject to chemical/photochemical damages, double-strand breaks, and errors that take place during DNA replication (Modrich 2016). Thus, the DNA destruction induced by exposure to NPs may vary based on microbe ecological conditions, representing only one aspect among the overall highly complicated processes that are at present not adequately understood.

3. Key Determinants of NP Bioactivity

The toxicity of NPs is governed by a range of intrinsic and environmental factors, such as material properties, their shape and size, crystallinity, surface chemistry and shell material, as well as ambient light, temperature, and pH, as well as the species of the target microorganism. This section will discuss some important features that influence the bactericidal activity of NPs.

3.1. The Effect of NP Size and Shape

The consensus for nanoparticle size is 1–100 nm, as mentioned in the current legislation, but there is no scientific evidence to support the appropriateness of this value (Brosset 2013). Ivask et al argued that the physical size threshold for novel biological effects of particles is < 20 nm (Ivask et al. 2014). However, the influence of particle size on the toxicity potential of NPs is yet to be fully defined given the relatively new field of nanotoxicology (Musee, Thwala, and Nota 2011). Nevertheless, it is agreed that understanding the impacts of the scale and shape of the constituent particles is critical to predicting their toxicity mechanisms (Al-jumaili 2019, Dolai, Mandal, and Jana 2021, Güell et al. 2023, Subhan, Neogi, and Choudhury 2022).

Indeed, a decrease in NP size generally leads to increase in the ratio of surface area to volume, thereby altering its physical-chemical properties (e.g., surface atom reactivity, optical property), which in turn allows the material to display phenomena not possible to observe in the bulk material counterpart (Fu and Yao 2001, Rao et al. 2002, Babayevska et al. 2022). In general, an increased toxicity with a decreasing particle size has been observed, since smaller NPs can more easily penetrate microbial membranes, thereby increasing their antibacterial efficacy (Gliga et al. 2014, Zahmatkesh et al. 2021). The size of NPs determines the nature of the membrane wrapping around them during internalization (Zhao et al. 2011). Ligand deficiency in NPs and receptor deficiency on cells lead to optimal endocytosis. The uptake of NPs with a diameter of 25-30 nm was suggested to be optimal, as shown by mathematical modelling (Yuan et al. 2010). The dependence of NP toxicity on its size can be attributed to its ability to interact with various macromolecules in a biological system and interfere with their function. Toxicity of NPs is triggered by the generation of oxidative stress by the formation of free radicals in which particle size plays a decisive role. Ag NP of various sizes (5-100 nm) fabricated by a common synthesis protocol were tested against bacterial strains of *E. coli, B.*

subtilis, and *S. aerues*. A substantial increase in bactericidal activity was observed with decreasing particle size, where particles in sub10 nm range showed the highest bactericidal efficiency (Agnihotri, Mukherji, and Mukherji 2014). Sirelkhatim et al concluded that size of ZnO NPs was a crucial factor for their bactericidal applications (Sirelkhatim et al. 2015). In another report, ZnO NPs of various shapes (e.g., branched rod, sphere and flower like) and sizes were synthesized and tested against *E. coli* and *S. aureus*. The bactericidal activity was found to increase with their decreasing size and was the highest for flower-like morphology of ZnO NPs (Talebian, Amininezhad, and Doudi 2013, Lallo da Silva et al. 2019).

The shape of NPs is another important factor affecting their bactericidal properties. Nanoparticles can be custom-engineered and come in various shapes like rod, spherical, triangular, truncated triangular, hexagonal, etc., as seen in figure 6. The shape of NPs influences their uptake by the cells. The uptake of Ag NPs by human mesenchymal stem cells showed no shape dependence whereas toxicity of the same NPs toward bacteria *S. aureus* was significant and shape dependent (Helmlinger et al. 2016). The bactericidal effect of NP decreased as platelets > spheres > rods > cubes (Helmlinger et al. 2016). Using mathematical modeling, it was demonstrated that the uptake of spherical nanoparticles was the highest, followed by cubic, rod, and disk-like NPs (Li, Kröger, and Liu 2015). The change in membrane energy during the internalization of NPs may account for the effect of their shape. The size of particles was in the sub-100 nm range. The change in membrane bending energy was minimal for spherical NPs and was considerably high for disk-like NPs (Li, Kröger, and Liu 2015). However, the sizes of nanodisk and nanorod particles were different. Gratton et al found a different trend in the uptake of NPs of size greater than 100 nm (Gratton et al. 2008). The uptake of rod-shaped NPs was the highest, followed by that for the sphere-, cylinder- and cube-shaped NPs (Gratton et al. 2008). The cellular uptake of longer Au NPs was less than that for short nanorods while their surface charges were similar (Qiu et al. 2010, Joudeh and Linke 2022). The uptake of spherical Au NPs was five times more than that of rod-shaped particles (Verma and Stellacci 2010, Saidin et al. 2021). The shorter membrane wrapping timing of spherical NPs was one of the possible explanations for this phenomenon (Verma and Stellacci 2010, Tsikourkitoudi, Henriques-Normark, and Sotiriou 2022). Atoms at the corners and the edge of a rough surface or uneven shapes are more biologically active because of their lower bonding coordination. Thus, the formation of NP ions and their dissolution is significantly enhanced at these sites, leading to higher bactericidal properties.

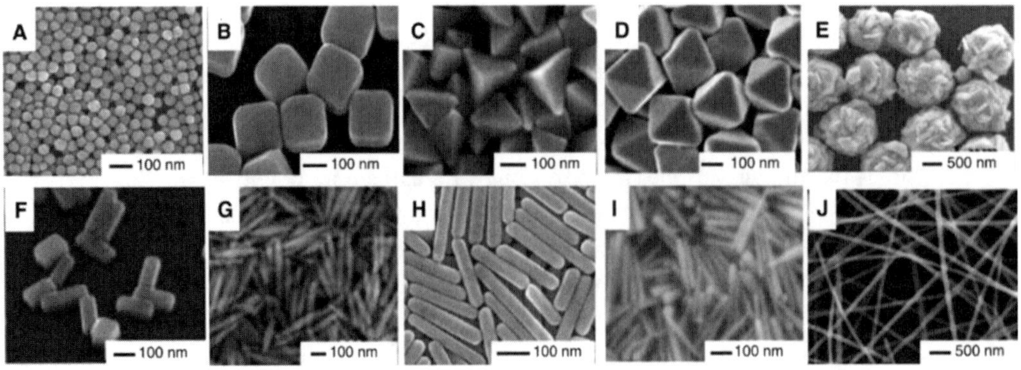

Figure 6. Silver nanostructures with a variety of morphologies. Reproduced with permission from (Liang et al. 2015).

3.2. The Effect of Surface Charge and Coating of NPs

Apart from their size and shape, the surface charge is an important factor dictating the cellular fate of NPs (Al-Jumaili, Alancherry, et al. 2019, Kumar et al. 2020). The adsorption of NPs on cells and their behavior in a biological medium is influenced by their surface charge (Pietroiusti et al. 2011). It has been found that positively charged NPs are internalized much more efficiently than neutral or negatively charged particles (Fröhlich 2012). Electrostatic interactions between cells and NPs favor their adhesion onto cells thus facilitating their uptake (He et al. 2010). Abbaszadegan et al tested the effect of surface charge of Ag NPs against four strains of bacteria (both gram-positive and gram-negative). The positively charged Ag NPs were found to have higher antibacterial effect against all tested species (Abbaszadegan et al. 2015). NPs are often encapsulated by a shell material, which acts as a stabilizing, reactive or biocompatibility agent. Encapsulation materials affect the nanoparticle charge, thus affecting its interaction with the cell surface. The release of ions from NPs is also affected by the encapsulating material. Consequently, the surface coating and charge greatly influence the NPs toxicity. Suresh et al showed a greater bactericidal effect of biogenically prepared Ag (zeta potential of -12 ± 2 mV) than colloidally prepared (zeta potential of -42 ± 5 mV) and oleate encapsulated (zeta potential of -45 ± 4 mV) NPs. The NPs were tested against strains of gram-negative (*E. coli* and *S. oneidensis*) and gram positive (*B. subtilis*) bacteria in the above study (Suresh et al. 2010). In another work, El Badawy et al tested four different surface-coated Ag NPs (having different surface charges) against B*acillus* species (El Badawy et al. 2010). They found that in general, positively charged NPs were most toxic because of greater electrostatic interactions with the cell surface. Fang et al synthesized two different surfaces with Au NPs of varying nanoparticle density and surface charge (sparse NPs with a zeta potential of -27 ± 3 mV, and dense NPs with zeta potential of -20 ± 3 mV) and tested them against *S. aureus* (Fang et al. 2015). The surface with more positive zeta potential was found to be more bactericidal and the bactericidal effect was not dependent on NP density. Likewise, a difference in overall surface charge because of encapsulating material was one of the important factors in determining the bactericidal effect of CdSe, CdTe and ZnS–$AgInS_2$ quantum dots (Wang et al. 2010). Radovic-Moreno et al. showed higher bacterial adhesion to poly (d,l-lactic-co-glycolic acid)-b-poly(l-histidine)-b-poly (ethylene glycol) (PLGA-PLH-PEG) encapsulated NPs than (PLGA-PEG) coated ones on account of higher zeta potential of the coating (Radovic-Moreno et al. 2012). The zeta potential of coating was found to be pH dependent in this study, thus affecting the bacterial adhesion.

However, non-specific protein adsorption on NPs and their agglomeration induced by the surface charge can limit their use in certain applications. Thus, NPs are frequently coated with neutral molecules or non-ionic polymers like PEG to overcome such issues. PEG-coated Fe_3O_4 NPs were found to show less agglomeration and less unspecific internalization by macrophage cells (Xie et al. 2007). Lee et al showed bactericidal effect of zero-valent (no surface charge) Fe nanoparticles against *E. coli* (Lee et al. 2008). They concluded that considerable cell wall disruption by Fe NP may be a possible reason for its bactericidal activity. In another study, it was demonstrated that Ag NPs coated with neutral glutathione showed antibacterial activity against gram positive and gram-negative bacteria (Taglietti et al. 2012). Coating of NPs with zwitterionic polymers to obtain neutral surface has also been studied (Rouhana, Jaber, and Schlenoff 2007, Estephan, Jaber, and Schlenoff 2010). CdSe/CdZnS NPs coated by zwitterionic ligands showed their non-adherence to some living molecules (Park et al. 2011).

These studies demonstrated that neutrally coated NPs are particularly useful when specific interactions between them and biological systems are required.

Negatively charged NPs interact quite differently with the cells. The adsorption of NPs on the negatively charged cell membrane is typically lower than that for the positively or neutrally charged species and consequently their uptake by cells is also lower (Al-Jumaili, Mulvey, et al. 2019). It is believed that binding of anionic NPs on cationic sites of cell membrane leads to their attachment and subsequent uptake by cells. Evidently, an increase in bactericidal effect of negatively charged citrate capped Ag NPs against *E. coli* was observed by Goli et al (Goli et al. 2013). These NPs were functionalized by protein-coated polypropylene surfaces. He et al demonstrated the non-phagocytic uptake of polymeric NPs to decrease with the increasing negative charge on NPs (He et al. 2010).

3.3. The Effect of Medium/Corona

The surface properties of NPs are altered or greatly influenced by the properties of the medium surrounding them. As NPs enter the biological medium, their surface gets instantaneously coated with biomolecules (e.g., lipids and protein). This biomolecule-NPs complex affects the biological fate and function (e.g., cellular uptake, bio-distribution) of the latter (Pietroiusti, Campagnolo, and Fadeel 2013, Gunawan et al. 2014, Kumar et al. 2020). Furthermore, the adsorption of biomolecules on NPs increases their effective particle size, subsequently altering their internalization process. Concerning the bactericidal activity of NPs, the bacterial growth medium can also have a profound effect on the functioning of NPs. Pelletier et al observed the growth inhibition of *E. coli* and *B. subtilis* by Ce NPs to occur more effectively in a minimal medium than the complex LB medium (Pelletier et al. 2010). Li et al showed the effect of water chemistry on the toxicity of ZnO NPs towards *E. coli* (Li, Lin, and Zhu 2013). An increase in pH and dissolved organic matter in the solution reduced the concentration of Zn^{+2} released from ZnO ions, thus lowering its toxicity. In addition, the presence of Ca^{+2} and Mg^{+2} was also found to decrease the toxicity of ZnO NPs to *E. coli*.

The crystalline structure, composition, surface roughness, and ligand arrangement of NPs also greatly influence their behavior in biological media. The effect of the crystalline structure of TiO_2 on their toxicity against algae, plants, and microcrustaceans has been demonstrated by Clément and colleagues (Clément, Hurel, and Marmier 2013). The anatase phase of TiO_2 was found to be more toxic than the rutile phase. The formation of aggregates by NPs with the rutile crystalline structure in an aqueous medium is believed to lower their toxicity (Clément, Hurel, and Marmier 2013). In another study, the bactericidal effect of Al_2O_3 and TiO_2 (two different crystalline structures) NPs were assessed against *E. coli* and *Cupriavidus metallidurans*. Al_2O_3 NPs were found to be more toxic than TiO_2 and no effect of the crystalline structure of TiO_2 on its toxicity was observed (Simon-Deckers, Loo, Mayne-L'hermite, Herlin-Boime, Menguy, Reynaud, Gouget, and Carriere 2009).

4. Advantages of Nanomaterial as Antimicrobial Agents

Antimicrobial resistance has emerged as a significant global health concern, with pathogens becoming increasingly resistant to traditional antibiotics and disinfectants. This has spurred the exploration of alternative antimicrobial strategies, including the use of nanomaterials as a promising solution. Nanomaterials possess several advantages over conventional antimicrobial agents. Their high surface-to-volume ratio and unique physicochemical properties, such as enhanced catalytic activity and improved stability, confer superior antimicrobial efficacy (Basavegowda and Baek 2021) .

Compared to bulk materials, nanomaterials can more effectively interact with and disrupt the cell membranes of microorganisms, leading to their inactivation or death (Nisar et al. 2019, Hoseinzadeh et al. 2017). Additionally, the small size of nanomaterials allows them to penetrate deeper into biofilms and reach otherwise inaccessible areas, enhancing their ability to combat resistant pathogens (Vimbela et al. 2017, Basavegowda and Baek 2021). In addition to their antimicrobial properties, nanomaterials can also be used to develop novel delivery systems for antimicrobial agents, improving their bioavailability and effectiveness. The versatility of nanomaterials enables the development of multifunctional antimicrobial agents that can be tailored to specific applications, such as targeted drug delivery or environmental remediation(Singh et al. 2019, Chandrakala, Aruna, and Angajala 2022, Roy et al. 2021).The antimicrobial properties of nanomaterials can be further enhanced through the use of combinations of different metal or metal oxide nanoparticles, exploiting their synergistic effects(Gold et al. 2018, Nikolova and Chavali 2020).

Importantly, nanomaterials can target the cell membranes of resistant microbes differently, circumventing the traditional resistance mechanisms developed by pathogens (Gupta et al. 2019, Makabenta et al. 2021). Furthermore, the small size and increased surface area of nanomaterials facilitate better interaction and penetration into microbial cells, enhancing their antimicrobial potency (Basavegowda and Baek 2021). Indeed, the versatility of nanomaterials allows for the development of multi-metallic nanoparticles with enhanced antimicrobial properties. These bimetallic or multi-metallic nanoparticles can exhibit synergistic effects, resulting in improved antimicrobial efficacy compared to single-metal nanoparticles (Arora, Thangavelu, and Karanikolos 2020, Basavegowda and Baek 2021). Unlike traditional antimicrobial agents, which often require harsh chemicals and energy-intensive synthesis, nanomaterials can be produced through more environmentally friendly and cost-effective methods, such as green synthesis using plant extracts. Additionally, the antimicrobial activity of nanomaterials can be further enhanced through the incorporation of active compounds or the development of hybrid systems, combining the unique properties of nanomaterials with the benefits of traditional antimicrobials.

The advantages of nanomaterial-based antimicrobial agents, including their broad-spectrum efficacy, and eco-friendly production, position them as a promising alternative to conventional antimicrobials in addressing the growing challenge of antibiotic resistance. In conclusion, the unique properties of nanomaterials, such as their high surface-to-volume ratio, enhanced catalytic activity, and improved stability, make them a promising alternative to conventional antimicrobial agents. Their versatility and ability to overcome resistance mechanisms offer significant advantages in the fight against infectious diseases.

5. Challenges and Future Perspectives

Within the vast number of studies on the antimicrobial activities of NPs, the lack of united/comprehensive standards makes it difficult to evaluate/compare the antibacterial efficiency of each type of nanoparticle. There is no single procedure to justify the experimental conditions (e.g., dissimilarity of microbial strains, temperature, exposure times, UV/Visible illumination, pH), NPs compositions (e.g., metal, metal oxide, modified NPs), characteristics of NPs (e.g., size, shape) and so on (Wang, Hu, and Shao 2017, Al-Jumaili, Kumar, et al. 2019). Indeed, efforts to develop new strategies/procedures that can cope with these experimental issues are much needed.

Although some types of NPs seem to be non-toxic to human cells, neurological and respiratory damage, circulatory issues, and other toxicity-related effects of these particles are the main concerns at this point in time (Dizaj et al. 2014). The toxicological implications of several important types of nanoparticles have been widely discussed (Ahamed, AlSalhi, and Siddiqui 2010, Hajipour et al. 2012). Just like the antibacterial performance, there is a lack of comprehensive knowledge about the overall mechanisms by which nanoparticles interact with eukaryotic cells. The cytotoxicity of nanoparticles is also reliant on various experimental conditions and their inherent characteristics (Stankic et al. 2016). In a recent report, a comparison of 11 types of metallic nanoparticles of similar diameter (< 20 nm) revealed that variations in their toxicological profiles might be elucidated by two main mechanisms. The first pathway is based on the release of metal ions that are observed in CuO and ZnO. The second pathway is related to ROS generation, detected for Mn_3O_4 and Co_3O_4. This study indicates that both mechanisms, i.e., via ion release and ROS generation, may be activated upon nanoparticle internalization by human cells (Ivask et al. 2015). However, the interaction mechanisms between metal nanoparticles and mammalian cells should be clearly understood for any safe clinical application of these materials. In addition to toxicity, employing NPs may have physical limitations. For instance, their small scale and great surface area can cause particle-particle aggregation, resulting in difficulties of dealing with nanoparticles in liquid and/or dry systems (Lara et al. 2011). Nonetheless, there is a strongly growing commercial usage of biocidal nanoparticles. For instance, wound dressings containing biocidal sputtered nano silver are now used in clinical applications to hinder bacterial contamination of burn wounds (Lok et al. 2007).

Engineered NPs are highly expected to continue their development as potential candidates for numerous bio-applications (Al-Jumaili, Duru, et al. 2023, Al-Jumaili, Al-Azzawi, et al. 2023, Al-Jumaili, Al-Jumaili, et al. 2022). However, their interaction with biological systems must be characterized fully before the era of nanomedicine can be fully realized. The surface properties of NPs play a pivotal role in their interaction with cells. The toxicity of these engineered NPs primarily depends on their size, shape, surface coating, charge, and the nature of the biological medium. Furthermore, NP crystallinity, composition, and ligand arrangement have also been found to affect NP toxicity. Thus, the fabrication of NPs with optimum surface characteristics and efficient control over their interaction with cells holds the future perspective of nanobiotechnology.

Conclusion

Ongoing advances in nanotechnology have led to the development of innovative biomaterials with the potential to improve the healthcare system. Inorganic nanoparticles have been presented as good antimicrobial agents, which can deliver significant benefits in several applications However, to be effectively employed in biomedical applications, a better understanding of the mechanisms that govern the interactions between NPs and living organisms is required.

Even though substantial progress has been achieved to clarify the toxicological profile of nanomaterials, the antibacterial mechanisms of metal and metal oxide nanoparticles are not fully understood. In this chapter, several key pathways responsible for cell-NP interactions are discussed, specifically the generation of metallic ions and ensuing damage to intracellular components of cells, ROS-mediated cell damage, physical deformation of cell membranes, modulation of intracellular and intercellular signaling, and direct DNA damage to halt cell replication.

Acknowledgments

Ahmed Al-Jumaili is thankful to College of Applied Science, University of Fallujah, Iraq for financial support.

References

State Government of Victoria, Department of Health & Human Services. 2015. *"Antibiotic resistant bacteria."* https://www.betterhealth.vic.gov.au/health/conditionsandtreatments/antibiotic-resistant-bacteria.

A, Ravishankar Rai V *and Jamuna Bai. 2015. "Nanoparticles and their potential application as antimicrobials." *Science against microbial pathogens: communicating current research and technological advances* 1:197-209.

Abbaszadegan, Abbas, Yasamin Ghahramani, Ahmad Gholami, Bahram Hemmateenejad, Samira Dorostkar, Mohammadreza Nabavizadeh, and Hashem Sharghi. 2015. "The effect of charge at the surface of silver nanoparticles on antimicrobial activity against gram-positive and gram-negative bacteria: a preliminary study." *Journal of Nanomaterials* 16 (1):53.

Abramova, Anna, Aharon Gedanken, Viktor Popov, Ean-Hin Ooi, Timothy J Mason, Eadaoin M Joyce, James Beddow, Ilana Perelshtein, and Vadim Bayazitov. 2013. "A sonochemical technology for coating of textiles with antibacterial nanoparticles and equipment for itsimplementation." *Materials Letters* 96:121-124.

Agarwal, Happy, and VenkatKumar Shanmugam. 2020. "A review on anti-inflammatory activity of green synthesized zinc oxide nanoparticle: Mechanism-based approach." *Bioorganic chemistry* 94:103423.

Agnihotri, Shekhar, Soumyo Mukherji, and Suparna Mukherji. 2014. "Size-controlled silver nanoparticles synthesized over the range 5–100 nm using the same protocol and their antibacterial efficacy." *RSC Advances* 4 (8):3974-3983.

Ahamed, Maqusood, Mohamad S AlSalhi, and MKJ Siddiqui. 2010. "Silver nanoparticle applications and human health." *Clinica chimica acta* 411 (23):1841-1848.

Ahmad, Razi, Mohd Mohsin, Tokeer Ahmad, and Meryam Sardar. 2015. "Alpha amylase assisted synthesis of TiO 2 nanoparticles: structural characterization and application as antibacterial agents." *Journal of hazardous materials* 283:171-177.

Ahmed, Bilal, Fuad Ameen, Asfa Rizvi, Khursheed Ali, Hana Sonbol, Almas Zaidi, Mohammad Saghir Khan, and Javed Musarrat. 2020. "Destruction of cell topography, morphology, membrane, inhibition of respiration, biofilm formation, and bioactive molecule production by nanoparticles of Ag, ZnO, CuO, TiO2, and Al2O3 toward beneficial soil bacteria." *ACS omega* 5 (14):7861-7876.

Al-jumaili, Ahmed. 2019. "Fabrication and characterization of antibacterial surfaces derived from geranium essential oil using PECVD." James Cook University.

Al-Jumaili, Ahmed, Surjith Alancherry, Kateryna Bazaka, and Mohan V Jacob. 2017a. "The Electrical Properties of Plasma-Deposited Thin Films Derived from Pelargonium graveolens." *Electronics* 6 (4):86.

Al-Jumaili, Ahmed, Surjith Alancherry, Kateryna Bazaka, and Mohan V Jacob. 2017b. "Review on the antimicrobial properties of carbon nanostructures." *Materials* 10 (9):1066.

Al-Jumaili, Ahmed, Surjith Alancherry, Daniel Grant, Avishek Kumar, Kateryna Bazaka, and Mohan V Jacob. 2019. "Plasma Treatment of Polymeric Membranes." In *Non-Thermal Plasma Technology for Polymeric Materials*, 211-240. Elsevier.

Al-Jumaili, Ahmed, Kateryna Bazaka, and Mohan V Jacob. 2017. "Retention of Antibacterial Activity in Geranium Plasma Polymer Thin Films." *Nanomaterials* 7 (9):270.

Al-Jumaili, Ahmed, Avishek Kumar, Kateryna Bazaka, and Mohan V Jacob. 2019. "Electrically Insulating Plasma Polymer/ZnO Composite Films." *Materials* 12 (19):3099.

Al-Jumaili, Ahmed, Peter Mulvey, Avishek Kumar, Karthika Prasad, Kateryna Bazaka, Jeffrey Warner, and Mohan V Jacob. 2019. "Eco-friendly nanocomposites derived from geranium oil and zinc oxide in one step approach." *Scientific reports* 9.

Al-Jumaili, Ahmed, Muhammad Adeel Zafar, Kateryna Bazaka, Janith Weerasinghe, and Mohan V Jacob. 2022. "Bactericidal vertically aligned graphene networks derived from renewable precursor." *Carbon Trends* 7:100157.

Al-Jumaili, Saif, Athar Al-Azzawi, Osman Nuri Uçan, and Adil Deniz Duru. 2023. "Classification of the Level of Alzheimer's Disease Using Anatomical Magnetic Resonance Images Based on a Novel Deep Learning Structure." In *Diagnosis of Neurological Disorders Based on Deep Learning Techniques*, 29-46. CRC Press.

Al-Jumaili, Saif, Ahmed Al-Jumaili, Salam Alyassri, Adil Deniz Duru, and Osman Nuri Uçan. 2022. "Recent Advances on Convolutional Architectures in Medical Applications: Classical or Quantum?" 2022 International Symposium on Multidisciplinary Studies and Innovative Technologies (ISMSIT).

Al-Jumaili, Saif, Adil Deniz Duru, Abdullahi Abdu Ibrahim, and Osman Nuri Uçan. 2023. "Investigation of epileptic seizure signatures classification in EEG using supervised machine learning algorithms."

Arora, Naman, Kavitha Thangavelu, and Georgios N Karanikolos. 2020. "Bimetallic nanoparticles for antimicrobial applications." *Frontiers in Chemistry* 8:412.

Babayevska, Nataliya, Łucja Przysiecka, Igor Iatsunskyi, Grzegorz Nowaczyk, Marcin Jarek, Ewa Janiszewska, and Stefan Jurga. 2022. "ZnO size and shape effect on antibacterial activity and cytotoxicity profile." *Scientific Reports* 12 (1):8148.

Barhoum, Ahmed, María Luisa García-Betancourt, Jaison Jeevanandam, Eman A Hussien, Sara A Mekkawy, Menna Mostafa, Mohamed M Omran, Mohga S. Abdalla, and Mikhael Bechelany. 2022. "Review on natural, incidental, bioinspired, and engineered nanomaterials: history, definitions, classifications, synthesis, properties, market, toxicities, risks, and regulations." *Nanomaterials* 12 (2):177.

Basavegowda, Nagaraj, and Kwang-Hyun Baek. 2021. "Multimetallic nanoparticles as alternative antimicrobial agents: challenges and perspectives." *Molecules* 26 (4):912.

Bazaka, K, MV Jacob, W Chrzanowski, and K Ostrikov. 2015. "Anti-bacterial surfaces: natural agents, mechanisms of action, and plasma surface modification." *RSC Advances* 5 (60):48739-48759.

Bindhu, MR, M Umadevi, M Kavin Micheal, Mariadhas Valan Arasu, and Naif Abdullah Al-Dhabi. 2016. "Structural, morphological and optical properties of MgO nanoparticles for antibacterial applications." *Materials Letters* 166:19-22.

Boonstra, Johannes, and Jan Andries Post. 2004. "Molecular events associated with reactive oxygen species and cell cycle progression in mammalian cells." *Gene* 337:1-13. doi: http://dx.doi.org/10.1016/j.gene.2004.04.032.

Brosset, Estelle. 2013. "The law of the European Union on nanotechnologies: comments on a paradox." *Review of European, Comparative & International Environmental Law* 22 (2):155-162.

Canaparo, Roberto, Federica Foglietta, Tania Limongi, and Loredana Serpe. 2020. "Biomedical applications of reactive oxygen species generation by metal nanoparticles." *Materials* 14 (1):53.

Chandrakala, V, Valmiki Aruna, and Gangadhara Angajala. 2022. "Review on metal nanoparticles as nanocarriers: Current challenges and perspectives in drug delivery systems." *Emergent Materials* 5 (6):1593-1615.

Chen, Hanqing, Bing Wang, Di Gao, Ming Guan, Lingna Zheng, Hong Ouyang, Zhifang Chai, Yuliang Zhao, and Weiyue Feng. 2013. "Broad-spectrum antibacterial activity of carbon nanotubes to human gut bacteria." *Small* 9 (16):2735-2746.

Chen, Kai Loon, and Geoffrey D Bothun. 2013. "Nanoparticles meet cell membranes: probing nonspecific interactions using model membranes." *Environmental science & technology* 48 (2):873-880.

Choi, Okkyoung, and Zhiqiang Hu. 2008. "Size dependent and reactive oxygen species related nanosilver toxicity to nitrifying bacteria." *Environmental science & technology* 42 (12):4583-4588.

Clément, Laura, Charlotte Hurel, and Nicolas Marmier. 2013. "Toxicity of TiO 2 nanoparticles to cladocerans, algae, rotifers and plants–effects of size and crystalline structure." *Chemosphere* 90 (3):1083-1090.

Czyżowska, Agnieszka, and Anna Barbasz. 2022. "A review: zinc oxide nanoparticles–friends or enemies?" *International journal of environmental health research* 32 (4):885-901.

D'Agostino, Ingrid B., and Joseph J. Kieber. 1999. "Phosphorelay signal transduction: the emerging family of plant response regulators." *Trends in Biochemical Sciences* 24 (11):452-456. doi: https://doi.org/10.1016/S0968-0004(99)01465-6.

Dakal, Tikam Chand, Anu Kumar, Rita S. Majumdar, and Vinod Yadav. 2016. "Mechanistic Basis of Antimicrobial Actions of Silver Nanoparticles." *Frontiers in Microbiology* 7:1831. doi: 10.3389/fmicb.2016.01831.

Damm, Cornelia, Helmut Münstedt, and Alfons Rösch. 2008. "The antimicrobial efficacy of polyamide 6/silver-nano-and microcomposites." *Materials Chemistry and Physics* 108 (1):61-66.

Dizaj, Solmaz Maleki, Farzaneh Lotfipour, Mohammad Barzegar-Jalali, Mohammad Hossein Zarrintan, and Khosro Adibkia. 2014. "Antimicrobial activity of the metals and metal oxide nanoparticles." *Materials Science and Engineering: C* 44:278-284. doi: http://dx.doi.org/10.1016/j.msec.2014.08.031.

Dolai, Jayanta, Kuheli Mandal, and Nikhil R Jana. 2021. "Nanoparticle size effects in biomedical applications." *ACS Applied Nano Materials* 4 (7):6471-6496.

Dwyer, Daniel J., Michael A. Kohanski, and James J. Collins. 2009. "Role of reactive oxygen species in antibiotic action and resistance." *Current Opinion in Microbiology* 12 (5):482-489. doi: http://dx.doi.org/10.1016/j.mib.2009.06.018.

El Badawy, Amro M, Rendahandi G Silva, Brian Morris, Kirk G Scheckel, Makram T Suidan, and Thabet M Tolaymat. 2010. "Surface charge-dependent toxicity of silver nanoparticles." *Environmental science & technology* 45 (1):283-287.

Estephan, Zaki G, Jad A Jaber, and Joseph B Schlenoff. 2010. "Zwitterion-stabilized silica nanoparticles: toward nonstick nano." *Langmuir* 26 (22):16884-16889.

Fang, Bing, Ying Jiang, Klaus Nüsslein, Vincent M Rotello, and Maria M Santore. 2015. "Antimicrobial surfaces containing cationic nanoparticles: How immobilized, clustered, and protruding cationic charge presentation affects killing activity and kinetics." *Colloids and Surfaces B: Biointerfaces* 125:255-263.

Feng, QL, Jian Wu, GQ Chen, FZ Cui, TN Kim, and JO Kim. 2000. "A mechanistic study of the antibacterial effect of silver ions on Escherichia coli and Staphylococcus aureus."

Foster, J Estelle, Qin Sheng, Jonathan R McClain, Mark Bures, Thalia I Nicas, Kenneth Henry, Malcolm E Winkler, and Raymond Gilmour. 2004. "Kinetic and mechanistic analyses of new classes of inhibitors of two-component signal transduction systems using a coupled assay containing HpkA–DrrA from Thermotoga maritima." *Microbiology* 150 (4):885-896.

Franco, Domenico, Giovanna Calabrese, Salvatore Pietro Paolo Guglielmino, and Sabrina Conoci. 2022. "Metal-based nanoparticles: Antibacterial mechanisms and biomedical application." *Microorganisms* 10 (9):1778.

Fröhlich, Eleonore. 2012. "The role of surface charge in cellular uptake and cytotoxicity of medical nanoparticles." *Int J Nanomedicine* 7 (1):5577-91.

Fu, Guifen, Patricia S. Vary, and Chhiu-Tsu Lin. 2005. "Anatase TiO2 Nanocomposites for Antimicrobial Coatings." *The Journal of Physical Chemistry B* 109 (18):8889-8898. doi: 10.1021/jp0502196.

Fu, Hong-Bing, and Jian-Nian Yao. 2001. "Size effects on the optical properties of organic nanoparticles." *Journal of the American Chemical Society* 123 (7):1434-1439.

Gao, Feng, Tianyi Shao, Yunpeng Yu, Yujie Xiong, and Lihua Yang. 2021. "Surface-bound reactive oxygen species generating nanozymes for selective antibacterial action." *Nature communications* 12 (1):745.

Gharpure, Saee, Aman Akash, and Balaprasad Ankamwar. 2020. "A review on antimicrobial properties of metal nanoparticles." *Journal of nanoscience and nanotechnology* 20 (6):3303-3339.

Gliga, Anda R, Sara Skoglund, Inger Odnevall Wallinder, Bengt Fadeel, and Hanna L Karlsson. 2014. "Size-dependent cytotoxicity of silver nanoparticles in human lung cells: the role of cellular uptake, agglomeration and Ag release." *Particle and fibre toxicology* 11 (1):11.

Godoy-Gallardo, Maria, Ulrich Eckhard, Luis M Delgado, Yolanda JD de Roo Puente, Mireia Hoyos-Nogués, F Javier Gil, and Roman A Perez. 2021. "Antibacterial approaches in tissue engineering using metal ions and nanoparticles: From mechanisms to applications." *Bioactive Materials* 6 (12):4470-4490.

Gold, Karli, Buford Slay, Mark Knackstedt, and Akhilesh K Gaharwar. 2018. "Antimicrobial activity of metal and metal-oxide based nanoparticles." *Advanced Therapeutics* 1 (3):1700033.

Goli, Kiran K, Nimish Gera, Xiaomeng Liu, Balaji M Rao, Orlando J Rojas, and Jan Genzer. 2013. "Generation and properties of antibacterial coatings based on electrostatic attachment of silver nanoparticles to protein-coated polypropylene fibers." *ACS applied materials & interfaces* 5 (11):5298-5306.

Gratton, Stephanie EA, Patricia A Ropp, Patrick D Pohlhaus, J Christopher Luft, Victoria J Madden, Mary E Napier, and Joseph M DeSimone. 2008. "The effect of particle design on cellular internalization pathways." *Proceedings of the National Academy of Sciences* 105 (33):11613-11618.

Gudkov, Sergey V, Dmitriy E Burmistrov, Dmitriy A Serov, Maxim B Rebezov, Anastasia A Semenova, and Andrey B Lisitsyn. 2021. "A mini review of antibacterial properties of ZnO nanoparticles." *Frontiers in Physics* 9:641481.

Güell, Frank, Andrés Galdámez-Martínez, Paulina R Martínez-Alanis, Ariadne Cristina Catto, Luís Fernando da Silva, Valmor R Mastelaro, Guillermo Santana, and Ateet na Dutt. 2023. "ZnO-based nanomaterials approach for photocatalytic and sensing applications: Recent progress and trends." *Materials Advances*.

Gunawan, Cindy, May Lim, Christopher P Marquis, and Rose Amal. 2014. "Nanoparticle–protein corona complexes govern the biological fates and functions of nanoparticles." *Journal of Materials Chemistry B* 2 (15):2060-2083.

Gupta, Akash, Shazia Mumtaz, Cheng-Hsuan Li, Irshad Hussain, and Vincent M Rotello. 2019. "Combatting antibiotic-resistant bacteria using nanomaterials." *Chemical Society Reviews* 48 (2):415-427.

Hajipour, Mohammad J., Katharina M. Fromm, Ali Akbar Ashkarran, Dorleta Jimenez de Aberasturi, Idoia Ruiz de Larramendi, Teofilo Rojo, Vahid Serpooshan, Wolfgang J. Parak, and Morteza Mahmoudi. 2012. "Antibacterial properties of nanoparticles." *Trends in Biotechnology* 30 (10):499-511. doi: http://dx.doi.org/10.1016/j.tibtech.2012.06.004.

He, Chunbai, Yiping Hu, Lichen Yin, Cui Tang, and Chunhua Yin. 2010. "Effects of particle size and surface charge on cellular uptake and biodistribution of polymeric nanoparticles." *Biomaterials* 31 (13):3657-3666.

Heinlaan, Margit, Angela Ivask, Irina Blinova, Henri-Charles Dubourguier, and Anne Kahru. 2008. "Toxicity of nanosized and bulk ZnO, CuO and TiO2 to bacteria Vibrio fischeri and crustaceans Daphnia magna and Thamnocephalus platyurus." *Chemosphere* 71 (7):1308-1316. doi: http://dx.doi.org/10.1016/j.chemosphere.2007.11.047.

Helmlinger, J, C Sengstock, C Groß-Heitfeld, C Mayer, TA Schildhauer, M Köller, and M Epple. 2016. "Silver nanoparticles with different size and shape: equal cytotoxicity, but different antibacterial effects." *RSC Advances* 6 (22):18490-18501.

Hernansanz-Agustín, Pablo, and José Antonio Enríquez. 2021. "Generation of reactive oxygen species by mitochondria." *Antioxidants* 10 (3):415.

Holban, Alina Maria, Monica Cartelle Gestal, and Alexandru Mihai Grumezescu. 2016. "Control of biofilm-associated infections by signaling molecules and nanoparticles." *International Journal of Pharmaceutics* 510 (2):409-418. doi: http://dx.doi.org/10.1016/j.ijpharm.2016.02.044.

Holister, Paul, Jan-Willem Weener, Cristina Román, and Tim Harper. 2003. "Nanoparticles." *Technology White Papers* 3 (1-11).

Hoseinzadeh, Edris, Pouran Makhdoumi, Parisa Taha, Hooshyar Hossini, John Stelling, Mohammad Amjad Kamal, and Ghulam Md Ashraf. 2017. "A review on nano-antimicrobials: metal nanoparticles, methods and mechanisms." *Current drug metabolism* 18 (2):120-128.

Hsueh, Yi-Huang, Kuen-Song Lin, Wan-Ju Ke, Chien-Te Hsieh, Chao-Lung Chiang, Dong-Ying Tzou, and Shih-Tung Liu. 2015. "The antimicrobial properties of silver nanoparticles in Bacillus subtilis are mediated by released Ag+ ions." *PloS one* 10 (12):e0144306.

Huang, Zhongbing, Xu Zheng, Danhong Yan, Guangfu Yin, Xiaoming Liao, Yunqing Kang, Yadong Yao, Di Huang, and Baoqing Hao. 2008. "Toxicological Effect of ZnO Nanoparticles Based on Bacteria." *Langmuir* 24 (8):4140-4144. doi: 10.1021/la7035949.

Ivask, Angela, Imbi Kurvet, Kaja Kasemets, Irina Blinova, Villem Aruoja, Sandra Suppi, Heiki Vija, Aleksandr Käkinen, Tiina Titma, and Margit Heinlaan. 2014. "Size-dependent toxicity of silver nanoparticles to bacteria, yeast, algae, crustaceans and mammalian cells *in vitro*." *PloS one* 9 (7):e102108.

Ivask, Angela, Tiina Titma, Meeri Visnapuu, Heiki Vija, Aleksandr Kakinen, Mariliis Sihtmae, Suman Pokhrel, Lutz Madler, Margit Heinlaan, Vambola Kisand, Ruth Shimmo, and Anne Kahru. 2015. "Toxicity of 11 Metal Oxide Nanoparticles to Three Mammalian Cell Types *In Vitro*." *Current Topics in Medicinal Chemistry* 15 (18):1914-1929.

J. Díaz-Visurragal, 2, C. Gutiérrez2, C. von Plessing1 and A. García3. 2011. "Metal nanostructures as antibacterial agents." *Science against microbial pathogens: communicating current research and technological advances*:210-218.

Jiang, Wei, Hamid Mashayekhi, and Baoshan Xing. 2009. "Bacterial toxicity comparison between nano-and micro-scaled oxide particles." *Environmental Pollution* 157 (5):1619-1625.

Joudeh, Nadeem, and Dirk Linke. 2022. "Nanoparticle classification, physicochemical properties, characterization, and applications: a comprehensive review for biologists." *Journal of Nanobiotechnology* 20 (1):262.

Kamel, Michael SA, Ahmed Al-Jumaili, Michael Oelgemöller, and Mohan V Jacob. 2022. "Inorganic nanoparticles to overcome efficiency inhibitors of organic photovoltaics: An in-depth review." *Renewable and Sustainable Energy Reviews* 166:112661.

Karimi, Ebrahim. 2019. "Antimicrobial Activities of Nanoparticles." In *Nanotechnology for Agriculture: Crop Production & Protection*, 171-206. Springer.

Kasemets, Kaja, Angela Ivask, Henri-Charles Dubourguier, and Anne Kahru. 2009. "Toxicity of nanoparticles of ZnO, CuO and TiO2 to yeast Saccharomyces cerevisiae." *Toxicology in Vitro* 23 (6):1116-1122. doi: http://dx.doi.org/10.1016/j.tiv.2009.05.015.

Kaweeteerawat, Chitrada, Chong Hyun Chang, Kevin R Roy, Rong Liu, Ruibin Li, Daniel Toso, Heidi Fischer, Angela Ivask, Zhaoxia Ji, and Jeffrey I Zink. 2015. "Cu nanoparticles have different impacts in Escherichia coli and Lactobacillus brevis than their microsized and ionic analogues." *ACS nano* 9 (7):7215-7225.

Klueh, U, V Wagner, S Kelly, A Johnson, and JD Bryers. 2000. "Efficacy of silver-coated fabric to prevent bacterial colonization and subsequent device-based biofilm formation." *Journal of Biomedical Materials Research Part A* 53 (6):621-631.

Knetsch, Menno L. W., and Leo H. Koole. 2011. "New Strategies in the Development of Antimicrobial Coatings: The Example of Increasing Usage of Silver and Silver Nanoparticles." *Polymers* 3 (4):340-366. doi: 10.3390/polym3010340.

Kumar, Ashutosh, Alok K. Pandey, Shashi S. Singh, Rishi Shanker, and Alok Dhawan. 2011. "Engineered ZnO and TiO2 nanoparticles induce oxidative stress and DNA damage leading to reduced viability of Escherichia coli." *Free Radical Biology and Medicine* 51 (10):1872-1881. doi: http://dx.doi.org/10.1016/j.freeradbiomed.2011.08.025.

Kumar, Avishek, Ahmed Al-Jumaili, Kateryna Bazaka, Peter Mulvey, Jeffrey Warner, and Mohan V Jacob. 2020. "In-situ surface modification of terpinen-4-ol plasma polymers for increased antibacterial activity." *Materials* 13 (3):586.

Kumar, Avishek, Daniel Grant, Surjith Alancherry, Ahmed Al-Jumaili, Kateryna Bazaka, and Mohan V Jacob. 2017a. "Plasma polymerization: Electronics and biomedical application." In *Plasma Science and Technology for Emerging Economies*, 593-657. Springer.

Kumar, Avishek, Daniel Grant, Surjith Alancherry, Ahmed Al-Jumaili, Kateryna Bazaka, and Mohan V Jacob. 2017b. "Plasma polymerization: Electronics and biomedical application." *Plasma science and technology for emerging economies: An AAAPT experience*:593-657.

Kumar, V. Siva, B. M. Nagaraja, V. Shashikala, A. H. Padmasri, S. Shakuntala Madhavendra, B. David Raju, and K. S. Rama Rao. 2004. "Highly efficient Ag/C catalyst prepared by electro-chemical deposition method in controlling microorganisms in water." *Journal of Molecular Catalysis A: Chemical* 223 (1–2):313-319. doi: http://dx.doi.org/10.1016/j.molcata.2003.09.047.

Lallo da Silva, Bruna, Marina Paiva Abuçafy, Eloisa Berbel Manaia, João Augusto Oshiro Junior, Bruna Galdorfini Chiari-Andréo, Rosemeire CL R Pietro, and Leila Aparecida Chiavacci. 2019. "Relationship between structure and antimicrobial activity of zinc oxide nanoparticles: An overview." *International journal of nanomedicine*:9395-9410.

Lara, Humberto H, Elsa N Garza-Treviño, Liliana Ixtepan-Turrent, and Dinesh K Singh. 2011. "Silver nanoparticles are broad-spectrum bactericidal and virucidal compounds." *Journal of nanobiotechnology* 9 (1):1.

Lee, Changha, Jee Yeon Kim, Won Il Lee, Kara L. Nelson, Jeyong Yoon, and David L. Sedlak. 2008. "Bactericidal Effect of Zero-Valent Iron Nanoparticles on Escherichia coli." *Environmental Science & Technology* 42 (13):4927-4933. doi: 10.1021/es800408u.

Lemire, Joseph A, Joe J Harrison, and Raymond J Turner. 2013. "Antimicrobial activity of metals: mechanisms, molecular targets and applications." *Nature Reviews Microbiology* 11 (6):371-384.

Li, Mei, Daohui Lin, and Lizhong Zhu. 2013. "Effects of water chemistry on the dissolution of ZnO nanoparticles and their toxicity to Escherichia coli." *Environmental pollution* 173:97-102.

Li, Qilin, Shaily Mahendra, Delina Y Lyon, Lena Brunet, Michael V Liga, Dong Li, and Pedro JJ Alvarez. 2008. "Antimicrobial nanomaterials for water disinfection and microbial control: potential applications and implications." *Water research* 42 (18):4591-4602.

Li, Ying, Martin Kröger, and Wing Kam Liu. 2015. "Shape effect in cellular uptake of PEGylated nanoparticles: comparison between sphere, rod, cube and disk." *Nanoscale* 7 (40):16631-16646.

Liang, Hongyan, Hong Wei, Deng Pan, and Hongxing Xu. 2015. "Chemically synthesized noble metal nanostructures for plasmonics." *Nanotechnology Reviews* 4 (3):289-302.

Lok, Chun-Nam, Chi-Ming Ho, Rong Chen, Qing-Yu He, Wing-Yiu Yu, Hongzhe Sun, Paul Kwong-Hang Tam, Jen-Fu Chiu, and Chi-Ming Che. 2006. "Proteomic analysis of the mode of antibacterial action of silver nanoparticles." *Journal of Proteome research* 5 (4):916-924.

Lok, Chun-Nam, Chi-Ming Ho, Rong Chen, Qing-Yu He, Wing-Yiu Yu, Hongzhe Sun, Paul Kwong-Hang Tam, Jen-Fu Chiu, and Chi-Ming Che. 2007. "Silver nanoparticles: partial oxidation and antibacterial activities." *JBIC Journal of Biological Inorganic Chemistry* 12 (4):527-534. doi: 10.1007/s00775-007-0208-z.

Lopes, Sílvia, Carlos Pinheiro, Amadeu M. V. M. Soares, and Susana Loureiro. 2016. "Joint toxicity prediction of nanoparticles and ionic counterparts: Simulating toxicity under a fate scenario." *Journal of Hazardous Materials* 320:1-9. doi: http://dx.doi.org/10.1016/j.jhazmat.2016.07.068.

Lu, Hoang D., Alina C. Spiegel, Amanda Hurley, Lark J. Perez, Katharina Maisel, Laura M. Ensign, Justin Hanes, Bonnie L. Bassler, Martin F. Semmelhack, and Robert K. Prud'homme. 2015. "Modulating Vibrio cholerae Quorum-Sensing-Controlled Communication Using Autoinducer-Loaded Nanoparticles." *Nano Letters* 15 (4):2235-2241. doi: 10.1021/acs.nanolett.5b00151.

Luef, Birgit, Sirine C Fakra, Roseann Csencsits, Kelly C Wrighton, Kenneth H Williams, Michael J Wilkins, Kenneth H Downing, Philip E Long, Luis R Comolli, and Jillian F Banfield. 2013. "Iron-reducing bacteria accumulate ferric oxyhydroxide nanoparticle aggregates that may support planktonic growth." *The ISME journal* 7 (2):338-350.

Macomber, Lee, Scott P. Elsey, and Robert P. Hausinger. 2011. "Fructose-1,6-bisphosphate aldolase (class II) is the primary site of nickel toxicity in Escherichia coli." *Molecular Microbiology* 82 (5):1291-1300. doi: 10.1111/j.1365-2958.2011.07891.x.

Makabenta, Jessa Marie V, Ahmed Nabawy, Cheng-Hsuan Li, Suzannah Schmidt-Malan, Robin Patel, and Vincent M Rotello. 2021. "Nanomaterial-based therapeutics for antibiotic-resistant bacterial infections." *Nature Reviews Microbiology* 19 (1):23-36.

Marambio-Jones, Catalina, and Eric M. V. Hoek. 2010. "A review of the antibacterial effects of silver nanomaterials and potential implications for human health and the environment." *Journal of Nanoparticle Research* 12 (5):1531-1551. doi: 10.1007/s11051-010-9900-y.

Martinez-Castanon, GA, N Nino-Martinez, F Martinez-Gutierrez, JR Martinez-Mendoza, and Facundo Ruiz. 2008. "Synthesis and antibacterial activity of silver nanoparticles with different sizes." *Journal of Nanoparticle Research* 10 (8):1343-1348.

Mendes, Carolina Rosai, Guilherme Dilarri, Carolina Froes Forsan, Vinícius de Moraes Ruy Sapata, Paulo Renato Matos Lopes, Peterson Bueno de Moraes, Renato Nallin Montagnolli, Henrique Ferreira, and Ederio Dino Bidoia. 2022. "Antibacterial action and target mechanisms of zinc oxide nanoparticles against bacterial pathogens." *Scientific Reports* 12 (1):2658.

Miller, Kristen P, Lei Wang, Yung-Pin Chen, Perry J Pellechia, Brian C Benicewicz, and Alan W Decho. 2015. "Engineering nanoparticles to silence bacterial communication." *Frontiers in microbiology* 6.

Mingeot-Leclercq, Marie-Paule, and Jean-Luc Décout. 2016. "Bacterial lipid membranes as promising targets to fight antimicrobial resistance, molecular foundations and illustration through the renewal of aminoglycoside antibiotics and emergence of amphiphilic aminoglycosides." *MedChemComm* 7 (4):586-611.

Modrich, Paul. 2016. "Mechanisms in E. coli and human mismatch repair (Nobel Lecture)." *Angewandte Chemie International Edition* 55 (30):8490-8501.

Moritz, Michał, and Małgorzata Geszke-Moritz. 2013. "The newest achievements in synthesis, immobilization and practical applications of antibacterial nanoparticles." *Chemical Engineering Journal* 228:596-613.

Mu, Haibo, Jiangjiang Tang, Qianjin Liu, Chunli Sun, Tingting Wang, and Jinyou Duan. 2016. "Potent antibacterial nanoparticles against biofilm and intracellular bacteria." *Scientific reports* 6:18877.

Mukherji, Ruchira, and Asmita Prabhune. 2015. "A new class of bacterial quorum sensing antagonists: glycomonoterpenols synthesized using linalool and alpha terpineol." *World Journal of Microbiology and Biotechnology* 31 (6):841-849.

Muñoz-Bonilla, Alexandra, and Marta Fernández-García. 2012. "Polymeric materials with antimicrobial activity." *Progress in Polymer Science* 37 (2):281-339.

Musee, Ndeke, Melusi Thwala, and Nomakhwezi Nota. 2011. "The antibacterial effects of engineered nanomaterials: implications for wastewater treatment plants." *Journal of Environmental Monitoring* 13 (5):1164-1183.

Naddeo, JJ, Matthew Ratti, Sean M O'Malley, Julianne C Griepenburg, Daniel M Bubb, and Eric A Klein. 2015. "Antibacterial Properties of Nanoparticles: A Comparative Review of Chemically Synthesized and Laser-Generated Particles." *Advanced Science, Engineering and Medicine* 7 (12):1044-1057.

Nazzaro, Filomena, Florinda Fratianni, Laura De Martino, Raffaele Coppola, and Vincenzo De Feo. 2013. "Effect of essential oils on pathogenic bacteria." *Pharmaceuticals* 6 (12):1451-1474.

Ng, Wai-Leung, and Bonnie L Bassler. 2009. "Bacterial quorum-sensing network architectures." *Annual review of genetics* 43:197-222.

Nikolova, Maria P, and Murthy S Chavali. 2020. "Metal oxide nanoparticles as biomedical materials." *Biomimetics* 5 (2):27.

Nisar, Parveen, Nasir Ali, Lubna Rahman, Muhammad Ali, and Zabta Khan Shinwari. 2019. "Antimicrobial activities of biologically synthesized metal nanoparticles: an insight into the mechanism of action." *JBIC Journal of Biological Inorganic Chemistry* 24:929-941.

Pacher, Pál, Joseph S. Beckman, and Lucas Liaudet. 2007. "Nitric Oxide and Peroxynitrite in Health and Disease." *Physiological Reviews* 87 (1):315-424. doi: 10.1152/physrev.00029.2006.

Pal, Sukdeb, Yu Kyung Tak, and Joon Myong Song. 2007. "Does the antibacterial activity of silver nanoparticles depend on the shape of the nanoparticle? A study of the gram-negative bacterium Escherichia coli." *Applied and environmental microbiology* 73 (6):1712-1720.

Palza, Humberto. 2015. "Antimicrobial polymers with metal nanoparticles." *International journal of molecular sciences* 16 (1):2099-2116.

Park, Joonhyuck, Jutaek Nam, Nayoun Won, Ho Jin, Sungho Jung, Sungwook Jung, So-Hye Cho, and Sungjee Kim. 2011. "Compact and Stable Quantum Dots with Positive, Negative, or Zwitterionic Surface: Specific Cell Interactions and Non-Specific Adsorptions by the Surface Charges." *Advanced Functional Materials* 21 (9):1558-1566.

Pelgrift, Robert Y., and Adam J. Friedman. 2013. "Nanotechnology as a therapeutic tool to combat microbial resistance." *Advanced Drug Delivery Reviews* 65 (13–14):1803-1815. doi: http://dx.doi.org/10.1016/j.addr.2013.07.011.

Pelletier, Dale A, Anil K Suresh, Gregory A Holton, Catherine K McKeown, Wei Wang, Baohua Gu, Ninell P Mortensen, David P Allison, David C Joy, and Martin R Allison. 2010. "Effects of engineered cerium oxide nanoparticles on bacterial growth and viability." *Applied and environmental microbiology* 76 (24):7981-7989.

Perelshtein, Ilana, Anat Lipovsky, Nina Perkas, Tzanko Tzanov, M Arguirova, M Leseva, and Aharon Gedanken. 2015. "Making the hospital a safer place by sonochemical coating of all its textiles with antibacterial nanoparticles." *Ultrasonics sonochemistry* 25:82-88.

Pietroiusti, Antonio, Luisa Campagnolo, and Bengt Fadeel. 2013. "Interactions of engineered nanoparticles with organs protected by internal biological barriers." *Small* 9 (9-10):1557-1572.

Pietroiusti, Antonio, Micol Massimiani, Ivana Fenoglio, Massimiliano Colonna, Federica Valentini, Giuseppe Palleschi, Antonella Camaioni, Andrea Magrini, Gregorio Siracusa, and Antonio Bergamaschi. 2011. "Low doses of pristine and oxidized single-wall carbon nanotubes affect mammalian embryonic development." *Acs Nano* 5 (6):4624-4633.

Pleskova, Svetlana N., Ekaterina N. Gorshkova, Viktor V. Novikov, and Marc Solioz. 2016. "Treatment by serum up-conversion nanoparticles in the fluoride matrix changes the mechanism of cell death and the elasticity of the membrane." *Micron* 90:23-32. doi: http://dx.doi.org/10.1016/j.micron.2016.08.005.

Pourhajibagher, Maryam, Ahmad Sodagar, and Abbas Bahador. 2020. "An *in vitro* evaluation of the effects of nanoparticles on shear bond strength and antimicrobial properties of orthodontic adhesives: A systematic review and meta-analysis study." *International Orthodontics*.

Prabhakaran, Pranesha, Muhammad Aqeel Ashraf, and Wan Syaidatul Aqma. 2016. "Microbial stress response to heavy metals in the environment." *RSC Advances* 6 (111):109862-109877.

Prabhu, Sukumaran, and Eldho K. Poulose. 2012. "Silver nanoparticles: mechanism of antimicrobial action, synthesis, medical applications, and toxicity effects." *International Nano Letters* 2 (1):32. doi: 10.1186/2228-5326-2-32.

Pramanik, Arindam, Dipranjan Laha, Debalina Bhattacharya, Panchanan Pramanik, and Parimal Karmakar. 2012. "A novel study of antibacterial activity of copper iodide nanoparticle mediated by DNA and membrane damage." *Colloids and Surfaces B: Biointerfaces* 96:50-55.

Qiu, Ruoyi, Miho Sakato, Elizabeth J Sacho, Hunter Wilkins, Xingdong Zhang, Paul Modrich, Manju M Hingorani, Dorothy A Erie, and Keith R Weninger. 2015. "MutL traps MutS at a DNA mismatch." *Proceedings of the National Academy of Sciences* 112 (35):10914-10919.

Qiu, Yang, Ying Liu, Liming Wang, Ligeng Xu, Ru Bai, Yinglu Ji, Xiaochun Wu, Yuliang Zhao, Yufeng Li, and Chunying Chen. 2010. "Surface chemistry and aspect ratio mediated cellular uptake of Au nanorods." *Biomaterials* 31 (30):7606-7619. doi: http://dx.doi.org/10.1016/j.biomaterials.2010.06.051.

Radovic-Moreno, Aleksandar F, Timothy K Lu, Vlad A Puscasu, Christopher J Yoon, Robert Langer, and Omid C Farokhzad. 2012. "Surface charge-switching polymeric nanoparticles for bacterial cell wall-targeted delivery of antibiotics." *ACS nano* 6 (5):4279-4287.

Radzig, M. A., V. A. Nadtochenko, O. A. Koksharova, J. Kiwi, V. A. Lipasova, and I. A. Khmel. 2013. "Antibacterial effects of silver nanoparticles on gram-negative bacteria: Influence on the growth and biofilms formation, mechanisms of action." *Colloids and Surfaces B: Biointerfaces* 102 (Supplement C):300-306. doi: https://doi.org/10.1016/j.colsurfb.2012.07.039.

Railsback, Justin G, Abhishek Singh, Ryan C Pearce, Timothy E McKnight, Ramón Collazo, Zlatko Sitar, Yaroslava G Yingling, and Anatoli V Melechko. 2012. "Weakly charged cationic nanoparticles induce DNA bending and strand separation." *Advanced materials* 24 (31):4261-4265.

Rao, CNR, GU Kulkarni, P John Thomas, and Peter P Edwards. 2002. "Size-dependent chemistry: properties of nanocrystals." *Chemistry–A European Journal* 8 (1):28-35.

Rapacka-Zdończyk, Aleksandra, Agata Woźniak, Klaudia Michalska, Michał Pierański, Patrycja Ogonowska, Mariusz Grinholc, and Joanna Nakonieczna. 2021. "Factors determining the susceptibility of bacteria to antibacterial photodynamic inactivation." *Frontiers in Medicine* 8:642609.

Rouhana, Layal L, Jad A Jaber, and Joseph B Schlenoff. 2007. "Aggregation-resistant water-soluble gold nanoparticles." *Langmuir* 23 (26):12799-12801.

Roy, Arpita, Apoorva Sharma, Saanya Yadav, Leta Tesfaye Jule, and Ramaswamy Krishnaraj. 2021. "Nanomaterials for remediation of environmental pollutants." *Bioinorganic Chemistry and Applications* 2021 (1):1764647.

Saidin, Syafiqah, Mohamad Amin Jumat, Nur Ain Atiqah Mohd Amin, and Abdullah Sharaf Saleh Al-Hammadi. 2021. "Organic and inorganic antibacterial approaches in combating bacterial infection for biomedical application." *Materials Science and Engineering: C* 118:111382.

Sathiyaseelan, Anbazhagan, Kumar Vishven Naveen, Xin Zhang, Kiseok Han, and Myeong-Hyeon Wang. 2023. "Research progress on chitosan-zinc oxide nanocomposites fabrication, characterization, biomedical and environmental applications." *Coordination Chemistry Reviews* 496:215398.

Sawai, J. 2003. "Quantitative evaluation of antibacterial activities of metallic oxide powders (ZnO, MgO and CaO) by conductimetric assay." *Journal of Microbiological Methods* 54 (2):177-182. doi: http://dx.doi.org/10.1016/S0167-7012(03)00037-X.

Shrivastava, Siddhartha, Tanmay Bera, Arnab Roy, Gajendra Singh, P Ramachandrarao, and Debabrata Dash. 2007. "Characterization of enhanced antibacterial effects of novel silver nanoparticles." *Nanotechnology* 18 (22):225103.

Siddiqui, Mohammad Nahid, Halim Hamid Redhwi, Dimitris S Achilias, Elena Kosmidou, Eythimia Vakalopoulou, and Maria D Ioannidou. 2018. "Green synthesis of silver nanoparticles and study of their antimicrobial properties." *Journal of Polymers and the Environment* 26 (2):423-433.

Simon-Deckers, Angelique, Sylvain Loo, Martine Mayne-L'hermite, Nathalie Herlin-Boime, Nicolas Menguy, Cecile Reynaud, Barbara Gouget, and Marie Carriere. 2009. "Size-, composition-and shape-dependent toxicological impact of metal oxide nanoparticles and carbon nanotubes toward bacteria." *Environmental science & technology* 43 (21):8423-8429.

Simon-Deckers, Angélique, Sylvain Loo, Martine Mayne-L'hermite, Nathalie Herlin-Boime, Nicolas Menguy, Cécile Reynaud, Barbara Gouget, and Marie Carrière. 2009. "Size-, Composition- and Shape-Dependent Toxicological Impact of Metal Oxide Nanoparticles and Carbon Nanotubes toward Bacteria." *Environmental Science & Technology* 43 (21):8423-8429. doi: 10.1021/es9016975.

Singh, Akhand Pratap, Arpan Biswas, Aparna Shukla, and Pralay Maiti. 2019. "Targeted therapy in chronic diseases using nanomaterial-based drug delivery vehicles." *Signal transduction and targeted therapy* 4 (1):33.

Singh, Braj R, Brahma N Singh, Akanksha Singh, Wasi Khan, Alim H Naqvi, and Harikesh B Singh. 2015. "Mycofabricated biosilver nanoparticles interrupt Pseudomonas aeruginosa quorum sensing systems." *Scientific reports* 5.

Singh, Mritunjai, Shinjini Singh, S Prasad, and IS Gambhir. 2008. "Nanotechnology in medicine and antibacterial effect of silver nanoparticles." *Digest Journal of Nanomaterials and Biostructures* 3 (3):115-122.

Sirelkhatim, Amna, Shahrom Mahmud, Azman Seeni, Noor Haida Mohamad Kaus, Ling Chuo Ann, Siti Khadijah Mohd Bakhori, Habsah Hasan, and Dasmawati Mohamad. "Review on Zinc Oxide Nanoparticles: Antibacterial Activity and Toxicity Mechanism." *Nano-Micro Letters*:1-24.

Sirelkhatim, Amna, Shahrom Mahmud, Azman Seeni, Noor Haida Mohamad Kaus, Ling Chuo Ann, Siti Khadijah Mohd Bakhori, Habsah Hasan, and Dasmawati Mohamad. 2015. "Review on zinc oxide nanoparticles: antibacterial activity and toxicity mechanism." *Nano-Micro Letters* 7 (3):219-242.

Sondi, Ivan, and Branka Salopek-Sondi. 2004. "Silver nanoparticles as antimicrobial agent: a case study on E. coli as a model for Gram-negative bacteria." *Journal of Colloid and Interface Science* 275 (1):177-182. doi: http://dx.doi.org/10.1016/j.jcis.2004.02.012.

Song, HY, KK Ko, IH Oh, and BT Lee. 2006. "Fabrication of silver nanoparticles and their antimicrobial mechanisms." *Eur Cells Mater* 11 (Suppl 1):58.

Stankic, Slavica, Sneha Suman, Francia Haque, and Jasmina Vidic. 2016. "Pure and multi metal oxide nanoparticles: synthesis, antibacterial and cytotoxic properties." *Journal of nanobiotechnology* 14 (1):73.

Subhan, Md Abdus, Newton Neogi, and Kristi Priya Choudhury. 2022. "Industrial manufacturing applications of zinc oxide nanomaterials: A comprehensive study." *Nanomanufacturing* 2 (4):265-291.

Suresh, Anil K, Dale A Pelletier, Wei Wang, Ji-Won Moon, Baohua Gu, Ninell P Mortensen, David P Allison, David C Joy, Tommy J Phelps, and Mitchel J Doktycz. 2010. "Silver nanocrystallites: biofabrication using

Shewanella oneidensis, and an evaluation of their comparative toxicity on gram-negative and gram-positive bacteria." *Environmental science & technology* 44 (13):5210-5215.

Szabo, Csaba, Harry Ischiropoulos, and Rafael Radi. 2007. "Peroxynitrite: biochemistry, pathophysiology and development of therapeutics." *Nat Rev Drug Discov* 6 (8):662-680. doi: http://www.nature.com/nrd/journal/v6/n8/suppinfo/nrd2222_S1.html.

Taglietti, Angelo, Yuri A Diaz Fernandez, Elvio Amato, Lucia Cucca, Giacomo Dacarro, Pietro Grisoli, Vittorio Necchi, Piersandro Pallavicini, Luca Pasotti, and Maddalena Patrini. 2012. "Antibacterial activity of glutathione-coated silver nanoparticles against gram positive and gram negative bacteria." *Langmuir* 28 (21):8140-8148.

Talebian, Nasrin, Seyedeh Matin Amininezhad, and Monir Doudi. 2013. "Controllable synthesis of ZnO nanoparticles and their morphology-dependent antibacterial and optical properties." *Journal of Photochemistry and Photobiology B: Biology* 120:66-73. doi: http://dx.doi.org/10.1016/j.jphotobiol.2013.01.004.

Thanbichler, Martin, Sherry C Wang, and Lucy Shapiro. 2005. "The bacterial nucleoid: a highly organized and dynamic structure." *Journal of cellular biochemistry* 96 (3):506-521.

Tobaldi, DM, C Piccirillo, N Rozman, RC Pullar, MP Seabra, A Sever Škapin, PML Castro, and JA Labrincha. 2016. "Effects of Cu, Zn and Cu-Zn addition on the microstructure and antibacterial and photocatalytic functional properties of Cu-Zn modified TiO 2 nano-heterostructures." *Journal of Photochemistry and Photobiology A: Chemistry* 330:44-54.

Troncarelli, M Zampoli, H Mello Brandão, J Carine Gern, A Sá Guimarães, and H Langoni. 2013. "Nanotechnology and antimicrobials in veterinary medicine." *Formatex*:543-556.

Tsikourkitoudi, Vasiliki, Birgitta Henriques-Normark, and Georgios A Sotiriou. 2022. "Inorganic nanoparticle engineering against bacterial infections." *Current Opinion in Chemical Engineering* 38:100872.

Turrens, Julio F. 2003. "Mitochondrial formation of reactive oxygen species." *The Journal of Physiology* 552 (Pt 2):335-344. doi: 10.1113/jphysiol.2003.049478.

Utsumi, Ryutaro. 2008. *Bacterial signal transduction: networks and drug targets*. Vol. 631: Springer Science & Business Media.

Vanlalveni, Chhangte, Samuel Lallianrawna, Ayushi Biswas, Manickam Selvaraj, Bishwajit Changmai, and Samuel Lalthazuala Rokhum. 2021. "Green synthesis of silver nanoparticles using plant extracts and their antimicrobial activities: A review of recent literature." *RSC advances* 11 (5):2804-2837.

Vasantharaj, Seerangaraj, Selvam Sathiyavimal, Palanisamy Senthilkumar, Felix LewisOscar, and Arivalagan Pugazhendhi. 2019. "Biosynthesis of iron oxide nanoparticles using leaf extract of Ruellia tuberosa: antimicrobial properties and their applications in photocatalytic degradation." *Journal of Photochemistry and Photobiology B: Biology* 192:74-82.

Vatansever, Fatma, Wanessa C. M. A. de Melo, Pinar Avci, Daniela Vecchio, Magesh Sadasivam, Asheesh Gupta, Rakkiyappan Chandran, Mahdi Karimi, Nivaldo A. Parizotto, Rui Yin, George P. Tegos, and Michael R. Hamblin. 2013. "Antimicrobial strategies centered around reactive oxygen species – bactericidal antibiotics, photodynamic therapy, and beyond." *FEMS Microbiology Reviews* 37 (6):955-989. doi: 10.1111/1574-6976.12026.

Verma, Ayush, and Francesco Stellacci. 2010. "Effect of surface properties on nanoparticle–cell interactions." *Small* 6 (1):12-21.

Vimbela, Gina V, Sang M Ngo, Carolyn Fraze, Lei Yang, and David A Stout. 2017. "Antibacterial properties and toxicity from metallic nanomaterials." *International journal of nanomedicine*:3941-3965.

Wang, Li, Yue Pan, Zhi-Hui Yuan, Huan Zhang, Bao-Yu Peng, Fang-Fang Wang, and Wei Qian. 2016. "Two-Component Signaling System VgrRS Directly Senses Extracytoplasmic and Intracellular Iron to Control Bacterial Adaptation under Iron Depleted Stress." *PLoS pathogens* 12 (12):e1006133.

Wang, Lingling, Huzhi Zheng, Yijuan Long, Mei Gao, Jianyu Hao, Juan Du, Xiaojiao Mao, and Dongbo Zhou. 2010. "Rapid determination of the toxicity of quantum dots with luminous bacteria." *Journal of hazardous materials* 177 (1):1134-1137.

Wang, Linlin, Chen Hu, and Longquan Shao. 2017. "The antimicrobial activity of nanoparticles: present situation and prospects for the future." *International journal of nanomedicine* 12:1227.

Wigley, Dale B. 2013. "Bacterial DNA repair: recent insights into the mechanism of RecBCD, AddAB and AdnAB." *Nature reviews. Microbiology* 11 (1):9.

Woon, Sze-Ann, and Dale Fisher. 2016. "Antimicrobial agents–optimising the ecological balance." *BMC medicine* 14 (1):114.

Wyszogrodzka, Gabriela, Bartosz Marszałek, Barbara Gil, and Przemysław Dorożyński. 2016. "Metal-organic frameworks: mechanisms of antibacterial action and potential applications." *Drug Discovery Today* 21 (6):1009-1018. doi: http://dx.doi.org/10.1016/j.drudis.2016.04.009.

Xie, Jin, Chenjie Xu, Nathan Kohler, Yanglong Hou, and Shouheng Sun. 2007. "Controlled PEGylation of monodisperse Fe3O4 nanoparticles for reduced non-specific uptake by macrophage cells." *Advanced Materials* 19 (20):3163-3166.

Xiu, Zong-ming, Qing-bo Zhang, Hema L. Puppala, Vicki L. Colvin, and Pedro J. J. Alvarez. 2012. "Negligible Particle-Specific Antibacterial Activity of Silver Nanoparticles." *Nano Letters* 12 (8):4271-4275. doi: 10.1021/nl301934w.

Yin, Jun-Jie, Baozhong Zhao, Qingsu Xia, and Peter P Fu. 2013. "Spectroscopy for Studying the Generation and Scavenging of Reactive Oxygen Species by Nanomaterials." *Nanopharmaceutics: The Potential Application of Nanomaterials*:375.

You, Chuangang, Chunmao Han, Xingang Wang, Yurong Zheng, Qiyin Li, Xinlei Hu, and Huafeng Sun. 2012. "The progress of silver nanoparticles in the antibacterial mechanism, clinical application and cytotoxicity." *Molecular Biology Reports* 39 (9):9193-9201. doi: 10.1007/s11033-012-1792-8.

Yuan, Hongyan, Ju Li, Gang Bao, and Sulin Zhang. 2010. "Variable nanoparticle-cell adhesion strength regulates cellular uptake." *Physical review letters* 105 (13):138101.

Zahmatkesh, Iman, Mikhail Sheremet, Liu Yang, Saeed Zeinali Heris, Mohsen Sharifpur, Josua P Meyer, Mohammad Ghalambaz, Somchai Wongwises, Dengwei Jing, and Omid Mahian. 2021. "Effect of nanoparticle shape on the performance of thermal systems utilizing nanofluids: A critical review." *Journal of Molecular Liquids* 321:114430.

Zhao, Feng, Ying Zhao, Ying Liu, Xueling Chang, Chunying Chen, and Yuliang Zhao. 2011. "Cellular uptake, intracellular trafficking, and cytotoxicity of nanomaterials." *Small* 7 (10):1322-1337.

Section 4

Chapter 4.1

Novel Advances in 3D Printing Nanomaterials for Tissue Engineering Applications

K. Vijayalakshmi[1,*]
Prabhakaran Arunachalam[2]
and P. N. Sudha[3]

[1]Department of Chemistry (SFS), Madras Christian College, East Tambaram, Chennai, India
[2]Electrochemical Sciences Research Chair (ESRC), Chemistry Department, College of Science, King Saud University, Riyadh, Saudi Arabia
[3]Department of Physiology, Saveetha Dental College and Hospitals, Saveetha Institute of Medical and Technical Sciences (SIMATS), Saveetha University, Chennai, Tamil Nadu, India

Abstract

In recent years, 3D printing technologies also termed fast prototyping or additive manufacturing involve the production of precise geometries of complex tissue in a layer-by-layer fashion by utilizing the top-down approach with the help of anatomically accurate 3D models of the tissue generated by computer graphics. The manufacturing of 3D printed scaffolds involves microscale/nanoscale printing with nanotechnological components and nanoscale textures. This 3D printing has an advantage like rapid fabrication and high precision in the manufacture of a scaffold which have gained recognition for their use with improved extent and rate of targeted tissue regeneration with physical and chemical guidance cues. To gain explicit knowledge, this chapter deals with the discussion of numerous applications of 3D bioprinting technology on the construction of various representative tissues and organs, including skin, cardiac, bone, cartilage which opens a new era of dynamic research.

Keywords: 3D printing, scaffolds, nanomaterials, tissue repair, regeneration

* Corresponding Author's Email: kvijayalakshmi@mcc.edu.in

In: Emerging Nanomaterials in Biomedical Science
Editors: Eepsita Priyadarshini and Ajith Manayil Parambil
ISBN: 979-8-89530-600-0

1. Introduction

Tissue engineering is an interdisciplinary field that combines life sciences and engineering to develop therapies to regenerate functional tissue (Noah Laird et al. 2021). To target the growth and regeneration of new tissue, tissue engineering relies on the three pillars of cells, bioactive factors, and biomaterial scaffolds. The three-dimensional (3D) printing process, which is also known as additive manufacturing, allows the fabrication of precise and complex structures through layer-by-layer material stacking. This technology has been primarily utilized for producing personalized biomaterials for tissue engineering and regenerative medicine applications (Liwen Zhang et al. 2022). The concept of 3D printing was introduced in 1896 by a scientist named Charles W. Hull, who invented stereolithography to form solid 3D structures by printing material thinly layer by layer. Materials produced layer-by-layer possess accurate control of three-dimensional scaffold architectures with high resolution, cost-effectiveness, and scalability. 3D printing technology enables the precise tuning of biocompatible scaffold density and pore size according to tissue requirements.

As scaffolds are constructed of biodegradable and biocompatible polymers, they mimic native tissue's extracellular matrix (ECM) by exhibiting chemical, biological, and mechanical cues that influence cell phenotypes and tissues' formation (Echeverria Molina et al. 2021). Structurally complex and compositionally complex scaffolds for tissue engineering can be made using 3D printing. In the biomedical field, personalized prosthetics, implants, anatomical models, and medical devices have been created with 3D printing technology, and this technology has also been used for reconstructing artificial tissues and organs (Kim et al. 2008), such as skin, nerves, blood vessels, cartilage, and bones. In recent years, many researchers have applied functionalization methods derived from nanotechnology research to further improve these scaffolds' capacity to promote regeneration and communicate with cells.

2. Scaffolds for Tissue Engineering

A scaffold is an essential part of the 3D framework for cell interactions which can act as a carrier to carry the right cytokines and growth factors to the site of repair (Mohammadi et al. 2007, Kawaguchi et al. 2004, Khojasteh et al. 2013). These systems can either serve as cellular systems or delivery systems for drugs and cells for the regeneration of cells and tissues. Specifically, tissue engineering scaffolds act as artificial ECMs to promote cell adhesion, growth, and transport, resulting in functional constructs that replace or mend damaged tissues (Amini et al. 2012).

Engineering scaffolds have the ability to produce the desirable cellular interactions for the formation of new tissues, particularly for the regeneration of tissues and drug delivery systems. These scaffolds possess ideal characteristics such as biocompatibility, resolvability, osteoinductivity, osteoconductivity, bioactivity, and a three-dimensionally porous, highly rough surface (Qu et al. 2019) that promotes cell proliferation, attachment, and differentiation. Additionally, the scaffold must match the mechanical properties of the tissues at the site of implantation.

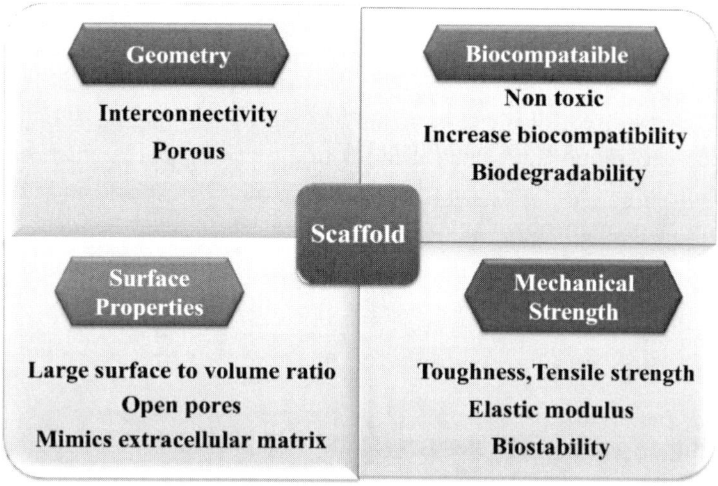

Figure 1. Ideal Characteristics of scaffold.

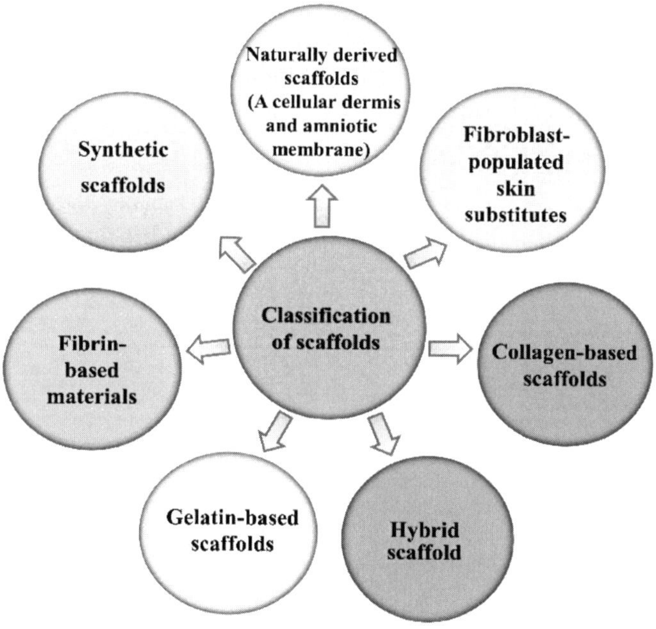

Figure 2. Classification of scaffold.

Some of the major characteristics of the biologically effective interconnective permeable scaffold were the well-defined porous three-dimensional physicochemical structure with a high surface-to-volume ratio and patient-specific geometries to promote nutrients, organize cells into 3D architecture, desired anisotropic tissue structure formation and to facilitate tissue in-growth during *in vitro* development of bone tissue or during *in vivo* implantation (Tarik Arafat et al. 2014, Trachtenberg et al. 2014). To regenerate and repair a cell, the cellular material will need to colonize the host cell sufficiently. Scaffolds can work as a framework to support cellular migration into the defect from surrounding tissues. These scaffolds have been shown to facilitate and regulate cellular processes such as synthesis, mitosis, and migration. It is possible to use scaffolds as matrixes for the adhesion of endogenous or exogenous cells, as well as delivery vehicles for genes, exogenous cells, and growth factors (Shivani Sachdeeva et al. 2022). In recent years, scaffold fabrication has played a vital role in the novel world of tissue engineering. 3D printed scaffolds can be manufactured by the incorporation of a combination of nanotechnological components, nanoscale texturing, and microscale/nanoscale printing. To assist three-dimensional tissue regeneration, scaffolds function as artificial ECM that can be used temporarily or permanently. Figure 1 and Figure 2 represent the ideal characteristics of the scaffold and its classification.

3. Materials for 3-D Print Scaffolds

Researchers in tissue engineering have created and employed a wide variety of 3D printing techniques. In 3D printing scaffolds, fabrication technology and material selection were crucial. To assist surgeons in their work and better educate patients about their problems, 3D models of the injured organ in a patient can be created (Melchels et al. 2010; Gross et al. 2014) and scaffolds and implants for regenerative medicine were produced by 3D printing techniques. An illustration of the different scaffold materials that are used in tissue engineering can be seen in Figure 3.

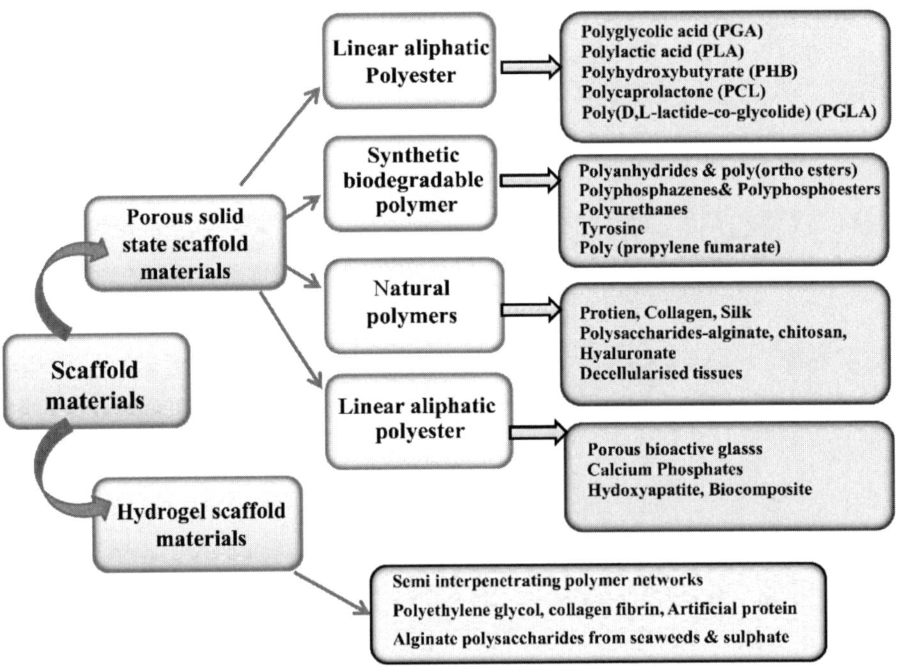

Figure 3. Various scaffold materials utilized for tissue engineering applications.

3D printing materials include stainless steel, magnesium, zinc, cobalt, tantalum, and titanium-based alloys, as well as nonmetallic materials such as bioceramics (tricalcium phosphate, calcium phosphate, and hydroxyapatite), polymers (collagen, chitosan, hyaluronic acid fibrin, sodium alginate) and composite materials (Lei Zhang et al.2019). Furthermore, a few active metal ions, such as Cu^{2+}, Mg^{2+}, and Co^{2+} can encourage vascularization in bone tissue in addition to being osteoinductive. Among the different materials evaluated, scaffolds made of chitosan, gelatin, sodium alginate, collagen, and platelet-rich plasma were selected for their potential to transport cells, growth factors, and genes. There has been a lot of interest in certain synthetic biodegradable polymers, such as polylactic acid (PLA), poly(caprolactone) (PCL), and polyglycolic acid (PGA) because ester linkages can be hydrolyzed to produce non-toxic byproducts.

4. Scaffold Fabrication Technologies

A number of traditional fabrication methods were employed to make 3D scaffolds, including particle leaching, phase separation, freeze drying, gas foaming, fiber meshes, and fiber bonding, melt molding, and solution casting (Kinoshita et al. 2013). However, these methods cannot eliminate heterogeneities in pore size, porosity, interconnectivity, and architecture, which may complicate studies assessing the effects of scaffold characteristics on developing tissues (Thimm et al. 2011).

Additionally, these methods might not be suitable for constructing a scaffold that fits a specific defect and has an architecture that mimics the complexity of native tissues. In recent years, rapid prototyping (RP), often referred to as solid-freeform fabrication (SFF) techniques was developed and this allowed for the creation of scaffolds with accurate internal morphology, exterior shape, and "reproducible" 3D architecture even in complex cases (Moroni et al. 2006). The various 3D scaffolding fabrication techniques employed in recent research are represented in Figure 4.

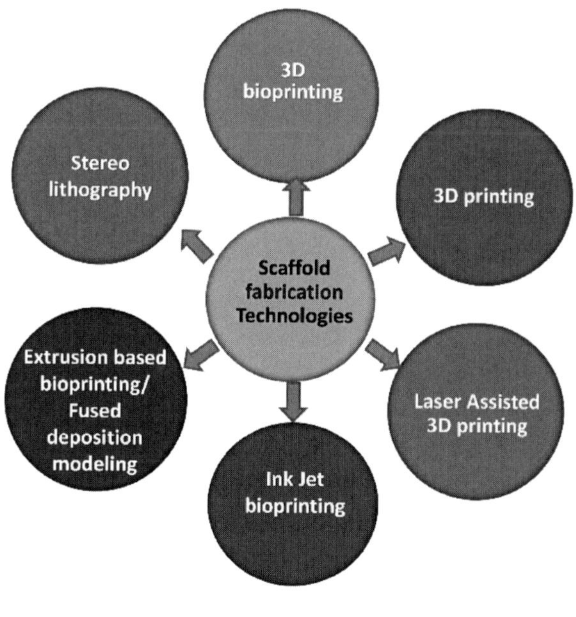

Figure 4. Various 3D scaffolding fabrication techniques.

4.1. Three-Dimensional Printing

3D printing is a relatively new method of creation of tissue engineering scaffolds with controlled architecture which was fabricated by the successive addition of consecutive 2D layers of a material (Angelika Zaszczynska et al. 2021). With the help of the segmentation software and computer-aided design (CAD), three-dimensional (3D) printing transforms two-dimensional (2D) medical pictures (such as CT, MRI, etc.) which were arranged sequentially into three-dimensional models (3D) models that are saved as digital files (such as STL, AMF) that may be printed into actual 3D structures (Ali et al. 2018, Murphy et al. 2020).

Numerous medical professionals are using 3D printing technology for implantable medical device manufacturing, surgical planning, instructional modeling, and other purposes (Ravnic et al. 2017). In the healthcare field, 3D printing technology has become a remarkable, promising technology with great progress in printing various complex bionic structures meanwhile, the combination of cell microenvironments, 3D printing technology, and biomaterials produces emerging strategies such as hybrid bioprinting which show wide application possibilities (Moor et al.2020).

4.2. Three Dimensional Bioprinting

In 3D bioprinting, structures are built layer by layer using bio-inks that mimic natural tissue composition and behavior. 3D bioprinting has excellent tissue engineering capabilities and has several applications in drug development, transplantation, and regenerative medicine. To recreate human tissues, 3D bioprinting relies on the intricacy of managing the dispersion of cells and biological components and in recent years wound healing has gained popularity with the use of 3D bioprinting bionic medical dressings (Kolesky et al. 2016). Bioprinting is already being utilized in prostheses, drug delivery and development, cancer research, and even patient and physician education (Ventola 2014, Matai et al. 2020, Mason et al. 2016). To advance the 3D bioprinting process, more cutting-edge approaches and methods have been investigated in recent months.

Romanazzo and his coworkers reported about the creation of calcium phosphate-incorporated ceramic-based bioink for bone tissue regeneration and these findings suggest that after being placed in water they were able to 3D bioprint bone-like tissue that hardens within minutes (Romanazzo et al. 2021). His novel strategy provides a fresh approach to the creation of bone-like constructions for bone tissue regeneration, with the potential to create specialized bone microenvironments for disease modeling, multiple cell delivery, and *in vivo* bone repair. In another study, Zhang and his coworkers constructed a pair of robotic arms that work together by using cement materials to create a large 3D-printed tissue model, for which they have now been awarded a patent (Zhang et al. 2019, U.S. Patent 0036337A1).

4.3. Laser-Assisted 3D Printing

Bioprinting using lasers (LAB), a game-changing technology, is a promising and emerging technology mainly used to fabricate complex bioengineered scaffolds that mimic tissues and organs (Reiza Dolendo Ventura et al. 2021). LAB reproduces tissues and organs with high accuracy, which makes it ideal for biological research, tissue engineering, and regenerative medicine. In laser-assisted bioprinting, the sacrificial layer is evaporated and propelled by the laser energy, resulting in droplets that are applied to the receiving substrate. Through a nozzle-free method, this device provides excellent resolution at the microscale. Yet, the technique is labor-intensive and makes it harder to fabricate useful 3D models because of the low overall flow rate produced by this high-resolution performance (Malda et al. 2013). When compared to other bio-printing techniques, LAB technique offers numerous key advantages and is available on Poietis' range of next-generation bioprinting (NGB) platforms. In conjunction with other complementary technologies like microvalve techniques and bioextrusion, these NGB platforms provide high degrees of versatility which is especially suitable for research (NGB-R) and clinical (NGB-C) applications.

4.4. Inkjet Bioprinting

Inkjet-based bioprinting is a non-contact printing method and bio-fabrication strategy in which the diluted solutions are distributed as individual droplets to produce the pattern of a substrate pattern with the help of thermal, piezoelectric, or microvalve technologies (Patrick Rider et al. 2018). With remarkable evolution, this technique can now produce intricate 3D structures in addition to 2-dimensional (2D) prints. Recently this technique has been utilized in the production of electronic components and has been successfully applied to a range of medicinal applications, including the creation of scaffolds, 3D tissues, and cell patterns. To create the 3D printed structures, the bioink solution was printed into the paper and when compared to scaffolds devoid of growth factors, the bioink—a cell solution supported by a growth factor—showed improved cell viability and proliferation for tissue growth. Currently, inkjet printing is frequently applied to develop nanocomposite materials in 3D structures. The advantages of inkjet printing include high resolution, high throughput, low cost, repeatability, and ease of use.

4.5. Extrusion-Based Bioprinting/Fused Deposition Modelling

An extrusion-based 3D printing technique, Fused Deposition Modeling (FDM), melts thermoplastic polymers in a semi-liquid state and then extrudes the material onto a platform to create solid geometries. In recent years, additive manufacturing (AM) based on extrusion has gained popularity for building detailed 3D prototypes and parts in layers. The extrusion method of 3D printing involves heating and extruding metal or plastic filaments through nozzles to form a 3D part over time. Pneumatic or mechanical piston- and screw-type operating modules are used in this system to discharge bioink through a nozzle. When compared to powder-based and liquid additive manufacturing, extrusion-based additive manufacturing provides more cost-effective production, higher building volume, and more design freedom. Currently, this technology utilizes ceramics, foodstuffs, metallic materials, elastomers (silicones), polymers, biomaterials, composites, smart materials, energy materials, glasses, photopolymers, cellulose, and wood. Aerospace, construction, food production, the medical, and electrical industries benefit from extrusion-based AM.

4.6. Stereolithography

Stereolithography (SLA) is a process in which the 3D scaffolds are produced from liquid polymer with the help of light-mediated chemical reactions. This method involves depositing a photosensitive (photocurable) polymer onto a surface medium and exposing it to light (UV range of 300–400 nm). Until the scaffold is entirely designed, this technique is repeated again and again, covering the first layer when it has cured. Polypropylene fumarate (PPF) containing photo--crosslinkable linkages and polyethylene glycol acrylate are two examples of biomaterials used in SLA. One major benefit of adopting SLA is that it allows the user to define and control the scaffold's defined geometries at a high resolution, very similar to the computer-aided design (Kim et al. 2010). Rats' dorsal skin was implanted with degradable crosslinked structures made by SLA using Poly (trimethylene carbonate-co-ε-caprolactone) resins, as reported by Mastuda and his coworkers (Matsuda et al. 2010). The observed results indicate that no negative effects were seen after a month. Stereolithography (SLA) is a process in which 3D scaffolds are produced from liquid polymers with light-mediated chemical reactions. This method involves depositing a photosensitive (photocurable) polymer onto a surface medium and exposing it to light (UV range 300-400 nm). Until the scaffold is entirely designed, this technique is repeated again and again, covering the first layer when it has cured. Polypropylene fumarate (PPF) containing photo-crosslinkable linkages and polyethylene glycol acrylate are two examples of biomaterials used in SLA. One major benefit of adopting SLA is that it allows the user to define and control the scaffold's defined geometries at a high resolution, very similar to computer-aided design (Kim et al.2010). The dorsal skin of rats was implanted with degradable crosslinked structures generated by SLA using polymers (trimethylene carbonate-co-caprolactone) (Matsuda et al. 2010). The observed results indicate that no negative effects were seen after a month.

5. Complex Heterogeneous 3D Tissue Construct

Combining nanotechnology with 3D printing techniques has led to several prospects and plenty of room for development across diverse application domains. The most promising advancements have been made in the fields of electronics and biomedicine. By incorporating nanoparticles, such as silver, carbon nanotubes (CNTs), and lithium compounds, into polymer, ceramic, or metallic matrixes, nanotechnology can be used in 3D printing applications. Nanocomposites are used in sensors, piezoelectric materials, electrically conductive systems such as circuits and screens, and energy storage devices such as micro-Li-ion batteries.

The metallic compounds incorporated into 3D printing nanomaterial matrices offer superior electrical and thermal conductivity. Materials such as these were used to make microcontrollers, diodes, capacitors, embedded wirings, resistors, sensors, actuators, microelectromechanical systems (MEMS), and transducers. Inkjet printing technology, for example, uses silver nanoparticle inks to boost the conductivity of screen-printing materials. Due to their semiconducting qualities, carbon nanoparticles like CNTs have also been deposited at precise quantities on a substrate for use in NO_2, humidity, and NH_3 sensing applications. Several researchers consider magnetic nanostructures produced with 3D printing to be excellent options for 3D data storage systems.

The fusion of nanomaterials with 3D bioprinting produces the optimal environment for multifunctional scaffolds (Bhatt et al.2017, Rana et al. 2017). Nanomaterials enhance the biomedical development of related fields such as tissue engineering, regenerative medicine, medical imaging, and cancer research in 3D printing technology. Improved specificity in scaffold design, as well as greater cell adhesion, material-cell interactions, and tissue formation, arise due to the incorporation of nanomaterials into three-dimensional bio-printed scaffolds.

In comparison to traditional biomaterials, nanomaterials can increase cell adhesion and proliferation due to their increased surface area to volume ratio and size range that more closely mimics natural tissues. Instead of using nanomaterials with high toxicity and low biocompatibility in 3D printing, more research should be done on natural nanomaterials like liposome nanoparticles and nano-hydroxyapatite. Bioactive glass, hydroxyapatite, and calcium phosphate are examples of nanomaterials that create a natural milieu for cell development and improve cell viability.

5.1. 3D Printed Bone

Bone is the second most transplanted tissue, and by 2027, the market is predicted to be worth $4 billion US (Grand View Research 2020, Faour et al. 2011). The two components that make up bone are the stiff cortical bone and the spongy cancellous bone which is composed of both inorganic mineral phases, such as hydroxyapatite, which gives the framework strength, and organic components, such as the protein collagen (Type I collagen makes up 97% of ECM protein). Millions of people worldwide receive bone grafts and artificial prosthesis transplants each year, resulting in the replacement or repair of native bone to address bone injuries and deformities.

New drug delivery systems based on nanoparticles can be used to promote osteo-immunomodulation and bone regeneration. Guided bone regeneration (GBR) has demonstrated the promise of nanomaterials, which have also demonstrated satisfactory mechanical properties, biocompatibility, vital barrier function, and increased osteogenesis and angiogenesis. Local use of nanoparticles in bone tissue engineering (BTE) can improve implant osseointegration, boost tissue regeneration, and stave off infections. Using gel nanofibers and gold nanoparticles (Au NPs), 3D nanocomposites can enhance both *in vitro* and *in vivo* bone regeneration (Samadian et al. 2021).

Several studies have shown that the hydroxyapatite/alginate nanocomposites display enhanced proliferation, cell adhesion, and osteogenic differentiation in comparison to pure alginate scaffolds (Chae et al. 2020, Barros et al. 2019, Benedini et al. 2020). According to Cottrill and his coworkers, osteogenesis and bone repair *in vivo* were successfully enhanced by the incorporation of non-phospholipid liposomes of stearyl amine (SA) and 20S-hydroxycholesterol into a methacrylate glycol chitosan (MeGC) hydrogel scaffold (Cotrill et al. 2020). An *in vitro* study by Calabrese et al. demonstrated that Ti scaffolds functionalized with TiO_2 and Fe_2O_3 were more effective in promoting cell proliferation and differentiation as well as improving antibacterial activity, suggesting that nano-functionalized Ti substrates could provide promising prototypes for BTE applications (Calabrese et al. 2021).

To assess their osteogenic activity on human dental pulp stem cells (hDPSCs), Xia and his coworkers created novel calcium phosphate cement scaffolds that incorporate iron oxide nanoparticles (IONP-CPC). The observed results indicate that in the seeded cells, the new CPC functionalized with IONPs significantly enhanced cell attachment, osteogenic differentiation, and bone mineralization (Xia et al. 2018). Predoi and his coworkers demonstrated that the wettability and corrosion resistance of Ti-based biomaterials have been increased by the integration of magnetic nanoparticles within coatings of hydroxyapatite and in addition this combination provides better cell viability, mineralization, and proliferation (Predoi et al. 2021). Using rGO hydrogels for BTE applications, Lu et al. found that rGO enhanced biocompatibility and osteoinductivity both *in vitro* and *in vivo* (Jiayu et al. 2013, Lu et al. 2016). Additionally, some research has shown that rGO with traditional BTE scaffolds—particularly those based on HA—improved bone development and mineralization *in vivo* and *in vitro* (Lee et al. 2015, Lee et al.2015, Liu et al. 2021).

5.2. 3D Printed Cartilage

An indispensable type of tissue in the human body is cartilage, which is composed of collagen and elastic fibers and is widely distributed in the trachea, auricles, and joints. Chondroitin sulfates, one of the polysaccharides derivatives present in cartilage complex with protein in the ground substance forming proteoglycan. Since cartilage tissue lacks blood vessels, it transfers nutrients weakly. As a result, once damaged, cartilage is difficult to repair. A new surgical technique known as matrix-associated autologous chondrocyte implantation has been presented in clinical settings as a superior way to cure cartilage abnormalities. Replacing a broken bone is the main goal of bone and cartilage tissue engineering. Scaffolds for mending tissues with a gradient composition, including tendon bone and bone cartilage tissue, can be made easily using 3D bioprinting technology. In scaffolds with a biomimetic composition and structure, subchondral bone and cartilage regeneration may be more rapid.

In order to replicate the anatomy of the trachea, Park and his coworkers created 3D-printed hollow bellows scaffolds and put functionalized sponges carrying chondrocytes within the scaffold's grooves. These tissue-engineered trachea demonstrated mechanical characteristics akin to natural trachea and were suitable for promoting cartilage regeneration *in vivo*, offering a novel tracheal replacement for the repair of circumferential tracheal defects. In addition to bone tissue regeneration, 3D cells bio-printed scaffolds have been tentatively expanded into cartilage tissue regeneration (Park et al. 2015).

Sun and his coworkers developed a scaffold loaded with bone marrow stromal cells (BMSCs) and conjugated it with growth differentiation factor 5 (GDF5) to heal articular cartilage damage. The printed BMSCs demonstrated accelerated chondrogenesis differentiation upon GDF5 addition. The scientists showed that the GDF5-conjugated BMSC-loaded scaffolds had superior results in cartilage regeneration by implanting them into rabbit knee cartilage lesions (Sun et al. 2019). Huang and his coworkers recently printed scaffolds using doped inorganic hydroxyapatite (HAP) in gelatin and then seeded MSCs for articular cartilage healing on the scaffolds. In addition to enhancing the scaffolds' rheological qualities, HAP aided in MSC chondrogenic differentiation as well as *in vivo* cartilage regeneration (Huang et al. 2021).

5.3. 3D-Printed Blood Vessel

The use of scaffold-based tissue engineering techniques helps create new blood vessels. The blood vessel serves as the body's biological blood transport pipeline, carrying blood from the heart to all of the body's tissues and organs to meet the different nutritional needs of bodily functions and returning metabolic end products, or waste, to the heart via the kidney, lung, and other organs that expel waste from the body. PCL/polydioxanone (PDO) small-diameter bilayer nanofibrous vascular structures were constructed by Emehebe and his coworkers using 3D printing, which can produce blood vessel printing at the necessary resolution (Datta et al. 2017, Emechebe et al. 2020).

With 3D printing, it is now possible to create vascular scaffolds more easily, which allows for a range of possibilities in the field of tissue vascularization. Researchers have published evaluations utilizing 3D-printed vascular tissue creation techniques (Hann et al. 2019, McCormack et al. 2020, Yu et al. 2020). Vascular regeneration is one method of treating the disorder, which involves regenerating blood vessels in injured tissues and organs to restore their structure and function (Nurden et al. 2011, Gu et al. 2017). The insertion of tissue-engineered vascular grafts, the implantation of non-biodegradable conduits, surgical bypass grafting, stent placement, and angioplasty are the current techniques for vascular regeneration (Carrabba et al. 2018). Vasculature scaffolds can be manufactured by 3D printing *in vivo* to form new blood vessel tissues.

Biomaterials like hydrogels are commonly used in vascular scaffolds and tissue development due to their biocompatibility and crosslinking properties. To achieve a specific function of vascular scaffolds, certain additive compounds such as magnetic nanoparticles and graphene (Wang et al. 2019) were utilized to form composite materials. Sodium alginate, poly (lactic-co-glycolic acid) (PLGA), and iron oxide nanoparticles were utilized by Joen and his coworkers (Jeon et al. 2018) to create surface-functionalized magnetic nanoparticles (MNPs). In this study, magnetic fields were found to enhance the intracellular delivery of surface-functionalized scaffolds. Zhu and his coworkers created a composite scaffold (MPHS) by physically doping polymer poly(3-hydroxybutyrate-co-3-hydroxyhexanoate), mesoporous bioactive glasses, and dimethylallyl glycine (DMOG). Composite scaffolds loaded with different DMOG dosages were evaluated using *in vitro* testing to assess their impact on vascularization and osteogenesis. The findings demonstrated that throughout four weeks, DMOG was regularly released from MPHS scaffolds, greatly enhancing angiogenesis and osteogenesis in the defects (Zhu et al. 2015).

To mimic cortical bone, Anada and his colleagues used a two-step 3D printing process to create an outer gel ring of octacalcium phosphate (OCP)-gelatin methacrylate (GelMA). Using human umbilical vein endothelial cells (HUVECs) spheroids and GelMA hydrogel as raw materials, they printed a central GelMA ring containing HUVECs spheroids to mimic the bone marrow space. Reported results illustrate the connection between vascularization and osteogenesis (Ananda et al. 2019).

5.4. 3D Printed Heart Valve

The heart, which is made up of several muscles that pump blood into the circulatory system, is one of the vital organs in human physiology. Scientists have investigated artificial heart valves made from PGA, PLA, collagen, and fibrin, among others. Hydrogels were considered a promising material for heart valve engineering due to their physicochemical and mechanical stability after hydration. Meshacrylated hyaluronic acid (Me-HA) and methacrylated gelatin (Me-Gel) hydrogels that are photo-crosslinkable were used by Duan and his coworkers (2014) to create heart valve conduits encapsulated with human aortic valvular interstitial cells (HAVIC). The initial creation of the collagen and glycosaminoglycan matrix was confirmed by the 3D bioprinted hydrogel conduits, which established cell survival and remodeling capability (Duan et al. 2014).

Artificial heart valve production could be completely revolutionized by the latest developments in 3D printing technology, which enable the creation of intricate, custom structures. The use of 3D bioprinting is restricted to the regeneration and repair of these problems, and it has been researched in human tissue structures, including the esophagus, urethra, and heart valves. For instance, prosthetic valve replacement is the only proven treatment for valvular heart disease (Cheung et al. 2015). To compensate for some of the disadvantages of prosthetic valve replacement, tissue engineering, and 3D bioprinting technologies can be used to construct customized functional valves based on patients' diseased valves (Vashistha et al. 2019).

Artificial polymeric heart valves with anatomically inspired properties have the potential to combine beneficial properties and advances in 3D printing have enabled us to develop silicon prototypes of novel designs quickly to learn more about their biophysical properties. TRISKELION is a prototype that resembles the architecture of the biological aortic valve with three symmetrically positioned cone-shaped cusps that serve as a core structure preserving the valve's thermodynamic and biophysical characteristics while maintaining its stability and long life.

Overall, silicone 3D printing makes it possible to quickly construct and investigate the ideal physical form of the new TRISKELION prototype, but it cannot compete with other valve options now available on the market. However, given its simplicity, the potential for low production costs, and exciting advancements in surface technology, polymers, nanomaterials, and biomaterials, it might be a worthwhile substitute (Tschorn et al. 2022).

5.5. 3D-Printed Wound Dressing

The development of skin substitutes opens new possibilities for skin regeneration and repair in tissue engineering. Among the many biomanufacturing techniques, 3D bioprinting has garnered particular attention due to its potent capacity to manipulate intricate 3D structures into functional 3D tissue scaffolds (Cubo et al. 2016). Wound healing with 3D bioprinted bionic medical dressings is a trend. To quicken the healing process, various cell types or essential macromolecules needed for wound healing were moved and released through the medical dressing's function.

A new medical-grade PCL fiber wound dressing loaded with human gingival mesenchymal stem cells created by certain researchers via 3D bioprinting process exhibits anisotropic mechanical properties. In the rat injury model, this dressing group's wound healing rate is noticeably higher than that of other groups and the full-thickness excision wound's scar formation is decreased (Shafiee et al. 2021). Skin tissue substitutes were created for substantial skin damage since simple dressings are insufficient to address patients' rehabilitation demands.

A novel technique for printing functional living skin (FLS) was created using 3D bioprinting and novel bionic bioinks. FLS stimulates skin appendages, such as hair follicles and sebaceous glands. It is a potential method for skin regeneration without scars. FLS can promote blood vessel and skin regeneration because of its physiological makeup. Its mechanical and bio-adhesive qualities also make it easy to prepare and implant into the surface of a lesion. This offers fresh hope for the future management of skin abnormalities involving large areas (Zhou et al. 2020).

5.6. Other 3D-Printed Organs

5.6.1. Lungs

The primary function of the lungs is to draw oxygen-rich air from the surrounding environment and transfer it into the bloodstream, where it is subsequently distributed throughout the body (Garcia-Mouton et al. 2019). By creating bioengineered air passageways and vascular gas exchange, tissue engineering provides physiological capabilities by attempting to replicate the whole range of specialized lung tissues. The body's defensive mechanism against infection and other dangerous environmental elements also depends on the lungs. The creation of a tissue that resembles normal lung tissue is the main objective of pulmonary tissue engineering. A research team from POSTECH has created a 3D lung model that includes four human alveolar cells: type I and II alveolar cells (NCI-H1703 and NCI-H441), lung fibroblasts (MRC5), and lung microvascular endothelial cells (HULEC-5a) utilizing ink jet bioprinting technique (Kang et al. 2021).

5.6.2. Kidney

A 3D printed kidney model is very useful when planning and supporting a partial nephrectomy for complex renal malignancies. 3D printed models designed for living donor kidney transplantation help prevent potential difficulties during the surgical procedure, lowering anxiety and boosting surgical team effectiveness. In order to depict the anatomical details and relationships between the anatomic structures, Kusaka and his coworkers constructed 3D models of kidney transplants based on the donor in two cases and a 3D model of the pelvis cavity based on the receiver in one case, all of which were created on the anatomic scale (Kusaka et al. 2015). According to Fan and his coworkers (Fan et al. 2019), when using 3D laparoscopic models during partial nephrectomies, patients achieved a reduced warm ischemic time and reduced intraoperative blood loss.

5.6.3. Liver

In the human body, the liver plays an important role in balancing biochemical conditions, as well as taking part in many different biochemical reactions. To create tissue constructs via 3D bioprinting, different cell types are encapsulated in a biomaterial solution or a mixture of biomaterials called bioink and the tissue constructs are then cross-linked or stabilized to create the final shape and structure of the planned constructions (Gungor-Ozkerim et al. 2018). To ensure correct bio-printed construct functionality, an ideal bioink with suitable mechanical and biological properties has been employed. Hydrogels comprising alginate, collagen, dECM, and gelatin, along with cells such as hepatic parenchymal cells and liver nonparenchymal cells, are frequently used in artificial liver models. In order to evaluate the organ-level response to hepatotoxicity brought on by a therapeutic medication, Nguyen et al. 3D-bioprinted liver tissue constructs utilizing nonparenchymal cells and patient-derived hepatocytes (Nguyen et al. 2016). Table 1 lists an overview of materials and nanomaterials used for 3D printed scaffolds for tissue engineering.

Table 1. Lists the overview of materials and molecules used for 3D printed scaffolds for various biomedical applications

Organs	Materials	Printer Type	Applications	References
Bone	Hydroyapatite	Selective Laser sintering (SLS)	Bone tissue engineering	Pei et al. 2017 Zhang et al. 2019
Cartilage	Polyurethane Polyethylene oxide	Fused deposition modeling (FDM)	Cartilage tissue repair	Hung et al. 2014 Hung et al. 2016
-	Silicon, alginate, silver nanoparticles	3D printing	Anti-infections Antimicrobial	Marnoor et al. 2013
Heart Valve	Collagen	3D printing	-	Lee et al. 2019
Blood vessel	Polypropylene fumarate	SLA	Vascular tissue regeneration	Melchiorri et al. 2016
Traches	Silk Fibroin	3D plotting	-	Kim et al. 2018
Liver	PCL, Collagen	Fused deposition modeling (FDM) and 3D plotting	-	Lee et al. 2016
Skin	Collagen	Ink Jet	Wound healing therapy	Koch et al. 2012
	Copper nanoparticles	Semi solid extrusion (SSE)	Antimicrobial	Guiterrez et al. 2019
	Zinc oxide nano particles	Semi solid extrusion (SSE)	Wound healing therapy	Cletus et al. 2020
Bone	Silver nanoparticles/rGO	Semi solid extrusion (SSE)	Antibacterial and osteogenic activity	Zhang et al. 2017
Bone	Polycaprolactone/silver nanoparticles		Bone Tissue engineering (BTE)	Radhakrishnan et al. 2021
Bone	3D-Printed Graphene and Graphene Quantum Dot-Reinforced Polycaprolactone Scaffolds	3D printing	Bone Tissue engineering (BTE)	Duo Meng et al. 2024

6. Future Prospects

Future research involving several sectors will examine the function mechanism of 3D-printed cell-delivery scaffolds in terms of precisely replicating intricate tissues, reconstructing organs, and even seeding cells *in vivo*. Even though 3D printing has been extensively studied for numerous biomedical applications, there is still significant space for advancement (Jakus et al. 2016, Zhu et al. 2016) in the optimization of printable inks, improved reproducibility with mass manufacturing, and standardization of printing techniques. 4D printing, a more contemporary technology that combines 3D printing with time has become more popular nowadays and is gaining a lot of interest in medical implant surgeries since the human body is a complicated environment with a variety of stimuli. With this technology, printed materials can alter their physical characteristics or functions in response to external stimuli like pH, water, temperature, or magnetism (Gao et al. 2016, Gladman et al. 2016). By enabling the production of more complex tissues and better medical applications, future developments in 3D bioprinting and related technologies hold great promise for advancement in the field of the tissue regeneration process. Specifically, certain research topics like "investigation of new bioink" "Extrusion-based bioprinting modification for cell viability and vascularization," "3D bioprinting in organoids and *in vitro* model" and "personalized and regenerative medicine research" were predicted to be hotspots for future research.

Conclusion

Integration of additive and conventional manufacturing techniques in hybrid 3D printing technologies has demonstrated encouraging advancements in printing resolution, stronger constructions, biological microenvironments that resemble their native habitats, and cell-related functions. The incorporation of nanomaterials or nanoparticles (NPs) with their unique physicochemical properties in the printed scaffolds is an emerging trend in 3D printed scaffolds which enhance the properties of printed systems, drug delivery capabilities, functionality, and drug release kinetics. In addition to improving mechanical and electrical qualities, scaffolds constructed using nanomaterials such as graphene oxide (GO), CNT, and Au NPs significantly affect the proliferation and differentiation of myogenic stem cells.

References

Ali, Anil Kumar, Lee, and Jackson. 2018. "Three-dimensional bioprinting for organ bioengineering: Promise and pitfalls." *Current Opinion in Organ Transplant* 23 (January): 649–656. 10.1097/MOT.0000000000000581.

Amini, Laurencin, Nukavarapu.2012. "Bone tissue engineering: recent advances and challenges." *Critical Reviews in biomedical engineering* 40 (September):363–408. 10.1615/critrevbiomedeng.v40.i5.10.

Anada, Pan, Stahl, Mori, Fukuda et al.2019. "Vascularized bone-mimetic hydrogel constructs by 3D bioprinting to promote osteogenesis and angiogenesis." *International Journal of molecular sciences* 20 (March):1096.10.3390/ijms20051096.

Angelika Zaszczynska, Maryla Moczulska-Heljak, Arkadiusz Gradys and Paweł Sajkiewicz.2021. "Advances in 3D Printing for Tissue Engineering." *Materials* 14 (June) :3149. https://doi.org/10.3390/ma14123149.

Barros, Ferraz, Azeredo, Fernandes, Gomes, Monteiro. 2019. "Alginate-nanohydroxyapatite hydrogel system: Optimizing the formulation for enhanced bone regeneration." *Material Science Engineering* 105 (December) :109985. https://doi.org/10.1016/j.msec.2019.109985.

Benedini, Laiuppa, Santillán, Baldini, Messina. 2020. "Antibacterial alginate/nano-hydroxyapatite composites for bone tissue engineering: Assessment of their bioactivity, biocompatibility, and antibacterial activity." *Material Science Engineering* 115 (October):11110.10.1016/j.msec.2020.111101.

Bhatt and Anbarasu. 2017. "Nanoscale Biomaterials for 3D Printing." I*OSR Journal of pharmacy and biological sciences* 12, (July): 80–86. 10.9790/3008-1203068086.

Calabrese, Franco, Petralia, Monforte, Condorelli, Squarzoni, Traina, Conoci. 2021. "Dual-Functional Nano-Functionalized Titanium Scaffolds to Inhibit Bacterial Growth and Enhance Osteointegration." *Nanomaterials* 11 (October) : 2634. 10.3390/nano11102634.

Carrabba, Madeddu. 2018. "Current strategies for the manufacture of small size tissue engineering vascular grafts." *Frontiers in Bioengineering and Biotechnology* 6, (April):41. 10.3389/fbioe.2018.00041.

Chae, Yang, Moon, Troczynski, Ko. 2020. "Biomimetically Mineralized Alginate Nanocomposite Fibers for Bone Tissue Engineering: Mechanical Properties and *in Vitro* Cellular Interactions." *ACS Applied Bio Materials* 3 (September): 6746–6755. https://doi.org/10.1021/acsabm.0c00692.

Cheung, Duan and Butcher.2015. "Current Progress in Tissue Engineering of Heart Valves: Multiscale Problems, Multiscale Solutions." *Expert Opinion on Biological Therapy* 15 (June): 1155–1172. 10.1517/14712598.2015.1051527.

Cleetus, Primo, Fregoso, Raveendran, Noveron, Spencer, Ramana and Joddar. 2020. "Alginate Hydrogels with Embedded ZnO Nanoparticles for Wound Healing Therapy." *International Journal of Nanomedicine* 15 (July): 5097-5111. 10.2147/IJN.S255937.

Cottrill, Lazzari, Pennington, Ehresman, Schilling, Dirckx, Theodore, Sciubba, Witham 2020. "Oxysterols as promising small molecules for bone tissue engineering: Systematic review." *World Journal of Orthopaedics*11 (July) : 328–344.10.5312/wjo.v11.i7.328.

Cubo, Garcia, Del Canizo et al. 2016. "3D bioprinting of functional human skin: production and *in vivo* analysis." *Biofabrication.* 9 (December) 015006. 10.1088/1758-5090/9/1/015006.

Datta, Ayan and Ozbolat. 2017. "Bioprinting for Vascular and Vascularized Tissue Biofabrication." *Acta Biomaterials* 51, (March): 1–20. https://doi.org/10.1016/j.actbio.2017.01.035.

Fan, Meng, Zhu, Ye, Li, Li, Ye, Liu, Z, Weiqin, Xie. 2019. "Three-Dimensional Printing for Laparoscopic Partial Nephrectomy in Patients with Renal Tumors." *Journal of International Medical Research* 47 (September): 4324–4332.10.1177/0300060519862058.

Faour, Dimitriou, Cousins, Giannoudis, 2011. "The use of bone graft substitutes in large cancellous voids: any specific needs?" *Injury* 42 (Suppl 2) (September): S87–S90. 10.1016/j.injury.2011.06.020.

Gao, Yang, Zhao, Jin, Ma and Xu. 2016. "4D bioprinting for biomedical applications." *Trends in Biotechnology* 34, (September):746–756. 10.1016/j.tibtech.2016.03.004.

Garcia-Mouton, Hidalgo, Cruz, Perez-Gil. 2019. "The lord of the lungs: the essential role of pulmonary surfactant upon inhalation of nanoparticles." *European Journal of Pharmaceutics and Biopharmaceutics* 144 (November):230–43. 10.1016/j.ejpb.2019.09.020.

Gladman, Matsumoto, Nuzzo, Mahadevan and Lews. "Biomimetic 4D printing." *Nature Materials* 15, (2016):413–418. 10.1038/nmat4544.

Grand View Research. Bone graft and substitutes market size, share & trends analysis report by product (allograft, synthetic), by application (dental, spinal fusion), By Region (APAC, North America), And Segment Forecasts, 2020 – 2027 (2020). htt ps://www.grandviewresearch.com/industry-analysis/bone-grafts-substitutes-market.

Gross, Erkal, Lockwood, Chen and Spence. 2014. "Evaluation of 3D printing and its potential impact on biotechnology and the chemical sciences." *Anal. Chem.* 86 (January) 3240–3253. https://doi.org/10.1021/ac403397r.

Gu, Hong, Potter et al.2017. "Mesenchymal stem cells and vascular regeneration." *Microcirculation* 24, (September):e12324. https://doi.org/10.1111/micc.12324.

Gungor-Ozkerim, Inci, Zhang, Khademhosseini, Dokmeci. 2018. "Bioinks for 3D bioprinting: An overview." *Biomaterial Science* 6, (March): 915–946. doi: https://doi.org/10.1039/C7BM00765E.

Gutierrez, Burdiles, Quero, Palma, Olate-Moya and Palza. 2019. "3D Printing of Antimicrobial Alginate/Bacterial-Cellulose Composite Hydrogels by Incorporating Copper Nanostructures" *ACS Biomaterial Science Engineering* 5 (October):6290–6299. https://doi.org/10.1021/acsbiomaterials.9b01048.

Hann, Cui, Esworthy et al.2019. "Recent advances in 3D printing: vascular network for tissue and organ regeneration." *Translational Research* 211 (September):46–63. 10.1016/j.trsl.2019.04.002.

Huang, Huang, Liang, Yuan, Bian, Duan, Rong, Xiong, Wang, Xia. 2021. "3D printed gelatin/hydroxyapatite scaffolds for stem cell chondrogenic differentiation and articular cartilage repair." *Biomaterial Science* 9 (February):2620–30. doi https://doi.org/10.1039/D0BM02103B.

Hung, Tseng and Hsu. 2014. "Synthesis and 3D printing of biodegradable polyurethane elastomer by a water-based process for cartilage tissue engineering applications." *Advanced Healthcare Materials* 3 (October):1578–1587. 10.1002/adhm.201400018.

Hung, Tseng, Dai and Hsu.2016. "Water-based polyurethane 3D printed scaffolds with controlled release function for customized cartilage tissue engineering." *Biomaterials* 83, (March):156–168. 10.1016/j.biomaterials.2016.01.019.

Jakus, Rutz and Shah.2016."Advancing the field of 3D biomaterial printing" *Biomedical Materials* 11, (January):014102. 10.1088/1748-6041/11/1/014102.

Jeon, Subbiah, Bonaedy et al. 2018. "Surface functionalized magnetic nanoparticles shift cell behavior with on/off magnetic fields." *Journal of Cell Physiology* 233 (February):1168–1178. 10.1002/jcp.25980.

Jiayu, Yu-Shi, Cheng, Wang, Qiu, Li, Zou. 2013. "Self-Supporting Graphene Hydrogel Film as an Experimental Platform to Evaluate the Potential of Graphene for Bone Regeneration." *Advanced Functional Materials* 23, (February): 3494–3502. https://doi.org/10.1002/adfm.201203637.

Kang, Park, Kim, Kim, Lee, Kim et al. 2021. "All-inkjet-printed 3D alveolar barrier model with physiologically relevant microarchitecture." *Advanced Science* 8 (March): 2004990. https://doi.org/10.1002/advs.202004990.

Kawaguchi, Hirachi, Hasegawa, Iwata, Hamaguchi, Shiba et al. 2004. "Enhancement of periodontal tissue regeneration by transplantation of bone marrow mesenchymal stem cells" *Journal of Periodontology*.75 (September):1281-7.10.1902/jop.2004.75.9.1281.

Khojasteh, Behnia, Naghdi, Esmaeelinejad, Alikhassy, Stevens. 2013. "Effects of different growth factors and carriers on bone regeneration: A systematic review." *Oral Surg Oral Med Oral Pathol Oral Radiol* 116 (December):e405-23.10.1016/j.oooo.2012.01.044.

Kim, Yeatts, Dean, Fisher. 2010. "Stereolithographic bone scaffold design parameters: osteogenic differentiation and signal expression." *Tissue Eng. Part B, Rev* 16 (October): 523-539.10.1089/ten.TEB.2010.0171.

Kim, Kim, Kim, Li, Wada, Leisk. 2008. "Bone Tissue Engineering with Premineralized Silk Scaffolds." *Bone* 42 (June): 1226–1234.10.1016/j.bone.2008.02.007.

Kim, Yeon, Lee, Chao, Lee, Seo et al.2018. "Precisely printable and biocompatible silk fibroin bioink for digital lightprocessing 3D printing." *Nature Communication* 9(April):1620. 10.1038/s41467-018-03759-y.

Kinoshita and Maeda. 2013. "Recent developments of functional scaffolds for craniomaxillofacial bone tissue engineering applications." *The Scientific World Journal* 21 ():1- 21.

Koch, Kuhn, Sorg, Gruene, Schlie, Gaebel, et al. 2010. "Laser printing of skin cells and human stem cells." *Tissue Engineering Part C Methods* 16 (September): 847–854.10.1155/2013/863157.

Kolesky, Homan, Skylar-Scott and Lewis. (2016)."Three-Dimensional Bioprinting of Thick Vascularized Tissues." *Proceedings of the National Academy of Sciences* 113 (March): 3179–3184. 10.1073/pnas.1521342113.

Kusaka, Sugimoto, Fukami, Sasaki, Takenaka, Anraku, Kenmochi, Shiroki, Hoshinaga K.2015. "Initial experience with a tailor-made simulation and navigation program using a 3-D printer model of kidney transplantation surgery." *Transplantation Proceedings* 47 (April) 596-599. 10.1016/j.transproceed.2014.12.045.

Lee, Shin, Lee, Jin, Kang, Hong, Jeong, Huh, Han. 2015. "Enhanced Osteogenesis by Reduced Graphene Oxide/Hydroxyapatite Nanocomposites." *Scientific Reports* 5, (Decemberr): 18833. https://doi.org/10.1038/srep18833.

Lee, Choi, Yong, Pati, Shim, Kang et al. 2016. "Development of a 3D cell printed construct considering angiogenesis for liver tissue engineering." *Biofabrication* 8(January):015007.10.1088/1758-5090/8/1/015007.

Lee, Hudson, Shiwarski, Tashman, Hinton and Yerneni, S. 2019. "3D bioprinting of collagen to rebuild components of the human heart." *Science* 365 (August): 482–487. 10.1126/science.aav9051.

Lee, Shin, Jin, Kang, Hwang, Park, Hong, Han. 2015. "Reduced graphene oxide-coated hydroxyapatite composites stimulate spontaneous osteogenic differentiation of human mesenchymal stem cells." *Nanoscale* 7, (May): 11642–11651.10.1039/C5NR01580D.

Lei Zhang, Guojing Yang, Blake N Johnson, Xiaofeng Jia. 2019. "Three-dimensional (3D) printed scaffold and material selection for bone repair." *Acta Biomaterialia*84 (January):16-33. 10.1016/j.actbio.2018.11.039.

Liu, Dong, Qin, Sui, Wang.2021. "Three-dimensional porous reduced graphene oxide/hydroxyapatite membrane for guided bone regeneration." *Colloids Surfaces B Biointerfaces* 208 (December): 112102.10.1016/j.colsurfb.2021.112102.

Liwen Zhang, Helen Forgham, Ao Shen, Jiafan Wang, Jiayuan Zhu, Xumin Huang, Shi-Yang Tang,Chun Xu, *de* Thomas P Davis*a* and Ruirui Qiao. 2022. "Nanomaterial integrated 3D printing for biomedical applications." *Journal of Material Chemistry B* 10 (August): 7473-7490. doi https://doi.org/10.1039/D2TB00931E.

Lu, Cheng, He, Lyu, Wang, Yu, Qiu, Zou, Li.2016. "Multilayered Graphene Hydrogel Membranes for Guided Bone Regeneration." *Advanced Materials* 28 (June): 4025–4031.10.1002/adma.201505375.

Malda, Visser, Melchels, Jüngst, Hennink, Dhert, Groll and Hutmacher. 2013. "25th anniversary article: engineering hydrogels for biofabrication." *Advanced Materials* 25 (September) 5011–5028. 10.1002/adma.201302042.

Mannoor, Jiang, James, Kong, Malatesta, Soboyejo et al. 2013. "3D printed bionic ears." *Nano Letters* 13, (May): 2634–2639. https://doi.org/10.1021/nl4007744.

Mason, Visintini, Quay. 2016. "An Overview of Clinical Applications of 3-D Printing and Bioprinting." In CADTH Issues in Emerging Health Technologies; Canadian Agency for Drugs and Technologies in Health: Ottawa, ON, Canada, (April):1–19. PMID: 31211545.

Matai, Kaur, Seyedsalehi, McClinton, Laurencin. 2020. "Progress in 3D bioprinting technology for tissue/organ regenerative engineering." *Biomaterials* 226 119536. https://doi.org/10.1016/j.biomaterials.2019.119536.

Matsuda, Mizutani. 2010. "Liquid acrylate-endcapped biodegradable poly (ϵ-caprolactone-co-trimethylene carbonate). II. Computer-aided stereolithographic microarchitectural surface photoconstructs." *Journal of Biomedical Material Research Part A* 62, (December):395–403.10.1002/jbm.10295.

McCormack, Highley, Leslie et al. 2020. "3D printing in suspension baths: keeping the promises of bioprinting afloat." *Trends in Biotechnology* 38 (June) :584–593. https://doi.org/10.1016/j.tibtech.2019.12.020.

Melchels, Feijen, and Grijpma. 2010. "A review on stereolithography and its applications in biomedical engineering." Biomaterials 31 (August) 6121–6130. https://doi.org/10.1016/j.biomaterials.2010.04.050.

Melchiorri, Hibino, Best, Yi, Lee, Kraynak et al. 2016. "3D-printed biodegradable polymeric vascular grafts." *Advanced Healthcare Materials* 5(February): 319–325. 10.1002/adhm.201500725.

Mohammadi, Shokrgozar, Mofid. 2007. "Culture of human gingival fibroblasts on a biodegradable scaffold and evaluation of its effect on attached gingiva: A randomized, controlled pilot study." *Journal of Periodontology* 78 (October):1897-903.10.1902/jop.2007.070083.

Moor, Fernandez, Vercruysse, Tytgat, Asadian, De Geyter. 2020. "Hybrid Bioprinting of Chondrogenically Induced Human Mesenchymal Stem Cell Spheroids." *Frontiers in Bioengineering and Biotechnology* 8 (May): 484. doi.org/10.3389/fbioe.2020.00484.

Moroni, DeWijn and Van Blitterswijk.2006. "3D fiberdeposited scaffolds for tissue engineering: influence of pores geometry and architecture on dynamic mechanical properties." *Biomaterials* 27 (March):974–985. https://doi.org/10.1016/j.biomaterials.2005.07.023.

Murphy, De Coppi, Atala. 2020. "Opportunities and challenges of translational 3D bioprinting." *Nature Biomedical Engineering* 4 (April): 370–380.10.1038/s41551-019-0471-7.

Nguyen, Funk, Robbins, Crogan-Grundy, Presnell, Singer, Roth. 2016. "Bioprinted 3D Primary Liver Tissues Allow Assessment of Organ-Level Response to Clinical Drug Induced Toxicity *In Vitro*." *PLoS ONE* 11 (July) e0158674. https://doi.org/10.1371/journal.pone.0158674.

Noah Z Laird, Timothy M Acri, Jaidev Chakka, Juliana Quarterman, Walla Malkawi, Satheesh Elangovan, Aliasger K Salem.2021. "Applications of nanotechnology in 3D printed tissue engineering scaffolds." *European Journal of Pharmaceutics and Biopharmaceutics* 161, (April):15-28. https://doi.org/10.1016/j.ejpb.2021.01.018.

Nurden. 2011. "Platelets, inflammation and tissue regeneration." *Thromb Haemost* 105(Suppl 1) (May):S13–S33. 10.1160/THS10-11-0720.

Park, Hong, Ju, Jung, Kang, Lee, Cho. 2015. A novel tissue-engineered trachea with a mechanical behavior similar to native trachea." *Biomaterials* 62 (September):106–115. 10.1016/j.biomaterials.2015.05.008.

Patrick Rider, Zeljka Peric Kacarevic, Said Alkildani, Sujith Retnasingh and Mike Barbeck. 2018. "Bioprinting of tissue engineering scaffolds." *Journal of Tissue Engineering* 9 (October): 1–16. 10.1177/2041731418802090.

Pei, Ma, Zhang, Sun, Sun, Fan et al. 2017. "Creating hierarchical porosity hydroxyapatite scaffolds with osteoinduction by threedimensional printing and microwave sintering." *Biofabrication* 9(November): 045008. 10.1088/1758-5090/aa90ed.

Predoi, Iconaru, Ciobanu, Predoi, Buton, Megier, Beuran.2021. "Development of Iron-Doped Hydroxyapatite Coatings." *Coatings* 11 (February):186. https://doi.org/10.3390/coatings11020186.

Qu, Fu, Han, Sun. 2019. "Biomaterials for bone tissue engineering scaffolds: A review." *RSC Advances* 9 (August): 26252–26262. doi:https://doi.org/10.1039/C9RA05214C.

Radhakrishnan, Nagarajan, Belaid, Farha, Iatsunskyi, Coy, Soussan, Huon, Bares, Belkacemi, Teyssier, Balme, Miele, Cornu, Kalkura, Cavailles, Bechelany. 2021. "Fabrication of 3D printed antimicrobial polycaprolactone scaffolds for tissue engineering applications." *Material Science Engineering C* 118(January): 111525. https://doi.org/10.1016/j.msec.2020.111525.

Rana, Kumar and Ramalingam. 2017. "Impact of Nanotechnology on 3D Bioprinting." *Journal of bionanoscienes* 11 (February): 1–6. doi: 10.1166/jbns.2017.1417.

Ravnic, Leberfinger, Koduru, Hospodiuk, Moncal, Datta, Dey, Rizk, Ozbolat. 2017. "Transplantation of Bioprinted Tissues and Organs: Technical and Clinical Challenges and Future Perspectives." *Annals of Surgery* 266 (July):48–58. 10.1097/SLA.0000000000002141.

Reiza Dolendo Ventura.2021. "An Overview of Laser-assisted Bioprinting (LAB) in Tissue Engineering Applications." *Medical Lasers* 10 (June): 76-81. https://doi.org/10.25289/ML.2021.10.2.76.

Romanazzo, Molley, Nemec, Lin, Sheikh, Gooding, Wan, Li, Kilian, Roohani. 2021. "Synthetic Bone-Like Structures Through Omnidirectional Ceramic Bioprinting in Cell Suspensions." *Advanced Functional Materials* 31 (January):2008216. https://doi.org/10.1002/adfm.202008216.

Samadian, Khastar, Ehterami, Salehi. 2021. "Bioengineered 3D Nanocomposite Based on Gold Nanoparticles and Gelatin Nanofibers for Bone Regeneration: *In Vitro* and *in Vivo* Study." *Scientific Reports* 11 (July):13877. https://doi.org/10.1038/s41598-021-93367-6.

Shafiee, Cavalcanti, Saidy et al. 2021."Convergence of 3D printed biomimetic wound dressings and adult stem cell therapy." *Biomaterials* 268 (January) 120558. https://doi.org/10.1016/j.biomaterials.2020.120558.

Shivani Sachdevaa, Amit Mania, Hiral Voraa, Harish Salujab, Nishant Mankac and Vipin Dehaned. 2022."Scaffolds in tissue engineering."*Journal of Cellular Biotechnology* 8 (September): 95–103.10. 3233/JCB-220070.

Sun, You, Jiang, Zhai, Dai. 2019. "3D-bioprinting a genetically inspired cartilage scaffold with GDF5-conjugated BMSC-laden hydrogel and polymer for cartilage repair." *Theranostics* 9 (September):6949-6961. 10.7150/thno.38061.

Tarik Arafat, Gibson, Li. 2014.State of the art and future direction of additive manufactured scaffolds-based bone tissue engineering." *Rapid Prototyping Journal* 20, (January):13–26. 10.1108/RPJ-03-2012-0023.

Thimm, Wust, Hofmann, Hagenm¨uller and R. Muller. 2011. "Initial cell pre-cultivation can maximize ECM mineralization by human mesenchymal stem cells on silk fibroin scaffolds." *Acta Biomaterialia* 7 (May): 2218–2228.10.1016/j.actbio.2011.02.004.

Trachtenberg, Mountziaris, Miller, Wettergreen, Kasper, Mikos. 2014."Open-source three-dimensional printing of biodegradable polymer scaffolds for tissue engineering." *Journal of Biomedical Material Research Part A* 102 (December):4326–433.10.1002/jbm.a.35108.

Tschorn, Schröter, Hartrumpf, Kühnel, Ostovar, Albes. 2022. "Engineering a New Polymeric Heart Valve Using 3D Printing—TRISKELION." Medicina 58, (November): 1695. 10.3390/medicina58111695.

Vashistha, Kumar, Dangi, Sharma, Chhabra and Shukla. 2019. "Quest for Cardiovascular Interventions: Precise Modeling and 3D Printing of Heart Valves." *Journal of biological engineering* 13, (February): 12. https://doi.org/10.1186/s13036-018-0132-5.

Ventola. 2014 "Medical Applications for 3D Printing: Current and Projected Uses." *Pharmacology and therapeeutics* 39, (October):704–711. PMID: 25336867, PMCID: PMC4189697.

Wang, Kang, Sun, Li, Cao, Jin. 2017."Mapping porous microstructures to yield desired mechanical properties for application in 3D printed bone scaffolds and orthopaedic implants." *Materials and Design* 135 (July): 62–68. 10.1016/j.matdes.2017.07.021.

Wang, Song, Zhang et al. 2019. "Silk-graphene hybrid hydrogels with multiple cues to induce nerve cell behavior." *ACS Biomaterial Science Engineering* 5, (February):613–622.10.1021/acsbiomaterials. 8b01481.

Xia, Chen, Zhang, Wang, Chen, Reynolds, Ma, Schneider, Gu, Xu. 2018. "Injectable calciumphosphate scaffold with iron oxide nanoparticles to enhance osteogenesis via dental pulp stem cells." *Artificial Cells Nanomedicine and Biotechnology* 46 (January): 423–433. 10.1080/21691401.2018.1428813.

Yu, Schimelman, Wang et al. 2020. "Photopolymerizable biomaterials and light-based 3D printing strategies for biomedical applications." *Chemical Reviews* 120, (April):10695–10743. https://doi.org/10.1021/acs.chemrev.9b00810.

Zhang, Zhai, Xu, Yao, Zhu, Chang and Wu. 2017. "3D-printed bioceramic scaffolds with antibacterial and osteogenic activity." *Biofabrication* 9 (June): 025037 10.1088/1758-5090/aa6ed6.

Zhang, Boca, Newkirk, Fuhlbrigge, Zhang, Li. 2019."System for Robotic 3D Printing." U.S. Patent 2019/0036337A1 (31st January) Zhang, Sun, Wu, Ma, Xing, Xing et al. 2019."3D printing of calcium phosphate bioceramic with tailored biodegradation rate for skull bone tissue reconstruction." *Biodesign and Manufacturig* 2, (August): 161–171.10.1007/s42242-019-00046-7.

Zhou, Li, Pei, Yang, Gu et al. 2020. "Dopamine-modified hyaluronic acid hydrogel adhesives with fast-forming and high tissue adhesion." *ACS Applied Material Interfaces*12 (March) 18225-18234.10.1021/acsami.9b22120.

Zhu, Ma, Gou, Mei, Zhang and Chen. 2016."3D printing of functional biomaterials for tissue engineering." *Current Opinion in Biotechnology* 40,(August):103–112.10.1016/j.copbio.2016.03.014.

Zhu, Zhao, Xin, Zhu, Zhang. 2015. "3D-printed dimethyloxallyl glycine delivery scaffolds to improve angiogenesis and osteogenesis." *Biomaterial Science* 3, (June): 1236–1244.https://doi.org/10.1039/c5bm00132c.

Chapter 4.2

The Use of Carbon Dots, Graphene, and Silica-Based Nanomaterial in Disease Diagnosis and Therapeutics

P. R. Kavya*
Samayeta Pramanik*
Mobasshara Uzma
Upama Majumder
and Debasish Kar†
Department of Biotechnology, Faculty of Life of Allied Health Sciences,
Ramaiah University of Applied Sciences, Bangalore, India

Abstract

Various fields of medicine and therapy have been invested in using nanoproducts and biosensors. The diagnosis in the early stage plays a critical role in their enhanced therapy and recovery. The problems of drug resistance, incidence, and relapse are increased when treatment is given with conventional therapy. Nanomaterials have seen significant growth in various fields, including biomedical applications. One of the areas of interest for therapeutic interventions is the interaction between nanomaterials and immune system components. This chapter classifies and characterizes different carbon-based nanomaterials (CNP) types like carbon dots, graphene & silica, enhancing efficiency in biosensors to diagnose diseases. The biocompatibility of silica in silicon nanomaterials has proved to have no toxigenicity, an important aspect of diagnosis. However, it shows great efficacy in targeted drug delivery and bio-imaging sensors, which are the major components of the diagnosis. Their extensive properties include fluorescence, effective cell membrane penetration, and therapeutic enzyme transport. Clinical translational challenges in the field are also discussed, and recent developments and future perspectives are explored.

Keywords: carbon dot, silica, graphene-based, immunotherapy, therapeutic enzymes

---

* Contributed equally to this work.
† Corresponding Author's Email: debasish.bios@gmail.com

In: Emerging Nanomaterials in Biomedical Science
Editors: Eepsita Priyadarshini and Ajith Manayil Parambil
ISBN: 979-8-89530-600-0

1. Introduction

After more than 20 years of basic nanoscience research, nanotechnology applications are fulfilling the promise that nanotechnology will benefit society in both expected and unexpected ways. Nanotechnology is helping to significantly improve, even revolutionize, many areas of technology and industry: information technology, homeland security, medicine, transportation, energy, food safety, and environmental sciences.

Nanomaterials are important in diagnosing and treating diseases due to their unique properties, such as large surface area, tunable size, and versatility. They increase the sensitivity and specificity of diagnostic techniques, allowing early and accurate disease detection. Nanomaterials facilitate targeted drug delivery, reduce side effects, and improve treatment efficacy. Their ability to penetrate biological barriers and interact at the molecular level makes them ideal for advanced medical applications such as imaging, biosensing, and cancer therapy (Faouzi et al. 2021). Nanomaterials are an important advance in personalized medicine and treating complex diseases.

2. Types of Carbon-Based Nanomaterials

2.1. Carbon Dots

2.1.1. Structure and Properties

Carbon dots (CDs) are new carbon-based nanomaterials extracted by isolating and purifying single-walled carbon nanotubes. CDs were randomly found, indicating that the carbon size was less than 10 nm. In the last decade, biological and biomedical applications of CDs have become more widespread due to their small size (110 nm), unique optical properties, ease of use, and bioactivity (Zhang et al. 2019) Because of their small size, they can cross several natural biological barriers *in vivo*, including ion channels, the blood-brain barrier (BBB), and glomerular filtration. CDs have been successfully grafted onto various silica nanostructures using different synthesis methods (e.g., gel chemistry, inverse microemulsion, surfactant modeling, and molecular imprinting technology (MIT)), which produce hybrid nanostructures with multimodal properties for various purposes (Figure 1). CDs have excellent photoluminescence, biocompatibility, and easy surface functionalization. Depending on the type of precursor, various functional groups such as -OH, -COOH, -CHO, -NH2, and -SH can be applied to the surface of the CDs during the preparation process. CDs with oxygen-containing functional groups tend to be negatively charged, while CDs with N-containing functional groups are positively charged (Mozhgani et al. 2020). The difference in charge gives the CDs different passivation properties. The carboxyl group of the surface promotes its water solubility, while other chemical groups facilitate surface functionalization and passivation. The number of amino groups is usually determined by ninhydrin colorimetry, where lysine is used as the standard solution, and a standard curve is created using the spectrum (Zhang et al. 2019). The resulting solution is measured with an ultraviolet spectrometer at a wavelength of 560 nm. The Boehm titration can determine the concentration of the carboxyl group by combining the C points with NaHCO3 solution and then titrating the mixed solution with HCl as the endpoint pH.

2.1.2. Applications

CDs are highly effective in visualizing and studying biological structures and processes with high sensitivity and resolution.

2.1.2.1. Fluorescence Imaging

DCDs absorb light at a specific wavelength (excitation) and emits light at a longer wavelength (emission), producing a bright fluorescent signal. The fluorescence properties of CDs can be tuned by modifying their size, surface functional groups, and composition, allowing for multicolor imaging. CDs stain and visualize cells and organelles, providing detailed images of cell morphology and function (Zhang et al. 2019). Due to their low toxicity and high photostability, CDs are suitable for real-time imaging of live cells, enabling the study of dynamic biological processes. CDs can penetrate tissues, making them useful for imaging complex tissue structures and detecting abnormalities at the tissue level.

2.1.2.2. Photoluminescence Imaging

When CDs are excited by a light source, they emit light through photoluminescence, similar to fluorescence but involving different mechanisms and timescales. CDs often have high quantum yields, meaning many absorbed photons are re-emitted as light, enhancing their brightness. CDs are used for non-invasive imaging of live animals, enabling the study of biological processes in real-time within a living organism (Zhang et al. 2019). CDs can be conjugated with specific biomolecules to target and illuminate cells, tissues, or disease markers, facilitating targeted imaging and diagnostics. CDs serve as contrast agents in various imaging modalities, improving the visibility and differentiation of structures within biological samples.

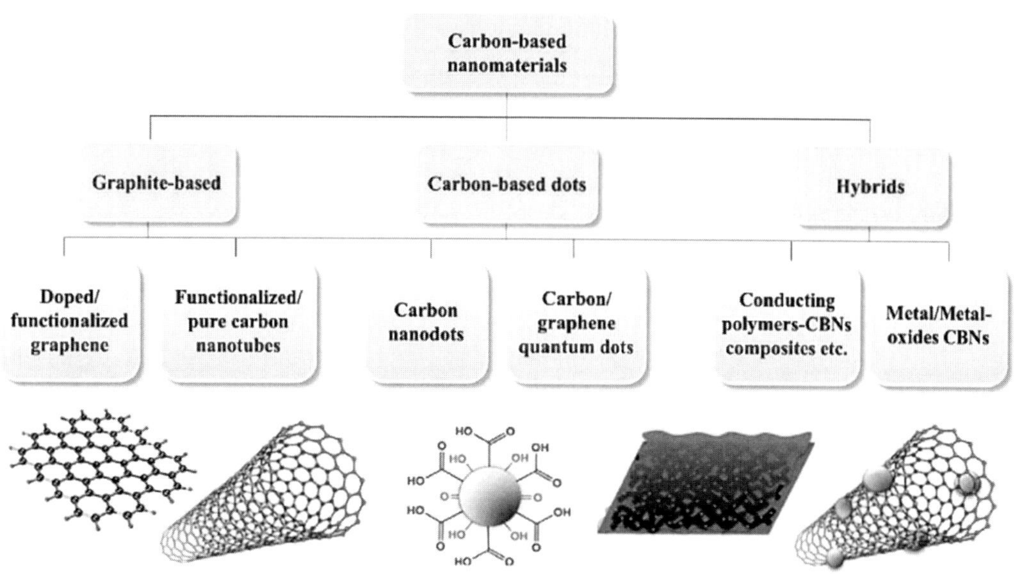

Figure 1. Types of Carbon-based nanomaterials. CBN that have been and have the potential to be used for biofueled fuel cells. Hybridization states of carbon-based nanomaterials. Many chemical and electronic properties of carbonaceous nanomaterials are determined by the dominant hybridization state of the carbon-carbon bonds. (Reproduced with permission from Sugano, 2011, Copyright 2011, RSC).

2.2. Graphene and Graphene-Based Nanomaterials

2.2.1. Structure and Properties

Graphene is a monatomic two-dimensional carbon sheet with hexagonally arranged sp2-hybridized carbon atoms. The graphene family varies in wall number (single/multiwall), surface chemistry, size (length and diameter), number of layers, sheet quality, and purity. The different types include monolayer graphene (one layer thick), multilayer graphene, ultrathin graphite (more than i3-5 nm thick but less than 100 nm), GO (- atom-thick carbon sheets with carboxylate groups), and reduced GO (Figure. 1). These new nanomaterials have unique properties such as high surface area, versatile surface, better biocompatibility, photothermal activity, pH sensitivity, etc. They can also be functionalized with various polymers through covalent or non-covalent bonds to minimize toxicity and develop specific/selective interactions (Zhang et al. 2019). Graphene and its nanocomposites have recently been widely used in biomedicine in bioimaging, biosensing, cancer therapy, drug/gene delivery, and tissue engineering.

2.2.2. Applications

Nanocarriers: Graphene-based nanomaterials can be functionalized to carry and release drugs at targeted sites, improving the efficacy and reducing side effects of treatments.

Fluorescence Imaging: Graphene quantum dots are used for bioimaging due to their strong and tunable fluorescence properties, aiding in cellular and molecular imaging.

Scaffolds: Graphene-enhanced scaffolds promote cell growth and differentiation, making them useful in tissue engineering and regenerative medicine.

Coatings and Materials: Graphene's antimicrobial properties are utilized in coatings and materials to prevent bacterial growth and infections.

2.3. Silica-Based Nanomaterials

2.3.1. Structure and Properties

Silica-based nanomaterials leverage silica's unique properties (silicon dioxide, SiO_2) at the nanoscale. Due to their versatility, stability, and biocompatibility, these materials have gained significant attention in various scientific and industrial fields. This introduction provides an overview of silica-based nanomaterials (Table 1), highlighting their fundamental characteristics, synthesis methods, and broad applications (Zhang et al. 2019). Silica is generally non-toxic and biocompatible, making it suitable for biomedical applications, including drug delivery and tissue engineering. These materials exhibit excellent thermal stability, allowing their use in high-temperature applications. Silica is chemically inert in many environments, providing durability and longevity in various applications.

2.3.2. Types and Application

In Table 1, we have summarized the various types and applications of silica-based nanomaterials (Zhang et al. 2019).

Table 1. Summary of different types and applications of silica-based nanomaterials

Types	Structure	Application
Mesoporous Silica	Array of pores between 2-50 nm	Used in drug delivery, catalysis, and as adsorbents due to high surface area and tunable pore sizes
Silica Nanoshells	Consists of a dielectric silica core surrounded by a thin metal shell	Used in bioimaging, photothermal therapy
Silica Nanoparticles	Solid or hollow spherical particles with sizes typically ranging from a few nm to several hundred nm	Employed in drug delivery, carriers for catalysts, cosmetics, and various industry
Silica Aerogels	Highly porous and lightweight with a large internal surface area	Used in thermal insulation and environmental remediation

3. Common Diseases

In this part of the chapter, we will focus on the diagnosis and therapeutic significance of carbon dots and graphene-based nanomaterials in common diseases like AIDS and cancer.

3.1. Acquired Immune Deficiency Syndrome (AIDS)

Unlike rare and infectious diseases, common diseases are more familiar among mankind and occur frequently. One such disease is Acquired Immune Deficiency Syndrome (AIDS) caused by Human Immunodeficiency Virus (HIV). HIV is a retrovirus with a high affinity to CD4 receptors of T cells, blocking the activity of T helper cells and causing immune deficiency (Mozhgani & Hamidreza. 2017).

The growing field of nanotechnology has many classes of nanomaterials, which have helped overcome the problem of designing sensors and enhancing their sensitivity. A few such techniques will be discussed below.

3.1.1. Electrochemical Impedance Spectroscopy (EIS)

Electrochemical impedance Spectroscopy is a sensitive technique that applies a periodic, low-amplitude AC pulse to measure the impedance within a suitable frequency range. Hu et al. demonstrated a diagnosis of the HIV-1 gene using graphene by the EIS method (Hu et al. 2011). In this method, 10-perylene tetracarboxylic acid was employed on the graphene along with NH2-IL gold nanoparticles, resulting in strong electrostatic interaction with single-stranded DNA. The single-stranded DNA was then hybridized with double-stranded DNA containing more negative charge, and the electrochemical impedance value was noted. The graphene platform containing PTCA and Au-IL particles was tested for selectivity towards DNA hybridization. This was compared using different mismatched DNA oligonucleotides, and the hybridization of 20-base oligonucleotide probes with the conserved sequence of the HIV-1 pol gene as the target. The ΔRet values for hybridization with cDNA were determined. Gong et al. designed an impedimetric DNA sensor using an electrode modified with a graphene-Nafion composite to detect the HIV-1 gene (Gong et al. 2017). The single-stranded DNA probes through π–π stacking were adsorbed onto the graphene surface. The probes hybridized with them in the presence of HIV gene oligonucleotides and constituted double-stranded DNA, which led to their exclusion from the graphene Nafion/glassy carbon electrode (GCE) surface.

This conformational transition of DNA caused changes in the interfacial charge monitored by EIS in a specific frequency.

3.1.2. Carbon Dots in Immunoassay

A highly specific microchip and microwell-plate-based immunoassay using fluorescent carbon dots was developed to detect HIV-1 p24 antigen, a protein biomarker (Chunduri et al. 2016). The carbon dots with high quantum yield are synthesized using ethylenediamine as nitrogen and citric acid as a carbon source by hydrothermal reaction in a single step. Then, using an amine-amine coupling reaction, the anime groups confirmed by FTIR on the carbon dots were coupled to streptavidin. The carbon dot-based immunoassay (CDIA) was discovered to have a linear dose-dependent detection range of 20–1000 pg/mL. When HIV-1 p24 antigen was detected by CDIA testing for HIV-negative plasma samples, no false positive results were observed. A microfluidic carbon dot immunoassay (μCDIA) was developed using the CDIA and demonstrated analytical sensitivity within the 30-1000 pg/mL range. It is relatively easy to adapt the CDIA and μCDIA to a lab-on-a-chip platform for usage in resource-constrained environments. It is also possible to multiplex the tests to detect other infections, such as TB and hepatitis.

3.2. Cancer

One of the main health issues affecting people worldwide is cancer, which has a high incidence and fatality rate (Montagnana M. & Lippi G., 2017). Cancer is a blatantly avoidable illness, just like many other chronic human disorders. The new era of cancer research may see the development of diagnostic techniques and chemotherapy delivery methods based on nanotechnologies, such as nanoparticles (NPs).

Nanomaterials have emerged as a powerful tool in cancer diagnosis due to their unique properties and functionalities (Nocito et al. 2021). The diverse physical-chemical characteristics of carbon-based nanomaterials, such as Carbon Dots (CDs), Quantum dots, Gold nanoparticles, Iron-oxide nanomaterials, etc. make these systems attractive for applications such as biosensing, bioimaging, and nano-carriers for drug delivery systems (Figure. 2). Not every type of cancer or every portion of the malignant tissue can be detected by every nanoparticle. The unique characteristics of nanoparticles and their ability to be designed to target certain cancer types and biomarkers determine how efficient they are in detecting cancer. On the other hand, various nanoparticles can be created or altered to target certain cancerous processes.

3.2.1. Graphene-Based Nanoparticles

Even with enormous efforts to develop novel treatment approaches and support screening initiatives to boost early detection, breast cancer remains one of the main causes that lead to women's mortality globally (Sitia et al. 2022). Recent studies have shown that these nanomaterials have played a major role in therapeutic and diagnostic aspects. Early diagnosis of breast cancer includes high levels of Human epidermal growth factor receptor-2 (HER-2), which requires the loading of anti-cancerous drugs onto graphene-based nanomaterials (Sitia et al. 2022).

Graphene's carbon structure is thought to be the reason behind its reported toxicity to cancer cells, while normal cells are predicted to be either hypoxic or harmless to graphene. To inhibit cancer, medicines, nucleic acid molecules, or genes were delivered to the cancer cells using materials based on graphene. Moreover, graphene-based materials can be employed in photothermal and photodynamic therapies since they can absorb light near the infrared spectrum. Graphene-based materials are expected to be used in batteries, supercapacitors, separation and electrocatalysis, anti-cancer medicine research, and other fields.

3.2.2. Carbon Dots (CDs) as Nanocarrier in Anticancer Drug Delivery

With a particle size of less than 10 nm, carbon dots (or CDs) are carbon-based nanomaterials with various uses, including innovative medication delivery in treating cancer, ophthalmic illnesses, and brain problems.

The World Health Organization predicts that by 2030, the number of deaths from cancer will reach approximately 13.1 million (Nocito et al. 2021). Our knowledge of CDs' physicochemical characteristics, diagnostic applications, and drug delivery features has greatly increased. As a result, we can now formulate and design CD-based theragnostic systems for treating a wide range of illnesses (Figure 2). It is possible to build CDs to identify and bind with biomarkers that are suggestive of malignancy. CDs can function as highly effective biosensors for timely cancer detection by identifying these biomarkers. The ability of CDs to transform light energy into heat makes them useful for photothermal therapy (Nocito et al. 2021). This method involves injecting CDs into tumors and shining light on them. The heat produced by the absorbed radiation of light kills cancer cells only, preserving healthy tissues. Photoacoustic imaging is a method that combines the tissue penetration ability of light with the precise spatial resolution of ultrasound. This method enables easy detection of deep-seated cancers by using CDs to absorb laser light and transform it into sonic waves.

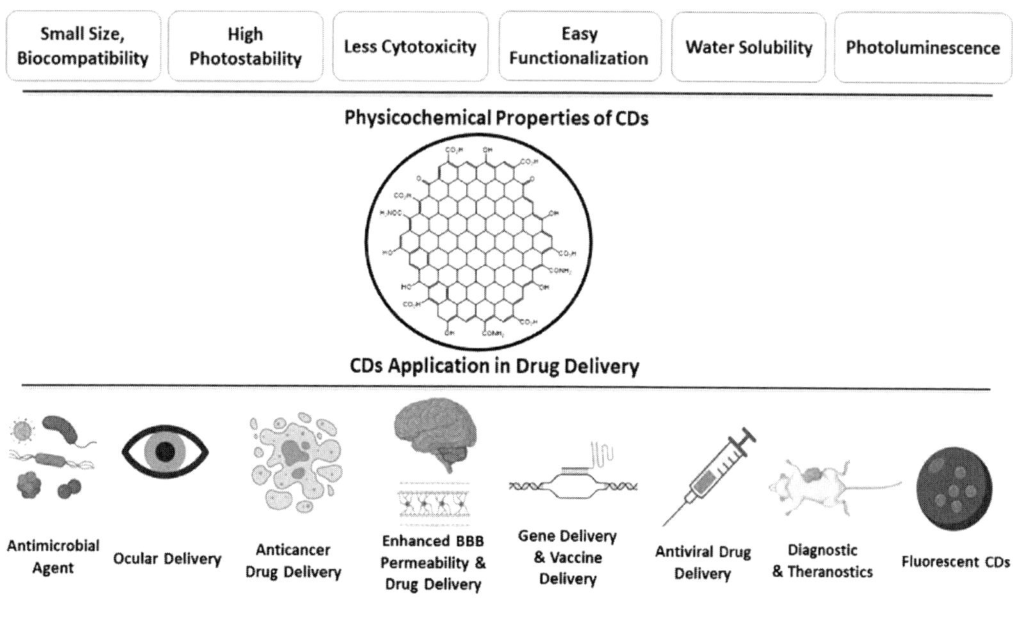

Figure 2. Carbon Dots (CDs) in drug delivery system. It shows various physiochemical properties and their application in drug delivery systems. It ranges from the micro molecular level to the high compatibility system. (Reproduced with permission from Kaurav et al. Copyright 2023).

4. Infectious Diseases

Infectious diseases are disorders caused by pathogenic microorganisms, including bacteria, viruses, fungi, and parasites. These pathogens can be transmitted directly or indirectly by various modes of transmission that depend on the microorganism, including skin contact, through vectors, or contaminated water or food. These disorders vary from mild to life-threatening. In this part, we will see how carbon dots, graphene, and silica-based nanoparticles are used for diagnosis and possible therapeutics for some of the most common infectious diseases.

4.1. SARS-COV-2

Severe Acute Respiratory Syndrome Coronavirus-2 is a variant of the family Corona viridae virus responsible for causing the COVID-19 pandemic. SARS-COV-2 is a positive sense single-stranded RNA virus that spreads through respiratory droplets produced from the affected or close contact with the patient or by skin contact with contaminated surfaces.

4.1.1. Use of Carbon Dots as Biosensors

Nanoparticle-based techniques can be orchestrated to manipulate the anti-virus system in the patient, as carbon-based nanomaterials show the most antiviral activity. This nanomaterial usage is critical in the targeted drug delivery system against SARS-COV-2 as the carbon-based nanomaterials do not trigger the immune system of the epithelial cells of the virus (Ahmadi et al. 2022). Carbon-based nanomaterials, including graphene and CDs, are used in bio-sensing and drug-delivery systems. These can be synthesized in top-down and bottom-up approaches.

Nanotechnology-based biosensors are used worldwide to detect the presence of microorganisms, which work by amplifying a small amount of signal based on transduction. CDs and graphene are widely used to design sensitive nanosensors, nanocarriers, and next-generation vaccines. Biosensors like GOs, CNTs, Rgos, and Graphene are applied to detect SARS-COV-2 by DNP/GO/AuNP (Ahmadi et al. 2022). Researchers have devised an optical nanosensor for the identification of SARS-CoV-2. This innovation involves enhancing the capacity of host proteins to bind with virion components, coupled with carbon nanotubes (CNTs) (Figure. 3) to deliver fluorescent readings indicative of protein activity. In conjunction with a biosensor based on single-walled carbon nanotubes (SWCNTs), a protein has been utilized (Figure. 3). This biosensor can interface with the spike protein's receptor, facilitating the detection process. The ACE2 proteins on the virus's surface were immobilized through a modification process, enabling the retention of fluorescence from the SWCNT. This was made possible by the strong binding affinity of the biosensor to the virus spike protein. This nanosensor demonstrates reduced sensitivity to the viral load of COVID-19, as evidenced by clinical trials. This characteristic could facilitate the development of rapid and point-to-point tests for detecting SARS-CoV. Quantum dots (QDs) are remarkable fluorescent materials known for their photochemical properties. QDs and CDs are utilized to label pathogens owing to their photoluminescence characteristics. Quantum dot nanobeads (QBs) offer the potential application in immunochromatographic assays (ICAs) as a point-of-care testing (POCT) method, featuring numerous advantages, including rapid processing speeds, which could prove advantageous in the detection of coronavirus. A fluorescent lateral flow immunoassay (LFIA) system employing QBs conjugated with S proteins exhibits remarkable sensitivity enhancement (Ahmadi et al. 2022). This assay illustrates that combining IgG/IgM antibodies with SiO2, Au, and QDs enables sensitive recognition of COVID-19 strains.

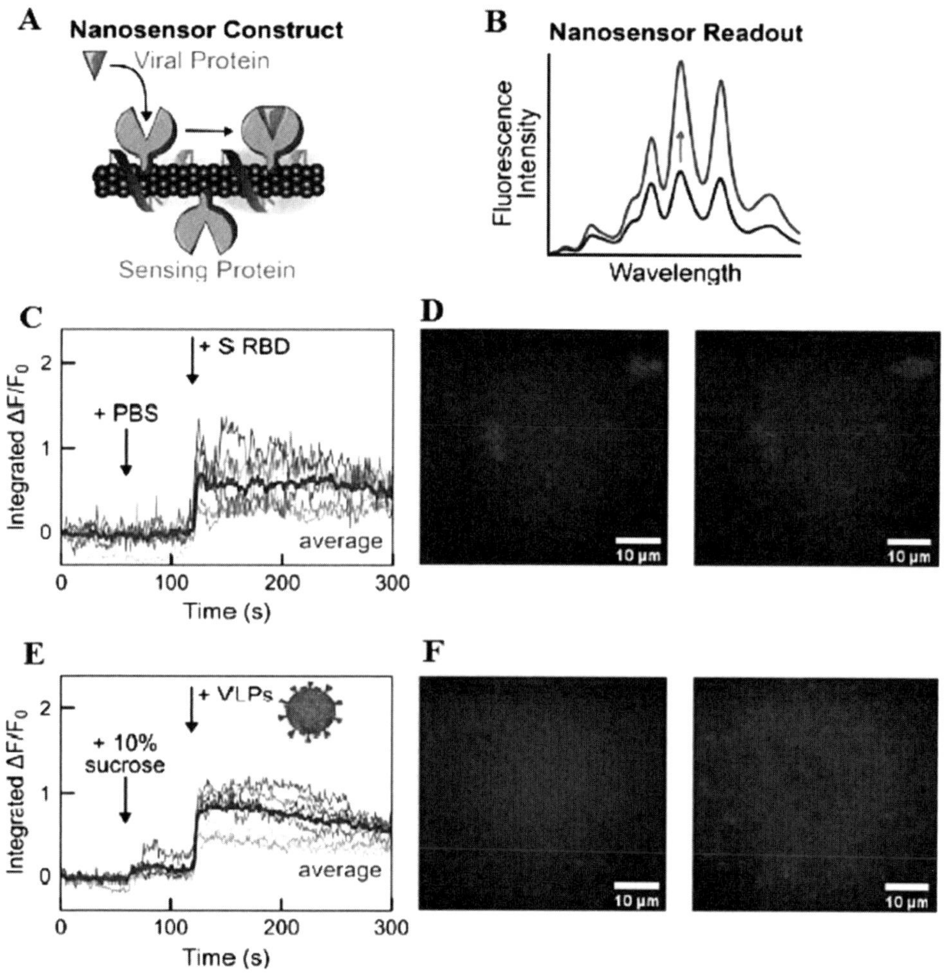

Figure 3. Nanosensor readings in COVID-19. (A, B) Rapid detection of virus spike proteins using carbon nanotube (CNT)-based Near-Infrared Nanosensors. Surface-immobilized ACE2-single-walled carbon nanotube (SWCNT) nanosensor (A-D) showed no change in fluorescence upon the addition of PBS at 1 minute. The introduction of SRBD resulted in a rapid turn-on fluorescence response. (D-F) Adding sucrose buffer at 1 minute led to a slight increase in fluorescence. (Reproduced with permission from Nano Lett et al. Copyright 2021, Science).

4.2. Malaria

Malaria is a life-threatening disease caused by the *Plasmodium* genus. It is transmitted through the bite of an infected female Anopheles mosquito that carries the pathogen to the bloodstream of the person being bit. Upon entering the bloodstream, this pathogen travels to the liver for maturation and reproduction, which then travels to the RBCs, where the onset of symptoms usually starts.

4.2.1. Use of Graphene against Malaria

Graphene is a 2D carbon-based nanomaterial with exceptional attributes that have led to its usage in inks, flexible electronics, and fabrics. Graphene can be synthesized on different surfaces and substrates through different processes to yield graphene-based products (Moutaouakil et al. 2020). Graphene biosensing against viral infection can serve both in malaria's diagnostic and preventive phases (Figure. 4).

Humans can be detected by their skin-associated molecular attractants by mosquitoes from a distance of 50m due to their developed olfactory and thermal senses. To prevent this, the use of graphene is critical. Graphene multilayer films, thanks to their 2D structure and ability to withstand penetration forces of about 50 μN can act as a physical and molecular barrier to mosquito bites. This can be done by using wearable technologies to embed graphene-based multilayer films into fabrics for protection against malaria. Dry GO and rGO are especially used as they interfere with the host-sensing process of mosquitos. This integration of graphene in fabrics is a huge step towards decreasing malaria. Its low cost and non-toxicity provide an extra layer of protection (Moutaouakil et al. 2020).

Besides wearable technology, using graphene in the bloodstream also caused a barrier-like behavior. In *Plasmodium falciparum*, it was shown that GO nanosheets serve as a physical obstruction between RBCs and the parasites. Here, the merozoites had no access to the healthy RBCs due to limited adhesion to the GO nanosheets. The growth and maturity of the parasites also got delayed when GO was introduced in infected RBCs, showing the anti-malarial activity of GO.

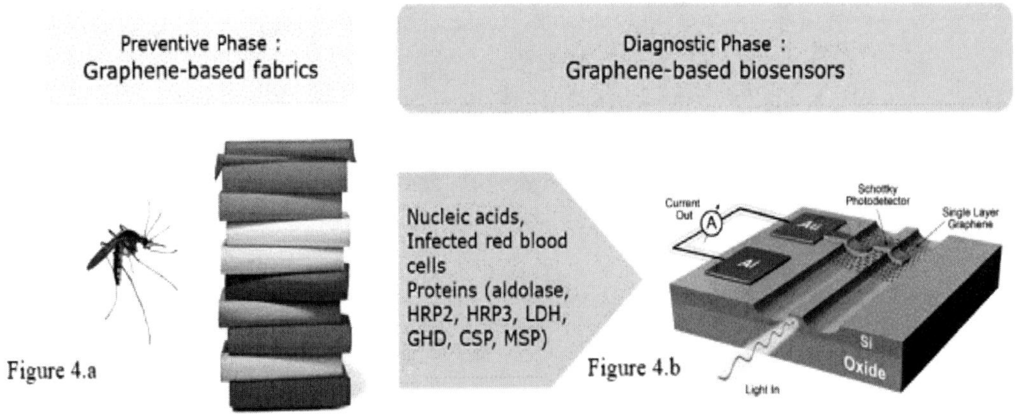

Figure 4. Utilization of graphene-based materials in malaria. (a) Graphene-based films serve as barriers in fabrics, effectively interfering with chemo-sensing mechanisms. This acts as a 'molecular barrier,' impeding mosquitoes from detecting skin-associated molecular attractants (b) Graphene-based biosensors are engineered for the detection of parasites (e.g., DNA) or host biomolecule markers (e.g., proteins). (Reproduced with permission from Amine El Moutaouakil et al. Copyright 2020, Science).

5. Rare Diseases

Rare diseases affect a total of approximately 400 million patients worldwide. However, this number is expected to increase due to the rapid development of sequencing technologies, which have identified over 6,000 to 10,000 rare diseases. However, most rare diseases lack specific treatments, and the clinical industry has traditionally focused on developing small-molecule drugs to alleviate symptoms rather than cure these diseases (Zhang et al. 2019). In this context, nanomaterials represent a therapeutic approach that is very suitable for treating rare diseases because they can correct the underlying genetic defects and provide a cure.

5.1. Gaucher Disease

Gaucher disease is classified as a lysosomal storage disease (LSD). Lysosomes are the most important digestive units of cells. Lysosome enzymes break down or "digest" nutrients, including certain complex carbohydrates and fats. In Gaucher disease, a type of fat-containing sugar (glucose), i.e., glycolipids, accumulate abnormally in the body due to a lack of glucocerebrosidase (Faouzi et al. 2021). Lipids' accumulation or "storage" causes the various symptoms or physical findings associated with lysosomal storage disease. Gaucher disease is the second most common type of lysosomal storage disease. (Recent publications indicate that Fabry disease is the most common LSD).

5.1.1. Properties

CDs can encapsulate therapeutic enzymes, such as glucocerebrosidase (an enzyme absent in Gaucher disease), within their structures. Functionalizing CDs with targeting ligands may enable targeting cells affected by Gaucher disease, such as macrophages. CDs can effectively penetrate cell membranes, allowing them to transport therapeutic enzymes into the lysosomes of diseased cells, where they are needed. CDs can act as fluorescent sensors for *in vivo* imaging and monitoring of the distribution of therapeutic agents. Their fluorescent properties allow real-time monitoring of enzymes on CDs facilitating the evaluation of therapy effectiveness (Faouzi et al. 2021). CDs are known for their low immunogenicity, which minimizes the risk of immune reactions when used in enzyme replacement therapy of patients with Gaucher disease.

The use of CDs in therapeutics majorly consists of:

- Enzyme replacement therapy (ERT): CDs filled with glucocerebrosidase (an enzyme deficient in Gaucher disease) are given to patients intravenously.
- Targeted delivery: Surface modification of CDs with mannose or other targeting ligands facilitates their macrophage uptake, the cells primarily affected by Gaucher disease.
- *In vivo* imaging: Fluorescently labeled CDs enable real-time imaging of enzyme distribution, confirming targeted delivery to affected organs.
- Therapeutic Efficacy: CD-mediated ERT improves enzyme levels in lysosomes, reducing substrate accumulation and alleviating symptoms of Gaucher disease.

5.2. Cystic Fibrosis

Cystic fibrosis (CF) is an inherited, life-threatening disease that affects many organs. This causes changes in the electrolyte transport system. People with CF have problems with the glands that produce sweat and mucus. CF makes mucus thicker. Symptoms begin in childhood. On average, people with CF live up to 30 years. But new treatments are extending life.

Treating CF using graphene-based nanomaterials involves utilizing the unique properties of graphene for targeted drug delivery, gene therapy, and imaging. Graphene-based vectors can deliver corrective genes, such as the cystic fibrosis transmembrane conductance regulator (CFTR) gene, to lung cells affected by CF. Graphene's unique structure allows for efficient penetration of cell membranes, facilitating the delivery of therapeutic genes into target cells. Graphene-based nanomaterials offer high surface area and loading capacity, allowing efficient loading of therapeutic agents such as antibiotics or CFTR modulators. Graphene with targeting ligands enables specific targeting of affected lung cells, minimizing off-target effects (Faouzi et al. 2021). Graphene-based scaffolds can mimic the extracellular matrix of lung tissues, promoting cell adhesion, proliferation, and differentiation. These scaffolds can aid in repairing damaged airway tissues caused by CF-associated inflammation and infection. Graphene-based nanomaterials can serve as contrast agents for imaging techniques such as MRI or CT scans, enabling visualization of lung function and disease progression. Their unique optical properties allow for real-time lung tissue imaging, facilitating treatment efficacy assessment (Faouzi et al. 2021). Graphene-based nanomaterials have been reported to exhibit anti-inflammatory properties, which can help alleviate inflammation in CF-affected lungs.

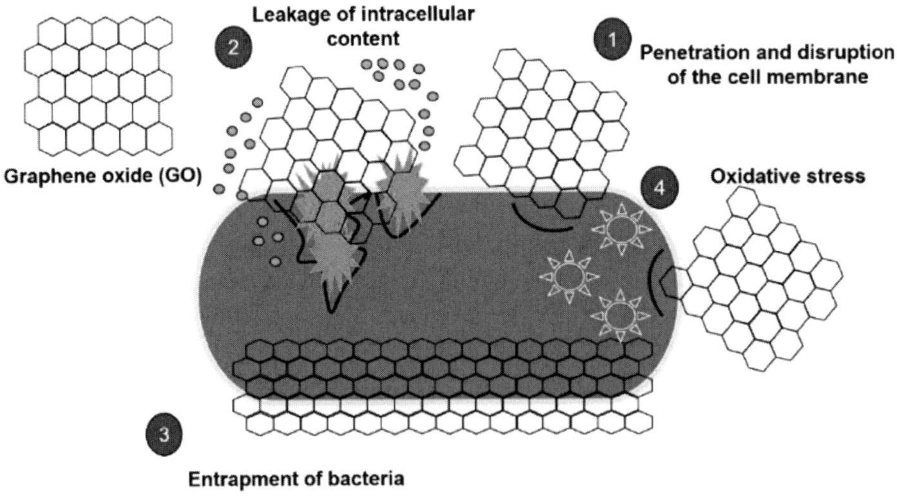

Figure 5. Antibacterial mechanism of GO in CF patients. Illustration of the antibacterial mechanism of GO in CF patients, which consists of disruption of the bacterial cell wall, leakage of intracellular content, oxidative stress, and bacterial entrapping. (Reproduced with permission from Vivas, Copyright 2019).

The use of graphene-based nanomaterials includes the following:

- Gene Editing: Graphene oxide (GO) nanosheets loaded with CFTR gene-editing tools are administered to CF patients via inhalation (Figure. 5).

- Targeted Delivery: Functionalizing GO with lung-targeting ligands facilitates their uptake by airway epithelial cells.
- Therapeutic Efficacy: Graphene-based nanomaterials deliver corrective genes or CFTR modulators to the affected lung cells, restoring chloride ion transport and reducing mucus viscosity.

5.3. Pompe Disease

Pompe disease (also known as acid maltase disease and glycogen storage disease II) is a rare genetic disorder that causes progressive weakness of the heart and skeletal muscles. It is caused by gene mutations that produce an enzyme called acid alpha-glucosidase (GAA), which the body uses to break down glycogen, a stored form of sugar used for energy. The enzyme functions in intracellular compartments called lysosomes, which act as the cell's clearing houses (B. P. Jiang et al. 2019). Lysosomes take in several substances, including glycogen, which GAA converts into glucose, a sugar that fuels muscles. In Pompe disease, mutations in the GAA gene reduce or eliminate this important enzyme that causes the buildup that most severely damages skeletal and heart muscles. The severity of the disease and the age of onset, which varies widely, are related to the degree of enzyme deficiency.

5.3.1. Structure and Application

Treating Pompe disease using silica nanoparticles involves leveraging their unique properties for enzyme replacement therapy and targeted drug delivery. Silica nanoparticles can encapsulate recombinant acid α-glucosidase (GAA), the deficient enzyme in Pompe disease, within their porous structure. Encapsulation protects the enzyme from degradation and enhances its stability during circulation. Silica nanoparticles efficiently deliver encapsulated enzymes into the lysosomes of affected cells. Mesoporous silica nanoparticles offer tunable pore sizes, allowing controlled loading and release of therapeutic enzymes or drugs (B.P. Jiang et al. 2019). Silica nanoparticles can provide sustained release of therapeutic agents, reducing the required treatment frequency. Functionalizing silica nanoparticles with targeting ligands (e.g., mannose) enables specific targeting of affected cells, such as muscle cells affected by Pompe disease. Targeted delivery minimizes off-target effects and enhances therapeutic efficacy. Silica nanoparticles can be labeled with fluorescent markers for imaging and monitoring the distribution of therapeutic agents *in vivo*. Their fluorescence properties allow real-time tracking of nanoparticles and the therapeutic cargo, facilitating the assessment of treatment efficacy (B. P. Jiang et al. 2019).

5.3.2. Use of Silica-Based Nanomaterial

- Enzyme Replacement Therapy: Mesoporous silica nanoparticles loaded with recombinant GAA enzyme are administered intravenously to Pompe disease patients.
- Targeted Delivery: Surface modification of silica nanoparticles with mannose facilitates their uptake by muscle cells affected by Pompe disease.

- Therapeutic Efficacy: Silica nanoparticle-mediated ERT improves enzyme levels in lysosomes, reducing glycogen accumulation and alleviating symptoms of Pompe disease.

6. Challenges and Future Prospectus

The nano-pharmaceuticals market, including the development of nanomedicine, has spread worldwide as it provides added value to traditional medicines. Since the definitions of nanomedicine vary from country to country, estimating the global market size for these products can take time and effort. The goal of developing generic clinical nanomedicine formulations is to reduce toxicity rather than improve drug efficacy. Nanomedicine has led to the development and progress of drug design of potent agents that can be manipulated by nanomedicine approaches, such as using nanocarriers so that these molecules can effectively cross the BBB (blood-brain barrier). Nanocarriers are perfect candidates for treating rare neurological diseases due to their useful properties, such as versatility, the ability to display three-dimensional shapes, and suitable dimensions that promote translocation across the BBB. Other properties of nanoparticles and nanomedicine include surface modification, in which their surface properties can be changed. This involves adjusting the surface charge or adding ligands that allow effective interaction with target cells to achieve a positive result.

Interestingly, nanoparticles and nanocarriers can also improve the sustained release of a drug, a critical property for effective therapy as it ensures drug availability over a longer period. This applies to common and rare brain diseases and more extensive diseases such as blood diseases. Without meaningful disease models for less-studied rare diseases, researchers are concerned with improving the predictability and effectiveness of preclinical studies *in vivo*. Despite the growing advancements in technologies for the diagnosis of various diseases, there are a few diseases that need to be drawn attention to, such as cardiovascular diseases, neurodegenerative diseases, and sexually transmitted diseases like chlamydia. Rather than imposing applications for specific materials currently in vogue, the future of nanomedicine will depend on the rational development of numerous materials for constructing nanotechnology and sophisticated tools based on an efficient and full understanding of biological processes. Although these technologies are proficient in detecting diseases, how promising is using nanomaterials in patient safety?

Achieving precision in targeting nanomaterials to specific cells or tissues presents a challenging task. Non-specific binding can result in background noise and inaccurate diagnostic outcomes. Overcoming the difficulty of developing efficient delivery systems capable of transporting nanomaterials to the intended site in the body without degradation or loss of function is crucial. Nanomaterials may face instability under certain conditions (e.g., temperature, pH, or light exposure), ultimately impacting their diagnostic properties. Achieving consistent quality and reproducibility of nanomaterials on a large scale poses a challenge, given that variations in size, shape, and surface chemistry can influence their performance.

Conclusion

This chapter discusses the unique properties of nanoparticles, specifically carbon-dots, graphene, and silica-based, and their applications in diagnosing various diseases and their prospective therapies. Nanotechnology is emerging as a promising field that can revolutionize industry and the health market. Due to their large surface area, biocompatibility, electrochemical properties, cost-effectiveness, and tunable sizes, they play a prominent role in diagnosis and therapeutics. These characteristics are also crucial for their development into biosensors and can be deployed as advanced nanosystems for targeted drug delivery after modifications and functionalization of their structures. Nanomaterials provide a promising avenue for rare diseases by addressing the underlying genetic defects and providing targeted therapies. In healthcare, nanomaterials play a pivotal role, offering unparalleled opportunities for disease diagnosis and prospects in therapeutics. The properties of nanomaterials allow for precise targeting, enhanced drug delivery, and minimizing cytotoxic effects, making the prospects of therapeutics achievable. Nano-sensors, drug delivery systems, and diagnostic tools offer unprecedented sensitivity and efficiency, enabling early detection, accurate diagnosis, and effective treatment strategies.

References

Ahmadi, Sepideh, Majid Ebrahimi Warkiani, Mohammad Rabiee, Siavash Iravani, and Navid Rabiee. "Carbon-based Nanomaterials Against SARS-CoV-2: Therapeutic and Diagnostic Applications." *OpenNano* 10 (March 1, 2023): 100121. https://www.sciencedirect.com/science/article/pii/S2352952023000014.

Amine El Moutaouakil, Mohamed Belmoubarik and Weng Kung Peng. "Graphene in the fight against malaria." *Applied Physics* (2020). https://arxiv.org/pdf/2008.13605.

Chunduri, L. A. Avinash. "Development of carbon dot-based microplate and microfluidic chip immunoassay for rapid and sensitive detection of HIV-1 p24 antigen. *Microfludics and Nanofluidics* (2016). https://link.springer.com/article/10.1007/s10404-016-1763-9.

Faouzi, Abdelfattah, and Valérie Gaëlle Roullin. "Think Big, Start Small: How Nanomedicine Could Alleviate the Burden of Rare CNS Diseases." *Pharmaceuticals* 14, no. 2 (January 30, 2021): 109. https://www.ncbi.nlm.nih.gov/pmc/articles/PMC7912386.

Gong, Qiaojuan, Yongdong Wang, and Haiying Yang. "A Sensitive Impedimetric DNA Biosensor for the Determination of the HIV Gene Based on graphene-Nafion Composite Film." *Biosensors & Bioelectronics/Biosensors & Bioelectronics (Online)* 89 (March 1, 2017): 565– 69. http://dx.doi.org/10.1016/j.jmat.2019.03.004.

Hu, Yuwei, Shucheng Hua, Fenghua Li, Yuanyuan Jiang, Xiaoxue Bai, Dan Li, and Li Niu. "Green-synthesized Gold Nanoparticles Decorated Graphene Sheets for Label-free Electrochemical Impedance DNA Hybridization Biosensing." B*iosensors & Bioelectronics/Biosensors & Bioelectronics (Online)* 26, no. 11 (July 1, 2011): 4355–61. https://doi.org/10.1016/j.bios.2011.04.037.

Jiang, B. P., B. Zhou, Z. Lin, H. Liang, X. C. Shen, Chem. Eur. J. "Recent Advancements in Carbon nanomaterials" *Current Opinion in Biomedical Engineering* (2019), 25, 3993. https://pubmed.ncbi.nlm.nih.gov/30328167/.

Montagnana, Martina, and Giuseppe Lippi. "Cancer Diagnostics: Current Concepts and Future Perspectives." *Annals of Translational Medicine* 5, no. 13 (July 1, 2017): 268. https://doi.org/10.21037/atm.2017.06.20.

Mozhgani, Sayed-Hamidreza. "Nanotechnology-based strategies for HIV-1 and HTLV-1 retroviruses." *Heliyon* 6 (2020) https://doi.org/10.1016/j.heliyon.2020.e04048.

Nocito, Giuseppe, Giovanna Calabrese, Stefano Forte, Salvatore Petralia, Caterina Puglisi, Michela Campolo, Emanuela Esposito, and Sabrina Conoci. "Carbon Dots as Promising Tools for Cancer Diagnosis and Therapy." *Cancers* 13, no. 9 (April 21, 2021): 1991. https://doi.org/10.3390/cancers13091991.

Sitia, Leopoldo, Marta Sevieri, Lorena Signati, Arianna Bonizzi, Arianna Chesi, Francesco Mainini, Fabio Corsi, and Serena Mazzucchelli. "HER-2-Targeted Nanoparticles for Breast Cancer Diagnosis and Treatment." *Cancers* 14, no. 10 (May 13, 2022): 2424.https://doi.org/10.3390/cancers14102424.

Zhang, Nan, Michael Ming-Yuan Wei, and Qiang Ma. "Nanomedicines: A Potential Treatment for Blood Disorder Diseases." *Frontiers in Bioengineering and Biotechnology* 7 (November 28, 2019) https://doi:10.3109/10717544.2015.1126769.

Chapter 4.3

Oral Drug Delivery: Nanomaterials-Based Strategies, Challenges, and Future Prospects

Jagriti Gupta[1,2]
Paulraj Rajamani[1]
and Jaydeep Bhattacharya[2,*]
[1]School of Environmental Sciences, Jawaharlal Nehru University, New Delhi, India
[2]School of Biotechnology, Jawaharlal Nehru University, New Delhi, India

Abstract

Nanomaterials have shown great potential for improving the effectiveness of oral drug delivery, giving rise to the possibility of resolving many issues related to conventional drug delivery techniques. This chapter explores nanomaterials' multifaceted role in improving oral therapeutics' bioavailability and biodistribution. Key challenges addressed include the harsh gastrointestinal environment, poor solubility, and limited permeability of many drugs. Nanocarriers, such as liposomes, solid lipid nanoparticles, polymeric nanoparticles, and inorganic nanoparticles, are explored for their ability to protect drug molecules, enhance absorption, and facilitate targeted delivery. The chapter also highlights innovative strategies to improve drug stability and release profiles, including surface modification and stimuli-responsive nanomaterials. The clinical translation of these advancements is examined, emphasizing the importance of rigorous preclinical and clinical trials to assess safety, efficacy, and scalability. Current perspectives on regulatory hurdles and integrating nanomaterials into established drug delivery systems (DDS) are discussed, underscoring the need for standardized protocols and interdisciplinary collaboration. This chapter addresses these aspects and provides a comprehensive overview of state-of-the-art nanomaterial-enabled oral drug delivery, offering insights into future directions and potential clinical applications. Integrating nanotechnology in oral DDS represents a significant leap forward, promising to enhance therapeutic outcomes and patient compliance.

Keywords: nanomaterials, oral drug delivery, bioavailability, biodistribution, regulatory hurdles

---

* Corresponding Author's Email: jaydpb@jnu.ac.in / jaydpb@gmail.com

In: Emerging Nanomaterials in Biomedical Science
Editors: Eepsita Priyadarshini and Ajith Manayil Parambil
ISBN: 979-8-89530-600-0

1. Introduction

Oral drug delivery is highly favored for its convenience, non-invasiveness, and patient compliance, with oral formulations comprising about 90% of the global pharmaceutical market (Lang et al. 2020; Prasad et al. 2017). The global market for oral drug delivery, valued at USD 120.44 billion in 2023, is projected to reach USD 187.56 billion by 2029, growing at a CAGR of 7.74% (https://www.techsciresearch.com/report/oral-drug-delivery-market/7236.html). Despite its popularity, oral delivery faces challenges with drugs that have poor solubility, stability, and bioavailability (Bertoni et al. 2020). Nanotechnology-based systems offer a promising solution, significantly enhancing drug potency and efficacy across various administration routes (Chauhan et al. 2024). It enables controlled release, improved bioavailability, and precise tissue or cell targeting through methods such as ligand-based targeting, pH-responsive targeting, etc., thereby maximizing therapeutic outcomes and minimizing side effects (Acharya et al. 2024; Vyas et al. 2023). Nanotechnology is poised to revolutionize healthcare and personalized medicine. This chapter delves into the challenges of oral drug delivery, the role of nanomaterials, strategies for enhancing biodistribution, and clinical trial perspectives in this field.

2. Challenges in Oral Drug Delivery

Oral drug administration continues to be widely utilized for its ease of use, strong patient adherence, and economic viability. However, several significant challenges persist, impacting oral medications' efficacy, bioavailability, and safety. These challenges primarily stem from the physiological and biochemical barriers within the human gastrointestinal (GI) tract, alongside formulation-related issues and patient-specific factors (Viswanathan et al. 2017) (Figure 1). Overcoming these obstacles is crucial for developing successful and efficient oral DDS.

2.1. Physiological Barriers

The GI tract poses significant physiological barriers that can hinder the absorption of orally administered drugs. Factors such as pH variations, enzymatic degradation, mucus barriers, intestinal transit time, epithelial cell layer, and the microbiota can affect drug absorption and bioavailability (Lou et al. 2023). Additionally, efflux transporters in the intestinal epithelium can actively pump drugs out of cells, reducing their effectiveness. Understanding these barriers is crucial for developing effective oral DDS.

2.1.1. Variation of pH in GI Tract

The GI tract has a critical role in oral drug delivery but faces challenges due to its pH levels. The stomach's acidic environment, ranging from 1 to 2.5 (Homayun et al. 2019), can degrade acid-labile drugs and reduce their therapeutic efficacy (Figure 1). Peptide-based drugs like insulin are prone to degradation under these conditions. Enteric coatings often protect drugs as they pass through the stomach (Brown et al. 2019). In contrast, the pH of the intestinal fluid ranges from 6 in the duodenum, to 7.4 in the terminal ileum (Fallingborg et al. 1990). These pH fluctuations can influence drug activity through pH-induced oxidation, deamidation, or hydrolysis (Singh et al. 2021).

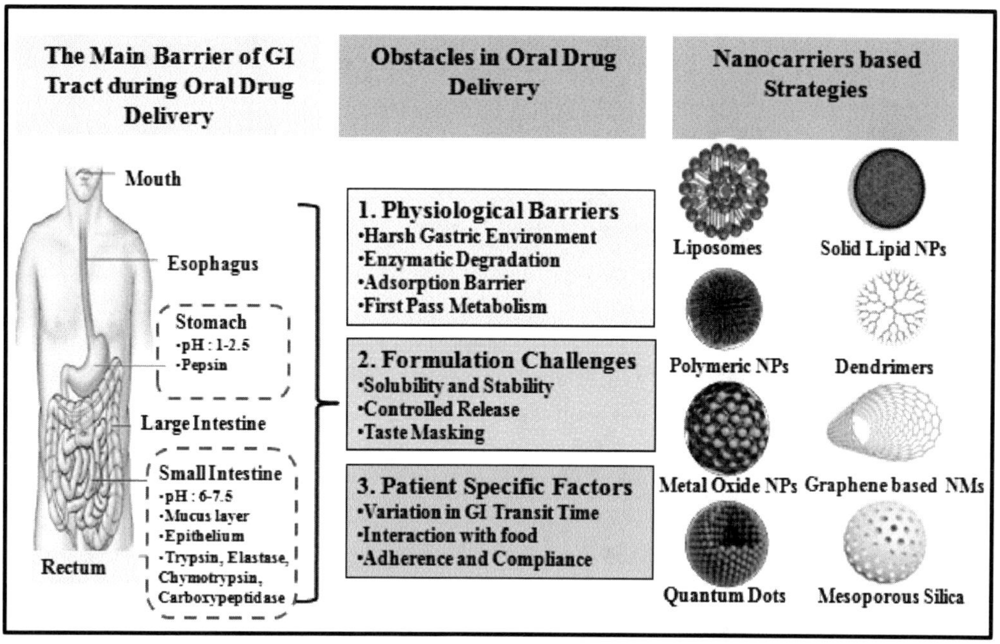

Figure 1. Challenges in oral drug delivery and nanocarriers-based strategies for overcoming these obstacles.

2.1.2. Enzymatic Degradation

Enzymes in the gastrointestinal tract are essential for food digestion but can also degrade drugs before absorption (Figure 1). The stomach contains proteases like pepsin that break down proteins into smaller peptides, which can degrade drugs before absorption. To counteract enzymatic degradation, strategies include enzyme inhibitors, nanoparticle encapsulation, or chemical modification of drugs. pH-responsive hydrogels can encapsulate drugs, protecting them from stomach acids and gastric enzymes. These materials stay intact under adverse conditions, safeguarding the drug until it encounters the target pH, at which point they release their contents. Yamagata et al. (2006) showed that pH-sensitive hydrogel microparticles can effectively protect delicate drugs such as insulin from degradation by gastric and intestinal enzymes. Cerchiara et al. (2016) formulated an oral pH-responsive microencapsulation system based on chitosan/carboxymethylcellulose polyelectrolytes, successfully protecting drugs against gastric acidity and enzymatic activity. Pancreatic enzymes like lipase, trypsin, amylase, and peptidases are plentiful at the entrance of the small intestine (duodenum) and can degrade nucleic acids, thus compromising the stability of biomolecules in the GI tract (O'Neill et al. 2011). However, these enzyme concentrations decrease significantly in the jejunum and further along the intestines. The short transit time of digested food in the duodenum often prevents these enzymes from fully degrading drugs (Davies et al. 2017). Additionally, the duodenum's lower pH compared to the later parts of the small intestine allows for controlled drug release by adjusting the pK_a of the delivery carriers. Lozoya-Agullo et al. (2018) demonstrated that using poly (lactic-co-glycolic acid) (PLGA) nanoparticles for colon delivery of ibuprofen can significantly reduce premature release in the duodenum due to the unsuitable pH environment.

Mechanical degradation, including osmotic stresses throughout the GI tract, peristalsis of GI muscles, and shear stresses from gastric juice flow, can also harm drug molecules (Valon and Levayer, 2019). Microencapsulation can address these issues by incorporating nanoparticles into hydrogels to enhance their properties (De France et al. 2016).

2.1.3. Mucus Barrier

The mucus layer lining the GI tract serves as a protective hurdle against pathogens and mechanical damage but also obstructs the absorption of drugs. Mucus is a viscoelastic gel composed of glycoproteins, water, electrolytes, and lipids (Figure 2a). It traps and slows the diffusion of drug molecules, reducing their availability at the epithelial surface for absorption. Nanoparticles and mucoadhesive DDS have been engineered to penetrate or adhere to the mucus layer, prolonging drug retention and absorption. Liu et al. (2018) devised an innovative liposome system composed of chitosan-thioglycolic acid-Pluronic F127 (CS-TGA-PF), designed for enhanced mucus adhesion and penetration, facilitating the oral delivery of paclitaxel.

2.1.4. Epithelial Cell Layer

The epithelial layer of the GI tract is the primary site for drug absorption. This layer consists of tightly packed cells that form a barrier to prevent the entry of harmful substances while allowing the selective absorption of nutrients and drugs (Figure 2a). The two main pathways for drug absorption across the epithelial barrier are transcellular (through the cells) and paracellular (between the cells) (Figure 2b). Transcellular transport often requires the drug to be lipophilic to pass through the cell membrane. In contrast, paracellular transport is limited by the tight junctions between epithelial cells, allowing only small hydrophilic molecules (Boegh et al. 2015). Permeation enhancers and nanoparticles can facilitate the opening of tight junctions or provide a more favorable route for transcellular transport.

2.1.5. Efflux Transporters

Efflux transporters of the multigene ATP binding cassette (ABC) transporter family, such as breast cancer resistance protein (BCRP), P-glycoprotein (P-gp), and multidrug resistance-associated protein 2 (MRP2), are proteins located in the epithelial cells of the GI tract that pump drugs back into the intestinal lumen, reducing their absorption (Chan et al. 2004). These transporters are part of the body's defense mechanism to limit the absorption of potentially harmful substances (Amin, 2013). However, they also affect the bioavailability of beneficial drugs (Levi et al. 2019). Inhibitors of efflux transporters and nanoparticles designed to bypass these transporters are being explored to improve drug absorption. Levi et al. (2019) fabricated the hydrogel microdevices to inhibit efflux transporters for better drug delivery and adsorption reversibly.

2.1.6. Gastrointestinal Motility and Transit Time

The motility and transit duration of the GI tract significantly impact how long a drug stays in various segments, thereby influencing its absorption. Swift passage through the stomach or small intestine can diminish the absorption window, especially for drugs with slow dissolution rates (Vinarov et al. 2021). Conversely, delayed transit can enhance the absorption of certain drugs. Controlled-release formulations and mucoadhesive systems are employed to regulate transit time and enhance drug bioavailability. In addition, several novel approaches are available, such as mucus-penetrating nanocarriers that can infiltrate deeply into the mucus gel layer of the stomach and intestinal mucosa, where they remain for extended periods. Within the mucus gel layer, charge-converting nanocarriers change their zeta potential from negative to positive, forming strong ionic interactions with anionic mucus glycoproteins. Furthermore, it has been shown that cyclodextrins and thiolated nanocarriers can even form covalent bonds with mucus glycoproteins' cysteine-rich subdomains (Kali et al. 2022).

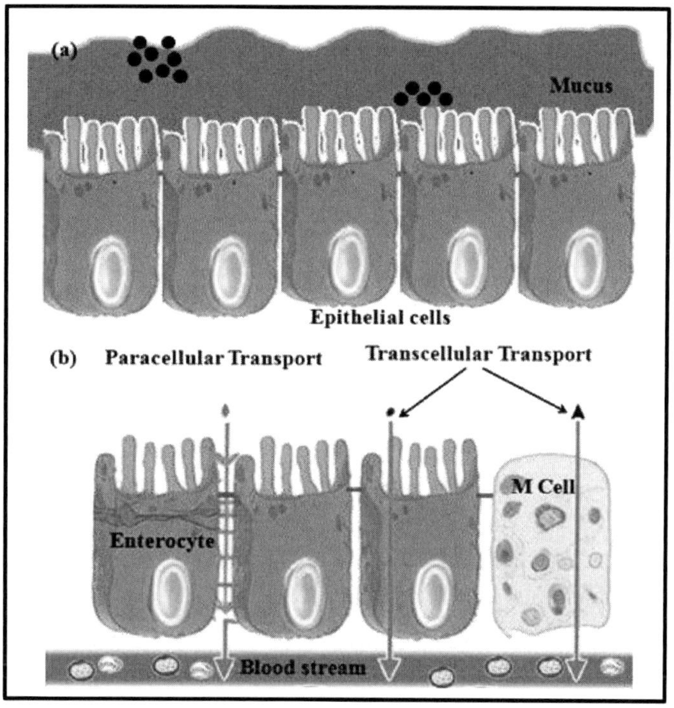

Figure 2. Physiological obstacles in oral drug delivery (a) mucus barrier and epithelial cell barrier (b) drug delivery mechanism employed by nanomaterials.

2.1.7. Microbiota

With trillions of microbes, the human gut microbiota is crucial to health and illness (Zhang et al. 2021). These microorganisms can also metabolize drugs, impacting their efficacy and safety. For example, certain bacteria in the gut can reduce drug bioavailability by metabolizing the active ingredient before it is absorbed (Colotti and Rinaldi, 2020). Understanding the interactions between the microbiota of the gut and drugs is vital for designing effective oral delivery systems. Probiotics, prebiotics, and antibiotics are sometimes used to modulate the gut microbiota to improve drug absorption and efficacy.

2.1.8. First-Pass Metabolism

First-pass metabolism refers to the pharmacological phenomenon where the concentration of a drug is significantly reduced before it reaches systemic circulation. After absorption from the GI tract, the drug enters the hepatic portal circulation and is transported to the liver, where it can be extensively metabolized. This reduces the bioavailability of many drugs, necessitating higher doses or alternative delivery methods (Herman and Santos, 2024). Prodrug strategies, where an inactive drug precursor is administered and then metabolized into the active form, can help overcome this barrier.

The physiological barriers of the GI tract pose substantial challenges to oral drug delivery. However, ongoing drug formulation and delivery technology advances are addressing these obstacles. Promising strategies include enteric coatings, enzyme inhibitors, mucoadhesive systems, nanoparticles, permeation enhancers, and modulation of the gut microbiota. These approaches aim to improve the bioavailability and efficacy of drugs administered orally. A thorough understanding of these barriers and the methods to overcome them is essential for the successful development of novel oral DDS.

2.2. Formulation Challenges

2.2.1. Drug Solubility and Stability

Poor water solubility in drugs hinders their absorption in the GI tract. Stability is crucial for oral medications, preventing degradation and potential toxicity (Vinarov et al. 2021). Factors such as pH, enzymatic activity, and interactions with food can significantly affect drug stability. Therefore, formulation strategies that enhance stability and permeability are essential for improving drug efficacy. Lipid-based DDS, particularly self-nano-emulsifying DDS, has effectively addressed these challenges (O'Driscoll et al. 2019).

2.2.2. Controlled Release

Designing formulations that allow for controlled drug release at specific sites in the GI tract is complex. Ensuring the drug releases its active ingredient at the desired location and rate requires sophisticated DDS and materials (Li et al. 2022).

2.2.3. Taste Masking

Oral medications must be palatable to ensure patient compliance, particularly for pediatric and geriatric populations. Masking the unpleasant taste of drugs without compromising their efficacy presents a significant challenge (Zhu et al. 2018).

2.3. Patient-Specific Factors

Patient compliance is a critical factor in the success of oral drug therapy. However, factors such as complex dosing regimens, large pill sizes, and unpleasant taste can affect patient adherence to oral medications. Developing formulations that are easy to administer, palatable, and convenient for patients can improve compliance and treatment outcomes.

2.3.1. Variability in GI Transit Time

The rate at which a drug moves through the GI tract can vary significantly between people, affected by age, dietary habits, and health conditions. This variability can affect the drug's absorption and effectiveness (Abuhelwa et al. 2016).

2.3.2. Interaction with Food

Food in the stomach can alter the drug's dissolution and absorption. Some drugs require an empty stomach for optimal absorption, while others are better absorbed with food, adding complexity to dosing schedules (Omachi et al. 2019).

2.3.3. Adherence and Compliance

Ensuring that patients adhere to their medication regimen is crucial for treatment efficacy. Factors like dosing frequency, regimen complexity, and side effects can impact patient compliance (Vinarov et al. 2021).

Overcoming the challenges in oral drug delivery necessitates a multidisciplinary approach, integrating pharmaceutical sciences, pharmacology, physiology, and materials science. Advances in formulation technology, drug design, and delivery systems offer promising solutions to these obstacles, potentially enhancing the efficacy and safety of oral medications. Addressing these challenges enables researchers to unlock the full potential of oral drug delivery, ultimately improving patient care and treatment outcomes.

3. Availability of Different Nanomaterials for Oral Drug Delivery

Drug delivery systems transport therapeutic substances into the body to achieve desired clinical effects. They regulate the timing, rate, and location of release, with oral delivery being the most preferred due to convenience and patient compliance. However, traditional oral systems face issues like poor bioavailability and uncontrolled release. Nanomaterials, including organic and inorganic types, enhance these systems by allowing better penetration and interaction with biological tissues, leading to controlled release and improved bioavailability. This precise targeting maximizes therapeutic outcomes and minimizes side effects. Several types of nanomaterials are being explored for this purpose.

3.1. Polymeric Nanoformulations

Polymeric nanoparticles (NPs) sized between 10 and 1000 nm are ideal for drug delivery due to their biocompatibility (Figure 3), simple formulation parameters, and precise control over multiple features (Wang et al. 2023). They can encapsulate therapeutics, entrap them in the polymer matrix, chemically conjugate them, or bind them to the NP surface (Al-hashimi et al. 2021). Various synthesis techniques yield different final products, such as emulsification, nanoprecipitation, ionic gelation, and microfluidics (Elmowafy et al. 2023). The two primary types of polymeric NPs are nanocapsules and nanospheres. These categories further divide into shapes like polymersomes, micelles, and dendrimers (Viswanathan et al. 2017). Polymersomes, resembling liposomes, have enhanced stability and efficiency in retaining cargo, rendering them potent for delivering substances into the cytosol. Polymeric micelles self-assemble into nanospheres, protecting drug cargo and improving circulation times. Dendrimers are hyperbranched polymers that may attach biomolecules or contrast agents to their surface while loading pharmaceuticals because of their highly regulated mass, size, shape, and surface chemistry (Ding et al. 2024). Despite challenges like particle aggregation and potential toxicity, many polymeric nanomedicines are undergoing clinical trials, demonstrating their potential in advanced therapeutic applications.

3.2. Lipid-Based Nanoformulations

Lipid-based nanoparticles (NPs) are amorphous platforms with an interior aqueous compartment encircled by at least one lipid bilayer (Figure 3). Their formulation simplicity, self-assembly, biocompatibility, high bioavailability, and capacity to transport big payloads are just a few of the many benefits they provide for drug administration. These are the most frequently approved classes of nanomedicines by the FDA (Mitchell et al. 2021).

Liposomes, a subset of lipid-based nanoparticles (NPs), are crafted from phospholipids, forming unilamellar and multilamellar vesicles. This versatile composition enables the concurrent delivery of hydrophilic, hydrophobic, and lipophilic drugs. Within this category, multilamellar vesicles (MLVs) feature large, concentric lipid bilayers, ideal for encapsulating drugs. Meanwhile, small unilamellar vesicles (SUVs), comprising a single lipid bilayer, offer heightened stability, enhanced drug encapsulation, and superior cellular uptake. Giant unilamellar vesicles (GUVs), distinguished by their expansive size, boast a singular, immense lipid bilayer enfolding a spacious aqueous compartment (Manca et al. 2024). These vesicles excel in targeted drug delivery, imaging, and vaccine systems. Their stability hinges upon factors including NP size, surface charge, lipid constitution, lamellar count, and surface modifications (Ferguson et al. 2023).

Lipid nanoparticles (LNPs) are mainly used to transport nucleic acids. They have four main constituents: phospholipids, cholesterol, cationic or ionizable lipids, and polyethylene glycolated (PEGylated) lipids. These constituents create micellar structures within the particle core. LNPs are essential in applications involving individualized genetic therapy because of these factors. However, LNPs face limitations, such as low drug loading and biodistribution issues, often resulting in high uptake in the liver and spleen (Seo et al. 2023). PEGylation improves LNPs performance by enhancing stability, bioavailability, circulation time, and immunogenicity, ensuring safer therapeutic outcomes. LNPs can be further categorized into solid lipid nanoparticles (SLNs) and nanostructured lipid carriers (NLCs), offering unique advantages in drug delivery, stability, and controlled release.

Lipid micelles (LMs) are self-assembled structures formed by the aggregation of amphiphilic molecules in aqueous solutions. Lipid micelles consist of a core-shell structure, with the hydrophobic tails of the lipids forming the core and the hydrophilic heads facing outward. They are used to solubilize hydrophobic drugs and enhance their bioavailability.

3.3. Inorganic Nanoparticles

Inorganic materials such as gold, iron, and silica are utilized to create nanostructured materials for drug delivery and imaging applications (Figure 3). These nanoparticles are engineered in various sizes, structures, and geometries, with gold nanoparticles (AuNPs) being the most extensively studied (Mitchell et al. 2021). AuNPs exhibit unique physical, electrical, magnetic, and optical properties derived from their base material (Gupta et al. 2020). Iron oxide nanoparticles, composed of magnetite or maghemite, possess superparamagnetic properties and are used as contrast agents, drug delivery vehicles, biosensors (Bharti et al. 2022) and in thermal-based therapies. Other common inorganic nanoparticles include calcium phosphate, mesoporous silicon, and porous particles employed for gene and drug delivery. Quantum dots, made from semiconductor materials, are primarily used in *in-vitro* imaging and hold potential for *in vivo* diagnostics (Gupta et al. 2022). Despite their advantageous properties, the clinical application of inorganic nanoparticles is restricted due to low solubility and toxicity concerns, particularly in formulations involving heavy metals (Gupta et al. 2021; Gupta and Rajamani, 2023).

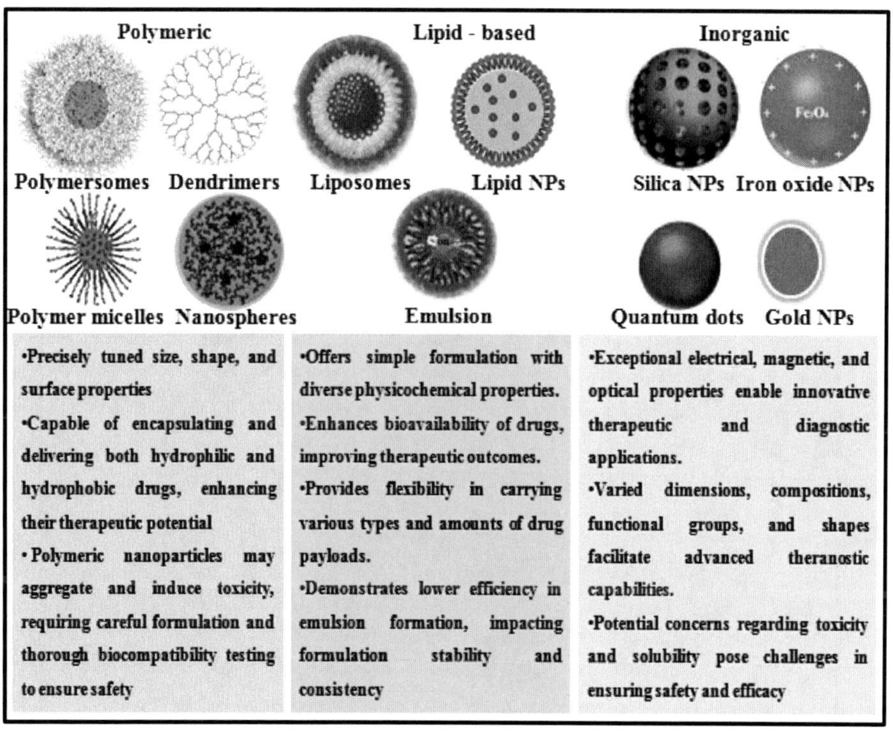

Figure 3. Classification of nanomaterials, advantages, and disadvantages for oral drug delivery.

3.4. Self-Emulsifying Drug Delivery Systems (SEDDS)

These formulations, blending oils, surfactants, and solvents, transform into finely dispersed oil-in-water emulsions upon encountering gastrointestinal fluids, thus augmenting drug solubility and absorption (Salawi et al. 2022).

Few other emulsifying DDS are double emulsions, which present intricate water-in-oil-in-water (W/O/W) structures, encapsulating hydrophilic drugs within lipophilic phases, ultimately elevating oral drug bioavailability. Conversely, nanoemulsions constitute finely dispersed droplets stabilized by surfactants or polymers, exhibiting enhanced oral drug solubility and absorption due to increased surface area and reduced droplet size (Wilson et al. 2022).

4. Strategies for Enhancing Biodistribution and Bioavailability of the Nanocarriers

Enhancing the biodistribution of orally delivered drugs is crucial to maximize the therapeutic efficacy of orally delivered drugs. Several strategies are employed to achieve this goal. The biodistribution and bioavailability of orally administered drugs are governed by particle size, surface charge, geometry, hydrophilicity/hydrophobicity, etc. (Figure 4). Nanomaterials can be engineered to optimize these parameters, significantly enhancing bioavailability (Banerjee et al. 2016). For instance, Banerjee et al. (2016)'s study found that rod-shaped polystyrene particles showed higher uptake and transport than spherical particles, and actively targeted rod-shaped particles accumulated more extensively within cells than their targeted spherical or disc-shaped counterparts. In this discussion, we delve into the various aspects and properties of nanomaterials that contribute to improved biodistribution and bioavailability of oral drugs.

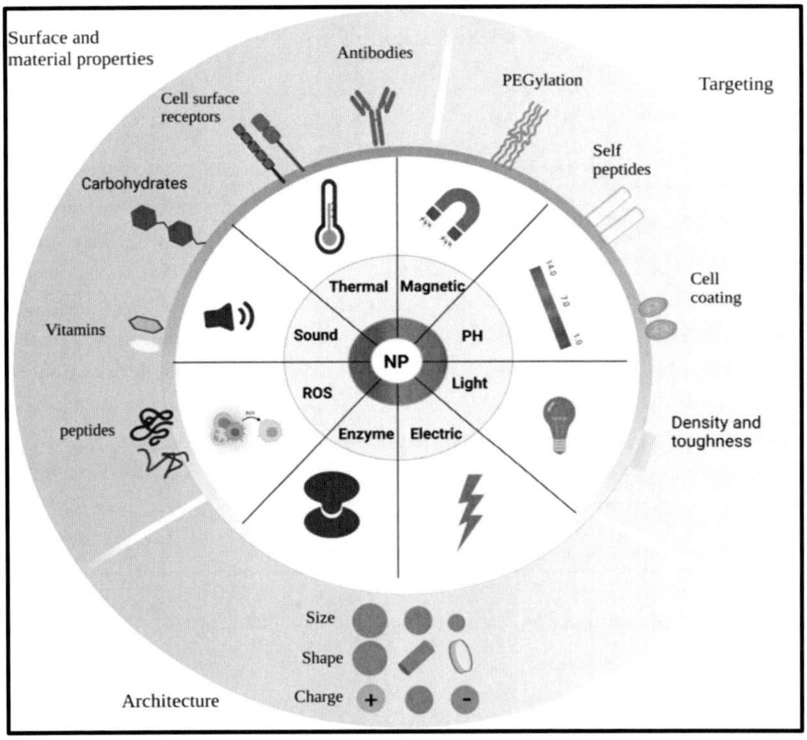

Figure 4. Frequently engineered nanoparticle surface properties enhance oral drug delivery, biodistribution, and bioavailability. Reproduced under the Creative Commons CC BY license, Copyright © 2022, Waheed et al.

4.1. Particle Size

The size of nanoparticles is critical for their adhesion, cellular uptake, and overall bioavailability in oral drug delivery. Research has demonstrated that reducing particle size can enhance adhesion, prolong residence time, and improve oral bioavailability. For instance, retinoic acid SLNs exhibited a threefold enhancement in oral bioavailability when particle size was reduced from 328.8 nm to 89.3 nm (Hu et al. 2020). Smaller particles, such as those measuring 40 nm and 100 nm, can penetrate deeper into the mucosal folds of the colorectal epithelium, thereby extending their residence time (Maisel et al. 2015). Additionally, 100 nm NLCs demonstrated higher cellular uptake efficiency and permeation ability in Caco-2 cells, with these NLCs showing the highest C_{max} and AUC (Li et al. 2016). Banerjee et al. (2016) also found that nanoparticle uptake by Caco-2 and Caco-2/HT-29 cells was inversely related to size, with 50 nm particles achieving the highest uptake. Consequently, further research is needed to determine the optimal particle size for oral drug delivery.

4.2. Surface Charge

The surface charge of nanoparticles significantly influences their interaction with the mucosal layer and cellular membranes. Positively charged nanoparticles interact more effectively with the negatively charged mucus and epithelial cells, enhancing absorption. Du et al. (2018) designed polyethylene glycol-block-polylactic acid (PEG-PLA) nanoparticles with different lipid charges to study their impact on oral absorption. They found that nanoparticles with a positive surface charge enhanced cellular uptake and transport through Caco-2 and small intestinal epithelial cells, improving oral bioavailability. Conversely, Czuba et al. (2018) discovered that negatively charged sodium dodecyl sulfate (SDS)-capped nanoparticles had higher *in-vitro* internalization than positively charged chitosan-coated nanoparticles in Caco-2 epithelial cultures. Additionally, they demonstrated that insulin SDS-NPs effectively reduced blood glucose levels in diabetic rats *in vivo*, suggesting that negatively charged nanoparticles are a promising approach to improving nanoparticle uptake and insulin bioavailability for oral delivery.

4.3. Surface Functionalization

Surface functionalization of nanoparticles significantly enhances their biodistribution and bioavailability in oral drug delivery. By modifying the surface by targeting ligands, polymers, or other functional groups, nanoparticles can achieve improved stability, mucosal adhesion, and cellular uptake. These modifications enable more precise targeting of specific tissues or cells, protect the drug from degradation, and facilitate controlled release, ultimately enhancing the therapeutic efficacy of the orally delivered drugs (Wang et al. 2020).

4.3.1. Hydrophilicity/Hydrophobicity

Balancing hydrophilicity and hydrophobicity are crucial for the stability and absorption of nanoparticles. Amphiphilic nanoparticles, which have hydrophilic and hydrophobic regions, can enhance drug solubility and interaction with biological membranes (Kumar et al. 2020).

4.3.2. Mucoadhesion

Enhancing the mucoadhesive properties of nanoparticles can significantly prolong their residence time in the GI tract, thereby increasing the likelihood of drug absorption. Polymers such as chitosan and polyacrylic acid are commonly employed for this purpose. Dyawanapelly et al. (2016) reported that PLGA nanoparticles with chitosan and oligosaccharide exhibited higher cellular uptake and superior mucoadhesive properties than uncoated PLGA nanoparticles.

4.3.3. PEGylation / Tannic Acid (TA) Coating

PEG is commonly used in nanoparticle surface modification to reduce interactions with luminal components and mucus in the gut, enhancing oral bioavailability. Studies show that doxorubicin (DOX)-loaded PEG-modified PLGA nanoparticles distribute uniformly on epithelial cells and improve oral bioavailability compared to DOX-S (Maisel et al. 2015). PEG-coated nanoparticles have been evaluated for their physicochemical properties, including molecular weight and surface coverage. Tannic acid (TA), a beneficial polyphenol, can improve the solubility of hydrophobic molecules by hydrogen bonding (Jackson and Letchford, 2016). These nanoparticles demonstrate high biological activity, reduce burst drug release, prevent stomach acidic degradation, and achieve intestinal site-specific drug release through pH stimulation, as confirmed under various pH conditions.

4.3.4. Specific Ligands

Nanoparticles coated with specific ligands, such as aptamers, antibodies, and cell surface receptors, have shown better intracellular uptake trends (Mitchell et al. 2021; Waheed et al. 2022) (Figure 3b). For example, PTX-PLGA nanoparticles with folate increased permeability in Caco-2 cells by 5-fold compared to free PTX (Roger et al. 2012). Functionalization with folic acid further increased transport capacity and uptake inside the cells.

4.3.5. Use of Absorption Enhancers

Lipophilic substances, especially macromolecules with large molecular weights, can enhance the lymphatic absorption of drugs. Lymphatic vessels exhibit higher permeability to nanoparticles than capillaries, thus bypassing liver metabolism (Wang et al. 2020). The absorption effects of fatty acids vary with chain length, with long-chain triglycerides promoting greater absorption. Additionally, certain phospholipids and surfactants can inhibit the P-glycoprotein (P-gp) efflux system, thereby increasing drug absorption (Ahmad et al. 2015).

4.4. Controlled Release

Nanomaterials enable controlled and sustained drug release, maintaining therapeutic levels over extended periods and reducing dosing frequency. For example, Li et al. (2022)

demonstrated the use of pH-responsive polyamine-modified cyclodextrin nanoparticles for the controlled release of anti-tumor drugs. Zhang et al. (2020) developed a polymeric vesicle loaded with the chemotherapeutic drug DOX and photothermal agents (IR780 iodide). They studied drug release dynamics under temperature, pH, and reducing conditions. Their findings illustrated that elevated temperatures, acidic pH, and heightened glutathione levels expedite the release of DOX and IR780. This study hints at the potential for precise modulation of drug release kinetics, offering a pathway towards achieving controlled release of encapsulated medications, thereby enhancing therapeutic outcomes.

5. Perspectives of Clinical Trials

Transitioning from preclinical research to clinical trials for nanomaterial-based oral DDS requires safety and toxicity assessments, biocompatibility evaluations, potential immune responses, and long-term effects. Detailed studies on pharmacokinetics and pharmacodynamics are necessary to understand nanomaterial-based formulations in the human body (Mitchell et al. 2021). Navigating the regulatory landscape is challenging, as agencies require evidence of safety, efficacy, and manufacturing quality (Shan et al. 2022). Developing scalable and reproducible manufacturing processes is essential for commercialization, ensuring consistency in particle size, surface characteristics, and drug loading (Homayun et al. 2019). Patient compliance and acceptance also require formulation acceptability, ease of administration, and potential side effects (Brown et al. 2019).

6. Case Studies and Clinical Examples

A study by Shan et al. (2022) illuminates the landscape of nanomedicine, revealing that over 500 formulations are currently undergoing clinical trials, primarily in phase I, with a significant emphasis on combating cancer and infections. Despite this promising trajectory, the journey to FDA approval for oral nanomedicines has been relatively restrained compared to other administration routes (Jia et al. 2023). While liposomes and lipid-based nanoformulations dominate the market, they're mainly administered intravenously due to challenges in oral delivery, including susceptibility to degradation in the gastrointestinal tract and poor absorption efficiency (Shan et al. 2022). The sanctioned oral nanomedicines are mostly nanocrystals and iron nanoparticles (Jia et al. 2023) and strive to augment the bioavailability, stability, and precision targeting of complex drugs, marking a pivotal advancement in therapeutic innovation. Notably, many nanomaterial-based oral DDS has traversed clinical trials and secured FDA endorsement, showcasing substantial promise in enhancing treatment outcomes (Table 1). Among these, two noteworthy natural polymer nanoparticle-based oral drugs, ORMD-0801 and HDV-I, have received FDA approval. However, their market entry has been deferred due to concerns surrounding their fluctuating toxicity and the spectrum of adverse effects observed across diverse clinical trials (Moroz et al. 2016). Additionally, a multitude of other formulations are currently under clinical evaluation or advancing through clinical trials, including curcumin-based nanoformulations for combating inflammation and cancer (Hafez Ghoran et al. 2022), as well as siRNA therapeutics ingeniously incorporated within nanocarriers for cancer treatment (Hattab et al. 2021).

Table 1. FDA-approved drug nanocrystals and iron nanoparticles based oral nanomedicines (Jia et al. 2023)

Trade Name	Drug Agent	Approval Year	Company	Clinical Applications
Gris-PEG®	Griseofulvin	1998	Recro Gainesville LLC (Gainesville, FL, USA)	Ringworm infections
Rapamune®	Rapamycin/ sirolimus	2000	Wyeth (Philadelphia, PA, USA)	Immunosuppressive therapy in kidney transplantation
Avinza®	Morphine sulfate	2002	King Pharma (Bristol, TN, USA)	Chronic pain
Ritalin LA®	Methylphenidate hydrochloride	2002	Novartis (Basel, Switzerland)	Attention-deficit-hyperactivity disorder
Emend®	Aprepitant	2003	Merck (Rahway, NJ, USA)	Nausea and vomiting due to chemotherapy
Tricor®	Fenofibrate	2004	Abbott (North Chicago, IL, USA)	Hypercholesterolemia
Triglide®	Fenofibrate	2005	Skye Pharma (San Diego, CA, USA)	Hypercholesterolemia
Megace®ES	Megestrol acetate	2005	Par Pharma (Petaluma, CA, USA)	Anorexia
Naprelan®	Naproxen sodium	2006	Wyeth (Philadelphia, PA, USA)	Inflammation
Cesamet®	Nabilone	2009	Lilly (Indianapolis, IN, USA)	Vomiting and nausea
Injectafer®	Iron nanoparticles Polynuclear iron (III) oxyhydroxide iron particles	2013	For Int. (Waltham, MA, USA)	
Monofer®	Iron nanoparticles Iron molecule with unbranched carbohydrate iron particle	2010	Pharmacosmos (Rorvangsvej, Holbæk, Denmark)	Anemia due to iron deficiency

Conclusion

Nanomaterial-based strategies for oral drug delivery have shown significant promise for enhancing therapeutic efficacy and patient compliance. These advanced delivery systems offer several advantages, including improved bioavailability, targeted delivery, and controlled release profiles. Integrating nanomaterials, such as nanoparticles, liposomes, and dendrimers, into oral drug formulations has the potential to overcome many of the limitations associated with conventional drug delivery methods. However, despite the promising advancements, several challenges remain. Issues related to the stability of nanomaterials in the harsh gastrointestinal environment, potential toxicity, large-scale manufacturing, and regulatory hurdles need to be addressed to fully realize these technologies' potential. Additionally, the interaction of nanomaterials with biological systems, particularly their long-term effects, requires further investigation to ensure safety and efficacy.

Prospects in this field are bright, with ongoing research focused on developing multifunctional nanomaterials capable of simultaneous diagnosis and therapy, improving targeting specificity, and enhancing patient-specific drug delivery through personalized medicine approaches. Advancements in materials science, coupled with a deeper understanding of the biological interactions of nanomaterials, will be crucial in overcoming current challenges. While nanomaterial-based oral drug delivery systems present a transformative approach to medicine, their successful translation from bench to bedside will depend on continued interdisciplinary research, innovation, and collaboration among scientists, clinicians, and regulatory bodies.

Acknowledgments

Jagriti gratefully acknowledges the University Grants Commission (India) for the fellowship provided during her doctoral studies.

Conflicts of Interest

The authors declare no conflict of interest.

References

Abuhelwa, Ahmad Y., David J. R. Foster, and Richard N. Upton. 2016. "A Quantitative Review and Meta-Models of the Variability and Factors Affecting Oral Drug Absorption—Part II: Gastrointestinal Transit Time." *The AAPS Journal* 18 (5): 1322–33. https://doi.org/10.1208/s12248-016-9953-7.

Acharya, Biswajeet, Amulyaratna Behera, Suchismeeta Behera, and Srikanta Moharana. 2024. "Recent Advances in Nanotechnology-Based Drug Delivery Systems for the Diagnosis and Treatment of Reproductive Disorders." *ACS Applied Bio Materials* 7 (3): 1336–61. https://doi.org/10.1021/acsabm.3c01064.

Ahmad, Javed, Saima Amin, Mahfoozur Rahman, Rehan Rub, Madhur Singhal, Mohammad Ahmad, Ziyaur Rahman, et al. 2015. "Solid Matrix Based Lipidic Nanoparticles in Oral Cancer Chemotherapy: Applications and Pharmacokinetics." *Current Drug Metabolism* 16 (8): 633–44. https://doi.org/10.2174/1389200216666150812122128.

Al-hashimi, Nihad, Mai Babenko, Maria Saaed, Negeen Kargar, and Amr ElShaer. 2021. "The Impact of Natural and Synthetic Polymers in Formulating Micro and Nanoparticles for Anti-Diabetic Drugs." *Current Drug Delivery* 18 (3): 271–88. https://doi.org/10.2174/1567201817666200810111726.

Amin, Md. Lutful. 2013. "P-Glycoprotein Inhibition for Optimal Drug Delivery." *Drug Target Insights* 7 (January):DTI.S12519. https://doi.org/10.4137/DTI.S12519.

Banerjee, Amrita, Jianping Qi, Rohan Gogoi, Jessica Wong, and Samir Mitragotri. 2016. "Role of Nanoparticle Size, Shape and Surface Chemistry in Oral Drug Delivery." *Journal of Controlled Release* 238 (September):176–85. https://doi.org/10.1016/j.jconrel.2016.07.051.

Bertoni, Serena, Nadia Passerini, and Beatrice Albertini. 2020. "Nanomaterials for Oral Drug Administration." In *Nanotechnology for Oral Drug Delivery*, 27–76. Elsevier. https://doi.org/10.1016/B978-0-12-818038-9.00004-1.

Bharti, Rajkamal, Jagriti Gupta, Paulraj Rajamani, Ranjita Ghosh Moulick, and Jaydeep Bhattacharya. 2022. "Iron Oxide Nanoparticles/PEDOT: PSS Nanocomposite-Based Modification of Both Glassy Carbon Electrode and Flexible Cotton Fiber OECT for Highly Sensitive Multi-Analytes Detection." *Applied Nanoscience*, August. https://doi.org/10.1007/s13204-022-02565-9.

Boegh, Marie, María García-Díaz, Anette Müllertz, and Hanne Mørck Nielsen. 2015. "Steric and Interactive Barrier Properties of Intestinal Mucus Elucidated by Particle Diffusion and Peptide Permeation." *European Journal of Pharmaceutics and Biopharmaceutics* 95 (September):136–43. https://doi.org/10.1016/j.ejpb.2015.01.014.

Brown, Tyler D., Kathryn A. Whitehead, and Samir Mitragotri. 2019. "Materials for Oral Delivery of Proteins and Peptides." *Nature Reviews Materials* 5 (2): 127–48. https://doi.org/10.1038/s41578-019-0156-6.

Cerchiara, T., A. Abruzzo, C. Parolin, B. Vitali, F. Bigucci, M. C. Gallucci, F. P. Nicoletta, and B. Luppi. 2016. "Microparticles Based on Chitosan/Carboxymethylcellulose Polyelectrolyte Complexes for Colon Delivery of Vancomycin." *Carbohydrate Polymers* 143 (June):124–30. https://doi.org/10.1016/j.carbpol.2016.02.020.

Chan, Lauretta M. S, Simon Lowes, and Barry H Hirst. 2004. "The ABCs of Drug Transport in Intestine and Liver: Efflux Proteins Limiting Drug Absorption and Bioavailability." *European Journal of Pharmaceutical Sciences* 21 (1): 25–51. https://doi.org/10.1016/j.ejps.2003.07.003.

Chauhan, Divya, Pavan K. Yadav, Nazneen Sultana, Arun Agarwal, Saurabh Verma, Manish K. Chourasia, and Jiaur R. Gayen. 2024. "Advancements in Nanotechnology for the Delivery of Phytochemicals." *Journal of Integrative Medicine*, April, S2095496424000529. https://doi.org/10.1016/j.joim.2024.04.005.

Colotti, Gianni, and Teresa Rinaldi. 2020. "The Central Role of Gut Microbiota in Drug Metabolism and Personalized Medicine." *Future Medicinal Chemistry* 12 (13): 1197–1200. https://doi.org/10.4155/fmc-2020-0023.

Czuba, Elodie, Mouhamadou Diop, Carole Mura, Anais Schaschkow, Allan Langlois, William Bietiger, Romain Neidl, et al. 2018. "Oral Insulin Delivery, the Challenge to Increase Insulin Bioavailability: Influence of Surface Charge in Nanoparticle System." *International Journal of Pharmaceutics* 542 (1–2): 47–55. https://doi.org/10.1016/j.ijpharm.2018.02.045.

Davies, Melanie, Thomas R. Pieber, Marie-Louise Hartoft-Nielsen, Oluf K. H. Hansen, Serge Jabbour, and Julio Rosenstock. 2017. "Effect of Oral Semaglutide Compared With Placebo and Subcutaneous Semaglutide on Glycemic Control in Patients With Type 2 Diabetes: A Randomized Clinical Trial." *JAMA* 318 (15): 1460. https://doi.org/10.1001/jama.2017.14752.

De France, Kevin J., Katelyn J. W. Chan, Emily D. Cranston, and Todd Hoare. 2016. "Enhanced Mechanical Properties in Cellulose Nanocrystal–Poly(Oligoethylene Glycol Methacrylate) Injectable Nanocomposite Hydrogels through Control of Physical and Chemical Cross-Linking." *Biomacromolecules* 17 (2): 649–60. https://doi.org/10.1021/acs.biomac.5b01598.

Ding, Ling, Prachi Agrawal, Sandeep K. Singh, Yashpal S. Chhonker, Jingjing Sun, and Daryl J. Murry. 2024. "Polymer-Based Drug Delivery Systems for Cancer Therapeutics." *Polymers* 16 (6): 843. https://doi.org/10.3390/polym16060843.

Du, Xiao-Jiao, Ji-Long Wang, Shoaib Iqbal, Hong-Jun Li, Zhi-Ting Cao, Yu-Cai Wang, Jin-Zhi Du, and Jun Wang. 2018. "The Effect of Surface Charge on Oral Absorption of Polymeric Nanoparticles." *Biomaterials Science* 6 (3): 642–50. https://doi.org/10.1039/C7BM01096F.

Dyawanapelly, Sathish, Uday Koli, Vimisha Dharamdasani, Ratnesh Jain, and Prajakta Dandekar. 2016. "Improved Mucoadhesion and Cell Uptake of Chitosan and Chitosan Oligosaccharide Surface-Modified Polymer Nanoparticles for Mucosal Delivery of Proteins." *Drug Delivery and Translational Research* 6 (4): 365–79. https://doi.org/10.1007/s13346-016-0295-x.

Elmowafy, Mohammed, Khaled Shalaby, Mohammed H. Elkomy, Omar Awad Alsaidan, Hesham A. M. Gomaa, Mohamed A. Abdelgawad, and Ehab M. Mostafa. 2023. "Polymeric Nanoparticles for Delivery of Natural Bioactive Agents: Recent Advances and Challenges." *Polymers* 15 (5): 1123. https://doi.org/10.3390/polym15051123.

Fallingborg, Jan, Lisbet A. Christensen, Margrethe Ingeman-Nielsen, Bent A. Jacobsen, Kirsten Abildgaard, Henrik H. Rasmussen, and Sten Nørby Rasmussen. 1990. "Measurement of Gastrointestinal pH and Regional Transit Times in Normal Children:" *Journal of Pediatric Gastroenterology and Nutrition* 11 (2): 211–14. https://doi.org/10.1097/00005176-199008000-00010.

Ferguson, Laura T., Xiaonan Ma, Jacob W. Myerson, Jichuan Wu, Patrick M. Glassman, Marco E. Zamora, Elizabeth D. Hood, et al. 2023. "Mechanisms by Which Liposomes Improve Inhaled Drug Delivery for Alveolar Diseases." *Advanced NanoBiomed Research* 3 (3): 2200106. https://doi.org/10.1002/anbr.202200106.

"Full Article: Extended-Release Morphine Sulfate in Treatment of Severe Acute and Chronic Pain." n.d. Accessed May 28, 2024. https://www.tandfonline.com/doi/full/10.2147/JPR.S6529#.

Gupta, Jagriti, Kishan Das, and Paulraj Rajamani. 2021. "Size-Responsive Differential Modulation in α-Amylase by MPA-CdSe QDs: Multispectroscopy and Molecular Docking Study." *Journal of Nanoparticle Research* 23 (8): 190. https://doi.org/10.1007/s11051-021-05298-y.

Gupta, Jagriti, Kishan Das, Ayushi Tanwar, Paulraj Rajamani, and Jaydeep Bhattacharya. 2022. "An Electrochemical Study of the Binding Interaction between Chitosan and MPA-CdSe QDs for the Development of Biocompatible Theranostic Nanoprobe." *Journal of Molecular Liquids* 358 (July):119193. https://doi.org/10.1016/j.molliq.2022.119193.

Gupta, Jagriti, Subhavna Juneja, and Jaydeep Bhattacharya. 2020. "UV Lithography-Assisted Fabrication of Low-Cost Copper Electrodes Modified with Gold Nanostructures for Improved Analyte Detection." *ACS Omega* 5 (7): 3172–80. https://doi.org/10.1021/acsomega.9b03125.

Gupta, Jagriti, and Paulraj Rajamani. 2023. "Size- and Surface Functionalization-Driven Molecular Interaction of CdSe Quantum Dots with Jack Bean Urease: Multispectroscopic, Thermodynamic, and AFM Approach." *Environmental Science and Pollution Research*, February. https://doi.org/10.1007/s11356-023-25356-3.

Hafez Ghoran, Salar, Andrea Calcaterra, Milad Abbasi, Fatemeh Taktaz, Kay Nieselt, and Esmaeil Babaei. 2022. "Curcumin-Based Nanoformulations: A Promising Adjuvant towards Cancer Treatment." *Molecules* 27 (16): 5236. https://doi.org/10.3390/molecules27165236.

Hattab, Dima, Amirah Mohd Gazzali, and Athirah Bakhtiar. 2021. "Clinical Advances of siRNA-Based Nanotherapeutics for Cancer Treatment." *Pharmaceutics* 13 (7): 1009. https://doi.org/10.3390/pharmaceutics13071009.

Herman, Timothy F., and Cynthia Santos. 2024. "First-Pass Effect." In *StatPearls*. Treasure Island (FL): StatPearls Publishing. http://www.ncbi.nlm.nih.gov/books/NBK551679/.

Homayun, Bahman, Xueting Lin, and Hyo-Jick Choi. 2019. "Challenges and Recent Progress in Oral Drug Delivery Systems for Biopharmaceuticals." *Pharmaceutics* 11 (3): 129. https://doi.org/10.3390/pharmaceutics11030129.

Hu, LianDong, Xing Tang, and FuDe Cui. 2010. "Solid Lipid Nanoparticles (SLNs) to Improve Oral Bioavailability of Poorly Soluble Drugs." *Journal of Pharmacy and Pharmacology* 56 (12): 1527–35. https://doi.org/10.1211/0022357044959.

Jackson, John K., and Kevin Letchford. 2016. "The Effective Solubilization of Hydrophobic Drugs Using Epigallocatechin Gallate or Tannic Acid-Based Formulations." *Journal of Pharmaceutical Sciences* 105 (10): 3143–52. https://doi.org/10.1016/j.xphs.2016.06.027.

Jia, Yuanchao, Yuxin Jiang, Yonglong He, Wanting Zhang, Jiahui Zou, Kosheli Thapa Magar, Hamza Boucetta, Chao Teng, and Wei He. 2023. "Approved Nanomedicine against Diseases." *Pharmaceutics* 15 (3): 774. https://doi.org/10.3390/pharmaceutics15030774.

Kali, Gergely, Patrick Knoll, and Andreas Bernkop-Schnürch. 2022. "Emerging Technologies to Increase Gastrointestinal Transit Times of Drug Delivery Systems." *Journal of Controlled Release* 346 (June):289–99. https://doi.org/10.1016/j.jconrel.2022.04.016.

Kumar, Rahul, Arvind Sirvi, Shamandeep Kaur, Sanjaya K. Samal, Sabyasachi Roy, and Abhay T. Sangamwar. 2020. "Polymeric Micelles Based on Amphiphilic Oleic Acid Modified Carboxymethyl Chitosan for Oral Drug Delivery of Bcs Class Iv Compound: Intestinal Permeability and Pharmacokinetic Evaluation." *European Journal of Pharmaceutical Sciences* 153 (October):105466. https://doi.org/10.1016/j.ejps.2020.105466.

Lang, Xuqian, Ting Wang, Mengjie Sun, Xiguang Chen, and Ya Liu. 2020. "Advances and Applications of Chitosan-Based Nanomaterials as Oral Delivery Carriers: A Review." *International Journal of Biological Macromolecules* 154 (July):433–45. https://doi.org/10.1016/j.ijbiomac.2020.03.148.

Levy, Elizabeth S., Karen E. Samy, Nicholas G. Lamson, Kathryn A. Whitehead, Deanna L. Kroetz, and Tejal A. Desai. 2019. "Reversible Inhibition of Efflux Transporters by Hydrogel Microdevices." *European Journal of Pharmaceutics and Biopharmaceutics* 145 (December):76–84. https://doi.org/10.1016/j.ejpb.2019.10.007.

Li, Bi-Lian, Jin Zhang, Wen Jin, Xiang-Yu Chen, Jian-Mei Yang, Shao-Ming Chi, Qiong Ruan, and Yan Zhao. 2022. "Oral Administration of pH-Responsive Polyamine Modified Cyclodextrin Nanoparticles for Controlled Release of Anti-Tumor Drugs." *Reactive and Functional Polymers* 172 (March):105175. https://doi.org/10.1016/j.reactfunctpolym.2022.105175.

Li, Huipeng, Minglei Chen, Zhigui Su, Minjie Sun, and Qineng Ping. 2016. "Size-Exclusive Effect of Nanostructured Lipid Carriers on Oral Drug Delivery." *International Journal of Pharmaceutics* 511 (1): 524–37. https://doi.org/10.1016/j.ijpharm.2016.07.049.

Liu, Yanhua, Tong Yang, Shijie Wei, Chengming Zhou, Yang Lan, Aichen Cao, Jianhong Yang, and Wenping Wang. 2018. "Mucus Adhesion- and Penetration-Enhanced Liposomes for Paclitaxel Oral Delivery." *International Journal of Pharmaceutics* 537 (1–2): 245–56. https://doi.org/10.1016/j.ijpharm.2017.12.044.

Lou, Jie, Hongli Duan, Qin Qin, Zhipeng Teng, Fengxu Gan, Xiaofang Zhou, and Xing Zhou. 2023. "Advances in Oral Drug Delivery Systems: Challenges and Opportunities." *Pharmaceutics* 15 (2): 484. https://doi.org/10.3390/pharmaceutics15020484.

Lozoya-Agullo, Isabel, Francisca Araújo, Isabel González-Álvarez, Matilde Merino-Sanjuán, Marta González-Álvarez, Marival Bermejo, and Bruno Sarmento. 2018. "PLGA Nanoparticles Are Effective to Control the Colonic Release and Absorption on Ibuprofen." *European Journal of Pharmaceutical Sciences* 115 (March):119–25. https://doi.org/10.1016/j.ejps.2017.12.009.

Maisel, Katharina, Laura Ensign, Mihika Reddy, Richard Cone, and Justin Hanes. 2015. "Effect of Surface Chemistry on Nanoparticle Interaction with Gastrointestinal Mucus and Distribution in the Gastrointestinal Tract Following Oral and Rectal Administration in the Mouse." *Journal of Controlled Release* 197 (January):48–57. https://doi.org/10.1016/j.jconrel.2014.10.026.

Manca, Maria Letizia, Matteo Aroffu, Federica Fulgheri, Matteo Perra, and Ines Castangia. 2024. "Conventional Methods for Preparing Liposomes of Various Types (MLVs, LUVs, SUVs)." In *Liposomes in Drug Delivery*, 461–88. Elsevier. https://doi.org/10.1016/B978-0-443-15491-1.00011-0.

Mitchell, Michael J., Margaret M. Billingsley, Rebecca M. Haley, Marissa E. Wechsler, Nicholas A. Peppas, and Robert Langer. 2021. "Engineering Precision Nanoparticles for Drug Delivery." *Nature Reviews Drug Discovery* 20 (2): 101–24. https://doi.org/10.1038/s41573-020-0090-8.

Moroz, Elena, Simon Matoori, and Jean-Christophe Leroux. 2016. "Oral Delivery of Macromolecular Drugs: Where We Are after Almost 100 Years of Attempts." *Advanced Drug Delivery Reviews* 101 (June):108–21. https://doi.org/10.1016/j.addr.2016.01.010.

O'Driscoll, Caitriona M., Andreas Bernkop-Schnürch, Julian D. Friedl, Véronique Préat, and Vincent Jannin. 2019. "Oral Delivery of Non-Viral Nucleic Acid-Based Therapeutics - Do We Have the Guts for This?" *European Journal of Pharmaceutical Sciences* 133 (May):190–204. https://doi.org/10.1016/j.ejps.2019.03.027.

Omachi, Fukiko, Masaki Kaneko, Ryosuke Iijima, Machiko Watanabe, and Fumio Itagaki. 2019. "Relationship between the Effects of Food on the Pharmacokinetics of Oral Antineoplastic Drugs and Their Physicochemical Properties." *Journal of Pharmaceutical Health Care and Sciences* 5 (1): 26. https://doi.org/10.1186/s40780-019-0155-1.

O'Neill, Martin J., Ludovic Bourre, Silvia Melgar, and Caitriona M. O'Driscoll. 2011. "Intestinal Delivery of Non-Viral Gene Therapeutics: Physiological Barriers and Preclinical Models." *Drug Discovery Today* 16 (5–6): 203–18. https://doi.org/10.1016/j.drudis.2011.01.003.

Prasad, Vinay, Kevin De Jesús, and Sham Mailankody. 2017. "The High Price of Anticancer Drugs: Origins, Implications, Barriers, Solutions." *Nature Reviews Clinical Oncology* 14 (6): 381–90. https://doi.org/10.1038/nrclinonc.2017.31.

Roger, Emilie, Stephen Kalscheuer, Ameya Kirtane, Bharath Raja Guru, Alex E. Grill, Judith Whittum-Hudson, and Jayanth Panyam. 2012. "Folic Acid Functionalized Nanoparticles for Enhanced Oral Drug Delivery." *Molecular Pharmaceutics* 9 (7): 2103–10. https://doi.org/10.1021/mp2005388.

Salawi, Ahmad. 2022. "Self-Emulsifying Drug Delivery Systems: A Novel Approach to Deliver Drugs." *Drug Delivery* 29 (1): 1811–23. https://doi.org/10.1080/10717544.2022.2083724.

Seo, Yoseph, Hayeon Lim, Hyunjun Park, Jiyun Yu, Jeongyun An, Hah Young Yoo, and Taek Lee. 2023. "Recent Progress of Lipid Nanoparticles-Based Lipophilic Drug Delivery: Focus on Surface Modifications." *Pharmaceutics* 15 (3): 772. https://doi.org/10.3390/pharmaceutics15030772.

Shan, Xiaoting, Xiang Gong, Jie Li, Jingyuan Wen, Yaping Li, and Zhiwen Zhang. 2022. "Current Approaches of Nanomedicines in the Market and Various Stage of Clinical Translation." *Acta Pharmaceutica Sinica B* 12 (7): 3028–48. https://doi.org/10.1016/j.apsb.2022.02.025.

Singh, Suruchi, Pankaj Bhatt, Satish Kumar Sharma, and S. Rabiu. 2021. "Digital Transformation in Healthcare: Innovation and Technologies." In *Blockchain for Healthcare Systems*, by Sheikh Mohammad Idrees, Parul Agarwal, and M. Afshar Alam, 1st ed., 61–79. Boca Raton: CRC Press. https://doi.org/10.1201/9781003141471-5.

Valon, Léo, and Romain Levayer. 2019. "Dying under Pressure: Cellular Characterisation and *in Vivo* Functions of Cell Death Induced by Compaction." *Biology of the Cell* 111 (3): 51–66. https://doi.org/10.1111/boc.201800075.

Vinarov, Zahari, Mohammad Abdallah, José A. G. Agundez, Karel Allegaert, Abdul W. Basit, Marlies Braeckmans, Jens Ceulemans, et al. 2021. "Impact of Gastrointestinal Tract Variability on Oral Drug Absorption and Pharmacokinetics: An UNGAP Review." *European Journal of Pharmaceutical Sciences* 162 (July):105812. https://doi.org/10.1016/j.ejps.2021.105812.

Viswanathan, Pragasam, Yuvashree Muralidaran, and Gokulakannan Ragavan. 2017. "Challenges in Oral Drug Delivery: A Nano-Based Strategy to Overcome." In *Nanostructures for Oral Medicine*, 173–201. Elsevier. https://doi.org/10.1016/B978-0-323-47720-8.00008-0.

Vyas, Kunj, Maharshsinh Rathod, and Mayur M. Patel. 2023. "Insight on Nano Drug Delivery Systems with Targeted Therapy in Treatment of Oral Cancer." *Nanomedicine: Nanotechnology, Biology and Medicine* 49 (April):102662. https://doi.org/10.1016/j.nano.2023.102662.

Waheed, Saquib, Zhibin Li, Fangyingnan Zhang, Anna Chiarini, Ubaldo Armato, and Jun Wu. 2022. "Engineering Nano-Drug Biointerface to Overcome Biological Barriers toward Precision Drug Delivery." *Journal of Nanobiotechnology* 20 (1): 395. https://doi.org/10.1186/s12951-022-01605-4.

Wang, Taoran, Erika Fleming, and Yangchao Luo. 2023. "An Overview of the Biochemistry, Synthesis, Modification, and Evaluation of Mucoadhesive Polymeric Nanoparticles for Oral Delivery of Bioactive Compounds." *Advanced Composites and Hybrid Materials* 6 (1): 6. https://doi.org/10.1007/s42114-022-00586-0.

Wang, Yuanyuan, Chao Pi, Xianhu Feng, Yi Hou, Ling Zhao, and Yumeng Wei. 2020. "The Influence of Nanoparticle Properties on Oral Bioavailability of Drugs." *International Journal of Nanomedicine* Volume 15 (August):6295–6310. https://doi.org/10.2147/IJN.S257269.

Wilson, Russell J., Yang Li, Guangze Yang, and Chun-Xia Zhao. 2022. "Nanoemulsions for Drug Delivery." *Particuology* 64 (May):85–97. https://doi.org/10.1016/j.partic.2021.05.009.

Yamagata, Tetsuo, Mariko Morishita, Nikhil J. Kavimandan, Koji Nakamura, Yu Fukuoka, Kozo Takayama, and Nicholas A. Peppas. 2006. "Characterization of Insulin Protection Properties of Complexation Hydrogels in Gastric and Intestinal Enzyme Fluids." *Journal of Controlled Release* 112 (3): 343–49. https://doi.org/10.1016/j.jconrel.2006.03.005.

Zhang, Xintong, Ying Han, Wei Huang, Mingji Jin, and Zhonggao Gao. 2021. "The Influence of the Gut Microbiota on the Bioavailability of Oral Drugs." *Acta Pharmaceutica Sinica B* 11 (7): 1789–1812. https://doi.org/10.1016/j.apsb.2020.09.013.

Zhang, Yu, Saji Uthaman, Wenliang Song, Kuen Hee Eom, Su Hyeon Jeon, Kang Moo Huh, Amal Babu, In-Kyu Park, and Il Kim. 2020. "Multistimuli-Responsive Polymeric Vesicles for Accelerated Drug Release in Chemo-Photothermal Therapy." *ACS Biomaterials Science & Engineering* 6 (9): 5012–23. https://doi.org/10.1021/acsbiomaterials.0c00585.

Zhu, Ying, Xinru You, Keqing Huang, Faisal Raza, Xin Lu, Yuejian Chen, Arvind Dhinakar, et al. 2018. "Effect of Taste Masking Technology on Fast Dissolving Oral Film: Dissolution Rate and Bioavailability." *Nanotechnology* 29 (30): 304001. https://doi.org/10.1088/1361-6528/aac010.

Chapter 4.4

Advancements in Nanotechnology for Age-Related Macular Degeneration: Innovations in Nanodiagnostics and Nanotherapeutics

Dissanayakage Dilshan Sampath Dissanayaka*
Department of Biomedical Sciences, International College of Business and Technology, Kandy, Sri Lanka

Abstract

Age-related macular degeneration (AMD) is a leading cause of irreversible vision loss in elderly people due to complications in the retinal pigment epithelium (RPE) and macular tissues, thus presenting a significant challenge to public health. AMD exists in two forms: dry (nonexudative) and wet (exudative), based on the progression of the disease. Conventional treatments such as anti-vascular endothelial growth factor (anti-VEGF) intravitreal medications offer therapeutic benefits but often come with challenges and side effects, such as endophthalmitis. Nanotechnology has shown promising results for various diseases, including AMD. However, novel therapies, including gene therapy, show promising outcomes with a single intravitreal injection. Over the years, nanotechnology has substantially progressed in treating diseases, including AMD, by offering precise drug delivery and immune modulation within the retina. This chapter of the book explains the recent advancements in nanotechnology-enabled immunotherapy for AMD, including the role of nano-drug delivery systems in AMD treatment. This chapter also examines nanodiagnostics' efficacy and their challenges in modulating the retinal immune microenvironment associated with AMD progression. Overall, this chapter provides a comprehensive overview of the current and future aspects of nanotechnology-enabled immunotherapy for AMD, highlighting its capability to improve vision care.

Keywords: age-related macular degeneration, nanotherapeutics, nanodiagnostics, drug delivery system, retinal pigment epithelium

* Corresponding Author's Email: dilshansampathdissanayaka93@gmail.com

In: Emerging Nanomaterials in Biomedical Science
Editors: Eepsita Priyadarshini and Ajith Manayil Parambil
ISBN: 979-8-89530-600-0

1. Introduction

One of the major causes of blindness is age-related macular degeneration (AMD), which accounts for approximately 9% of all vision impairment cases and affects 200 million people globally (Vyawahare and Shinde 2022). The retina, a light-sensitive tissue lining the back of the eye, is divided into two major areas: the macula and the peripheral retina. The peripheral retina provides wide-angle vision, while the macula controls central vision (Boyd 2023). The macula, a slightly oval-shaped area with diameter about 5 mm, is located below the center of the optic disc. Like the rest of the retina, the macula converts light energy into electrical signals that travel through the optic nerve to the brain (Deng et al. 2022). AMD is a medical condition that primarily affects people over the age of 50 and is caused by the progressive deterioration of the macula. Based on the progression of the pathology in the macular, AMD is grouped into four stages, starting from normal aging changes to the advanced atrophic stage. Stage 1 involves small drusen (<63 μm) without pigment abnormalities. Stage 2 (early AMD) features some intermediate-sized drusen (63–124 μm) but no abnormalities in the Retinal Pigment Epithelium (RPE). Stage 3 (intermediate AMD) is characterized by a significant number of moderate drusen, at least one large drusen (≥125 μm), and abnormalities in the RPE. Stage 4 (Advanced AMD) is defined by geographic atrophy (GA) of any neovascular changes associated with macular degeneration, resulting in visual loss (Keenan, Cukras, and Chew 2021; Deng et al. 2022). Advanced AMD is classified into two categories based on the progression: neovascular (wet AMD) and non-vascular or geographic atrophy (dry AMD) (Lim et al. 2012). Dry AMD is common, whereas wet AMD is uncommon, but its severity is much higher than dry AMD. Dry AMD is characterized by the growth of tiny clumps of protein called drusen in the RPE. Wet AMD is characterized by abnormal blood vessels growing under the retina that may leak blood or other fluids, resulting in scarring of the macula (Ammar et al. 2020) (Table 1).

Table 1. Differences between dry and wet AMD

Factor	Dry AMD (nonexudative or non-neovascular)	Wet AMD (exudative or neovascular)
Visual impairment	Gradually loss the central vision	Rapid and severely loss the central vision
Prevalence	Common (80-90% of all cases)	Less common (10-20% of all cases), but most common form of advanced AMD
Stages	Early, Intermediate, and Advanced	Always advanced stage
Pathophysiology	Featured by presence of protein deposits (drusen) in early or intermediate stages and GA in advanced stages	Characterized by abnormal CNV, leading to haemorrhaging, fluid leakage, and fibrosis in the retinal layers or subretinal space
Symptoms	No prominent symptoms in the early stage. The intermediate stage includes symptoms such as minor blurriness in the central vision. Symptoms in the advance stage are like wet AMD.	Straight lines appearing curved or distorted, decreased central vision in one or both eyes, requiring brighter light for reading or close-up tasks, trouble adjusting to low light conditions, a distinct blurry spot or blind spot in the visual field.

1.1. Epidemiology

Epidemiological information explains that AMD is considered the third leading cause (8 million) of distance vision impairment, followed by glaucoma (7.7 million) and diabetic retinopathy (3.9 million). Cataract is the highest contributor to global distance vision impairment, accounting for 94 million cases among 1 billion people in 2020 (Steinmetz et al. 2021) (Figure 1).

The likelihood of vision loss due to AMD increases with age. It is a leading cause of legal blindness in the U.S., impacting over 14% of white Americans aged 80 and older, according to the National Eye Institute (NEI). Caucasian Americans are the most at risk for developing AMD, making up 89% of cases, while black and Hispanic Americans each represent 4% of cases. NEI's data projects that as the U.S. population aged 65 and older grows, the number of AMD cases is expected to rise from 2.07 million to 5.44 million by 2050 (Nashine 2021).

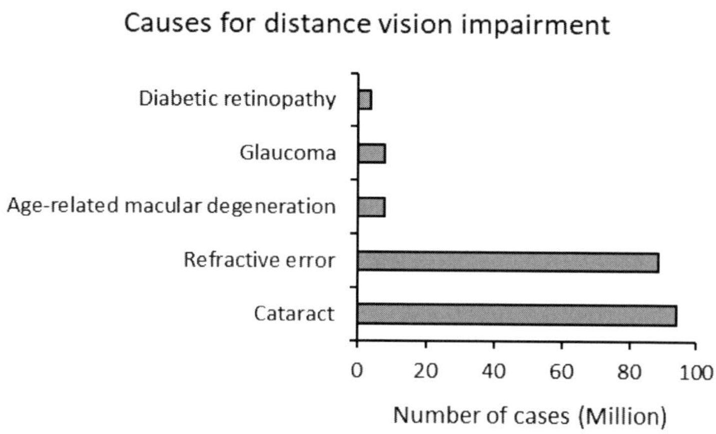

Figure 1. Causes for global prevalence of distance vision impairment or blindness: information extracted from (Steinmetz et al. 2021).

1.2. General Risk Factors

Factors associated with AMD are classified into two broad areas: personal and environmental factors, with genetic predisposition, age, sex, and iris color falling under the former, whereas nicotine consumption, nutritional factors, and sunlight exposure are categorized as environmental factors (Stahl 2020; Gheorghe, Mahdi, and Musat 2015). However, according to Nowak (2006), there are four major contributors other than genetic predisposition involved in the development of AMD: lipofuscinogenesis (formation and accumulation of lipofuscin within the RPE), drusogenesis (drusen protein deposit formation and accumulation), local inflammation, and neovascularization. Further, genes such as CFH, ARMS2-HTRA1, C2-CFB-SKIV2L, and C3 have a significant association with AMD risk, according to consortium case-control studies (Fritsche et al. 2013). Additionally, compounds such as high-density lipoprotein cholesterol, omega-3 fatty acids (docosahexaenoic acid and eicosapentaenoic acid), and carotenoid pigment have a significant risk impact on the progression of AMD (Mrowicka et al. 2022). Moreover, regular usage of some medications, such as aspirin, which is used for cardiovascular diseases, has a significant association with augmenting the risk of getting wet AMD. However, the effects of non-steroid anti-inflammatory drugs (acetaminophen) on wet AMD progression are negligible (Liew et al. 2013; Wang, Wang, and Chan 2011; Dissanayaka et al. 2023).

2. Pathophysiology

The pathogenesis of AMD is quite complex and elucidated via histopathologic dissection and genetic tests (Warwick and Lotery 2018). The aging process of the eye is characterized by the accumulation of cellular debris, originating from the RPE, at the junction where the RPE meets Bruch's membrane and the neurosensory retina. These drusen are often the earliest observable indicators of AMD detectable via ophthalmoscopy, preceding any significant impact on visual function (Khan et al. 2016). Drusen are complex structures predominantly composed of lipids, proteins, and carbohydrates and manifest as small white or yellowish deposits on the macula (Figure 2). The deposition of drusen within Bruch's membrane, alongside other structural and biochemical alterations associated with AMD, such as chronic activation of the complement system and inflammation, results in the thickening and reduced permeability of the membrane (Algvere, Kvanta, and Seregard 2016). This obstruction hinders the transport of nutrients to the retina and the exchange of waste products to the choroid, and it is accompanied by the thinning of the choroidal vasculature. These processes, along with neurodegenerative changes within the photoreceptor-RPE complex, lead to pigmentary abnormalities in the RPE, such as hypo- or hyperpigmentation, which are characteristic of the early or intermediate stages of the disease (Flores et al. 2021).

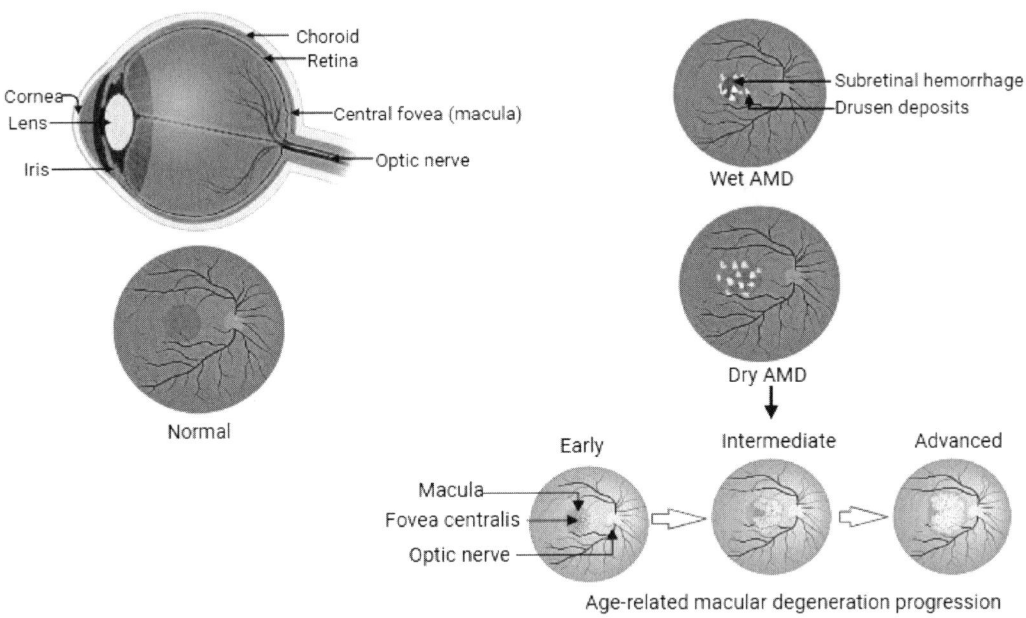

Figure 2. Pathology and progression of AMD. Structure of a healthy eye (left side), main types of AMD (right side-top), and the progression of AMD (right side-bottom). The image was created in BioRender.com.

Dry AMD is marked by a gradual decline in visual function, caused by the degeneration of the choriocapillaris, atrophy of the outer retina, and disruption and eventual death of the photoreceptor layer. During the progression of AMD, drusen accumulates and undergoes morphological changes. Small, well-defined drusen particles, known as hard drusen, are frequently observed in younger individuals and do not indicate an increased risk of AMD (Jhingan et al. 2021).

The development of wet AMD involves the recruitment of immune cells to the damaged macula and the release of proinflammatory and proangiogenic cytokines, particularly Vascular Endothelial Growth Factor (VEGF). VEGF plays a crucial role as an angiogenic factor for vascular endothelial cells. VEGF is produced by numerous cells, including cancer cells and keratinocytes (Duffy, Bouchier-Hayes, and Harmey 2013, Liu et al. 2021, Chen, Chan, et al. 2021). Treatments are designed to target the pro-angiogenesis function of VEGF. Various retinal cell types, including RPE cells, can produce and secrete VEGF. VEGF promotes endothelial cell proliferation and migration, leading to angiogenesis and increased vascular permeability. The newly formed blood vessels leak fluid, which disrupts and damages the photoreceptor layer, impairing vision. While these abnormal vessels typically arise from choroidal circulation, neovascularization and leakage can also originate from the retinal vasculature. Without treatment, this subretinal exudation leads to fibrosis and atrophy of the macula, resulting in irreversible central vision loss within a few months (Garcia-Garciaetal. 2022).

3. Nanomaterial-Based Diagnosis of AMD

Often, imprecise diagnoses can prolong treatment and increase the risk of additional complications, prompting the need for combined therapies. Today's diagnostic methods predominantly utilize imaging technologies, transitioning from traditional film-based to modern digital techniques. These modalities furnish various diagnostic features essential for precise evaluations and timely interventions (Sarkar and Dyawanapelly 2021). The initial stage in diagnosing AMD involves gathering a thorough medical history and conducting a thorough eye examination, including tests for visual acuity and Amsler grid assessment (Apte 2021). The following table summarizes the development of diagnostic approaches in diagnosing AMD (Table 2).

3.1. Nanodiagnostics

Nanotechnology is characterized as technology that operates on a scale ranging from 1 to 100 nanometers in at least one dimension. Moreover, the term 'nanoparticle' is quite extensive and can encompass both organic (such as lipids and biopolymers) and inorganic nanomaterials (like metals, oxides, and carbon) that come in various shapes like cubes, stars, needles, spheroids, or custom-designed geometries, all of which have an aerodynamic diameter of less than 100 nm (Contera, Bernardino de la Serna, and Tetley 2020). Nanomaterials can be created from various substances: silicates, non-oxide ceramics, and metal oxides. These materials have unique features, such as an optimal size, improved solubility, easy access to cell barriers, and increased reactivity due to their distinct structural properties (Rezaei et al. 2019). Nanoparticles are designed to enhance drug effectiveness by improving drug delivery efficiency while minimizing side effects. They can also enhance imaging techniques, facilitating a more accurate diagnosis of diseases. Furthermore, nanomaterials are applied to develop scaffolds for tissue regeneration, improving quicker and more effective healing of damaged tissues (Lin et al. 2015). Delivering drugs using nanotechnology in the body provides benefits such as controlled release, injectable and sterilizable formulations, and extended shelf life after lyophilization. Moreover, incorporating drugs and genetic materials into nanoparticles can shield them from immediate dilution and degradation and help overcome issues related to drug solubility. Nanoparticles can be engineered to transport a wide range of molecules, including proteins, DNA, lipids, and various organic and inorganic substances (Prow et al. 2006; Dharmapriya et al. 2023). Polymer-based nanoparticles, constructed from biocompatible materials like poly-lactic-co-glycolic acid, chitosan, polyethylene glycol, and hyaluronic acid, offer non-toxic and biodegradable options for drug delivery. Incorporating bio- or mucoadhesive polymers like chitosan can further enhance drug bioavailability by extending residence time in the mucosal layer. Additionally, lipid-based nanoparticles, including liposomes, solid lipid nanoparticles, and nanostructured lipid carriers, offer versatile platforms for encapsulating therapeutic cargo, thereby expanding the possibilities for targeted drug delivery in ocular applications (Torrecillas-Picazo et al. 2017).

Table 2. Conventional diagnostic methods for identifying AMD

Diagnosis method	Principle and application	Limitations	Reference
Amsler grid	The grid consists of lines and a central dot, crucial for engaging the macula for clear vision; distortions or missing areas may reveal retinal abnormalities	It identifies macular issues but may overlook peripheral retinal problems, especially since early-stage retinal diseases can present changes too subtle for patients to notice	(Bjerager et al. 2023, Loewenstein et al. 2003)
Fundus autofluorescence	Natural fluorescence from lipofuscin and other compounds in the RPE is detected under specific light wavelengths, typically obtained at 488 nm (blue light), emitting yellow-green light in the 510-560 nm range. This fluorescence is used to create images that depict the condition and spatial distribution of RPE cells	Autofluorescence patterns can vary among different diseases, and some conditions may present similar autofluorescence characteristics, requiring additional diagnostic tests for accurate diagnosis	(Yung, Klufas, and Sarraf 2016, Frampton et al. 2017)
Fundus Photography	Captures detailed images of the retina and optic nerve, highlighting damage indicators like lipid exudates, microaneurysms, and pigmentary changes caused by retinal cell damage	Cannot produce accurate images in the presence of macular edema without stereoscopic imaging, high labor demands, long duration, and limited precision	(Kanagasingam et al. 2014, Wintergerst et al. 2020)
Fluorescein Angiography	Injecting sodium fluorescein to visualize CNV and retinal complications	Invasive, time-consuming and can cause nausea and anaphylaxis	(Roisman and Goldhardt 2017, Tomi, Marin, and life 2014)
Optical Coherence Tomography	Generates 3D cross-sectional images of the retina, estimating layer thickness and visualizing drusen as RPE deformations and vitreoretinal interface abnormalities	Inability to visualize certain structures such as the RPE and choroid in detail due to their high scattering properties, challenges in imaging through media opacities such as cataracts or vitreous hemorrhage	(Huang et al. 2018, Schmidt-Erfurth et al. 2017)
Deep Learning technology in Artificial Intelligence	Identify early AMD biomarkers accurately quantify drusen load by analyzing layer positions	Potential biases in training data, struggle in generalizing across diverse populations, or adapting to variations in imaging quality	(Yildirim et al. 2023, Morelle et al. 2023)

3.2. Applications and Challenges of Nanodiagnostics for AMD

Nanotechnology in biomedical science is an emerging field that provides advancements in drug delivery, diagnostic imaging, and tissue engineering. Most conventional ocular treatments focus on minimizing macular edema, or CNV, which occurs due to the development of vitreous VEGF; periodic intraocular injections may increase the need for surgery. Pharmacotherapies for AMD often involve periocular or intraocular administration to improve therapeutic concentrations in target tissues. However, rapid clearance of agents remains a challenge, necessitating sustained drug release and coupling drugs with liposomes, microparticles, or nanoparticles (Birch and Liang 2007). When administering a nanoparticle-encapsulated drug via injection, it must traverse several barriers within the eye. Initially, it encounters vitreous humor, a gel-like matrix comprising collagen fibers and hyaluronic acid. The movement of nanoparticles within this environment relies on their diffusion rate and interactions with binding sites. Subsequently, the nanoparticles confront the inner limiting membrane, which consists of an extracellular matrix and serves as a barrier separating the retina from the vitreous. Finally, the nanoparticles reach Müller cells, from the vitreal border to the outer nuclear membrane, where photoreceptor cells are located, presenting the final barrier to drug delivery within the eye (Zhang et al. 2024).

Nanodiagnostics tools have become increasingly essential in the early identification, continuous monitoring, and assessment of treatment effectiveness for AMD. Numerous nanodiagnostics, including AMD, are used for disease diagnosis, as depicted in the diagram (Figure 3).

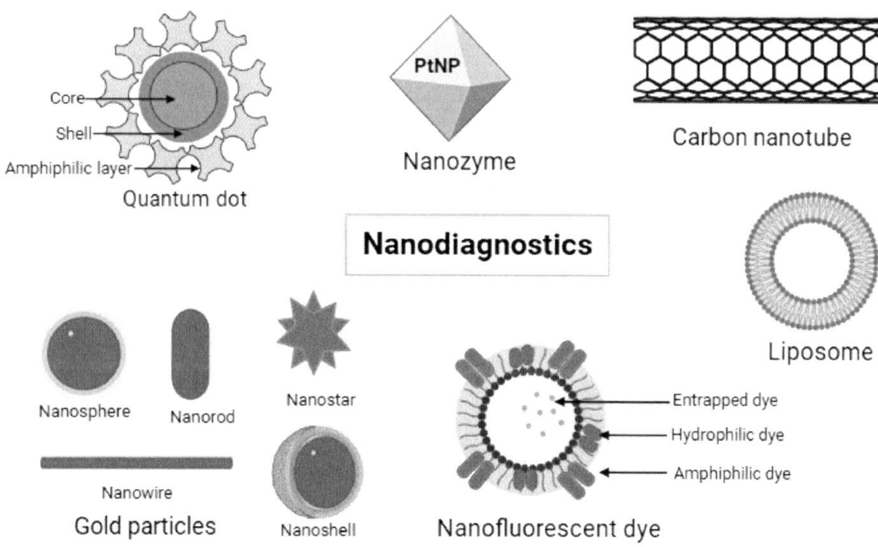

Figure 3. Different nanodiagnostics are used for AMD diagnosis. The image was created in BioRender.com.

3.2.1. Quantum Dots

Quantum dots (QDs) have diverse uses in ophthalmology, from bioimaging to gene/stem cell tracking and monitoring ocular lymphatics. These semiconductor nanoparticles offer unique optical properties, enabling high-resolution imaging of retinal cells and early AMD detection.

Smaller QDs are often more effective. The most common biocompatible QDs are cadmium-selenium QDs with a zinc sulfide core, which are commercially available in various sizes. Their core-shell structure allows them to cross biological barriers like the blood-retinal barrier without disrupting their constituents. At the same time, the shell enhances inertness and reduces harmful interactions with living tissues (Sarwat et al. 2019). Complications of QDs include potential toxicity due to heavy metal components such as cadmium, limitations in long-term stability and biocompatibility, and concerns regarding environmental impact and disposal (Jamieson et al. 2007).

3.2.2. Gold Nanoparticles

Gold nanoparticles have emerged as promising tools in ophthalmology, offering enhanced imaging and potential treatment for conditions such as corneal and CNV, as well as nAMD. In its nanoparticle form, gold exhibits a unique property called surface plasmon resonance, generating strong electromagnetic fields on its surface and enhancing its radiative properties (Chen, Si, et al. 2021). These nanoparticles can be delivered to the eye through various methods, including intravenous, intravitreal, and topical administration. Notably, gold nanoparticles around 20 nm in size can effectively cross the blood-retinal barrier and disperse throughout the retinal layers without causing harm. Recent research has found that smaller gold nanoparticles, specifically those sized 3–5 nm, are non-toxic to cells but effectively inhibit the migration of VEGF-induced choroid-retina endothelial cells *in vitro* (Chan et al. 2019). Gold nanoparticles have drawbacks, including potential toxicity, particularly due to the heavy metal component, and limitations in long-term stability and biocompatibility (Arvizo, Bhattacharya, and Mukherjee 2010).

3.2.3. Liposome

Liposomes are versatile lipid vesicles that encapsulate hydrophilic and hydrophobic agents, enhancing medical diagnostics and treatments. For AMD, liposomes can carry imaging agents like fluorescent dyes and MRI contrast agents, which accumulate in retinal tissues for improved imaging and diagnosis (Alavi and Hamidi 2019). Functionalized with specific ligands or antibodies, they target AMD biomarkers, boosting diagnostic accuracy. Though generally biocompatible, challenges include maintaining stability and effective delivery to the retina. Despite these hurdles, liposomes show great potential for protecting nano- capsulated drug molecules from degradation and prolonging their duration in the vitreous matrix (Zhang et al. 2024).

3.2.4. Carbon Nanotubes

Carbon nanomaterials stand out among nanoparticles due to their high chemical resistance, excellent mechanical properties, light weight, and electrical conductivity. The most commonly used carbon nanomaterials are nanodiamonds, carbon nanotubes, and graphene-related materials (GRM) (Simon, Flahaut, and Golzio 2019). Carbon nanotubes (CNTs) can enhance imaging techniques like MRI and CT scans by targeting retinal tissues affected by AMD, providing improved contrast and resolution. Additionally, CNTs can be conjugated with molecules that bind to AMD biomarkers for precise localization and used in highly sensitive biosensors to detect AMD-related biomarkers at low concentrations. Despite biocompatibility and delivery efficiency challenges, CNTs hold promise for advancing early and accurate AMD diagnosis. However, the production of CNTs can be expensive, limiting their widespread use in diagnostics (Lu et al. 2013).

3.2.5. Nanozymes

Nanozymes offer advantages over natural enzymes, including ease of synthesis, cost-effective scalability, and robust stability in diverse environmental conditions. They encompass various types, such as ROS-scavenging and metal-based nanozymes. For example, cerium (Ce)-based nanozymes protect against oxidative stress in the eyes and potentially slow diabetic cataract progression. Iron (Fe)-based nanozymes, like those containing Prussian blue (PB), reduce ROS production in AMD (Wong, Wong, and Liu 2024). Platinum-based nanoparticles (PtNPs) have emerged as powerful nanozymes that mimic natural antioxidant enzymes such as peroxidase, catalase, and superoxide dismutase. Biocompatible PtNPs, coated with a biocorona, are internalized by cells and exhibit enhanced enzymatic activity in acidic lysosomal environments, demonstrating superior stability, biocompatibility, and anti-inflammatory potential (Cupini et al. 2023). Challenges include concerns over biocompatibility and long-term effects, difficulties in targeted delivery, regulatory hurdles, costs of production, and the need for optimization in catalytic efficiency and stability under physiological conditions (Yadav and Maurya 2021).

4. Treatments for AMD

First-generation neovascular AMD (nAMD) treatments, including laser photocoagulation, radiotherapy, and angiostaticsteroids aim to prevent visual sharpness loss rather than enhance visual function. Laser photocoagulation targets the extrafoveal lesion by destroying the CNV that grows under the macular, resulting in blind spots (scotomas) in the central visual field (Virgili and Bini 2007). Photodynamic therapy with Verteporfin (as a photosensitizer) also depicts promising results in treating wet AMD disrupting CNV via targeting newly synthesized blood vessels in CNV lesions without interfering with other choroidal vessels and retinal pigment epithelium (Amoaku 2010). The discovery that the Vascular Epithelial Growth Factor drives CNV and that its inhibition can counteract this process has been a significant advancement in ophthalmology (Pożarowska and Pożarowski 2016). Anti-VEGF agents targeting VEGF or its receptors have outperformed previous treatments by improving and maintaining visual function and are now the primary treatment for wet AMD. Annually, around 1 million patients receive VEGF inhibitors for nAMD and other retinal diseases. Current VEGF inhibitors administered via intravitreal injection include Ranibizumab (brand name: Lucentis), Bevacizumab (brand name: Avastin), Aflibercept (brand name: Eylea), and Brolucizumab (brand name: Beovu). The recommended dosing schedules are every 4–8 weeks for Ranibizumab and Bevacizumab, every 12 weeks for Aflibercept, and every 8–12 weeks for Brolucizumab (Flaxel et al. 2020). In contrast, dry AMD is more common, responsible for fewer severe vision loss cases, and lacks preventive or therapeutic options (Flores et al. 2021).

There is a significant association between antioxidants and the progression of AMD. The accumulation of reactive oxygen species (ROS) in the retina generates oxidative stress, directly impacting dry AMD progression. However, the ocular antioxidant defense system, consisting of enzymatic and non-enzymatic substances such as melatonin, melanin, and zeaxanthin, plays a crucial role in scavenging free radicals and mitigating this oxidative damage (Čolak and Žorić

2019). A research group working on age-related eye diseases, conducted a project to ascertain the protective effects of antioxidants on dry AMD progression through a multicenter, randomized, double-masked trial. Participants were categorized into four AMD severity levels (from category 1: no AMD to category 4: advanced AMD) and assigned to receive oral daily supplementation with antioxidants (vitamin C 500 mg, vitamin E 400 IU, β-carotene 15 mg), zinc (zinc oxide 80 mg and copper 2 mg), a combination of both, or a placebo. The study found that progression to advanced AMD over five years of control patients in category 2 was 1.3%, 18% for category 3, and 43% for category 4. The combination of antioxidants and zinc significantly reduced the risk of progression in patients in categories 3 and 4 by 27% (Age-Related Eye Disease Study Research Group 2001). Also, omega-3 depicted significant therapeutic effects in visual parameters while decreasing drusen cover in early AMD patients (Feher et al. 2005). However, antioxidants such as anthocyanin supplements didn't significantly improve the development of macular pigment optical density (Olmedilla-Alonso et al. 2018). Further, Cho et al. (2004) stated that the effects of antioxidants such as vitamins and carotenoid supplements had no significant association with AMD progression. Antioxidant supplements have both benefits and drawbacks. It may be prudent to prescribe them only to patients who meet the specific criteria, carefully accounting for the risk-to-benefit ratio for everyone (Banerjee, Chawla, and Kumar 2021).

SS-31 is a water-soluble antioxidant targeting mitochondria and is part of the Szeto-Schiller family. It shows antioxidant and anti-apoptotic effects in various diseases, including neurodegeneration (Nashine 2021). Preclinical studies indicate that SS-31 improves retinal function, reduces TUNEL-positive cells, downregulates BAX, inhibits cytochrome c release, and mitigates mitochondrial dysfunction in the retina. According to Chen et al. (2011), SS-31 offers significant cytoprotection against hydrogen peroxide-induced oxidative stress in ocular cell lines, preventing mitochondrial depolarization, reducing ROS and LDH release, alleviating apoptosis, inhibiting cytochrome c release, suppressing Caspase-3 activation, and maintaining cellular integrity. SS-31 accumulates in the inner mitochondrial membrane without entering the mitochondrial matrix, making it safe without effects on normal mitochondria. Clinical studies reported that the main adverse effects were mild injection site reactions, including erythema (57%), pruritus (47%), pain (20%), urticaria (20%), and irritation (10%) (Zhu et al. 2022).

4.1. Challenges in Current Ocular Treatments

Systemic administration, via oral and intravenous routes, delivers drugs to the posterior eye segment through choroidal capillaries. However, the blood-retinal barrier and blood-aqueous barrier limit drug penetration, often necessitating higher doses for efficacy. This increase in dosage heightens the risk of drug toxicity, potentially causing adverse systemic effects (Begines et al. 2020). On the topical route, conventional formulations like ointments, eye drops, and suspensions offer good patient compliance but face significant challenges. Effective drug delivery is hindered by lacrimation, which can wash away drugs before they are absorbed, and the corneal epithelium barrier, which drugs must penetrate. Additionally, interactions with tear film enzymes, proteins, and anterior chamber molecules can reduce drug efficacy (Bonilla et al. 2021). Sub-conjunctival injection delivers drugs to the anterior eye segment, achieving higher concentrations compared to topical application. Administered beneath the conjunctival membrane, it bypasses the conjunctiva-cornea barrier, facilitating direct access to the transscleral route. However, drawbacks include limited drug concentration reaching the retina due to potential systemic elimination and decreased bioavailability caused by tear drainage (Afarid, Mahmoodi, and Baghban 2022). Though not currently utilized clinically, the suprachoroidal administration route holds promise for delivering drugs to the posterior eye segment. This method involves administering drugs via microneedles into the supracervical space under the inner surface of the sclera. However, drawbacks include potential complications such as choroidal detachment and suprachoroidal hemorrhage, as well as challenges related to accessing the suprachoroidal space. Additionally, the method is invasive and complex, limiting its widespread adoption (Ranta et al. 2010, Liu et al. 2012). The Subtenon route delivers drugs between Tenon's capsule and the sclera, providing extended contact time. Safer than intravitreal injection, it avoids needle-associated risks. However, concerns over increased intraocular pressure, worsening cataracts, and reduced efficacy compared to intravitreal steroids have decreased usage, with alternative methods gaining popularity (Ghate et al. 2007; Thorne et al. 2019). Intravitreal injection involves injecting ophthalmic drugs directly into the vitreous through the orbicularis ciliaris using a 30G injection needle, enhancing drug absorption compared to topical and systemic delivery. Targeting a specific site allows for greater drug delivery to the posterior eye segment than systemic administration. However, intravitreal administration may lead to technical complications such as intravitreal hemorrhages, endophthalmitis, trauma to the lens, and retinal detachment (Awwad et al. 2020; Mehta et al. 2018).

4.2. Nanotherapeutics for AMD

Nanotherapeutics for AMD offer prodigious benefits over conventional treatments, including targeted drug delivery, protection from degradation, sustained release, and deeper tissue penetration, potentially leading to more effective and patient-friendly treatment options. They are available in numerous forms and demonstrate different physiological mechanisms (Figure 4). The biocompatibility and biodegradability of polymeric nanoparticles like polylactic acid (PLA) and polylactic co-glycolic acid (PLGA) have been thoroughly investigated to deliver medicinal molecules, especially proteins and peptides. PLA- and PLGA-nanoparticles are spherical, with sizes between 300 and 400 nm for PLA and 200 and 300 nm for PLGA (Cruz et al. 2017). Natural metabolites like lactic and glycolic acids, which are quickly excreted from the body, can be produced by hydrolyzing PLGA. Bhatt et al. (2020) investigated the sustained release properties and therapeutic potential of resveratrol, a naturally occurring antioxidant and anti-inflammatory polyphenol, by loading it into PLGA polymeric nanoparticles. The entrapment efficiency of the nanoparticles was determined to be 65.21%. The *in vitro* study demonstrated that these resveratrol-loaded nanoparticles effectively reduced the expression of VEGF protein, indicating their potential for treating nAMD.

Sirolimus, also known as rapamycin, is an immunosuppressive drug widely used to treat and manage various diseases. Suri's group in 2021 aimed to develop sirolimus-loaded PLGA nanoparticles (SIR-PLGA-NP) and chitosan-decorated PLGA nanoparticles (SIR-CH-PLGA-NP) for non-invasive administration via the sub-conjunctival route. The modification with the cationic polymer chitosan demonstrated sustained in-vitro release of sirolimus and enhanced ex-vivo scleral permeation and penetration. Additionally, the SIR-CH-PLGA-NP showed improved cellular uptake and effectively reduced lipopolysaccharide-induced free-radical generation. This study suggests that chitosan-decorated PLGA nanoparticles could be a promising nanocarrier for sirolimus in treating AMD (Suri et al. 2021).

Melanin is a crucial radical scavenger and photoprotective agent, but its effectiveness in protecting against oxidative stress decreases with age. This reduction is associated with the onset of AMD (Kushwah et al. 2023). To counteract this decline, Kwon's group developed a therapeutic strategy using PEGylated (polyethylene glycol) synthetic melanin-like nanoparticles (MNPs) to replenish melanin in the RPE, aiming to treat AMD. These MNPs are designed to be biocompatible and specifically target ROS, offering robust antioxidant capabilities. They efficiently migrate to and accumulate within the RPE, localizing exclusively in the cytosol without penetrating the nucleus or mitochondria. According to the findings, MNPs can effectively substitute for natural melanin in the RPE, suggesting their potential as powerful natural radical scavengers to combat oxidative stress and treat ROS-related diseases such as AMD (Kwon et al. 2022).

Polyion complex micelles, measuring tens of nanometers, are created through electrostatic interactions. These micelles have shown a tendency to accumulate in and be retained at CNV lesions, making them highly effective for targeted drug delivery. Utilizing this property, a novel treatment for AMD was designed by encapsulating a dendritic photosensitizer within a polymeric-micelle formulation. This innovative approach allows for precise photodynamic therapy, selectively targeting CNV lesions. The treatment has demonstrated exceptional efficacy in occluding CNV while minimizing adverse phototoxic effects, highlighting its potential as a highly effective and safe therapeutic option for AMD (Tamaki 2009).

Elevated oxidative stress and the accumulation of ROS are crucial drivers in the onset and progression of AMD. To counteract these damaging effects, Wang's group utilized glycol chitosan-coated cerium oxide nanoparticles (GCCNP), which are antioxidative and water-soluble. GCCNP delivery effectively protects the retina from ongoing oxidative damage. It also enhanced this approach by combining GCCNP with an alginate-gelatin-based injectable hydrogel, resulting in synergistic antioxidant effects. This combination significantly accelerated the recovery of the RPE and photoreceptor cells (Wang et al. 2019).

In another study, an innovative implantable intraocular device was specifically engineered to adsorb, and trap complement factors implicated in AMD. This approach utilizes biocompatible implants composed of a nanopore polyacrylonitrile polymer membrane to dialyze proteins from the vitreous. These implants exhibit a remarkable affinity and capacity for complement protein absorption. This device showed excellent biocompatibility, with no inflammation and the preservation of normal retinal physiology and histology. These results indicate that prolonged intraocular suppression of complement factors is achievable with this nanopore polymer membrane, offering a promising and effective strategy for the treatment of AMD (Olson et al. 2020).

Anti-VEGF therapies, such as Ranibizumab, are effective in inhibiting the growth of new blood vessels, but repeated monthly intravitreal injections can lead to side effects. To address this, Gao's group has developed an innovative injectable supramolecular nanofiber hydrogel by combining betamethasone phosphate (BetP), a clinically approved anti-inflammatory drug, with the gold-standard anti-VEGF agent for AMD treatment, and CaCl2. This BetP-based hydrogel (BetP-Gel) is designed to be administered via intravitreal injection, providing a sustained release of anti-VEGF over an extended period. This prolonged release effectively inhibits retinal vascular proliferation and reduces CNV. Additionally, BetP-Gel can scavenge ROS, thereby diminishing local inflammation. Since hydrogel is composed entirely of clinically approved agents, it is well-positioned for rapid adoption in clinical settings for AMD treatment, offering a potential replacement for current anti-VEGF therapies (Gao et al. 2023).

A biocompatible nanomedicine delivery system called RGD@LP-2-DG, constructed by loading RGD peptide (Arginine-Glycine-Aspartic sequence specifically binds to integrins found in endothelial cells involved in angiogenesis)-modified liposomes (as a career) with 2-deoxy-D-glucose (2-DG), showed promising results in inhibiting neovascularization in wet AMD by targeting endothelial cell metabolism. Treatments targeting VEGF often result in inadequate responses in more than half of the patients, highlighting the need for more effective therapies. RGD@LP-2-DG has shown significant promise, demonstrating precise targeting of endothelial cells and robust inhibition of neovascularization in both *in vitro* and *in vivo* studies. Mechanistic studies have revealed that 2-DG disrupts N-glycosylation, which inhibits VEGFR2 and its downstream signaling pathways, critical in developing abnormal blood vessels. The combination of high inhibitory effects on neovascularization and excellent biocompatibility makes RGD@LP-2-DG a highly promising therapeutic candidate for the treatment of wet AMD, potentially addressing the shortcomings of current VEGF-targeting treatment (Chen et al. 2024).

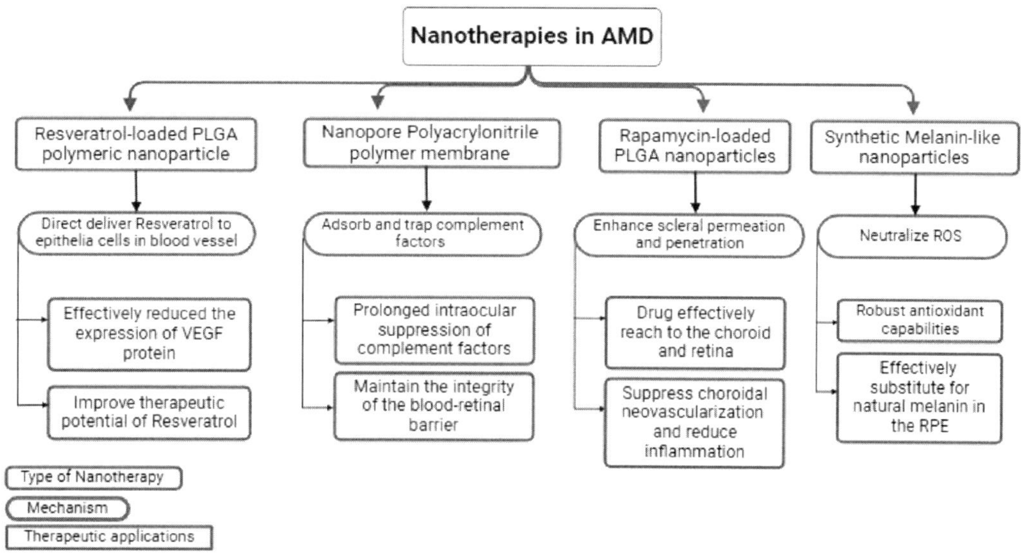

Figure 4. A summary of selected Nanotherapeutics used in AMD and their mechanisms and therapeutic applications. The image was created in BioRender.com.

Conclusion

Age-related macular degeneration significantly impacts elderly individuals, leading to visual impairments. Conventional treatments, such as anti-VEGF medications, offer therapeutic benefits but often come with challenges and side effects. Nanotechnology has shown promising results for various diseases, including AMD. This book chapter critically examines the current challenges of traditional treatments and the recent advancements in nano-derived therapies for AMD. Nanoparticles have advanced in ways that enhance drug delivery efficiency, stability, and accuracy, addressing many of the issues conventional treatments face. Nanodiagnostic tools, such as carbon tubes, gold particles, and nanozymes, play a crucial role in improving the diagnosis of ocular deformities by minimizing harmful interactions with ocular tissues and enhancing biocompatibility and stability. Additionally, this chapter discusses the practical barriers to nanomedicine and the pros and cons of nanotechnology-derived immunotherapy, highlighting novel pathways for treating AMD.

Acknowledgments

The author thanks the support of the Department of Biomedical Science, International College of Business and Technology, Sri Lanka. Further, the author thanks Mr. U. B. Dissanayaka, Mrs. P. Chandralatha, Ms. T. N. Dharmapriya, Mr. Buddhika Dissanayaka, Mrs. J. B. Dissanayaka, and Mr. Lakmal Priyankara for their valuable support.

References

Afarid Mehrdad, Shirin Mahmoodi, and Roghayyeh Baghban. 2022. "Recent achievements in nano-based technologies for ocular disease diagnosis and treatment, review and update." *Nanobiotechnology* 20 (1):361. doi: 10.1186/s12951-022-01567-7.

Alavi Mehran, and Mehrdad Hamidi. 2019. "Passive and active targeting in cancer therapy by liposomes and lipid nanoparticles." *Drug metabolism personalized therapy* 34 (1):20180032.

Algvere Peep V, Anders Kvanta, and Stefan Seregard. 2016. "Drusen maculopathy: a risk factor for visual deterioration." *Acta ophthalmologica* 94 (5):427-433.

Ammar Michael J, Jason Hsu, Allen Chiang, Allen C Ho, and Carl D Regillo. 2020. "Age-related macular degeneration therapy: a review." *Current opinion in ophthalmology* 31 (3):215-221.

Amoaku Winfried MK. 2010. "Treatments for Age-related Macular Degeneration." *Treatments for Age-related Macular Degeneration* 4:92–100. doi: 10.17925/EOR.2010.04.01.92.

Apte Rajendra S. 2021. "Age-related macular degeneration." *New England Journal of Medicine* 385 (6):539-547.

Arvizo Rochelle, Resham Bhattacharya, and Priyabrata Mukherjee. 2010. "Gold nanoparticles: opportunities and challenges in nanomedicine." *Expert opinion on drug delivery* 7 (6):753-763.

Awwad Sahar, Christin Henein, Nkiruka Ibeanu, Peng T Khaw, and Steve Brocchini. 2020. "Preclinical challenges for developing long acting intravitreal medicines." *European Journal of Pharmaceutics Biopharmaceutics* 153:130-149.

Banerjee Mousumi, Rohan Chawla, and Atul Kumar. 2021. "Antioxidant supplements in age-related macular degeneration: are they actually beneficial?" *Therapeutic Advances in Ophthalmology* 13:25158414211 030418.

Begines Belén, Ortiz Tamara, María Pérez-Aranda, Martínez Guillermo, Merinero Manuel, Argüelles-Arias Federico, and Alcudia Ana. 2020. "Polymeric Nanoparticles for Drug Delivery: Recent Developments and Future Prospects." *Nanomaterials (Basel)* 10 (7). doi: 10.3390/nano10071403.

Bhatt Priyanka, Fnu Gulimirerouzi, Bhatia Deepak, Shahid Amna, and Sutariya Vijaykumar. 2020. "Nanodelivery of Resveratrol-Loaded PLGA Nanoparticles for Age-Related Macular Degeneration." *AAPS PharmSciTech* 21 (8):291. doi: 10.1208/s12249-020-01836-4.

Birch David G, and Fong Qi Liang. 2007. "Age-related macular degeneration: a target for nanotechnology derived medicines." *International journal of nanomedicine* 2 (1):65-77.

Bjerager, Jakob, Miklos Schneider, Ivan Potapenko, Elon HC van Dijk, Carsten Faber, Jakob Grauslund, Kristina Pfau, Josef Huemer, Danson V Muttuvelu, and Marie LR Rasmussen. 2023. "Diagnostic accuracy of the Amsler grid test for detecting neovascular age-related macular degeneration: a systematic review and meta-analysis." *JAMA ophthalmology*.

Bonilla Lorena, Espina Marta, Severino Patricia, Cano Amanda, Ettcheto Miren, Camins Antoni, García Maria Luisa, Souto Eliana B, and Elena Sánchez-López . 2021. "Lipid Nanoparticles for the Posterior Eye Segment." *Pharmaceutics* 14 (1). doi: 10.3390/pharmaceutics14010090.

Boyd, Kierstan. 2023. "What is macular degeneration?." American Academy of Ophthalmology, accessed 02/06/2024. https://www.aao.org/eye-health/diseases/amd-macular-degeneration.

Chan, Chi-Ming, Chien-Yu Hsiao, Hsin-Ju Li, Jia-You Fang, Der-Chen Chang, and Chi-Feng Hung. 2019. "The inhibitory effects of gold nanoparticles on VEGF-A-induced cell migration in choroid-retina endothelial cells." *International Journal of Molecular Sciences* 21 (1):109.

Chen Fang, Si Peng, Zerda Adam de la, Jokerst Jesse V, and Myung David. 2021. "Gold nanoparticles to enhance ophthalmic imaging." *Biomaterial Science* 9 (2):367-390. doi: 10.1039/d0bm01063d.

Chen, Min, Bingqian Liu, Qianying Gao, Yehong Zhuo, and Jian Ge. 2011. "Mitochondria-targeted peptide MTP-131 alleviates mitochondrial dysfunction and oxidative damage in human trabecular meshwork cells." *Investigative ophthalmology visual science* 52 (10):7027-7037.

Chen Tzu-Ju, Chan Ti-Chun, Li Chien-Feng, Dissanayaka Dilshan Sampath, Kianpour Maryam, He Hong Lin, Huang Steven K, Li Wan-Shan, Chen Nai Yu, and Shiue Yow-Ling. 2021. "High glycosyltransferase 8 domain containing two protein levels contribute to poor prognosis in urothelial carcinoma." *International Journal Urology* 28 (11):1178-1187. doi: 10.1111/iju.14656.

Chen, XiRui, Liu SiWei, Chen MoXin, Ni Ni, Zhou Rong, Wang YiQi, Xu Yang, Wang YuanHui, HuiQin Gao, Zhang DanDan, Tang ZhiMin, Shu Qin, Zhang Jing, Li Lin, Ju YaHan, Gu ping. 2024. "Novel therapeutic perspectives for wet age-related macular degeneration: RGD-modified liposomes loaded with 2-deoxy-D-glucose as a promising nanomedicine." *Biomedicine Pharmacotherapy* 175:116776.

Cho, Eunyoung, Johanna M Seddon, Bernard Rosner, Walter C Willett, and Susan E Hankinson. 2004. "Prospective study of intake of fruits, vegetables, vitamins, and carotenoidsand risk of age-related maculopathy." *Archives of Ophthalmology* 122 (6):883-892.

Čolak, Emina, and Lepša Žorić. 2019. "6 - Antioxidants and Age-Related Macular Degeneration." In *Handbook of Nutrition, Diet, and the Eye (Second Edition)*, edited by Victor R. Preedy and Ronald Ross Watson, 85-106. Academic Press.

Contera, Sonia, Jorge Bernardino de la Serna, and Teresa D Tetley. 2020. "Biotechnology, nanotechnology and medicine." 4 (6):551-554.

Cruz, J., J. Flórez, R. Torres, M. Urquiza, J. A. Gutiérrez, F. Guzmán, and C. C. Ortiz. 2017. "Antimicrobial activity of a new synthetic peptide loaded in polylactic acid or poly(lactic-co-glycolic) acid nanoparticles against Pseudomonas aeruginosa, Escherichia coli O157:H7 and methicillin resistant Staphylococcus aureus (MRSA)." *Nanotechnology* 28 (13):135102. doi: 10.1088/1361-6528/aa5f63.

Cupini, S., S. Di Marco, L. Boselli, A. Cavalli, G. Tarricone, V. Mastronardi, V. Castagnola, E. Colombo, P. P. Pompa, and F. Benfenati. 2023. "Platinum Nanozymes Counteract Photoreceptor Degeneration and Retina Inflammation in a Light-Damage Model of Age-Related Macular Degeneration." *ACS Nano* 17 (22):22800-22820. doi: 10.1021/acsnano.3c07517.

Deng, Yanhui, Lifeng Qiao, Mingyan Du, Chao Qu, Ling Wan, Jie Li, Lulin Huang, and diseases. 2022. "Age-related macular degeneration: Epidemiology, genetics, pathophysiology, diagnosis, and targeted therapy." *Genes* 9 (1):62-79.

Dharmapriya, Thakshila Nadeeshani, Po-Jung Huang, Dissanayakage Dilshan Sampath Dissanayaka, Jinendra Jayasanka Dharmapriya, and Paichi Pat Shein. 2023. "Survey on social awareness towards energy conservation and ESD lesson on biofuel as a renewable energy source." *Journal of Environmental Studies and Sciences* 13 (4):658-667. doi: 10.1007/s13412-023-00848-3.

Dissanayaka Dissanayakage Dilshan Sampath, SHNP Gunawickrama, Thakshila Nadeeshani Dharmapriya, and KBS Gunawickrama. 2023. "Effects of acetaminophen exposure on behavior and erythrocyte nuclear morphology of juvenile Oreochromis niloticus." *Drug Chem Toxicology* 1-10. doi:10.1080/01480545.2023.2291983.

Duffy, Angela M, David J Bouchier-Hayes, and Judith H Harmey. 2013. "Vascular endothelial growth factor (VEGF) and its role in non-endothelial cells: autocrine signalling by VEGF." In *Madame Curie Bioscience Database [Internet]*. Landes Bioscience.

Feher, J, B Kovacs, I Kovacs, M Schveoller, A Papale, and Corrado Balacco Gabrieli. 2005. "Improvement of visual functions and fundus alterations in early age-related macular degeneration treated with a combination of acetyl-L-carnitine, n-3 fatty acids, and coenzyme Q10." *Ophthalmologica* 219 (3):154-166.

Flaxel, Christina J, Ron A Adelman, Steven T Bailey, Amani Fawzi, Jennifer I Lim, G Atma Vemulakonda, and Gui-shuang Ying. 2020. "Age-related macular degeneration preferred practice pattern®." *Ophthalmology* 127 (1):P1-P65.

Flores, Rita, Ângela Carneiro, Miguel Vieira, Sandra Tenreiro, and Miguel C Seabra. 2021. "Age-related macular degeneration: pathophysiology, management, and future perspectives." *Ophthalmologica* 244 (6):495-511.

Frampton, Geoff K, Neelam Kalita, L Payne, JL Colquitt, Emma Loveman, Susan M Downes, and AJ Lotery. 2017. "Fundus autofluorescence imaging: systematic review of test accuracy for the diagnosis and monitoring of retinal conditions." *Eye* 31 (7):995-1007.

Fritsche, L. G., W. Chen, M. Schu, B. L. Yaspan, Y. Yu, G. Thorleifsson, D. J. Zack, S. Arakawa, V. Cipriani, S. Ripke, R. P. Igo, Jr., G. H. Buitendijk, X. Sim, D. E. Weeks, R. H. Guymer, J. E. Merriam, P. J. Francis, G. Hannum, A. Agarwal, A. M. Armbrecht, I. Audo, T. Aung, G. R. Barile, M. Benchaboune, A. C. Bird, P. N. Bishop, K. E. Branham, M. Brooks, A. J. Brucker, W. H. Cade, M. S. Cain, P. A. Campochiaro, C. C. Chan, C. Y. Cheng, E. Y. Chew, K. A. Chin, I. Chowers, D. G. Clayton, R. Cojocaru, Y. P. Conley, B. K. Cornes, M. J. Daly, B. Dhillon, A. O. Edwards, E. Evangelou, J. Fagerness, H. A. Ferreyra, J. S. Friedman, A. Geirsdottir, R. J. George, C. Gieger, N. Gupta, S. A. Hagstrom, S. P. Harding, C. Haritoglou, J. R. Heckenlively, F. G. Holz, G. Hughes, J. P. Ioannidis, T. Ishibashi, P. Joseph, G. Jun, Y. Kamatani, N. Katsanis, N. Keilhauer C, J. C. Khan, I. K. Kim, Y. Kiyohara, B. E. Klein, R. Klein, J. L. Kovach, I. Kozak, C. J. Lee, K. E. Lee, P. Lichtner, A. J. Lotery, T. Meitinger, P. Mitchell, S. Mohand-Saïd, A. T. Moore, D. J. Morgan, M. A. Morrison, C. E. Myers, A. C. Naj, Y. Nakamura, Y. Okada, A. Orlin, M. C. Ortube, M. I. Othman, C. Pappas, K. H. Park, G. J. Pauer, N. S. Peachey, O. Poch, R. R. Priya, R. Reynolds, A. J. Richardson, R. Ripp, G. Rudolph, E. Ryu, J. A. Sahel, D. A. Schaumberg, H. P. Scholl, S. G. Schwartz, W. K. Scott, H. Shahid, H. Sigurdsson, G. Silvestri, T. A. Sivakumaran, R. T. Smith, L. Sobrin, E. H. Souied, D. E. Stambolian, H. Stefansson, G. M. Sturgill-Short, A. Takahashi, N. Tosakulwong, B. J. Truitt, E. E. Tsironi, A. G. Uitterlinden, C. M. van Duijn, L. Vijaya, J. R. Vingerling, E. N. Vithana, A. R. Webster, H. E. Wichmann, T. W. Winkler, T. Y. Wong, A. F. Wright, D. Zelenika, M. Zhang, L. Zhao, K. Zhang, M. L. Klein, G. S. Hageman, G. M. Lathrop, K. Stefansson, R. Allikmets, P. N. Baird, M. B. Gorin, J. J. Wang, C. C. Klaver, J. M. Seddon, M. A. Pericak-Vance, S. K. Iyengar, J. R. Yates, A. Swaroop, B. H. Weber, M. Kubo, M. M. Deangelis, T. Léveillard, U. Thorsteinsdottir, J. L. Haines, L. A. Farrer, I. M. Heid, and G. R. Abecasis. 2013. "Seven new loci associated with age-related macular degeneration." *Nat Genet* 45 (4):433-9, 439e1-2. doi: 10.1038/ng.2578.

Gao, Huiqin, Muchao Chen, Yan Liu, Dandan Zhang, Jingjing Shen, Ni Ni, Zhimin Tang, Yahan Ju, Xiaochan Dai, and Ai Zhuang. 2023. "Injectable anti-inflammatory supramolecular nanofiber hydrogel to promote anti-VEGF therapy in age-related macular degeneration treatment." *Advanced Materials* 35 (2):2204994.

Garcia-Garcia, J., R. Usategui-Martin, M. R. Sanabria, E. Fernandez-Perez, J. J. Telleria, and R. M. Coco-Martin. 2022. "Pathophysiology of Age-Related Macular Degeneration: Implications for Treatment." *Ophthalmic Research* 65 (6):615-636. doi: 10.1159/000524942.

Ghate, D., W. Brooks, B. E. McCarey, and H. F. Edelhauser. 2007. "Pharmacokinetics of intraocular drug delivery by periocular injections using ocular fluorophotometry." *Invest Ophthalmol Vis Sci* 48 (5):2230-7. doi: 10.1167/iovs.06-0954.

Gheorghe, A., L. Mahdi, and O. Musat. 2015. "Age-related macular degeneration." *Rom J Ophthalmol* 59 (2):74-7.

Huang, Xiaoli, Zhengwei Zhang, Jie Wang, Xiaomei Meng, Tiantian Chen, and Zhifeng Wu. 2018. "Macular assessment of preoperative optical coherence tomography in ageing Chinese undergoing routine cataract surgery." *Scientific reports* 8 (1):5103.

Jamieson, Timothy, Raheleh Bakhshi, Daniela Petrova, Rachael Pocock, Mo Imani, and Alexander M Seifalian. 2007. "Biological applications of quantum dots." *Biomaterials* 28 (31):4717-4732.

Jhingan, Mahima, Sumit Randhir Singh, Anindya Samanta, Supriya Arora, Davide Tucci, Sohani Amarasekera, Carlo Cagini, Marco Lupidi, and Jay Chhablani. 2021. "Drusen ooze: Predictor for progression of dry age-related macular degeneration." *Graefe's Archive for Clinical Experimental Ophthalmology* 259:2687-2694.

Kanagasingam, Yogesan, Alauddin Bhuiyan, Michael D Abràmoff, R Theodore Smith, Leonard Goldschmidt, and Tien Y Wong. 2014. "Progress on retinal image analysis for age related macular degeneration." *Progress in retinal eye research* 38:20-42.

Keenan, Tiarnán DL, Catherine A Cukras, and Emily Y Chew. 2021. "Age-related macular degeneration: epidemiology and clinical aspects." *Age-related Macular Degeneration: From Clinic to Genes Back to Patient Managemen*:1-31.

Khan, Kamron N, Omar A Mahroo, Rehna S Khan, Moin D Mohamed, Martin McKibbin, Alan Bird, Michel Michaelides, Adnan Tufail, and Anthony T Moore. 2016. "Differentiating drusen: Drusen and drusen-like appearances associated with ageing, age-related macular degeneration, inherited eye disease and other pathological processes." *Progress in retinal eye research* 53:70-106.

Kushwah, N., K. Bora, M. Maurya, M. C. Pavlovich, and J. Chen. 2023. "Oxidative Stress and Antioxidants in Age-Related Macular Degeneration." *Antioxidants (Basel)* 12 (7). doi: 10.3390/antiox12071379.

Kwon, Y. S., M. Zheng, A. Y. Zhang, and Z. Han. 2022. "Melanin-like Nanoparticles as an Alternative to Natural Melanin in Retinal Pigment Epithelium Cells and Their Therapeutic Effects against Age-Related Macular Degeneration." *ACS Nano* 16 (11):19412-19422. doi: 10.1021/acsnano.2c09087.

Liew, Gerald, Paul Mitchell, Tien Yin Wong, Elena Rochtchina, and Jie Jin Wang. 2013. "The Association of Aspirin Use With Age-Related Macular Degeneration." *JAMA Internal Medicine* 173 (4):258-264. doi: 10.1001/jamainternmed.2013.1583.

Lim, Laurence S, Paul Mitchell, Johanna M Seddon, Frank G Holz, and Tien Y Wong. 2012. "Age-related macular degeneration."*The Lancet* 379 (9827):1728-1738.

Lin, Tai-Chi, Kuo-Hsuan Hung, Chi-Hsien Peng, Jorn-Hon Liu, Lin-Chung Woung, Ching-Yao Tsai, Shih-Jen Chen, Yan-Ting Chen, and Chih-Chien Hsu. 2015. "Nanotechnology-based drug delivery treatments and specific targeting therapy for age-related macular degeneration." *Journal of the Chinese Medical Association* 78 (11):635-641. doi: 10.1016/j.jcma.2015.07.008.

Liu, Guihong, Tao Chen, Zhenyu Ding, Yang Wang, Yuquan Wei, and Xiawei Wei. 2021. "Inhibition of FGF-FGFR and VEGF-VEGFR signalling in cancer treatment." *Cell proliferation* 54 (4):e13009.

Liu, S., W. Liu, Y. Ma, K. Liu, and M. Wang. 2012. "Suprachoroidal injection of ketorolac tromethamine does not cause retinal damage." *Neural Regen Res* 7 (35):2770-7. doi: 10.3969/j.issn.1673-5374.2012.35.004.

Loewenstein, Anat, Rafael Malach, Michaela Goldstein, Igal Leibovitch, Adiel Barak, Eli Baruch, Yair Alster, Omer Rafaeli, Isaac Avni, and Yuval Yassur. 2003. "Replacing the Amsler grid: a new method for monitoring patients with age-related macular degeneration." *Ophthalmology* 110 (5):966-970.

Lu, Haibao, Weilong Yin, Wei Min Huang, and Jinsong Leng. 2013. "Self-assembled carboxylic acid-functionalized carbon nanotubes grafting onto carbon fiber for significantly improving electrical actuation of shape memory polymers." *RSC advances* 3 (44):21484-21488.

Mehta, Hemal, Adnan Tufail, Vincent Daien, Aaron Y Lee, Vuong Nguyen, Mehmet Ozturk, Daniel Barthelmes, and Mark C Gillies. 2018. "Real-world outcomes in patients with neovascular age-related macular degeneration treated with intravitreal vascular endothelial growth factor inhibitors." *Progress in retinal eye research* 65:127-146.

Morelle, Olivier, Maximilian WM Wintergerst, Robert P Finger, and Thomas Schultz. 2023. "Accurate drusen segmentation in optical coherence tomography via order-constrained regression of retinal layer heights." *Scientific Reports* 13 (1):8162.

Mrowicka, M., J. Mrowicki, E. Kucharska, and I. Majsterek. 2022. "Lutein and Zeaxanthin and Their Roles in Age-Related Macular Degeneration-Neurodegenerative Disease." *Nutrients* 14 (4). doi: 10.3390/nu1404 0827.

Nashine, S. 2021. "Potential Therapeutic Candidates for Age-Related Macular Degeneration (AMD)." *Cells* 10 (9). doi: 10.3390/cells10092483.

Nowak, Jerzy Z. 2006. "Age-related macular degeneration (AMD): pathogenesis and therapy." *Pharmacological reports* 58 (3):353.

Olmedilla-Alonso, Begoña, Rocío Estévez-Santiago, José-Manuel Silván, Milagros Sánchez-Prieto, and Sonia de Pascual-Teresa. 2018. "Effect of long-term xanthophyll and anthocyanin supplementation on lutein and zeaxanthin serum concentrations and macular pigment optical density in postmenopausal women." *Nutrients* 10 (8):959.

Olson, J. L., R. Velez-Montoya, A. Mackin, N. Manoharan, N. Mueller, and M. Erlanger. 2020. "Vitreous protein binding with nanopore dialysis membrane device." *Biomed Microdevices* 22 (3):46. doi: 10.1007/s10544-020-00500-9.

Pożarowska, Dorota, and Piotr Pożarowski. 2016. "The era of anti-vascular endothelial growth factor (VEGF) drugs in ophthalmology, VEGF and anti-VEGF therapy." *Cent Eur J Immunol* 41 (3):311-316.

Prow, Tarl, Jacob N Smith, Rhonda Grebe, Jose H Salazar, Nan Wang, Nicholas Kotov, Gerard Lutty, and James Leary. 2006. "Construction, gene delivery, and expression of DNA tethered nanoparticles." *Molecular vision* 12 (67-69):606-615.

Ranta, V. P., E. Mannermaa, K. Lummepuro, A. Subrizi, A. Laukkanen, M. Antopolsky, L. Murtomäki, M. Hornof, and A. Urtti. 2010. "Barrier analysis of periocular drug delivery to the posterior segment." *J Control Release* 148 (1):42-48. doi: 10.1016/j.jconrel.2010.08.028.

Rezaei, Razieh, Mohsen Safaei, Hamid Reza Mozaffari, Hedaiat Moradpoor, Sara Karami, Amin Golshah, Behroz Salimi, and Hossein Karami. 2019. "The role of nanomaterials in the treatment of diseases and their effects on the immune system." *Open access Macedonian journal of medical sciences* 7 (11):1884.

Roisman, Luiz, and Raquel Goldhardt. 2017. "OCT Angiography: an upcoming non-invasive tool for diagnosis of age-related macular degeneration." *Current ophthalmology reports* 5:136-140.

Sarkar, Aira, and Sathish Dyawanapelly. 2021. "Nanodiagnostics and Nanotherapeutics for age-related macular degeneration." *Journal of controlled release* 329:1262-1282.

Sarwat, S., F. Stapleton, M. Willcox, and M. Roy. 2019. "Quantum Dots in Ophthalmology: A Literature Review." *Current Eye Research* 44 (10):1037-1046. doi: 10.1080/02713683.2019.1660793.

Schmidt-Erfurth, U, S Klimscha, SM Waldstein, and H Bogunović. 2017. "A view of the current and future role of optical coherence tomography in the management of age-related macular degeneration." *Eye* 31 (1):26-44.

Simon, Juliette, Emmanuel Flahaut, and Muriel Golzio. 2019. "Overview of carbon nanotubes for biomedical applications." *Materials* 12 (4):624.

Stahl, A. 2020. "The Diagnosis and Treatment of Age-Related Macular Degeneration." *Dtsch Arztebl Int* 117 (29-30):513-520. doi: 10.3238/arztebl.2020.0513.

Steinmetz, Jaimie D, Rupert RA Bourne, Paul Svitil Briant, Seth R Flaxman, Hugh RB Taylor, Jost B Jonas, Amir Aberhe Abdoli, Woldu Aberhe Abrha, Ahmed Abualhasan, and Eman Girum Abu-Gharbieh. 2021. "Causes of blindness and vision impairment in 2020 and trends over 30 years, and prevalence of avoidable blindness in relation to VISION 2020: the Right to Sight: an analysis for the Global Burden of Disease Study." *The Lancet Global Health* 9 (2):e144-e160.

Suri, R., Y. R. Neupane, N. Mehra, M. Nematullah, F. Khan, O. Alam, A. Iqubal, G. K. Jain, and K. Kohli. 2021. "Sirolimus loaded chitosan functionalized poly (lactic-co-glycolic acid) (PLGA) nanoparticles for potential treatment of age-related macular degeneration." *Int J Biol Macromol* 191:548-559. doi: 10.1016/j.ijbiomac.2021.09.069.

Tamaki, Y. 2009. "Prospects for nanomedicine in treating age-related macular degeneration." *Nanomedicine (Lond)* 4 (3):341-52. doi: 10.2217/nnm.09.10.

Thorne, J. E., E. A. Sugar, J. T. Holbrook, A. E. Burke, M. M. Altaweel, A. T. Vitale, N. R. Acharya, J. H. Kempen, and D. A. Jabs. 2019. "Periocular Triamcinolone vs. Intravitreal Triamcinolone vs. Intravitreal Dexamethasone Implant for the Treatment of Uveitic Macular Edema: The PeriOcular vs. INTravitreal corticosteroids for uveitic macular edema (POINT) Trial." *Ophthalmology* 126 (2):283-295. doi: 10.1016/j.ophtha.2018.08.021.

Tomi, A, I Marin, and life. 2014. "Angiofluorographic aspects in age-related macular degeneration." *Journal of medicine* 7 (Spec Iss 4):4.

Torrecillas-Picazo, R, M Cerdà-Ibáñez, I Almor Palacios, JM Hervás Hernandis, R Ramón-Cosín, N Ruiz Del Rio, and A Duch-Samper. 2017. "Analysis and follow-up of type 1 choroidal neovascularisation with optical coherence tomography–angiography after antiangiogenic treatment." *Archivos de la Sociedad Española de Oftalmología* 92 (6):265-272.

Virgili, Gianni, and Alessandro Bini. 2007. "Laser photocoagulation for neovascular age-related macular degeneration." *Cochrane Database of Systematic Reviews* (3).

Vyawahare, H., and P. Shinde. 2022. "Age-Related Macular Degeneration: Epidemiology, Pathophysiology, Diagnosis, and Treatment." *Cureus* 14 (9):e29583. doi: 10.7759/cureus.29583.

Wang, K., M. Zheng, K. L. Lester, and Z. Han. 2019. "Light-induced Nrf2(-/-) mice as atrophic age-related macular degeneration model and treatment with nanoceria laden injectable hydrogel." *Sci Rep* 9 (1):14573. doi: 10.1038/s41598-019-51151-7.

Wang, Y, VM Wang, and CC Chan. 2011. "The role of anti-inflammatory agents in age-related macular degeneration (AMD) treatment."*Eye* 25 (2):127-139.

Warwick, A., and A. Lotery. 2018. "Genetics and genetic testing for age-related macular degeneration." *Eye* 32 (5):849-857. doi: 10.1038/eye.2017.245.

Wintergerst, M. W. M., D. K. Mishra, L. Hartmann, P. Shah, V. K. Konana, P. Sagar, M. Berger, K. Murali, F. G. Holz, M. P. Shanmugam, and R. P. Finger. 2020. "Diabetic Retinopathy Screening Using Smartphone-Based Fundus Imaging in India." *Ophthalmology* 127 (11):1529-1538. doi: 10.1016/j.ophtha. 2020.05.025.

Wong, Ka-Ying, Man-Sau Wong, and Juewen Liu. 2024. "Nanozymes for Treating Ocular Diseases." *Advanced Healthcare Materials*:2401309.

Yadav, S., and P. K. Maurya. 2021. "Biomedical applications of metal oxide nanoparticles in aging and age-associated diseases." *3 Biotech* 11 (7):338. doi: 10.1007/s13205-021-02892-8.

Yildirim, Kemal, Sami Al-Nawaiseh, Sophia Ehlers, Lukas Schießer, Michael Storck, Tobias Brix, Nicole Eter, and Julian Varghese. 2023. "U-Net-Based Segmentation of Current Imaging Biomarkers in OCT-Scans of Patients with Age Related Macular Degeneration." *Stud Health Technol Inform*:947-951.

Yung, Madeline, Michael A Klufas, and David Sarraf. 2016. "Clinical applications of fundus autofluorescence in retinal disease." *International journal of retina vitreous* 2:1-25.

Zhang, Yuhang, Stephanie Watson, Yogambha Ramaswamy, and Gurvinder Singh. 2024. "Intravitreal therapeutic nanoparticles for age-related macular degeneration: Design principles, progress and opportunities." *Advances in Colloid Interface Science*:103200.

Zhu, Yuexin, Manyu Luo, Xue Bai, Jicui Li, Ping Nie, Bing Li, and Ping Luo. 2022. "SS-31, a mitochondria-targeting peptide, ameliorates kidney disease." *Oxidative Medicine and Cellular Longevity* 2022.

Chapter 4.5

Microbial Nanocellulose-Based Biocomposites for Soft Bone Regeneration: From the Present to the Future

Probal Basu[1,2,*]
Nabanita Saha[2,3,†]
Tomáš Sáha[3]
and Petr Sáha[2,3]

[1]Department of Biomaterials, Institute of Clinical Dentistry, University of Oslo, Oslo, Norway
[2]Center of Polymer Systems, University Institute, Tomas Bata University in Zlin, Zlín, Czech Republic
[3]Footwear Research Center, University Institute, Tomas Bata University in Zlin, Zlín, Czech Republic

Abstract

Microbial nanocellulose (MC) draws special attention from the research community due to its unique nano-bio fabrication. This nanofiber-constructed biomaterial is biosynthesized by certain microbes in the purest form (devoid of lignin, pectin, and hemicellulose). Due to its striking similarities to animal collagen fibers, MC nanofibers have been used to develop osteoconductive biomaterial for soft tissue regeneration. High surface area, notable rheological properties, and biocompatibility make this material attractive for tissue regeneration. Concurrently, their bio-friendly characteristics are due to various bio-ceramic materials like calcium phosphate (CP) employed in conjunction with MC (for the development of scaffold/bio-composite). The significant interaction between MC's reinforcement nanofibrous matrix and the particulate fillers, such as CP, renders this bio-composite particularly intriguing. Considering the above perspectives, an effort has been made to explore and discuss the preparation and characterization of MC-CP-based bio-composites in soft bone regeneration applications. This book chapter provides a comprehensive insight into the significance of MC and biphasic calcium phosphate (BCP) based bio-composite.

Keywords: microbial nanocellulose, calcium phosphate, bio-composites, material properties, soft bone tissue regeneration

---

* Corresponding Author's Email: probal.basu@odont.uio.no
† Corresponding Author's Email: nabanita@utb.cz

In: Emerging Nanomaterials in Biomedical Science
Editors: Eepsita Priyadarshini and Ajith Manayil Parambil
ISBN: 979-8-89530-600-0

1. Introduction

A bone fracture is a traumatic event in the human body (Hou et al. 2022; Nunes et al. 2022). Substantial traumatic fractures result in notable morbidity, with the case incidence around 1.0 and 2.9 million on a global scale (Ghouri et al. 2021; Agarwal-Harding et al. 2015). The conventional fracture healing/repair approach typically involves grafting techniques utilizing various implant materials (Kuo et al. 2022). These methods also have limitations like secondary surgery and microbial infections. Alternatively, stem cell therapy has been notably used for bone fracture repair, but this approach is also recognized as costly (Yao et al. 2021; Koons et al. 2020; Basu et al. 2019a; Hao et al. 2002). Hydrogel-based bio-composites have been utilized notably in different applications like tissue engineering, wound healing, and anti-biofouling applications (Gatabi et al. 2022; Zhao et al. 2022; Keshk, 2014).

Microbial nanocellulose (MC) is considered suitable for preparing soft bone implants. MC is a promising biopolymer that can be obtained through lab-scale synthesis and is devoid of lignin, pectin, and hemicellulose (Hospodiuk-Karwowski et al. 2022; Basu et al. 2019a; Keshk, 2014). This eco-friendly nanocellulose material is being synthesized by *Acetobacter, Rhizobium, Agrobacterium, Pseudomonas*, and *Sarcina* bacterial strains through a fermentation procedure (Jain et al. 2024; Barja, 2021). Different properties like high surface area, notable rheological properties, and biocompatibility make this biomaterial exciting for tissue regeneration (Grøndahl, 2019; Dufresne, 2013). MC typically forms reversible hydrogels at low solid content, which include numerous -OH groups interacting through hydrogen bonds and van der Waals forces. The property of the MC-based hydrogels can be enhanced by applying different polymers like polyethylene glycol (PEG) that crosslinks with the nanofibers to form mechanically stable gels (Grøndahl, 2019; Syverud, 2017). In a hydrogel matrix, MC typically acts as a reinforcement fiber filler material, supporting the scaffold matrix and providing biocompatibility for osteoconduction. This osteogenic property of the developed scaffold can be further enhanced by using different bio-ceramic materials like calcium phosphates (CP).

Calcium phosphate, an essential human bone component, is a bio-stimulant for osteoblast cells, promoting bone fracture healing and regeneration (Zaszczyńska, et al. 2021). Various calcium phosphate phases like hydroxyapatite (HAp), β tri-calcium phosphate (β-TCP), octacalcium phosphate were extensively utilized to develop bone implant preparation. These bioactive calcium phosphates have been preferred due to their biocompatibility, ability to induce fracture healing, and bio-resorbability (Hussin et al. 2021). Interestingly, biphasic calcium phosphate (BCP: where HAp and β-TCP are present in different concentration ratios), has been significantly recommended for the development of bone substitute biomaterial due to its biocompatibility and degradability (Choi et al. 2021; Hong et al. 2017; Kaur et al. 2014; Kim et al. 2012; Hench, 1998).

Given the perspectives outlined above, an attempt has been made to explore and discuss the significance of MC/CP bio-composites for soft bone tissue regeneration applications. This book chapter provides a thorough knowledge regarding the significance of MC and biphasic calcium phosphate (BCP) based bio-composite.

2. Short Overview of Calcium-Based Polymeric Bio-Composites and Their Challenges for Bone Regeneration

Polymer-ceramic-based bio-composites have been utilized successfully in various soft bone tissue regeneration. Calcium phosphate-based bioceramic materials like hydroxyapatite, tricalcium phosphate, octacalcium phosphate have been exciting (Bharadwaza, and Jayasuriya, 2020). Moreover, calcium carbonate has also been utilized to prepare calcium-filled bio-composite for bone tissue engineering (Basu et al. 2018; Basu et al. 2019b). For the development of inorganic calcium-based polymeric biocomposites, various synthetic polymers like polyvinyl alcohol (PVA), polyvinylpyrrolidone (PVP), polyethylene glycol (PEG), Poly-lactic acid (PLA), Poly-ε- caprolactone (PCL) have been extensively used. Natural polymers like chitosan, alginate, and collagen have also been used due to their high tissue compatibility and biological efficiency (Qi et al. 2021, Basu et al. 2019b, Basu et al. 2018).

The idea behind developing an efficient tissue engineering-based bio-composite is to mimic the extracellular matrix, which usually facilitates tissue regeneration. The advantages of the polymeric bio-composite lie in the fact that these biomaterials are flexible in their structures. However, different challenges are still there for these materials, like inadequate elasticity, decreased dimensional stability, non-uniform structural properties, poor wettability, and biocompatibility (Trivedi et al. 2023). These challenges facilitate the limited osseointegration of different biomaterials, thereby limiting their functionality. In this context, cellulose-based bio-composites have shown great promise due to their notable cytocompatibility and significant mechanical properties (i.e., compressive strength) (Basu et al. 2019b, Basu et al. 2018).

3. MC-CP Bio-Composite Hydrogel Preparation and Characteristics

3.1. Microbial Nanocellulose Biosynthesis

MC is a biopolymer developed with the D-glucose units joined by β, 1 to 4-glycosidic bond. The MC fibrillar matrix comprises three-dimensional nanofibers, forming a hydrogel sheet-like structure. *Gluconobacter* is the most common bacterium utilized in the production of MC. This bacterium synthesizes ribbon-like cellulose I and 'thermodynamically stable' cellulose-II polymers under suitable carbon-nitrogen nutrient conditions and temperature (28-30°C) and pH (4.0-7.0). The MC nanocellulose is formed through a biosynthetic pathway. The bacteria initially develop 2-4 nm protofibrils composed of glucose chains that are then secreted by bacterial cell wall and produce microfibrils, and then the protofibrils are aggregated to form 80 nm cellulose fiber ribbons (Suresh, 2018; Esa et al. 2014).

Two fermentation methods have been reported for MC production: (i) static and (ii) dynamic. The maintenance of aerobic conditions in the static culture method to produce MC facilitates the bacteria's metabolic process, thereby promoting nanofiber biosynthesis. On the other hand, the dynamic process employs a 'forced oxygen' supply to the bacterial culture medium by applying a bioreactor environment. The disadvantage of the latter process is that MC's overall appearance and properties are changed for dynamic culture (Shavyrkina et al. 2021).

3.2. Preparation of the MC-CP Biocomposite Hydrogel

Both MC and CP are present as 'filler' material in MC-CP biocomposite hydrogels. In one way, MC acts as a 'polymer,' contributing to the formation of the overall 3D structure of the hydrogel; on the other hand, it also provides mechanical stability to the hydrogel by acting as a 'fibrous filler' material. CP plays an important role as a bioactive agent and a 'particulate filler' (Figure 1a).

In our standard procedure, the MC was first synthesized as a pellicle. Then, it was treated with NaOH (0.5 N NaOH, 80°C, 1 hour) solution to detach the bacterial mass. After that, the MC pellicle was washed in distilled water until the pH of the MC pellicle reached neutral, i.e., pH 7.0. (Basu et al. 2019a; Basu et al. 2019b).

A suspension of MC was then prepared through grinding (particle size: 158-159 nm; 10-12 minutes) using a mixture grinder. The ground MNC suspension was used in the preparation of MC-CP hydrogels.

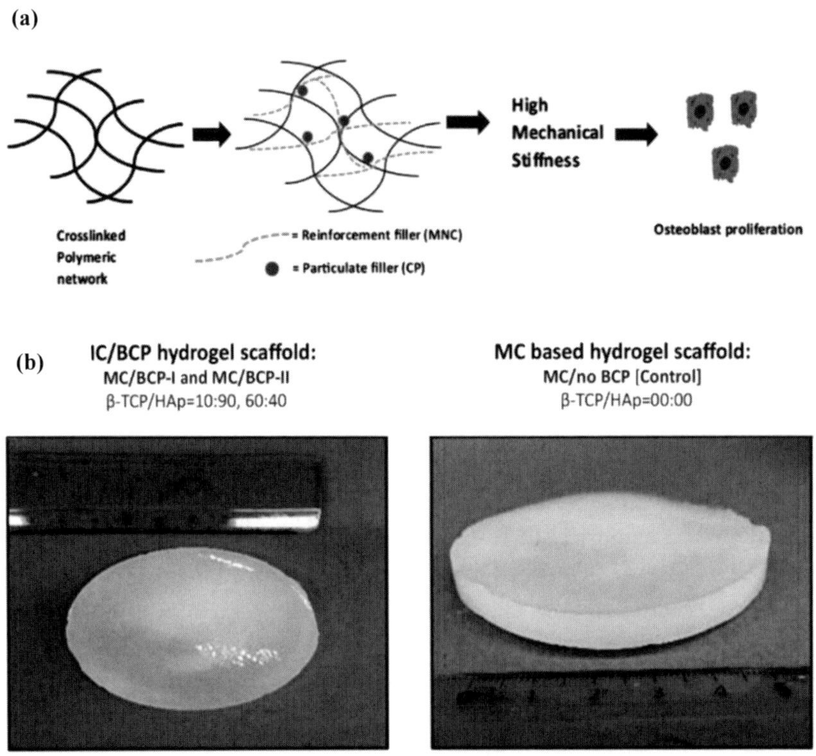

Figure 1. (a) Outline of mechanistic insights of MC/BCP based polymeric bio-composite in soft bone formation (b) A visual image of MC/BCP based hydrogel scaffold (before dry).

The two phases of CPs were selected, i.e., hydroxyapatite (HAp) and β-tri calcium phosphate (β-TCP) as biphasic calcium phosphate (BCP) for their bioactive potential related to soft bone tissue engineering. To achieve the MC-CP hydrogel, two polymer solutions (for "MC/BCP-I hydrogel" and "MC/BCP-II hydrogel") were prepared with MC (0.5g). Other hydrophilic polymers like Polyvinyl pyrrolidone (PVP, 0.5g) and Polyethylene glycol (PEG,1g) agar (2g) were also utilized. Glycerin was also given as a humectant. The biphasic components (MC/BCP-I hydrogel: 0.1 g β-TCP/ 0.9 g HAp and MC/BCP-II hydrogel: 0.6 g β-TCP/0.4 g HAp) were mixed thoroughly. The final volume of all the polymer solutions was 100 mL. A control was prepared without β-TCP and HAp and designated as "MC/ no BCP."

The physical cross-linking occurred within the polymer solutions due to moist heat and pressure treatment (120°C temperature for 20 minutes at 107 KPa pressure) (Basu et al. 2019a; Basu et al. 2019b). 100 mL MC/BCP physically crosslinked polymer solutions were solvent-cast into 75 mm petri plates. The petri plates were then cooled at room temperature (28°C). Finally, round-shaped hydrogels MC/ no BCP (Diameter: 75 mm; Thickness: 6.0-6.3 mm) and MC/BCP-I and II (Diameter: 72-75 mm; Thickness: 5.9-6.2 mm) were obtained. The visual image of the MC/BCP hydrogel scaffold is depicted in Figure 1b.

4. Properties and Characteristics of MC-CP Bio-Composites Hydrogel

The efficiency of a bio-composite for tissue engineering depends on their physico-chemical properties like porosity and chemical properties. These properties will indicate the promise of the developed bio-composites and provide an idea about the prospect of the biomaterial being applied for a specific application like cortical tissue regeneration or cartilage tissue regeneration. The properties of the developed MC/CP hydrogels, i.e., MC/BCP-I and MC/BCP-II, are discussed below regarding their efficiency and prospects in soft bone regeneration.

4.1. Porosity and Void Fraction

The porosity of MC-based scaffolds is an important factor in the evaluation of the performance of the biomaterial. Notably, porous biomaterial demonstrates significant cell adhesion and proliferation and facilitates notable cell viability (Lin et al. 2023; Basu et al. 2018). The porosity (%) of MC/BCP-II hydrogel is higher than MC/BCP-I (Figure 2 (a)), which indicates the higher efficiency of MC/BCP-II hydrogels. This further indicates that MC/BCP hydrogels can efficiently allow the diffusion of essential bioactive agents and drugs to the bone cells. As β-TCP and HAp are two different phases of the same calcium phosphate filler, the interactions between β-TCP and HAp might also influence the porosity of the scaffolds. On the other hand, the *void fraction* study indicates a higher void fraction of MC/no BCP than the other hydrogel scaffolds. The result of the void fraction analysis (Figure 2(b)) is comparable with the porosity of the scaffolds, i.e., MC/no BCP > MC/BCP-II > MC/BCP-I.

4.2. Thermal Property

Thermal characterization of the hydrogel scaffold indicates its compositional integrity. Figure 2(c) demonstrates the thermal characteristics of MC/BCP hydrogels, i.e., MC/BCP-I, MC/BCP-II. A significant endothermal peak between 50°C-100°C can be seen, possibly due to moisture loss (Pandit et al. 2012). Another prominent endothermal peak can be seen at around 250°C, which might result from the melting of the cellulose crystalline region (Oliveira et al. 2015). On the other hand, an exothermic peak can also be found around 400°C, corresponding to cellulose's thermal breakdown. Interestingly, the peaks have broadened from MC/no BCP to MC/BCP-II, possibly due to polymer-filler interaction within the biomaterial.

4.3. Fourier Transform Infrared Spectroscopy

The chemical characteristics of the hydrogels were confirmed through FT-IR spectroscopy. Significant results are found within the wavenumber, 2000 cm^{-1}. The bond vibration at 1044 cm^{-1} shows the C-O stretching of the C-O-C linkage of cellulose (Beltramino et al. 2016). In addition, the vibration at wavenumber 1107 cm^{-1} is because of the C-O-C stretching of the glycosidic bond of MC (Pavitra et al. 2015). 1342 cm^{-1} shows the C-H bending of PEG (León et al. 2017). Meanwhile, the bond vibration at 1654 cm^{-1} demonstrates the C=O stretching of PVP (Basu et al. 2019a). Finally, 1038 cm^{-1} bond vibration indicates the presence of calcium phosphate in the MC/BCP sample (Gadaleta et al. 1996).

4.4. X-Ray Diffraction

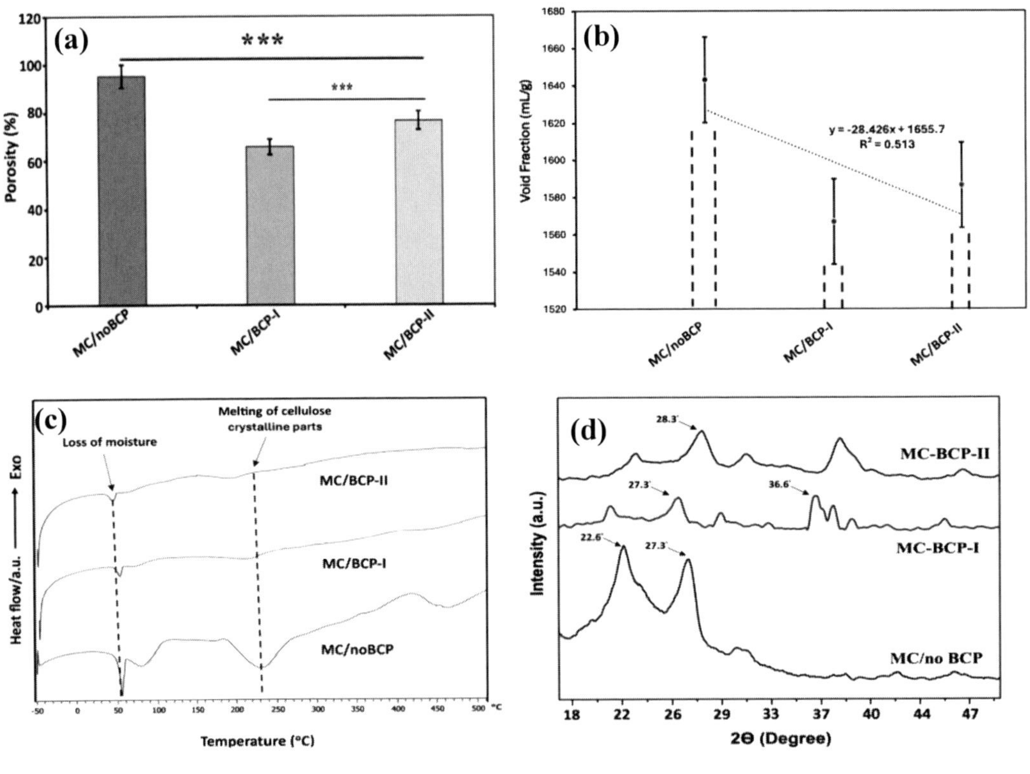

Figure 2. Physico-chemical characteristics of MC/BCP biocomposite hydrogels. Porosity (%) (a) and void fraction (b) ***P < 0.0001; DSC scan (c) for thermal property (d) analysis of MC/BCP hydrogel scaffolds.

Figure 2(d) exhibits the XRD (X-ray diffractograms) of MC/BCP hydrogels, where MC/no BCP hydrogel is kept as a control. The crystallinity index (CI%) of the hydrogel scaffolds are in the following order: MC/ no BCP (49.09%) > MC/BCP-I (48.08%) > MC/BCP-II (33.12%). This difference in CI% between MC/ no BCP and MC/BCP-I and II reflects a notable interaction between BC and BCP. Earlier research reported that a high degree of crystallinity will be present where the XRD peaks are sharper and stronger (Khamrai et al. 2017). The diffraction peak (2θ) at 22.6°is attributed to cellulose (Gong et al. 2017). The full width half maximum values (FWHM) analysis is performed (Panda et al. 2021). The FWHM of the peaks around 22° are found to be higher for MC/no BCP hydrogels (i.e., FWHM MC/no BCP = 11.19 (±0.149)) than MC/BCP I (i.e., FWHM MC/BCP I = 1.2 (±0.067)) and MC/BCP-II (i.e., FWHM MC/BCP II = 6.69 (±0.25)). These results further indicate a strong interaction between the cellulose and other compositional components like β-TCP and HAp in MC/BCP I and II scaffolds. The 27.3°peak is found due to atmospheric moisture (Klechkovskaya et al. 2003), and the 28.3°and 36.6° peaks correspond to calcium phosphate in BCP-based samples (Su et al. 2013).

4.5. Scanning Electron Microscopy (SEM) and Elemental Analysis

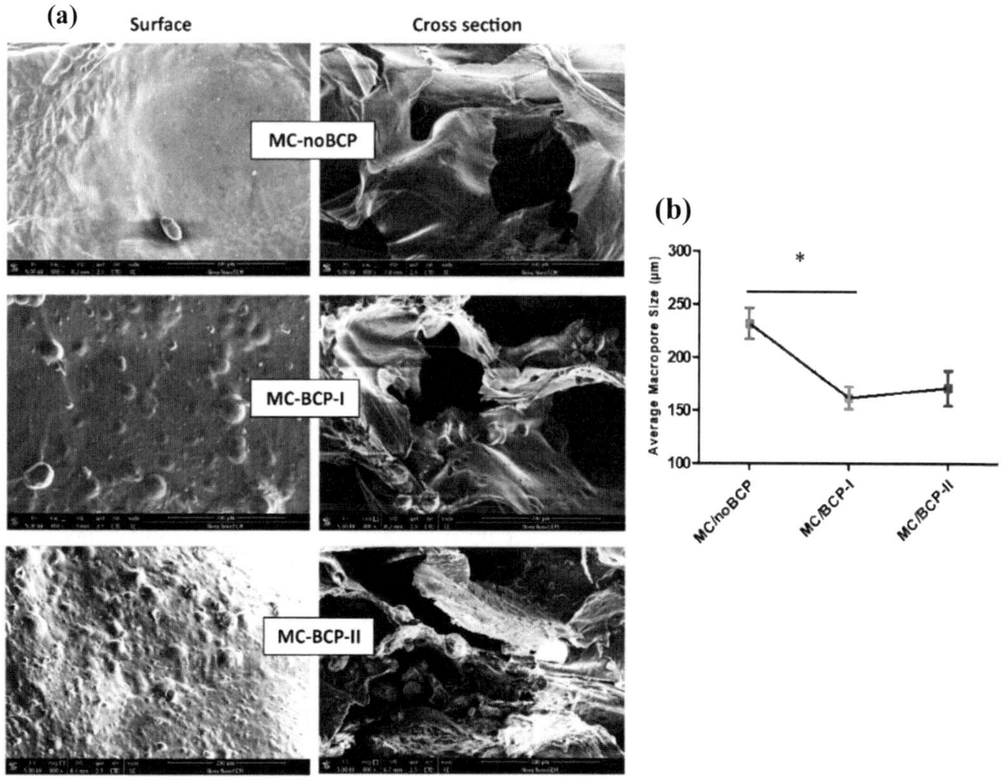

Figure 3. Morphological characterization and elemental analysis of MC/BCP hydrogel. (a) SEM morphological structure; (b) average macro pore size (* indicates P<0.0001).

Table 1. Calcium and Phosphate (%) with ± SD in MC/BCP hydrogel scaffolds in EDX

Elements (%)	MC/BCP-I	MC/BCP-II
Ca	5.42 (±0.11)	7.90 (±0.13)
P	1.99 (±0.11)	±0.18)

In Figure 3, the morphological characteristics of MC/BCP hydrogel are indicated through SEM analysis. The surface roughness of the hydrogels gradually increases from MC/no BCP to MC/BCP-II (Figure. 3(a)). On the other hand, it can be seen from the cross-sectional structure that all the hydrogels contain notable porous structures and confirmed that MC/BCP-I and MC/BCP-II are composite film/ biomaterial. Earlier research reported that macroporous structures in biomaterial have significance in bone regeneration (Mirhadi et al. 2018; Tolba et al. 2010). MC/BCP-based hydrogel (Figure 3(b)) contain significant macroporous structures (150-200 μm), which is an important characteristic for fracture healing.

Table 1 exhibits the existence of calcium (Ca) and phosphate (P) in MC/ BCP-based hydrogels. *EDX (energy dispersive X-ray analysis)* confirmed that MC/BCP-I and MC/BCP-II have maintained their structural integrity by not losing the notable concentrations of Ca and P elements from the polymeric structure.

4.6. Mechanical Property

The mechanical property of implant materials is a significant factor in the proper functioning of biomaterials. The proper analysis of the mechanical properties of the material is a challenging task. After being implanted, the implant materials generally experience a compressive stress environment. On the other hand, the inner structural attributes will also be influenced by the physiological viscous contents (e.g., physiological fluid).

4.6.1. Viscoelastic Property

The viscoelastic behavior of MC/BCP hydrogel is analyzed at a 1% strain rate at room temperature (28°C), maintaining the linear viscoelastic region (LVR). Figure 4 (a) exhibits the relationship between complex viscosity and shear stress of the hydrogels, which indicates a notable shear thinning behavior of the MC/BCP hydrogels. This behavior is observed in the following order: MC-BCP-II > MC-BCP-I and MC/no BCP. This suggests that the hydrogels maintain their gel-like consistency during the study duration (Kokol et al. 2021; Stanic et al. 2020). This shear thinning behavior of the MC/BCP hydrogels might be because of polymer-filler and filler-filler interaction. The interaction of MC with BCP is also a complex phenomenon where MC interacts with β-TCP and HAp with polymer-filler interaction and fibrous filler-particulate filler interaction. Hence, these differential levels of interactions influence the viscoelastic gel-like properties of the polymeric scaffold (Sheenoy, 1999). The abovementioned hypothesis can be further supported by observing the significant elastic behavior of MC/BCP hydrogels between 0.1-100 rad/s angular frequencies within the LVR. The MC/BCP I and II hydrogels show low damping in the initial stage compared to MC/ no BCP (Figure 4(b)). However, overall, all the MC/BCP-based hydrogels exhibit significant viscoelastic properties at 1% strain and 28°C.

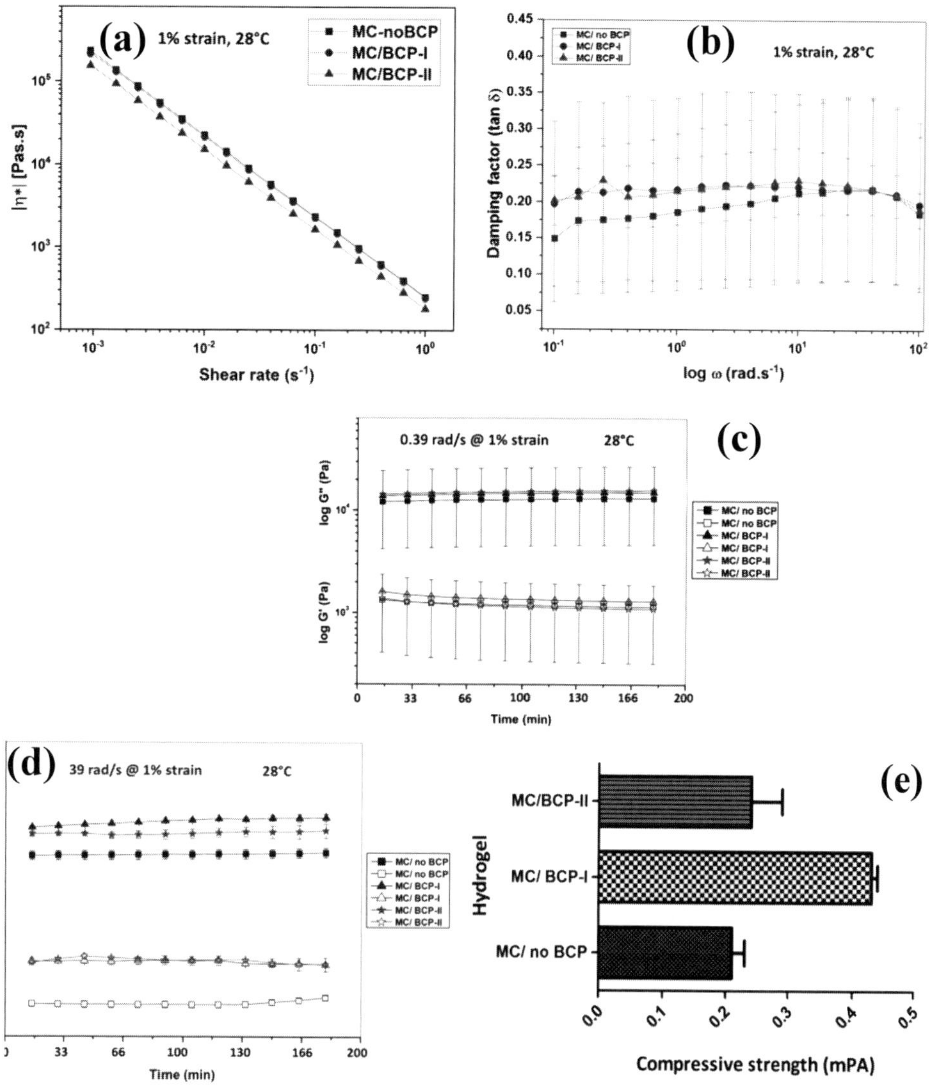

Figure 4. Mechanical behavior of MC/BCP biocomposite hydrogel. (a) viscosity vs shear rate and (b) damping factor between 0.1 to 100 radian/s angular frequency at 1% strain, 28°C. (c) The storage modulus (filled symbols) and loss modulus (open symbols) at 0.39 rad/s, 3.9 rad/s, 39 rad/s angular frequencies at 1 % strain, 28°C; (d) Compressive strength (mPA) values of MC/BCP hydrogel scaffolds. Error bars indicate the standard deviation (±) of the mean value.

The mechanical behavior of a polymeric material is generally also influenced by the deformation rate. It has been reported that materials behave liquid-like in slow deformations, and the same material could behave elastically like in fast deformations (Vliet et al. 1993). To understand the elastic and viscous behavior in low and fast deformation of MC/BCP hydrogels, the measurement is performed at 0.39 rad/s, 3.9 rad/s, and 39 rad/s angular frequencies as the function of time. This study was performed at a 1% strain rate and room temperature (28°C) (Figure 4(c)). MC/ no BCP is studied as a control. All the hydrogel maintains a higher storage modulus in 0.39, 3.9, and 39 rad/s than the loss modulus throughout the study time. Interestingly, the trend of the storage modulus for MC/ BCP-I was observed to be different in high deformation (39 rad/s), i.e., MC/BCP-I >MC/BCP-II> MC/ no BCP. This might be because of the interaction between two different ratios of β-TCP and HAp, i.e., 10:90 and 60:40, as particulate filler material within the polymer-filler system (Sheenoy, 1999).

4.6.2. Compressive Property

In Figure 4(d) indicates the compressive strengths of MC/BCP-I and II hydrogel scaffolds. These are comparable with the mechanical properties of human cancellous bone (Misch et al. 1999; Will et al. 2013). The compressive strength of the MC/BCP hydrogels is as follows: MC/BCP-I > MC/ BCP-II > MC/ no BCP. The higher compressive strength of MC/BCP-I may have resulted from significantly different polymer-filler interactions. In this context, it could also be mentioned that the interaction of differential concentration of HAp and β-TCP in MC/BCP-I and II might also influence the compression property of the studied bio-composites.

Conclusion and Prospects of MC-CP Bio-Composites in Soft Bone Regeneration

MC-based bio-composites are an interesting biomaterial due to their notable properties like high biocompatibility and significant bioactivity. Moreover, the mechanical properties, like the stiffness of MC nanofiber, are related to human soft tissue. This characteristic of the MC makes it important to utilize it as an efficient extracellular matrix-like material for soft tissue regeneration (Kesharwani et al. 2024). The nano-fibrous nature of this polymer provides additional support to the complex biocomposite structure. MC-based biomaterial loaded with bioactive agents like other bioactive polymers like collagen hyaluronic acid could promote cell proliferation and facilitate tissue engineering (Saleh Ahmed et al. 2024). Our work suggests that MC/BCP hydrogels contain important properties like porosity, significant macro-porous structures (150-240 μm), and chemical properties, which are important for facilitating bone fracture healing. The rheological property indicates these MC/BCP hydrogel scaffolds maintain notable gel-like properties and viscoelasticity. In addition, the compressive strengths of MC/BCP-I and II hydrogels are comparable with the human soft bone. This result suggests that MC/BCP-I and II hydrogels have the necessary physicochemical properties to become efficient osteoconductive materials and can be further suggested for their application for bone fracture healing. However, biological studies (*in vitro* and *in vivo*) of MC/BCP hydrogels are also recommended before their successful utilization for bone fracture repair, specifically soft-bone regeneration application.

Acknowledgments

The experimental parts of this work have been performed at the Centre of Polymer Systems, Tomas Bata University in Zlin, Czech Republic. This work also involves unpublished experimental data from the doctoral work of the first author (P.B.). The authors acknowledge the Ministry of Education, Youth & Sports of the Czech Republic—DKRVO (RP/CPS/2020/005). The first author would like to acknowledge the dean, head of the department, and the other members of the Faculty of Dentistry, Department of Biomaterials,

University of Oslo, Norway, for providing necessary additional support for successfully finishing the data analysis (through Origin software) and original draft writing & editing (by providing technological and infrastructural support). The first author dedicates this work to his amiable wife and late childhood friend.

References

Agarwal-Harding, Kiran, John Meara, Sarah Greenberg, Lars Hagander, David Zurakowski, George Dyer. "Estimating the Global Incidence of Femoral Fracture from Road Traffic Collisions." *Journal of Bone and Joint Surgery* 97, no. 6 (2015): e31. doi: 10.2106/JBJS.N.00314.

Barja, Franois. "Bacterial nanocellulose production and biomedical applications." *Journal of Biomedical Research* 35, no. 4(2021): 310- 317. doi: 10.7555/JBR.35.20210036

Basu, Probal, Nabanita, Saha, Radostina, Alexandrova, Petr, Saha. "Calcium Phosphate Incorporated Bacterial Cellulose-Polyvinylpyrrolidone Based Hydrogel Scaffold: Structural Property and Cell Viability Study for Bone Regeneration Application." *Polymers* 11, no. 11(2019a), 1821. https://doi.org/10.3390/polym11111821.

Basu Probal, Nabanita Saha & Petr Saha. "Inorganic calcium filled bacterial cellulose based hydrogel scaffold: novel biomaterial for bone tissue regeneration." *International Journal Polymeric Materials and Polymeric Biomaterials* 68, no. 1-3 (2019b):134-144. https://www.tandfonline.com/doi/full/10.1080/00914037.2018.1525733.

Basu, Probal, Nabanita Saha, Radostina Alexandrova, Boika Andonova-Lilova, Milena Georgieva, George Miloshev, Petr Saha. "Biocompatibility and Biological Efficiency of Inorganic Calcium Filled Bacterial Cellulose Based Hydrogel Scaffolds for Bone Bioengineering." *International Journal of Molecular Sciences* 19 (2018): 3980. https://doi.org/10.3390/ijms19123980.

Beltramino, Facundo, M. Blanca Roncero, Antonio L. Torres, Teresa Vidal, Cristina Valls. "Optimization of sulfuric acid hydrolysis conditions for preparation of nanocrystalline cellulose from enzymatically pretreated fibers." *Cellulose* 23, (2016):1777–1789. https://doi.org/10.1007/s10570-016-0897-y.

Bharadwaza, Angshuman and Jayasuriya, Ambalangodage. "Recent trends in the application of widely used natural and synthetic polymer nanocomposites in bone tissue regeneration." *Materials Science & Engineering C* 110 (2020): 1106. doi: https://doi.org/10.1016/j.msec.2020.110698.

Choi, Ji-Bong, Yu-Kyoung Kim, Seon-Mi Byeon, Jung-Eun Park, Tae-Sung Bae, Yong-Seok Jang, and Min-Ho Lee. 2021. "Fabrication and Characterization of Biodegradable Gelatin Methacrylate/Biphasic Calcium Phosphate Composite Hydrogel for Bone Tissue Engineering." *Nanomaterials* 11, no. 3: 617. https://doi.org/10.3390/nano11030617.

Dufresne, A. "Nanocellulose: a new ageless bionanomaterial." *Materials Today* 16, no. 6(2013), 220–227.

Esa, Faezah, Siti Masrinda Tasirin, Abd Rahman Norliza. "Overview of Bacterial Cellulose Production and Application." *Agriculture and Agricultural Science Procedia* 2, (2014): 113-119. doi: 10.1016/j.aaspro.2014.11.017.

Gadaleta, SJ, E. P. Paschalis, F. Betts, R. Mendelsohn, A.L Boskey. "Fourier Transform Infrared Spectroscopy of the Solution-Mediated Conversion of Amorphous Calcium Phosphate to Hydroxyapatite: New Correlations Between X-Ray Diffraction and Infrared Data." *Calcified Tissue International* 58, (1996): 9-16. DOI: 10.1007/BF02509540.

Gatabi, Zahra Rezanejad, Niloofar Heshmati, Mehri Mirhoseini, Maedeh Dabbaghianamiri. "The Application of Clay-Based Nanocomposite Hydrogels in Wound Healing." *Arabian Journal for Science and Engineering* 48, (2023): 8481–8494. https://doi.org/10.1007/s13369-022-06959-3.

Ghouri, Syed Imran, Mohammad Asim, Fuad Mustafa, Kanbar, Ellabib Ahad, Mohamed Ellabib, Hisham Al Jogol, Mohammed Muneer, Nuri Abdurraheim, et al. "Patterns, Management, and Outcome of Traumatic Femur Fracture: Exploring the Experience of the Only Level 1 Trauma Center in Qatar." *International Journal of Environmental Research and Public Health* 18, no. 11(2021): 5916. doi: 10.3390/ijerph18115916.

Gong, Jie, Jun Li, Jun Xu, Zhouyang Xiang, Lihuan Mo. "Research on cellulose nanocrystals produced from cellulose sources with various polymorphs." *RSC Advance* 7, (2017):33486 -33493. DOI: 10.1039/c7ra06222b.

Grøndahl, M. "Cytocompatibility and Functionalization of Nanocellulose for Bone Tissue Engineering Applications." Master of Science, Norwegian University of Science and Technology. Trondheim, 2019. http://hdl.handle.net/11250/2630946.

Hao, Zhuowen, Hanke Li, Yi Wang, Yingkun Hu, Tianhong Chen, Shuwei Zhang, Xiaodong Guo, Lin Cai, Jingfeng Li. "Supramolecular Peptide Nanofiber Hydrogels for Bone Tissue Engineering: From Multihierarchical Fabrications to Comprehensive Applications." *Advanced Science* 2103820, (2002): 1-36. https://doi.org/10.1002/advs.202103820.

Hench, Larry "Bioceramics," *Journal of the American Ceramic Society* 81, No. 7(1998): 1705-1728. doi:10.1111/j.1151-2916.1998.tb02540.x.

Hong, M-H, S-M Kim, YK Lee. "Mechanical and Biological Properties of Biphasic Calcium Phosphate Scaffold depending on Different Nanoparticle Fabrication Methods." *Journal of Ceramic Science and Technology* 08, no. 4(2017): 541-546. DOI: 10.4416/JCST2017-00033.

Hospodiuk-Karwowski, Monika, Syed Bokhari, Kai Chi, Kazim Moncal, Veli Ozbolat, Ibrahim Ozbolat, Jeffrey Catchmark "Dual-charge bacterial cellulose as a potential 3D printable material for soft tissue engineering." *Composites Part B: Engineering* 231(2022): 109598. https://doi.org/10.1016/j.compositesb.2021.109598.

Hou, Xiaodong, Yixing Chen, Feng Chen, Junfeng Liu, Tianlong Wang, Yiping Luo, et al. "The hydroxyapatite microtubes enhanced GelMA hydrogel scaffold with inner "pipeline framework" structure for bone tissue regeneration." *Composites Part B: Engineering* 228 (2022): 109396 https://doi.org/10.1016/j.compositesb.2021.109396.

Hussin, Mohamed Saiful Firdaus, Aludin Mohd Serah, Khairul Azri Azlan, Hasan Zuhudi Abdullah, Maizlinda Izwana Idris, Ihwan Ghazali, Amir Husni Mohd Shariff, Nurul Huda, and Azrul Abidin Zakaria. 2021. "A Bibliometric Analysis of the Global Trend of Using Alginate, Gelatine, and Hydroxyapatite for Bone Tissue Regeneration Applications. " *Polymers* 13, no. 4: 647. https://doi.org/10.3390/polym13040647.

Jain, Pooja, Ruby Yu-Tong Lin, Keerti Mishra, Harish Handral, Nileshkumar Dubey. "Three-dimensional eco-friendly bacterial nanocellulose (BNC) scaffold for regenerative dentistry: Characterization, cytocompatibility and differentiation potential." *Dental Materials*. 40, no. 1(2024):151-157. doi: 10.1016/j.dental.2023.11.001.

Kaur, Gurbinder, Pandey OP, Singh K, Homa Dan, Scott Brian, Pickrell Gary. "A review of bioactive glasses: Their structure, properties, fabrication, and apatite formation." *Journal of Biomedical Materials Research* 102, no.1 (2014): 254–274. doi: 10.1002/jbm.a.34690.

Kesharwani, Payal, Amit Alexander, Rahul Shukla, Smita Jain, Akansha Bisht, Kajal Kumari, Kanika Verma, Swapnil Sharma. "Tissue regeneration properties of hydrogels derived from biological macromolecules: A review." *International Journal of Biological Macromolecules* 271, (2024): 132280. https://doi.org/10.1016/j.ijbiomac.2024.132280.

Keshk, Sherif MAS. "Bacterial Cellulose Production and its Industrial Applications, Bacterial Cellulose Production and its Industrial Applications." *Journal of Bioprocessing and Biotechniques* 4 (2014): 1000150. DOI: 10.4172/2155-9821.1000150.

Khamrai, Moumita, Sovan Lal Banerjee, Patit P Kundu. "Modified bacterial cellulose based self-healable polyeloctrolyte film for wound dressing application." *Carbohydrate Polymer* 174, (2017): 580-590. doi: 10.1016/j.carbpol.2017.06.094.

Kim, Tae-Wan, Hyeong-Shin Lee, Dong-Hyun Kim, Hyeong-Ho Jin, Kyu-Hong Hwang, Jong Kook Lee, et al. "In situ synthesis of magnesium-substituted biphasic calcium phosphate and *in vitro* biodegradation." *Materials Research Bulletin* 47, no. 9(2012): 2506–2512. https://doi.org/10.1016/j.materresbull.2012.05.011.

Klechkovskaya, VV, YG Baklagina, ND Stepina, AK Khripunov, PA Buffat, EI Suvorova, IS Zanaveskina, AA Tkachenko, SV Gladchenko. "Structure of cellulose Acetobacter xylinum." *Crystallography Reports* 48, no. 5 (2003): 755–762. doi: 10.1134/1.1612596.

Kokol, Vanja, Yasir Beeran Pottathara, Mohor Mihelčič, Lidija Slemenik Perše. "Rheological properties of gelatine hydrogels affected by flow- and horizontally induced cooling rates during 3D cryo-printing."

Colloids and Surfaces A: Physicochemical and Engineering Aspects 616, (2021): 126356. https://doi.org/10.1016/j.colsurfa.2021.126356.

Koons, Gerry, Mani Diba, Antonios Mikos. "Materials design for bone-tissue engineering." *Nature Reviews Materials* 5 (2020): 584–603. https://doi.org/10.1038/s41578-020-0204-2.

Kuo, Yi-Jie, Chia-Hsien Chen, Pranjyan Dash, Yu-Chien Lin, Chih-Wei Hsu, Shao-Ju Shih, and Ren-Jei Chung. "Angiogenesis, Osseointegration, and Antibacterial Applications of Polyelectrolyte Multilayer Coatings Incorporated with Silver/Strontium Containing Mesoporous Bioactive Glass on 316L Stainless Steel." *Frontiers in Bioengineering and Biotechnology* 2022, no. 10 (2022): 818137. doi: 10.3389/fbioe.2022.818137.

León, Andrea, Patricia Reuquen, Carolina Garín, Rodrigo Segura, Patricio Vargas, Paula Zapata, and Pedro A. Orihuela. "FTIR and Raman Characterization of TiO2 Nanoparticles Coated with Polyethylene Glycol as Carrier for 2-Methoxyestradiol." *Applied Sciences* 7, no. 1 (2017): 49. https://doi.org/10.3390/app7010049.

Lin, Shin-Ping, Stephanie Singajaya, Tsui-Yun Lo, Shella Permatasari Santoso, Hsien-Yi Hsu, and Kuan-Chen Cheng. "Evaluation of Porous Bacterial Cellulose Produced from Foam Templating with Different Additives and Its Application in 3D Cell Culture." *International Journal of Biological Macromolecules* 234 (2023): 123680. https://doi.org/10.1016/j.ijbiomac.2023.123680.

Mirhadi, Seyed Mehdi, Nahid Hassanzadeh Nemati, Fariborz Tavangarian, Morteza Daliri Joupari. "Fabrication of hierarchical meso/macroporous TiO2 scaffolds by evaporation-induced self-assembly technique for bone tissue engineering applications." *Materials Characterization* 144(2018): 35–41. https://doi.org/10.1016/j.matchar.2018.06.035.

Misch, Carl, Zhimin Qu MS, Martha W Bidez. "Mechanical properties of trabecular bone in the human mandible: Implications for dental implant treatment planning and surgical placement." *Journal of Oral and Maxillofacial Surgery* 57, no. 6 (1999): 700–706. doi: 10.1016/s0278-2391(99)90437-8.

Nunes, Tiago, Pedro Armelim Almiro, Joana Coelho, Catarina Caetano, José P. Figueiredo, Ana Corte-Real. "Orofacial trauma injuries and disability assessment in adolescents: A Portuguese study." *Dental Traumatology* 38 (2022): 41– 47. https://doi.org/10.1111/edt.12700.

Oliveira, Rafael, Júlia G. Vieira, Hernane S. Barud, Rosana M. N. Assunção, Guimes Rodrigues Filho, Sidney J. L. Ribeiro, Younes Messadeqq. "Synthesis and Characterization of Methylcellulose Produced from Bacterial Cellulose under Heterogeneous Condition." *Journal of the Brazilian Chemical Society* 26, no. 9 (2015):1861–1870. https://doi.org/10.5935/0103-5053.20150163.

Pandit, Vinay, Roopa S Pai, Kusum Devi, Suresh Sarasija. "*In vitro-in vivo* evaluation of fast-dissolving tablets containing solid dispersion of pioglitazone hydrochloride." *Journal of Advanced Pharmaceutical Technology & Research* 3, no. 3(2012):160-170. doi: 10.4103/2231-4040.101008.

Panda, Pradeep Kumar, Jen-Ming Yang, Yen-Hsiang Chang. "Water-induced shape memory behavior of poly (vinyl alcohol) and p-coumaric acid-modified water-soluble chitosan blended membrane." *Carbohydrate Polymer* 257, (2021): 117633. https://doi.org/10.1016/j.carbpol.2021.117633.

Pavitra, Raja Marthandam, Sethu Gunasekaran, Ethirajulu Sailatha, Kamatchi Sundaram. "Investigations on Paper Making Raw Materials and Determination of Paper Quality by FTIR-UATR and UV-Vis DRS Spectroscopy." *International Journal of Current Research and Academic Review* 3, (2015): 42-59.

Qi, Jiangi, Yu, Tianqi, Hu, Bangyan, Wu, Hangwei and Ouyang, Hongwei. "Current Biomaterial-Based Bone Tissue Engineering and Translational Medicine." *International Journal of Molecular Sciences* 22, no. 19 (2021): 10233. doi: 10.3390/ijms221910233.

Saleh Ahmed, Julie Basu Ray, Mohamed H El-Sayed, Adel I Alalawy, Noha Omer, Mahmoud A Abdelaziz, Ragab Abouzeid. "Functionalization of bacterial cellulose: Exploring diverse applications and biomedical innovations: A review." *International Journal of Biological Macromolecules* 264, no. 1 (2024):130454. doi: 10.1016/j.ijbiomac.2024.130454.

Shavyrkina NA, EA Skiba, AE Kazantseva, EK Gladysheva, VV Budaeva, NV Bychin, YA Gismatulina, EI Kashcheyeva, GF Mironova, AA Korchagina, IN Pavlov, GV Sakovich. "Static Culture Combined with Aeration in Biosynthesis of Bacterial Cellulose." *Polymers* (Basel) 13, no. 23 (2021):4241. doi: 10.3390/polym13234241.

Sheenoy, A.V. "*Rheology of Filled Polymer Systems*," 1999, Elsevier: U.K.

Stanic, Sascha, Gergö Gottlieb, Thomas Koch, Lukas Göpperl, Klaus Schmid, Simone Knaus, Vasiliki-Maria Archodoulaki. "Influence of Different Types of Peroxides on the Long-Chain Branching of PP via Reactive Extrusion." *Polymers* 12, (2020): 886. https://doi.org/10.3390/polym12040886.

Suresh, Sumathi. "Biosynthesis and Assemblage of Extracellular Cellulose by Bacteria." Springer, Cham, Handbook of Environmental Materials Management (2018): 1-43. ISBN 978-3-319-58538-3. Available at: https://link.springer.com/referenceworkentry/10.1007%2F978-3-319-58538-3_71-1.

Su, Yingchao, Liyuan Niu, Yanbo Lu, Jianshe Lian, Guangyu Li. "Preparation and Corrosion Behavior of Calcium Phosphate and Hydroxyapatite Conversion Coatings on AM60 Magnesium Alloy." *Journal of The Electrochemical Society* 160, no. 11 (2013): C536-C541.

Syverud, K. "Tissue Engineering Using Plant-Derived Cellulose Nanofibrils (CNF) as Scaffold Material." In: Agarwal, U. P., Atalla, R. H. and Isogai, A. (editors). Nanocelluloses: Their preparation, Properties, and Applications. Washington DC: American Chemical Society; 2017.

Tolba, E, BM Abd-Elhady, B Elkholy, H Elkady, M. Eltonsi. "Biomimetic Synthesis of Guided-Tissue Regeneration Hydroxyapatite/Polyvinl Alcohol Nanocomposite Scaffolds: Influence Of Alignate on Mechanical and Biological Properties." *Journal of American Science* 6, (2010): 239-249.

Trivedi, Alok Kumar, Gupta, Manoj Kumar, Singh, Harinder. "PLA Based Biocomposites for Sustainable Products: A Review." *Advanced Industrial and Engineering Polymer Research* 6, no.4 (2023): 382-395. https://doi.org/10.1016/j.aiepr.2023.02.002.

Vliet, TV, H Luyten, P Walstra. In Food Colloids and Polymers: Stability & Mechanical Properties; Dickinson, E.; Walstra, P., Eds., Elsevier: U.K., 1993.

Will, Julia, Rainer Detsch, Aldo R Boccaccini, A. R. Characterization of Biomaterials. In Structural and Biological Characterization of Scaffolds; Bandyopadhyay, A., Bose, S., Eds.; Academic Press: Cambridge, MA, USA, 2013, Chapter 7.1; pp. 299–310.

Yao, Guang, Lei Kang, Cuicui Li, Sihong Chen, Qian Wang, Junzhe Yang, Yin Long, et al. "self-powered implantable and bioresorbable electrostimulation device for biofeedback bone fracture healing." *Proceedings of the National Academy of Sciences* 118, no. 28(2021): e2100772118.https://doi.org/10.1073/pnas.2100772118.

Zaszczyńska, Angelika, Maryla Moczulska-Heljak, Arkadiusz Gradys, and Paweł Sajkiewicz. 2021. "Advances in 3D Printing for Tissue Engineering" *Materials* 14, no. 12: 3149. https://doi.org/10.3390/ma14123149.

Zhao, Yue, Shanliang Song, Xiangzhong Ren, Junmin Zhang, Quan Lin, Quan Zhao. "Supramolecular Adhesive Hydrogels for Tissue Engineering Applications." *Chemical Reviews* 122, no. 6(2022): 5604-5640. doi: 10.1021/acs.chemrev.1c00815.

Section 5

Chapter 5.1

The Mechanistic Aspects of *In Vitro* Toxicity of Organic and Inorganic Nanoparticles

Induni Nayodhara Weerarathna[1,*]
Hellen Yayra Dzoagbe[2]
Elijah Skarlus Doelakeh[1]
Sanvi Arora[3]
Kanishk Shukla[3]
and Praveen Kumar[4]

[1]School of Allied Health Sciences, Datta Meghe Institute of Higher Education and Research, Sawangi, Meghe, Wardha, Maharashtra, India
[2]College of Pharmacy, Datta Meghe Institute of Higher Education and Research, Sawangi, Meghe, Wardha, Maharashtra, India
[3]Jawaharlal Nehru Medical College, Datta Meghe Institute of Higher Education and Research, Sawangi, Meghe, Wardha, Maharashtra, India
[4]Faculty of Engineering, Datta Meghe Institute of Higher Education and Research, Sawangi, Meghe, Wardha, Maharashtra, India

Abstract

Nanotechnology has grown exponentially, allowing nanoparticles (NPs) to be incorporated into many aspects of daily life, especially in the medical field. However, the toxicity of NPs presents serious obstacles to their safe use, showing up as a range of health issues such as lymphopenia, hypokalemia, and neurodegenerative diseases. The mechanistic features of nanoparticle toxicity are still poorly understood, despite their tremendous promise, which calls for targeted study initiatives. This chapter explores the complex field of *in vitro* toxicity of both inorganic and organic NPs, providing details on their intracellular destiny, cellular absorption processes, and resulting reactions. To understand the complex nature of NP toxicity, factors that impact it, such as biological variables, environmental conditions, and physicochemical features, are examined. Additionally, instances of NP toxicity through case studies highlight how crucial it is to comprehend the various pathways and results associated with multiple NP kinds. To enable safer uses of NPs, strategies for reducing their toxicity such as dose optimization and surface modification techniques are

* Corresponding Author's Email: induni.weerarathna@dmihr.edu.in

In: Emerging Nanomaterials in Biomedical Science
Editors: Eepsita Priyadarshini and Ajith Manayil Parambil
ISBN: 979-8-89530-600-0

investigated. In the end, this chapter emphasizes how critical it is to have a thorough understanding of NP toxicity to ensure their safe incorporation in different disciplines. It also calls for future research endeavors to fill up the knowledge gaps that now exist and promote breakthroughs in the field of nanotoxicology.

Keywords: nanoparticles, Toxicity, mechanisms, nanotoxicology, mitigation strategies

1. Introduction

A particle with a diameter of one to one hundred nanometers (nm) is called a nanoparticle. They are instrumental in electronics, energy, medical and environmental protection. Nanoparticles are mainly of two types- organic nanoparticles and inorganic particles (Virlan et al. 2016). Organic nanoparticles are carbon-based molecules; the most prevalent forms of organic nanoparticles are polymeric nanoparticles, liposomes, nano-gels, micelles, and dendrimers. They benefit bioanalysis and basic science as a model system to research molecular recognition and single molecule behavior signal transduction. They can also be used in drug delivery systems, targeted therapy, imagining, diagnosis/therapy, and biosensors (Feracci et al. 2012). Inorganic nanoparticles, which include gold (AuNPs), silver (AgNPs), platinum, titanium oxide (TiO2), zinc oxide (ZnO), and iron oxide (Fe3O4), are non-carbon-containing, water-soluble, and biocompatible with living systems.

Magnetic nanoparticles hold great significance since they cover a wide range of magnetic recording media and biomedical uses such as magnetic resonance, angiography, and cancer therapy (H. Alshammari et al. 2023). The study of engineering nanomaterials' harmful effects is known as neurotoxicology. There are several uses for nanoparticles, such as medical fields, drug and gene delivery, and the development of new screening techniques. They are also used in engineering sections like pipelines, materials like polymers and metals, and electronics (Egbuna et al. 2021a).

There are various mechanisms by which nanoparticles produce toxic effects, i.e., oxidative stress, inflammation, genotoxicity, membrane damage, and protein corona formation (Sukhanova et al. 2018a). The nanoparticles generate oxygen species, which leads to oxidative stress. It could damage some components like lipids, proteins, and DNA. TiO_2 and ZnO_2 are primarily responsible for generating oxygen species (Fu et al. 2014). This may result in chronic diseases such as cancer, neurodegenerative disorders, and cardiovascular disease. It is also necessary to carefully evaluate their biological fate and immunity to use inorganic nanoparticles as carriers for drug delivery and other biomedical applications (Mohammapdour and Ghandehari 2022). Exposure of nanoparticles to immune cells could lead to chronic inflammation, which can cause various diseases. Both inorganic and organic particles cause inflammatory responses. The immune system can either detect or bypass nanoparticles. Once identified, they can either trigger defensive (inflammatory) or anti-inflammatory reactions, or they can be tolerated (Ernst et al. 2021). Some nanoparticles interact with the genetic material, leading to gene mutation and chromosomal fragmentation. Metals like Ag and ZnO are primarily responsible for causing such an effect (Shukla et al. 2021). Some nanoparticles also communicate with the cell membranes, which would cause cell destruction. Lipid bilayer interaction can compromise the integrity and regular operations of cells (X. Zhang, Ma, and Wei 2021).

Some may also possess toxicological risks, such as altering the cellular uptake seen during surface coating (Awashra and Młynarz 2023). Others may include accumulation in the organs such as the liver, spleen, and lung, causing systemic toxicity and chronic disease (Wassenaar and Verbruggen 2021). During manufacture, use, and disposal, nanoparticles may be released into the environment, posing a risk to public health and worker exposure. For example, breathing in airborne nanoparticles presents severe dangers to respiratory health, such as lung cancer, fibrosis, and inflammation (Ray, Yu, and Fu 2009). This chapter will explain how organic and inorganic particles express toxicity and describe their potential adverse effects.

2. Overview of Nanoparticles

Nanoparticles are tiny particles ranging from 1 to 100 nanometers (Jeevanandam et al. 2018). In context to biomedicine, they are particles engineered or natural particles at the nanometer scale with unique physical and chemical properties due to their high surface-area-to-volume ratio, quantum effects, and the large surface energy of these particles that could provide them the opportunity to interact with biological systems at both the molecular and cell level (Albanese, Tang, and Chan 2012). Their properties show diversity and use fullness in a wide range of applications such as drug delivery, biochemical imaging or sensing, and diagnostic and therapeutic applications (J. Xu and Chow 2023). Nanoparticles can be made from a variety of materials, including metals, polymers, lipids, or ceramics, nanoparticles can be designed to be biocompatible, and functionalized with molecules of interest to deliver and release therapeutic agents to specific cells or tissues within the body (Altammar 2023). They have been widely used as carriers of a variety of therapeutics due to their ability to penetrate biological barriers (cell membrane), since they are minute, their surface could be modified to make them more stable, soluble, and able to target specific tissues or cells (McNamara and Tofail 2017).

2.1. Organic and Inorganic Nanoparticles

Numerous materials are available. and structures of nanoparticles, classifying them into organic and inorganic nanoparticles, and they have specific characteristics and applications (Khan, Saeed, and Khan 2019). Among organic nanoparticles, carbon materials are the most studied, which are applied in the form of carbon nanotubes, fullerenes, and polymeric nanoparticles such as polylactic-co-glycolic acid (PLGA) and liposomes that are biocompatible nanomaterials and have been well established for drug delivery (Abd Ellah and Abouelmagd 2017). Among these lipid nanoparticles, liposomes and solid lipid nanoparticles are commonly used to pack therapeutic agents for targeted delivery. Their branching structure gives dendrimers fine-tuned control over drug loading and release (Hagens et al. 2007). Inorganic nanoparticles, on the other hand, consist of such materials as metals (e.g., gold, silver), magnetic nanoparticles (e. g. iron oxide), quantum dots, and ceramic nanoparticles (e. g. silica, titania) (*Magnetic Nanoparticles* 2020). Metal nanoparticles, particularly in the optical and catalytic fields, are convenient for both imaging and treatment, magnetic nanoparticles are beneficial for targeted drug delivery, diagnosis such as medical resonance imaging (MRI). (Joseph et al. 2023). Quantum dots provide changeable luminescent properties that are perfect for imaging, while ceramic nanoparticles are suitable for drug delivery and imaging uses (Sim and Wong 2021).

2.2. Properties Affecting Toxicity

Size, shape, composition, and surface chemistry are the most apparent properties that affect the level of toxicity of nanoparticles. Smaller dimensions ranging from 1 to 100 nm cause objects to possess a larger surface area to volume ratio hence high reactivity and toxicity due to interactivity with biological systems (Bahadar et al. 2016). Further, some shapes like nanotubes or nanowires could have a higher toxicity in comparison to round nanoparticles due to the enhanced ability to penetrate cell membranes or exert oxidative stress (De Berardis et al. 2010). Nanoparticles having functional groups on the surface or a surface coating may trigger immune responses or cellular damage. Surface charge plays a significant role in determining how nanoparticles, interact with cell membranes and proteins to get internalized and the subsequent toxicity (González-García et al. 2022). However, the nanoparticles themselves, including both organic and inorganic compounds, may also be toxic depending on their content; for example, heavy metals of silver or cadmium may cause cytotoxicity or inflammation (Xuan et al. 2023).

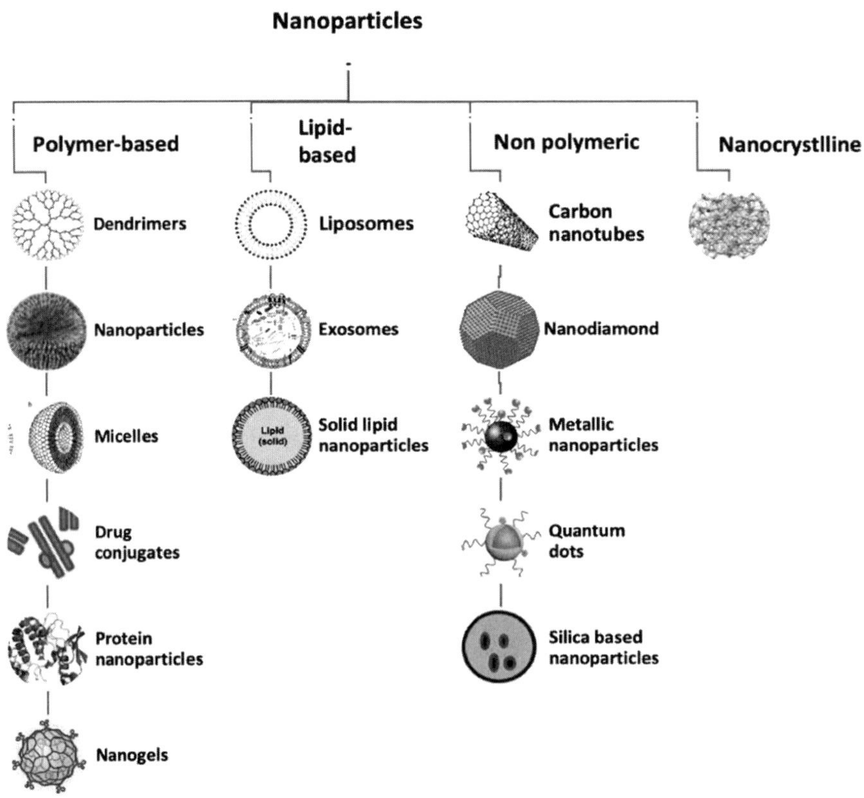

Figure 1. Diagrammatic representation of the various nanoparticle types resulting from different formulation .(Reproduced with permission from Abdellatif et al. Copyright, 2021, Nanotechnology Reviews) (Abdellatif and Alsowinea 2021).

2.3. Common Applications

Nanoparticles are versatile in bio-medical applications because of their peculiar characteristics (Wang and Wang 2014). In drug delivery, they improve the efficiency of the therapy and localization to the specific cells or tissues minimizing side effects due to other cells' damage (Elumalai, Srinivasan, and Shanmugam 2024). In medical applications, medical imaging techniques like MRI and fluorescence microscopy are aided by nanoparticles like gold nanoparticles, quantum dots, and other particles to enhance the noise ratio hence enhancing the diagnosis and monitoring of diseases at an early stage (Abdelmalak and Doyle 2020; Han et al. 2019). . Nanoparticles are also used in diagnostic applications, enabling the detection of biomolecules associated with diseases such as cancer and infections. Further, due to their antimicrobial activity, these compounds have many applications in wound healing as well as in killing bacteria when used as a disinfectant (Thwala et al. 2023). In therapy, the application of magnetic nanoparticles makes it possible to employ hyperthermia treatment under the principle that heat is selectively applied to destroy the cancerous cells, which of course, leads to the death of those cells, it proves that the nanoparticles are multi-functional and can offer several opportunities to enhance different aspects of the field of healthcare (Fatima, Charinpanitkul, and Kim 2021). Figure 1 shows the various nanoparticle forms resulting from different formulation types.

3. Factors Influencing Nanoparticle Toxicity

3.1. Physicochemical Properties

3.1.1. Size, Shape, and Surface Charge

The toxicity of nanoparticles (NPs) is greatly affected by their size and shape. The dimensions of nanoparticles play a role in the production of reactive oxygen species (ROS) and the ion release from silver nanoparticles (AgNP). Studies have shown an inverse relationship between the levels of ROS, the balance of apoptosis to necrosis, cell viability, and the size of PVP-coated AgNPs. Smaller AgNPs demonstrate increased activity because of their extensive surface area, which affects oxidative stress, and their reliance on interfacial interactions that influence the rate of AgNP dissolution into ions. It has been documented that AgNPs can cause significant endothelial cell damage and dysfunction in addition to causing a size-dependent increase in the permeability of the microvascular endothelial cell monolayer, which in turn triggers significant inflammatory responses. Smaller AgNPs build up inside the cell and induce cytotoxicity, alter endothelial cell monolayer shape, and stimulate the release of cytokines. NPs with smaller sizes exhibit greater toxicity compared to those with bigger sizes. For example, AgNPs measuring 10 nm are significantly more poisonous than AgNPs with larger sizes (20-100 nm) (T. Zhang et al. 2014).

It is, therefore, clear that the immunological and poisonous effects of AgNPs are highly dependent on the shape of the nanoparticles. Common forms include sheets, cylinders, spheres, rods, cubes, and ellipsoids. The shape is highly related to toxicity. This has been illustrated for NPs of diverse structural form and elemental composition. For example, the tendency of spherical NPs to undergo endocytosis is higher compared to nanotubes and nanofibers. Several studies have established that spherical fullerenes are not as good at preventing calcium channel blockage as single-walled carbon nanotubes (Sukhanova et al. 2018b). Stoehr and colleagues found that cytotoxicity was highest in Ag nanowires when tested on human lung epithelial A549 cells while spherical Ag particles were not cytotoxic to the cells (Stoehr et al. 2011).

Moreover, the particles' surface charge might determine the particles' ability to penetrate cell membranes or react with biomolecules and organelles. This could influence the cytotoxicity of NPs directly suggesting that toxicity increases as the surface charge increases. An investigation was conducted on the human hepatoma cell line (BEL-7202) using several iron NPs of different surface charges, showing that the NP with higher positive charge had high electrostatic interaction hence; high endocytic uptake (Kai et al. 2011). This was in line with another study performed by Baek and his colleagues who used similar size and shape ZnONPs, but positively charged ZnONPs showed more cytotoxicity to A549 cells than negatively charged (Baek et al. 2011). The same could be said when NPs, which bear a positive charge are associated with negatively charged DNA and destroy it in the process.

3.1.2. Surface Chemistry

Surface chemistry may have an impact on how NPs interact with cells. NPs coated on the surface alter the dissolution ratio, aggregation, and morphology. The toxicity of colloidal AgNPs can be reduced and their stability enhanced by citrate coatings. Citrate-coated AgNPs lose their immaculate surface coating when they are dried into a powder, but their cytotoxicity remains evident when they dissolve again. On the other hand, when the dried NP powder is re-dissolved into the water, PVP-modified AgNPs exhibit good stability and have minimal harmful effects on human skin HaCaT keratinocytes (Lu et al. 2010). In a different investigation, it was shown that iron oxide nanoparticles coated with polyethyleneimine (PEI) exhibited notably more absorption in cancer cells and macrophages than PEGylated iron oxide nanoparticles, leading to considerable cytotoxicity (Feng et al. 2018).

3.2. Biological Factors

3.2.1. Cell Type

To ascertain the acute toxicity of different NPs on the chosen cells, Joris et al. studied six different types of cells; Human stem cells (hNSC), progenitor cell line ReNcell, cancer cell line LA-N-2, and mouse-derived stem cells (mNSC), progenitor cell line C17.2, and cancer cell line euro-2a). Silver NPs (AgNPs), iron oxide NPs (IONPs), and AuNPs were employed. The results from the study indicate that all NP cell type combinations analyzed showed a general concentration-dependent drop in ATP signal after 24 hours of exposure. AuNP treatment caused the most severe effect, whereas IONPs had the least impact on the cells (Joris et al. 2016).

AgNPs elicited different responses from fibroblast cells (NIH3T3) and colon cancer epithelial cells (HCT116). To guard against apoptotic stimuli, the anti-apoptotic protein Bcl-2 can be produced in excess, which explains why the epithelial cells reacted to AgNPs less strongly. For fibroblast cells (IMR-90) and glioblastoma cells (U251), starch-coated AgNPs can have varying genotoxicity. AgNPs can cause chromosomal abnormalities, DNA damage, cell cycle arrest, and mitochondrial malfunction in both forms of cells by diffusing into the nucleus and mitochondria (T. Zhang et al. 2014). They can also raise the quantity of ROS in the cell.

3.2.2. Dose and Exposure Duration

Cell viability may be disproportionately reduced by high NP dosages. To ascertain the toxicity of spherical silica nanoparticles (SNPs) with diameters ranging from 20 to 200 nm on NIH/3T3 fibroblasts, A549 and HepG2 epithelial cells, Kim et al. conducted a study. SNPs were found to reduce the viability of A549 cells at concentrations ≥50 μg/ml. Compared to all other SNP dosages and sizes evaluated, 60 nm particles produced dramatically reduced viability at high doses of 200 and 500 μg/ml. When exposed to SNPs at doses ≥50 μg/ml, NIH/3T3 cells demonstrated size- and dose-dependent reductions in viability at 24 and 72 hours. In contrast, HepG2 cells did not exhibit size-dependent alterations to functionality in response to SNP concentrations above 50 μg/ml, except for the largest dosage of 60 nm SNPs, which drastically decreased functioning compared to any other SNP treatment (Kim et al. 2015).

According to research on the *in vitro* impact of cobalt NPs on cell viability, excessive metal-on-metal implant wear can increase cobalt NP concentration *in vivo*, which can have detrimental biological implications locally. TiO_2 nanoparticles (NPs) were also tested for toxicity against human lung and colon cells at various dose ranges and diameters (18, 30, and 87 nm, as confirmed by TEM). Even at the lowest dosage, these smaller NPs (18 nm) had a notable detrimental impact. (Abbasi et al. 2023).

3.2.3. Interactions with Biomolecules

One important biological element that greatly affects the toxicity of nanoparticles is their interaction with biomolecules. These interactions can trigger a range of biological reactions that ultimately impact the overall toxicity of the nanoparticles, such as oxidative stress, protein binding, and cellular absorption. Reactive nitrogen species (RNS) overproduction is a prominent factor in nanoparticle-associated genotoxicity. The generation of ROS and RNS by NPs may result from an inflammatory response, contact with the cell target, or inherent production. Consequently, there may be either immediate or subsequent primary clastogenic disruption to the genetic material, as well as secondary genotoxicity (aneugenic and DNA adduct formation) (Egbuna et al. 2021b).

When NPs interact with DNA, they cause primary toxicity. However, when NPs create or transport ROS or RNS, they cause secondary genotoxicity, which results in genetic damage. Under the indirect primary clastrogenic pathway, ROS-mediated primary lipid oxidation results in the production of unsaturated aldehydes, which in turn create exocyclic DNA adducts. The idea that nanoparticles can cause genotoxicity is supported by numerous scientific studies (Egbuna et al. 2021b). In their research, Kisin and colleagues, for instance, found that Chinese hamster fibroblasts (V79 cell line) were susceptible to single- and double-strand DNA lesions produced by single-wall carbon nanotubes (SWCNTs) at 96 μg/cm`(Kisin et al. 2007). AgNPs were discovered to significantly damage DNA in *S. cerevisiae*, mulberry silkworm larvae, adult *C. cornuta* microcrustaceans, and zebrafish (D. rerio) micronuclei and nucleus at varying doses in an independent investigation (Egbuna et al. 2021b).

3.3. Environmental Conditions

3.3.1. pH, Temperature, and Presence of Other Chemicals

Cupi et al. investigated the impact of pH on the toxicity of Ag, ZnO, and TiO2-engineered nanoparticles on Daphnia magna neonates. The neonates' capacity to navigate unimpeded in the medium for a minimum of 15 seconds was used to gauge how hazardous these NPs were. They found that maintaining a steady pH throughout the test was problematic, and adding buffer did not make the medium more stable; rather, it made the TiO2 NPs more prone to agglomeration and, ultimately, toxic at pH 7. They also stated that ZnO NPs' toxicity increased when their solubility in the media, which was kept at pH 7 (Cupi, Hartmann, and Baun 2016).

Given that the aggregates' mean size decreased at high temperatures, it is possible to relate the impact of temperature on the distribution of Cu NPs. Additionally, the temperature changed the ion dissolution and aggregation size of ZnO NPs, which changed their nanotoxicity (Wong and Leung 2014). The heightened toxicity of Ag NPs at higher temperatures can be partially attributed to the greater metal uptake rate at elevated temperatures. Furthermore, it has been observed that Ag NPs' increased nanotoxicity is correlated with higher Ag+ release rates at temperatures between 0 and 37°C (J. Liu and Hurt 2010; Kittler et al. 2010). Temperature was observed as a significant factor in the toxicity of zero-valent copper nanoparticles (ZVCN) towards Escherichia coli (Rispoli et al. 2010).

3.3.2. Synergistic Effects with Other Pollutants

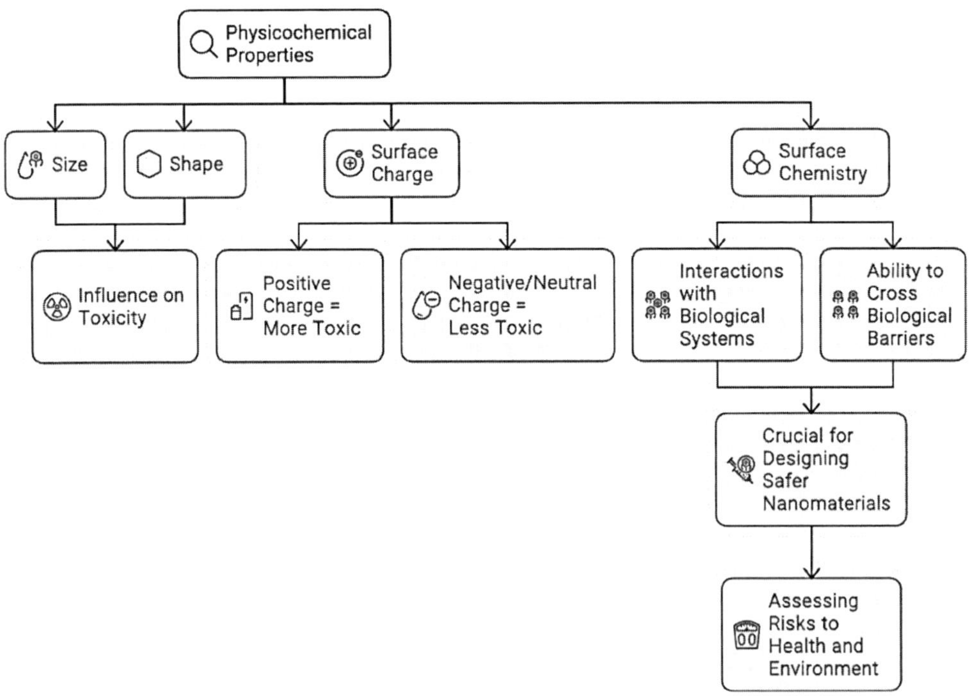

Figure 2. Influence of physicochemical properties on nanoparticle toxicity.

The concurrent presence of several toxicants can have unanticipated toxic effects on cells or organisms exposed to them, even though the individual toxicities of these substances are well established. A few studies looked at the toxicity and bioavailability of NPs to plants and animals, as well as the coexistence of contaminants. There was evidence of both antagonistic and synergistic effects. Cu^{2+} buildup in zebrafish was aided by cadmium telluride quantum dots (CdTe QDs), and synergistic effects were noted. RNS, a phenomenon similar to a Trojan horse was found in the interactions between metal ions and nanoparticles. Metal ions were transported into the cells using NPs (Ren, Hu, and Zhou 2016). The impact of physicochemical characteristics on nanoparticle toxicity is illustrated in Figure 2.

4. Mechanisms of *In Vitro* Toxicity

4.1. Cellular Uptake Mechanisms

One important factor influencing nanoparticle toxicity *in vitro* is their cellular absorption, which is controlled by several mechanisms that affect how the particles interact with biological systems. The main mechanism by which nanoparticles can enter cells is endocytosis, which includes receptor-mediated endocytosis, phagocytosis, and pinocytosis (Behzadi et al. 2017a). Pinocytosis is the non-specific absorption of extracellular fluid and its solutes, whereas phagocytosis is usually used by specialized cells such as macrophages to absorb bigger particles. The extremely selective process of receptor-mediated endocytosis, on the other hand, depends on the binding of nanoparticles to cell surface receptors, which in turn causes the plasma membrane to invade and vesicle formation to occur (Panariti, Miserocchi, and Rivolta 2012). By designing nanoparticles with surface ligands that specifically target different cell types, an increase in uptake efficiency and decrease off-target effects can be achieved (Yang et al. 2022).

Depending on their size, surface charge, and coating, internalized nanoparticles can localize inside the cytosol, nucleus, endosomes, and lysosomes, among other intracellular compartments. These characteristics control not only the absorption route but also the nanoparticles' eventual intracellular trafficking and destiny (Behzadi et al. 2017a). However, damage to the integrity of the membrane and disturbance of regular cellular processes can also result in greater toxicity. Comprehending these pathways is crucial to anticipate and alleviate the possible harmful consequences of nanoparticles, hence steering the creation of more secure nanomaterials intended for biomedical uses ('Preferential Binding of Positive Nanoparticles on Cell Membranes Is Due to Electrostatic Interactions: A Too Simplistic Explanation That Does Not Take into Account the Nanoparticle Protein Corona - ScienceDirect,' n.d.).

4.2. Intracellular Fate of Nanoparticles

When nanoparticles enter, they are frequently encased in endosomes. These endosomes can mature and fuse with lysosomes, exposing the nanoparticles to an acidic environment, full of hydrolytic enzymes. Certain nanoparticles, particularly those intended to be biodegradable, may be broken down by this endo-lysosomal pathway, releasing their payloads in a regulated fashion (Qiu et al. n.d.). Not all nanoparticles, nevertheless, travel down this route. Some can break out from endosomes and enter the cytoplasm by mechanisms including endosomal escape. These processes are triggered by the proton sponge effect, which causes osmotic swelling and vesicle rupture (Z. P. G. Xu 2022).

Depending on their shape and function, nanoparticles in the cytosol can interact with different organelles and may even make it to the nucleus or mitochondria. Nuclear localization of signal-functionalized nanoparticles, for example, can translocate into the nucleus, where they may alter gene expression and perhaps have genotoxic effects (Behzadi et al. 2017b). Oxidative stress and apoptosis can result from nanoparticles that target the mitochondria and impair their ability to function. The toxicity and therapeutic efficacy of nanoparticles are significantly influenced by their intracellular trafficking and destination that also determines their biological impacts. Thus, to optimize the design of nanoparticles for biomedical applications and minimize unfavorable consequences, a thorough understanding of these processes is required (Manke, Wang, and Rojanasakul 2013).

4.3. Cellular Responses to Nanoparticles

Cellular responses to nanoparticles vary based on their size, shape, composition, and surface properties (Behzadi et al. 2017c). Cells can internalize nanoparticles through endocytosis, leading to potential impacts on cellular functions (Abdal Dayem et al. 2017). Nanoparticles may cause oxidative stress, and inflammation, or disrupt cellular membranes, affecting viability and function (Ma and Yang 2016). Additionally, they can interact with cellular organelles, altering metabolic pathways and gene expression. The specific response depends on the type of nanoparticle and the cell type involved, highlighting the need for a thorough evaluation of nanoparticle-cell interactions to ensure safety in biomedical applications (Egbuna et al. 2021c).

4.3.1. Oxidative Stress

When the body's antioxidant reserves are exceeded by producing reactive oxygen species (ROS), oxidative stress results. ROS are very reactive molecules that can cause damage to cells, proteins, and DNA (Pizzino et al. 2017). These damages are related to different diseases and aging mechanisms. Although, under normal conditions, the body can counterbalance the amount of ROS produced by using antioxidants, an overproduction of ROS relative to the antioxidant capacity leads to oxidative stress (Correia et al. 2017). This stress can cause inflammation and death of cells and may promote disorders like cancer, cardiovascular disease, neurodegeneration, or diabetes (Y.-Z. Liu, Wang, and Jiang 2017). It is paramount that oxidative stress be managed to support health and prevent disease, both with lifestyle and potentially antioxidant supplementation (Sharifi-Rad et al. 2020).

Thus, the nanoparticles affect the cells by disrupting their structure and function, including cellular membranes, generating ROS, and causing the loss of cells' functioning. Nanoparticles, especially those of metals like silver and zinc oxide, have been known to cause oxidative stress that leads to cell destruction (Lewinski, Colvin, and Drezek 2008). Recent studies have also established that nanoparticles can initiate inflammation by stimulating certain immune cells to action. This may include a definition of pro-inflammatory cytokines and chemokines, subsequently causing inflammation. Surface chemical effects and endotoxin content of the nanoparticles are the key factors dictating the extent of inflammation (Swartzwelter et al. 2020).

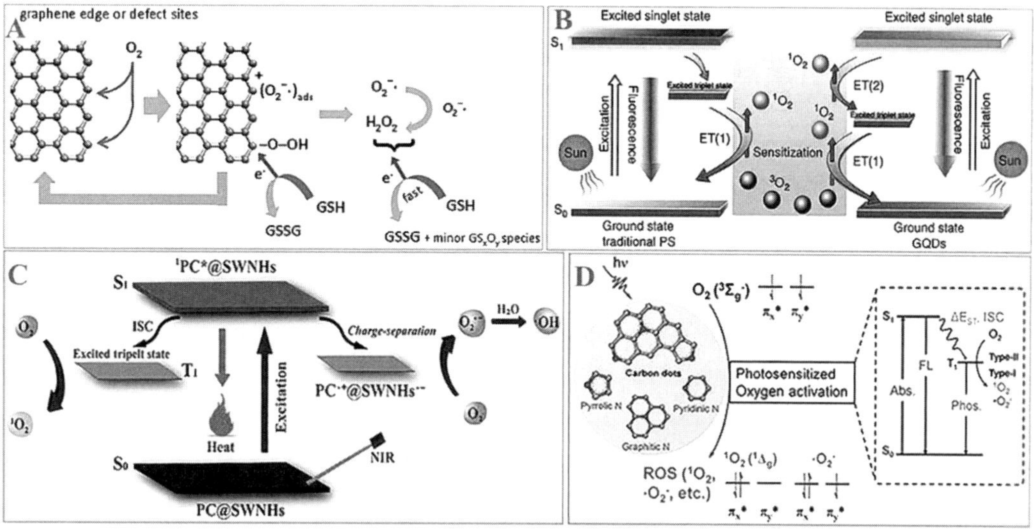

Figure 3. (A) Defect sites in carbon nanomaterials and production of ROS. Reprinted from Liu et al. with permission. [54] Copyright (2011) Wiley-VCH. (B) Diagram showing how GQDs (on the right) and conventional PDT agents (on the left) produce 1O2. Reprinted from Ge et al. with permission. [55]. Copyright (2014) Nature Publishing Group. (C) Diagram showing how ROS is produced by PC@SWNHs when exposed to a 650 nm laser. Reprinted from Lin et al. with permission. [56]. Copyright (2019) Elsevier. (D) Nitrogen-infused CDs for oxygen activation that is photosensitized. Reprinted from Wu et al. with permission. [57]. Copyright (2020) The Royal Society of Chemistry.

Liu et al. examined the connection between surface areas of different graphenic carbon compounds and oxidative stress. (Figure 3A) ('Antioxidant Deactivation on Graphenic Nanocarbon Surfaces - Liu - 2011 - Small - Wiley Online Library,' n.d.). The GQDs used a multistate sensitization technique to show a 1O2 quantum yield greater than 1.3. (Figure 3B) ('A Graphene Quantum Dot Photodynamic Therapy Agent with High Singlet Oxygen Generation | Nature Communications,' n.d.). The functionalized SWNHs showed a similar •OH production pathway in another experiment. In the meantime, 1O2 was produced from T1 to S0 using the traditional energy transfer method. (Figure 3C) ('Synthesis of Copper Peroxide Nanodots for H2O2 Self-Supplying Chemodynamic Therapy | Journal of the American Chemical Society,' n.d.). Heteroatom doping is a useful method for enhancing CD surface functionalization, which is crucial to produce ROS. The photosensitization properties of CDs were shown to be affected by graphitic N and pyrrolic N, among other N types. The energy difference (ΔEST) between S1 and T1 can be lowered by the graphitic N's presence in N-CDs to promote ISC and improve photosensitization effectiveness. (Figure 3D) ('Highly Efficient Oxygen Photosensitization of Carbon Dots: The Role of Nitrogen Doping - Nanoscale (RSC Publishing),' n.d.).

4.3.2. Genotoxicity

Some nanoparticles cause DNA damage either directly by interacting with the DNA molecules or indirectly by producing ROS. This genotoxic effect raises concerns about the potential long-term impacts, including carcinogenesis (Landsiedel et al. 2009). Figure 4 shows the possible mechanism of TiO_2-NP genotoxicity.

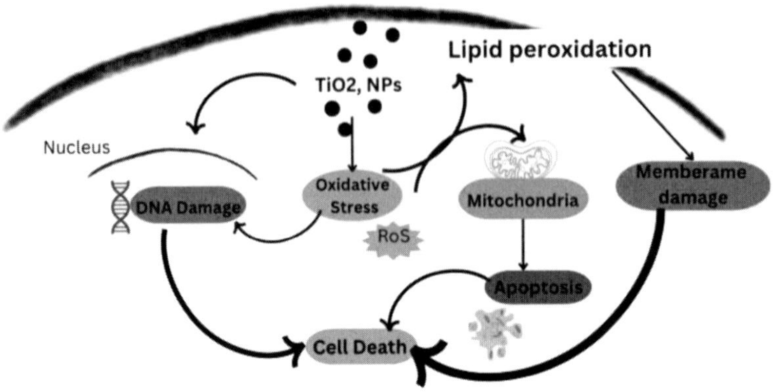

Figure 4. Possible mechanism of TiO2-NPs genotoxicity.

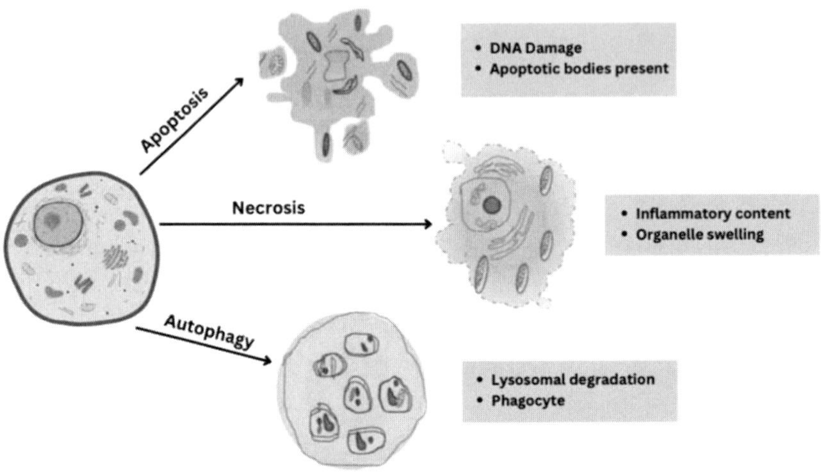

Figure 5. Major pathways of death cells through apoptosis, necrosis, and autophagy.

Table 1. Summary of cellular responses to nanoparticles

Sl. No.	Nanoparticles	Cellular Response	Description	References
1	Silver Nanoparticles (AgNPs) Zinc Oxide Nanoparticles (ZnO NPs) Titanium Dioxide Nanoparticles (TiO2 NPs) Cerium Oxide Nanoparticles (CeO2 NPs)	Oxidative Stress	Production of reactive oxygen species (ROS) leads to potential damage to cellular components like lipids, proteins, and DNA.	(Afzal et al. 2023)
2	Carbon Nanotubes (CNTs) Silica Nanoparticles (SiO2 NPs) Iron Oxide Nanoparticles (Fe3O4 NPs) Polystyrene Nanoparticles (PS NPs)	Inflammation	Activation of inflammatory pathways can result in chronic inflammation and associated tissue damage.	(Chen et al. 2018)
3	Gold Nanoparticles (AuNPs) Quantum Dots (QDs) AgNPs Cobalt Nanoparticles (Co NPs)	Genotoxicity	Damage to the genetic material within a cell causes mutations, which can lead to cancer and other genetic disorders.	(Phillips and Arlt 2009)
4	Fullerene Nanoparticles (C60) Copper Oxide Nanoparticles (CuO NPs) Graphene Oxide Nanoparticles (GO NPs) Nickel Nanoparticles (Ni NPs)	Apoptosis and Necrosis	Induction of programmed cell death (apoptosis) or uncontrolled cell death (necrosis), both of which can disrupt normal tissue function and health.	(Fink and Cookson 2005)

4.3.3. Apoptosis and Necrosis

Nanoparticles, depending on the exposure, result in the précised cell suicide (apoptosis) or random cell death (necrosis). Another important factor that regulates the mode of cell death involves the type and concentration of nanoparticles as well as the period of exposure. Apart from apoptosis, nanoparticles that generate high levels of oxidative stress enhance apoptosis (Hussain et al. 2012). Table 1 shows the summary of cellular responses to nanoparticles and Figure 5 illustrates the main processes by which autophagy, necrosis, and apoptosis because cell death.

5. Case Studies

5.1. Specific Examples of Organic and Inorganic Nanoparticle Toxicity

Four kinds of white-rot fungus were studied by Galindo et al. Over three to eight days, they evaluated the toxic impact of two organic NPs, sodium dodecyl sulfate, and didodecyldimethylammonium bromide (SDS/DDAB), monoolein and sodium oleate (Mo/NaO) NPS, and a few inorganic NPs on four species of white-rot fungi: *Phanerochaete chrysosporium* (Burdsall 38388), *Trametes versicolor* (Pilát 38412), *Lentinus sajor caju* (Fries), and *Pleurotus streams* (Jacquin) P. Kummer. Most soil biota ecosystems depend on these fungi for their well-being, productivity, and other operations.

From day 3 to the completion of the experiment, it was found that the growth of *P. ostreatus* and *L. major caju*n was inhibited by all Mo/NaO concentrations. Only the two highest doses hindered the development of *P. chrysosporium*, whereas, at concentrations of 9.8, 13.9, and 20.0 g/L, T. versicolor exhibited a decline in growth on days 4 and 5. The growth of the four fungus species was inhibited by all SDS/DDAB doses. These effects became noticeable on day two for L. sajor caju, P. chrysosporium, and T. versicolor, but they didn't become noticeable on day three for P. ostreatus. The effective doses causing 20 and 50% growth inhibition were calculated for each NP after the experiments (Galindo et al. 2013b).

Using human and rat intestinal cells, Jo et al. assessed the toxicity of chemically synthesized gold NPs supplied orally over 14 days. They also measured intestinal absorption and tissue distribution biokinetics. Based on the inhibition of cell growth, membrane damage, and oxidative stress, the gold nanoparticles were shown to be harmless for 24 hours. However, after 14 days (about 2 weeks), they discovered that nanosized gold particles were poisonous, displaying long-term and highly concentration-dependent toxic responses (Jo et al. 2015). Gaillet and Rouanet investigated the toxicity of AgNPs when they were orally exposed to humans. It was shown that oral exposure to AgNPs primarily results in harmful side effects in the liver and intestinal system. Remarkably, the cellular oxidative stress caused by the AgNPs generates free radicals and causes oxidative damage, which results in apoptosis or necrosis and inflammatory reaction-triggered toxicity. Compared to other nanosized metal oxides, carbon nanoparticles, and MWCNTs, the findings demonstrated that copper oxide nanoparticles are particularly hazardous to lung cells by inducing oxidative lesions and degrading DNA (Karlsson et al. 2008).

After a literature study, Valdiglesias et al. documented how iron oxide NPs affected normal, healthy human cells. It was stressed that short-term exposure to magnetic nanoparticles in humans is extremely cytotoxic, genotoxic, and as well as led to developmentally toxicity, and neurotoxipity and that more research is needed to assess the long-term exposure consequences (Valdiglesias et al. 2015).

6. Strategies for Mitigating Nanoparticle Toxicity

Nanotechnology has recently seen a huge rise in the therapeutical industry (Mitragotri and Patrick 2014). It is seen to be useful in many healthcare applications like vascular aging and related diseases, diabetes, food nanotechnology, nanomedicine, and so on. However, the rapid development in the field of nanoparticles has led to its environmental exposure. It has forced us to shed light on nanoparticle-related toxicity (Galindo et al. 2013a). The size of nanoparticles has a significant relation to their toxicity. If the size is decreased, the reactivity increases. Thus, smaller particles are more toxic (Johnston et al. 2010). The shape of the nanoparticles also helps determine the toxicity. Spherical nanoparticles tend to be taken up in a greater amount and the star-shaped particles end up being the most toxic (Bacchetta et al. 2018).

Surface coating is one way by which we can alter the properties of these particles. The process of surface coating helps to alter various parameters like stability, and dissolution and make the toxic particle nontoxic (Egbuna et al. 2021d). Taking the example of copper oxide nanoparticles, they had been seen to showcase toxicity towards lung epithelial cells due to their size. In a clearer picture, it was seen that after internalizing these particles, the larger particles would disintegrate and ultimately show dose-dependent toxicity (Egbuna et al. 2021d). The accumulation of nanoparticles in the body can ultimately lead to carcinogenesis or cardiovascular diseases (Mohammadpour et al. 2019). Table 2 shows the factors affecting toxicity and their mitigating strategies.

Table 2. Factors affecting toxicity and their mitigating strategies

Sl. No.	Factors affecting toxicity	Strategies to mitigate the toxicity	References
1	Size	Avoid using smaller-sized nanoparticles	(Johnston et al. 2010)
2	Shape	Avoid using star-shaped nanoparticles	(Bacchetta et al. 2018)
3	Dosing	Avoid higher doses or frequent doses of nanoparticles	(Mohammadpour et al. 2019)
4	Surface coating	Helps to alter the properties of nanoparticles and make them more stable	(Egbuna et al. 2021d)

Conclusion and Future Prospects

Understanding the mechanics of *in vitro* toxicity in organic and inorganic particles is essential for ensuring their safe use across various fields, including medicine, environmental science, and materials engineering. The detrimental impacts of organic and inorganic nanoparticles *in vitro* arise from a variety of molecular and cellular interactions. Key mechanisms through which nanoparticles induce harmful effects include oxidative stress, inflammation, genotoxicity, membrane damage, protein corona formation, and intracellular trafficking disruptions. Nanoparticles can generate oxidative stress by producing ROS, leading to cellular damage and inflammation. Inflammation, in turn, can cause further cellular damage and exacerbate toxic effects. Membrane damage disrupts cellular integrity and function, while via protein corona formation, proteins can change the way nanoparticles behave and interact with cells when they attach to their surface. Intracellular trafficking disruptions interfere with the normal functioning of cellular processes. By rigorously studying these pathways, researchers and scholars can develop targeted therapeutic interventions and design safer nanoparticles. This chapter also highlights the importance of understanding nanoparticle safety and outlines the necessary protocols for their evaluation. Further research is crucial for elucidating the detailed mechanics of nanoparticle interactions and contributing to developing enhanced safety measures and innovative applications in the future. This continuous investigation will support the safe and effective use of nanoparticles, ensuring their benefits outweigh any potential risks.

In future, there can be integration of advancing technologies and interdisciplinary methods for further understanding nanoparticle behavior and toxicity. The key focus in the future should be advancing analytical techniques, green nanotechnology, recycling of nanoparticles, and ensuring easy access to the technology. By innovating computational modeling, screening methods, personalized medicine, and developing standard protocol, the future of nanoparticle safety and its function plays a significant role. Continuous innovation in this area is required for the development of creative applications that would improve the environment and benefit society.

References

'A Graphene Quantum Dot Photodynamic Therapy Agent with High Singlet Oxygen Generation | Nature Communications.' n.d. Accessed 15 June 2024. https://www.nature.com/articles/ncomms5596.

Abbasi, Reza, Ghazal Shineh, Mohammadmahdi Mobaraki, Sarah Doughty, and Lobat Tayebi. 2023. 'Structural Parameters of Nanoparticles Affecting Their Toxicity for Biomedical Applications: A Review.' *Journal of Nanoparticle Research* 25 (3): 43. https://doi.org/10.1007/s11051-023-05690-w.

Abd Ellah, Noura H., and Sara A. Abouelmagd. 2017. 'Surface Functionalization of Polymeric Nanoparticles for Tumor Drug Delivery: Approaches and Challenges.' *Expert Opinion on Drug Delivery* 14 (2): 201–14. https://doi.org/10.1080/17425247.2016.1213238.

Abdal Dayem, Ahmed, Mohammed Hossain, Soo Lee, Kyeongseok Kim, Subbroto Saha, Gwang-Mo Yang, Hye Choi, and Ssang-Goo Cho. 2017. 'The Role of Reactive Oxygen Species (ROS) in the Biological Activities of Metallic Nanoparticles.' *International Journal of Molecular Sciences* 18 (1): 120. https://doi.org/10.3390/ijms18010120.

Abdellatif, Ahmed A. H., and Abdullah Fahad Alsowinea. 2021. 'Approved and Marketed Nanoparticles for Disease Targeting and Applications in COVID-19.' *Nanotechnology Reviews* 10 (1): 1941–77. https://doi.org/10.1515/ntrev-2021-0115.

Abdelmalak, Basem B., and D John Doyle. 2020. 'Recent Trends in Airway Management.' *F1000Research* 9 (May):355. https://doi.org/10.12688/f1000research.21914.1.

Afzal, Sheryar, Aimi Syamima Abdul Manap, Ali Attiq, Ibrahim Albokhadaim, Mahmoud Kandeel, and Sameer M. Alhojaily. 2023. 'From Imbalance to Impairment: The Central Role of Reactive Oxygen Species in Oxidative Stress-Induced Disorders and Therapeutic Exploration.' *Frontiers in Pharmacology* 14 (October):1269581. https://doi.org/10.3389/fphar.2023.1269581.

Albanese, Alexandre, Peter S. Tang, and Warren C.W. Chan. 2012. 'The Effect of Nanoparticle Size, Shape, and Surface Chemistry on Biological Systems.' *Annual Review of Biomedical Engineering* 14 (1): 1–16. https://doi.org/10.1146/annurev-bioeng-071811-150124.

Altammar, Khadijah A. 2023. 'A Review on Nanoparticles: Characteristics, Synthesis, Applications, and Challenges.' *Frontiers in Microbiology* 14 (April):1155622. https://doi.org/10.3389/fmicb.2023.1155622.

'Antioxidant Deactivation on Graphenic Nanocarbon Surfaces - Liu - 2011 - Small - Wiley Online Library.' n.d. Accessed 15 June 2024. https://onlinelibrary.wiley.com/doi/abs/10.1002/smll.201100651.

Awashra, Mohammad, and Piotr Młynarz. 2023. 'The Toxicity of Nanoparticles and Their Interaction with Cells: An *in Vitro* Metabolomic Perspective.' *Nanoscale Advances* 5 (10): 2674–2723. https://doi.org/10.1039/D2NA00534D.

Bacchetta, Renato, Nadia Santo, Irene Valenti, Daniela Maggioni, Mariangela Longhi, and Paolo Tremolada. 2018. 'Comparative Toxicity of Three Differently Shaped Carbon Nanomaterials on Daphnia Magna: Does a Shape Effect Exist?' *Nanotoxicology* 12 (3): 201–23. https://doi.org/10.1080/17435390.2018.1430258.

Baek, M, M K Kim, H J Cho, J A Lee, J Yu, H E Chung, and S J Choi. 2011. 'Factors Influencing the Cytotoxicity of Zinc Oxide Nanoparticles: Particle Size and Surface Charge.' *Journal of Physics: Conference Series* 304 (July):012044. https://doi.org/10.1088/1742-6596/304/1/012044.

Bahadar, Haji, Faheem Maqbool, Kamal Niaz, and Mohammad Abdollahi. 2016. 'Toxicity of Nanoparticles and an Overview of Current Experimental Models.' *Iranian Biomedical Journal* 20 (1): 1–11. https://doi.org/10.7508/ibj.2016.01.001.

Behzadi, Shahed, Vahid Serpooshan, Wei Tao, Majd A. Hamaly, Mahmoud Y. Alkawareek, Erik C. Dreaden, Dennis Brown, Alaaldin M. Alkilany, Omid C. Farokhzad, and Morteza Mahmoudi. 2017a. 'Cellular Uptake of Nanoparticles: Journey Inside the Cell.' *Chemical Society Reviews* 46 (14): 4218–44. https://doi.org/10.1039/c6cs00636a.

———. 2017b. 'Cellular Uptake of Nanoparticles: Journey Inside the Cell.' *Chemical Society Reviews* 46 (14): 4218–44. https://doi.org/10.1039/c6cs00636a.

———. 2017c. 'Cellular Uptake of Nanoparticles: Journey inside the Cell.' *Chemical Society Reviews* 46 (14): 4218–44. https://doi.org/10.1039/C6CS00636A.

Chen, Linlin, Huidan Deng, Hengmin Cui, Jing Fang, Zhicai Zuo, Junliang Deng, Yinglun Li, Xun Wang, and Ling Zhao. 2018. 'Inflammatory Responses and Inflammation-Associated Diseases in Organs.' *Oncotarget* 9 (6): 7204–18. https://doi.org/10.18632/oncotarget.23208.

Correia, Bruno, Joana Lourenço, Sérgio Marques, Verónica Nogueira, Ana Gavina, Maria Da Graça Rasteiro, Filipe Antunes, Sónia Mendo, and Ruth Pereira. 2017. 'Oxidative Stress and Genotoxicity of an Organic and an Inorganic Nanomaterial to Eisenia Andrei : SDS/DDAB Nano-Vesicles and Titanium Silicon Oxide.' *Ecotoxicology and Environmental Safety* 140 (June):198–205. https://doi.org/10.1016/j.ecoenv.2017.02.035.

Cupi, Denisa, Nanna B. Hartmann, and Anders Baun. 2016. 'Influence of pH and Media Composition on Suspension Stability of Silver, Zinc Oxide, and Titanium Dioxide Nanoparticles and Immobilization of Daphnia Magna under Guideline Testing Conditions.' *Ecotoxicology and Environmental Safety* 127 (May):144–52. https://doi.org/10.1016/j.ecoenv.2015.12.028.

De Berardis, Barbara, Gabriele Civitelli, Maria Condello, Pasquale Lista, Roberta Pozzi, Giuseppe Arancia, and Stefania Meschini. 2010. 'Exposure to ZnO Nanoparticles Induces Oxidative Stress and Cytotoxicity in Human Colon Carcinoma Cells.' *Toxicology and Applied Pharmacology* 246 (3): 116–27. https://doi.org/10.1016/j.taap.2010.04.012.

Egbuna, Chukwuebuka, Vijaykumar K. Parmar, Jaison Jeevanandam, Shahira M. Ezzat, Kingsley C. Patrick-Iwuanyanwu, Charles Oluwaseun Adetunji, Johra Khan, et al. 2021a. 'Toxicity of Nanoparticles in Biomedical Application: Nanotoxicology.' *Journal of Toxicology* 2021 (July):9954443. https://doi.org/10.1155/2021/9954443.

———. 2021b. 'Toxicity of Nanoparticles in Biomedical Application: Nanotoxicology.' *Journal of Toxicology* 2021 (July):9954443. https://doi.org/10.1155/2021/9954443.

———. 2021c. 'Toxicity of Nanoparticles in Biomedical Application: Nanotoxicology.' Edited by You-Cheng Hseu. *Journal of Toxicology* 2021 (July):1–21. https://doi.org/10.1155/2021/9954443.

———. 2021d. 'Toxicity of Nanoparticles in Biomedical Application: Nanotoxicology.' *Journal of Toxicology* 2021 (July):9954443. https://doi.org/10.1155/2021/9954443.

Elumalai, Karthikeyan, Sivaneswari Srinivasan, and Anandakumar Shanmugam. 2024. 'Review of the Efficacy of Nanoparticle-Based Drug Delivery Systems for Cancer Treatment.' *Biomedical Technology* 5 (March):109–22. https://doi.org/10.1016/j.bmt.2023.09.001.

Ernst, Lena M., Eudald Casals, Paola Italiani, Diana Boraschi, and Victor Puntes. 2021. 'The Interactions between Nanoparticles and the Innate Immune System from a Nanotechnologist Perspective.' *Nanomaterials* 11 (11): 2991. https://doi.org/10.3390/nano11112991.

Fatima, Hira, Tawatchai Charinpanitkul, and Kyo-Seon Kim. 2021. 'Fundamentals to Apply Magnetic Nanoparticles for Hyperthermia Therapy.' *Nanomaterials* 11 (5): 1203. https://doi.org/10.3390/nano11051203.

Feng, Qiyi, Yanping Liu, Jian Huang, Ke Chen, Jinxing Huang, and Kai Xiao. 2018. 'Uptake, Distribution, Clearance, and Toxicity of Iron Oxide Nanoparticles with Different Sizes and Coatings.' *Scientific Reports* 8 (1): 2082. https://doi.org/10.1038/s41598-018-19628-z.

Feracci, Helene, Berta Saez Gutierrez, William Hempel, and Isabel Segura Gil. 2012. 'Chapter 8 - Organic Nanoparticles.' In *Frontiers of Nanoscience*, edited by Jesus M. de la Fuente and V. Grazu, 4:197–230. Nanobiotechnology. Elsevier. https://doi.org/10.1016/B978-0-12-415769-9.00008-X.

Fink, Susan L., and Brad T. Cookson. 2005. 'Apoptosis, Pyroptosis, and Necrosis: Mechanistic Description of Dead and Dying Eukaryotic Cells.' *Infection and Immunity* 73 (4): 1907–16. https://doi.org/10.1128/IAI.73.4.1907-1916.2005.

Fu, Peter P., Qingsu Xia, Huey-Min Hwang, Paresh C. Ray, and Hongtao Yu. 2014. 'Mechanisms of Nanotoxicity: Generation of Reactive Oxygen Species.' *Journal of Food and Drug Analysis*, Nanomaterials - Toxicology and Medical Applications, 22 (1): 64–75. https://doi.org/10.1016/j.jfda.2014.01.005.

Gaillet, Sylvie, and Jean-Max Rouanet. 2015. 'Silver Nanoparticles: Their Potential Toxic Effects after Oral Exposure and Underlying Mechanisms – A Review.' *Food and Chemical Toxicology* 77 (March):58–63. https://doi.org/10.1016/j.fct.2014.12.019.

Galindo, T. P. S., R. Pereira, A. C. Freitas, T. A. P. Santos-Rocha, M. G. Rasteiro, F. Antunes, D. Rodrigues, et al. 2013a. 'Toxicity of Organic and Inorganic Nanoparticles to Four Species of White-Rot Fungi.' *Science of The Total Environment* 458–460 (August):290–97. https://doi.org/10.1016/j.scitotenv.2013.04.019.

Galindo, T. P. S., R. Pereira, A. C. Freitas, T. A. P. Santos-Rocha, M.G. Rasteiro, F. Antunes, D. Rodrigues, et al. 2013b. 'Toxicity of Organic and Inorganic Nanoparticles to Four Species of White-Rot Fungi.' *Science of The Total Environment* 458–460 (August):290–97. https://doi.org/10.1016/j.scitotenv.2013.04.019.

González-García, Laura E., Melanie N. MacGregor, Rahul M. Visalakshan, Artur Lazarian, Alex A. Cavallaro, Svenja Morsbach, Agnieszka Mierczynska-Vasilev, Volker Mailänder, Katharina Landfester, and Krasimir Vasilev. 2022. 'Nanoparticles Surface Chemistry Influence on Protein Corona Composition and Inflammatory Responses.' *Nanomaterials* 12 (4): 682. https://doi.org/10.3390/nano12040682.

Hagens, Werner I., Agnes G. Oomen, Wim H. De Jong, Flemming R. Cassee, and Adriënne J.A.M. Sips. 2007. 'What Do We (Need to) Know about the Kinetic Properties of Nanoparticles in the Body?' *Regulatory Toxicology and Pharmacology* 49 (3): 217–29. https://doi.org/10.1016/j.yrtph.2007.07.006.

H. Alshammari, Basmah, Maha M. A. Lashin, Muhammad Adil Mahmood, Fahad S. Al-Mubaddel, Nasir Ilyas, Nasir Rahman, Mohammad Sohail, Aurangzeb Khan, Sherzod Shukhratovich Abdullaev, and Rajwali Khan. 2023. 'Organic and Inorganic Nanomaterials: Fabrication, Properties and Applications.' *RSC Advances* 13 (20): 13735–85. https://doi.org/10.1039/D3RA01421E.

Han, Xiangjun, Ke Xu, Olena Taratula, and Khashayar Farsad. 2019. 'Applications of Nanoparticles in Biomedical Imaging.' *Nanoscale* 11 (3): 799–819. https://doi.org/10.1039/C8NR07769J.

'Highly Efficient Oxygen Photosensitization of Carbon Dots: The Role of Nitrogen Doping - Nanoscale (RSC Publishing).' n.d. Accessed 15 June 2024. https://pubs.rsc.org/en/content/articlelanding/2020/nr/c9nr10986b/unauth.

Hussain, Salik, Faris Al-Nsour, Annette B. Rice, Jamie Marshburn, Brenda Yingling, Zhaoxia Ji, Jeffrey I. Zink, Nigel J. Walker, and Stavros Garantziotis. 2012. 'Cerium Dioxide Nanoparticles Induce Apoptosis and Autophagy in Human Peripheral Blood Monocytes.' *ACS Nano* 6 (7): 5820–29. https://doi.org/10.1021/nn302235u.

Jeevanandam, Jaison, Ahmed Barhoum, Yen S Chan, Alain Dufresne, and Michael K Danquah. 2018. 'Review on Nanoparticles and Nanostructured Materials: History, Sources, Toxicity and Regulations.' *Beilstein Journal of Nanotechnology* 9 (April):1050–74. https://doi.org/10.3762/bjnano.9.98.

Jo, Mi-Rae, Song-Hwa Bae, Mi-Ran Go, Hyun-Jin Kim, Yun-Gu Hwang, and Soo-Jin Choi. 2015. 'Toxicity and Biokinetics of Colloidal Gold Nanoparticles.' *Nanomaterials* 5 (2): 835–50. https://doi.org/10.3390/nano5020835.

Johnston, Helinor J., Gary Hutchison, Frans M. Christensen, Sheona Peters, Steve Hankin, and Vicki Stone. 2010. 'A Review of the *in Vivo* and *in Vitro* Toxicity of Silver and Gold Particulates: Particle Attributes and Biological Mechanisms Responsible for the Observed Toxicity.' *Critical Reviews in Toxicology* 40 (4): 328–46. https://doi.org/10.3109/10408440903453074.

Joris, Freya, Daniel Valdepérez, Beatriz Pelaz, Stefaan J. Soenen, Bella B. Manshian, Wolfgang J. Parak, Stefaan C. De Smedt, and Koen Raemdonck. 2016. 'The Impact of Species and Cell Type on the Nanosafety Profile of Iron Oxide Nanoparticles in Neural Cells.' *Journal of Nanobiotechnology* 14 (1): 69. https://doi.org/10.1186/s12951-016-0220-y.

Joseph, Tomy, Debarshi Kar Mahapatra, Amin Esmaeili, Łukasz Piszczyk, Mohamed Hasanin, Mashhoor Kattali, Józef Haponiuk, and Sabu Thomas. 2023. 'Nanoparticles: Taking a Unique Position in Medicine.' *Nanomaterials* 13 (3): 574. https://doi.org/10.3390/nano13030574.

Kai, Wei, Xu Xiaojun, Pu Ximing, Hou Zhenqing, and Zhang Qiqing. 2011. 'Cytotoxic Effects and the Mechanism of Three Types of Magnetic Nanoparticles on Human Hepatoma BEL-7402 Cells.' *Nanoscale Research Letters* 6 (1): 480. https://doi.org/10.1186/1556-276X-6-480.

Karlsson, Hanna L., Pontus Cronholm, Johanna Gustafsson, and Lennart Möller. 2008. 'Copper Oxide Nanoparticles Are Highly Toxic: A Comparison between Metal Oxide Nanoparticles and Carbon Nanotubes.' *Chemical Research in Toxicology* 21 (9): 1726–32. https://doi.org/10.1021/tx800064j.

Khan, Ibrahim, Khalid Saeed, and Idrees Khan. 2019. 'Nanoparticles: Properties, Applications and Toxicities.' *Arabian Journal of Chemistry* 12 (7): 908–31. https://doi.org/10.1016/j.arabjc.2017.05.011.

Kim, In-Yong, Elizabeth Joachim, Hyungsoo Choi, and Kyekyoon Kim. 2015. 'Toxicity of Silica Nanoparticles Depends on Size, Dose, and Cell Type.' *Nanomedicine: Nanotechnology, Biology and Medicine* 11 (6): 1407–16. https://doi.org/10.1016/j.nano.2015.03.004.

Kisin, Elena R., Ashley R. Murray, Michael J. Keane, Xiao-Chun Shi, Diane Schwegler-Berry, Olga Gorelik, Sivaram Arepalli, et al. 2007. 'Single-Walled Carbon Nanotubes: Geno- and Cytotoxic Effects in Lung Fibroblast V79 Cells.' *Journal of Toxicology and Environmental Health, Part A* 70 (24): 2071–79. https://doi.org/10.1080/15287390701601251.

Kittler, S., C. Greulich, J. Diendorf, M. Köller, and M. Epple. 2010. 'Toxicity of Silver Nanoparticles Increases during Storage Because of Slow Dissolution under Release of Silver Ions.' *Chemistry of Materials* 22 (16): 4548–54. https://doi.org/10.1021/cm100023p.

Landsiedel, Robert, Maike Diana Kapp, Markus Schulz, Karin Wiench, and Franz Oesch. 2009. 'Genotoxicity Investigations on Nanomaterials: Methods, Preparation and Characterization of Test Material, Potential Artifacts and Limitations—Many Questions, Some Answers.' *Mutation Research/Reviews in Mutation Research* 681 (2–3): 241–58. https://doi.org/10.1016/j.mrrev.2008.10.002.

Lewinski, Nastassja, Vicki Colvin, and Rebekah Drezek. 2008. 'Cytotoxicity of Nanoparticles.' *Small* 4 (1): 26–49. https://doi.org/10.1002/smll.200700595.

Liu, Jingyu, and Robert H. Hurt. 2010. 'Ion Release Kinetics and Particle Persistence in Aqueous Nano-Silver Colloids.' *Environmental Science & Technology* 44 (6): 2169–75. https://doi.org/10.1021/es9035557.

Liu, Yun-Zi, Yun-Xia Wang, and Chun-Lei Jiang. 2017. 'Inflammation: The Common Pathway of Stress-Related Diseases.' *Frontiers in Human Neuroscience* 11 (June):316. https://doi.org/10.3389/fnhum.2017.00316.

Lu, Wentong, Dulal Senapati, Shuguang Wang, Oleg Tovmachenko, Anant Kumar Singh, Hongtao Yu, and Paresh Chandra Ray. 2010. 'Effect of Surface Coating on the Toxicity of Silver Nanomaterials on Human Skin Keratinocytes.' *Chemical Physics Letters* 487 (1–3): 92–96. https://doi.org/10.1016/j.cplett.2010.01.027.

Ma, Dan-Dan, and Wan-Xi Yang. 2016. 'Engineered Nanoparticles Induce Cell Apoptosis: Potential for Cancer Therapy.' *Oncotarget* 7 (26): 40882–903. https://doi.org/10.18632/oncotarget.8553.

Magnetic Nanoparticles. 2020. MDPI. https://doi.org/10.3390/books978-3-03928-269-2.

Manke, Amruta, Liying Wang, and Yon Rojanasakul. 2013. 'Mechanisms of Nanoparticle-Induced Oxidative Stress and Toxicity.' *BioMed Research International* 2013:942916. https://doi.org/10.1155/2013/942916.

McNamara, Karrina, and Syed A. M. Tofail. 2017. 'Nanoparticles in Biomedical Applications.' *Advances in Physics: X* 2 (1): 54–88. https://doi.org/10.1080/23746149.2016.1254570.

Mitragotri, Samir, and Stayton Patrick. 2014. 'Organic Nanoparticles for Drug Delivery and Imaging.' *MRS Bulletin* 39 (3): 219–23. https://doi.org/10.1557/mrs.2014.11.

Mohammadpour, Raziye, Marina A. Dobrovolskaia, Darwin L. Cheney, Khaled F. Greish, and Hamidreza Ghandehari. 2019. 'Subchronic and Chronic Toxicity Evaluation of Inorganic Nanoparticles for Delivery Applications.' *Advanced Drug Delivery Reviews* 144 (April):112–32. https://doi.org/10.1016/j.addr.2019.07.006.

Mohammapdour, Raziye, and Hamidreza Ghandehari. 2022. 'Mechanisms of Immune Response to Inorganic Nanoparticles and Their Degradation Products.' *Advanced Drug Delivery Reviews* 180 (January):114022. https://doi.org/10.1016/j.addr.2021.114022.

Panariti, Alice, Giuseppe Miserocchi, and Ilaria Rivolta. 2012. 'The Effect of Nanoparticle Uptake on Cellular Behavior: Disrupting or Enabling Functions?' *Nanotechnology, Science and Applications* 5 (September):87–100. https://doi.org/10.2147/NSA.S25515.

Phillips, David H., and Volker M. Arlt. 2009. 'Genotoxicity: Damage to DNA and Its Consequences.' In *Molecular, Clinical and Environmental Toxicology*, edited by Andreas Luch, 99:87–110. Experientia Supplementum. Basel: Birkhäuser Basel. https://doi.org/10.1007/978-3-7643-8336-7_4.

Pizzino, Gabriele, Natasha Irrera, Mariapaola Cucinotta, Giovanni Pallio, Federica Mannino, Vincenzo Arcoraci, Francesco Squadrito, Domenica Altavilla, and Alessandra Bitto. 2017. 'Oxidative Stress: Harms and Benefits for Human Health.' *Oxidative Medicine and Cellular Longevity* 2017:1–13. https://doi.org/10.1155/2017/8416763.

'Preferential Binding of Positive Nanoparticles on Cell Membranes Is Due to Electrostatic Interactions: A Too Simplistic Explanation That Does Not Take into Account the Nanoparticle Protein Corona - ScienceDirect.' n.d. Accessed 15 June 2024. https://www.sciencedirect.com/science/article/pii/S0928493116311894?via%3Dihub.

Qiu, Chong, Fei Xia, Junzhe Zhang, Qiaoli Shi, Yuqing Meng, Chen Wang, Huanhuan Pang, et al. n.d. 'Advanced Strategies for Overcoming Endosomal/Lysosomal Barrier in Nanodrug Delivery.' *Research* 6:0148. https://doi.org/10.34133/research.0148.

Ray, Paresh Chandra, Hongtao Yu, and Peter P. Fu. 2009. 'Toxicity and Environmental Risks of Nanomaterials: Challenges and Future Needs.' *Journal of Environmental Science and Health. Part C, Environmental Carcinogenesis & Ecotoxicology Reviews* 27 (1): 1–35. https://doi.org/10.1080/10590500802708267.

Ren, Chaoxiu, Xiangang Hu, and Qixing Zhou. 2016. 'Influence of Environmental Factors on Nanotoxicity and Knowledge Gaps Thereof.' *NanoImpact* 2 (April):82–92. https://doi.org/10.1016/j.impact.2016.07.002.

Rispoli, Fred, Angel Angelov, Daniel Badia, Amit Kumar, Sudipta Seal, and Vishal Shah. 2010. 'Understanding the Toxicity of Aggregated Zero Valent Copper Nanoparticles against *Escherichia Coli*.' *Journal of Hazardous Materials* 180 (1): 212–16. https://doi.org/10.1016/j.jhazmat.2010.04.016.

Sharifi-Rad, Mehdi, Nanjangud V. Anil Kumar, Paolo Zucca, Elena Maria Varoni, Luciana Dini, Elisa Panzarini, Jovana Rajkovic, et al. 2020. 'Lifestyle, Oxidative Stress, and Antioxidants: Back and Forth in the Pathophysiology of Chronic Diseases.' *Frontiers in Physiology* 11 (July):694. https://doi.org/10.3389/fphys.2020.00694.

Shukla, Ritesh K, Ashish Badiye, Kamayani Vajpayee, and Neeti Kapoor. 2021. 'Genotoxic Potential of Nanoparticles: Structural and Functional Modifications in DNA.' *Frontiers in Genetics* 12 (September): 728250. https://doi.org/10.3389/fgene.2021.728250.

Sim, Serjay, and Nyet Wong. 2021. 'Nanotechnology and Its Use in Imaging and Drug Delivery (Review).' *Biomedical Reports* 14 (5): 42. https://doi.org/10.3892/br.2021.1418.

Stoehr, Linda C, Edgar Gonzalez, Andreas Stampfl, Eudald Casals, Albert Duschl, Victor Puntes, and Gertie J Oostingh. 2011. 'Shape Matters: Effects of Silver Nanospheres and Wires on Human Alveolar Epithelial Cells.' *Particle and Fibre Toxicology* 8 (1): 36. https://doi.org/10.1186/1743-8977-8-36.

Sukhanova, Alyona, Svetlana Bozrova, Pavel Sokolov, Mikhail Berestovoy, Alexander Karaulov, and Igor Nabiev. 2018a. 'Dependence of Nanoparticle Toxicity on Their Physical and Chemical Properties.' *Nanoscale Research Letters* 13 (February):44. https://doi.org/10.1186/s11671-018-2457-x.

———. 2018b. 'Dependence of Nanoparticle Toxicity on Their Physical and Chemical Properties.' *Nanoscale Research Letters* 13 (1): 44. https://doi.org/10.1186/s11671-018-2457-x.

Swartzwelter, Benjamin J., Alexandra C. Fux, Litty Johnson, Elmer Swart, Sabine Hofer, Norbert Hofstätter, Mark Geppert, et al. 2020. 'The Impact of Nanoparticles on Innate Immune Activation by Live Bacteria.' *International Journal of Molecular Sciences* 21 (24): 9695. https://doi.org/10.3390/ijms21249695.

'Synthesis of Copper Peroxide Nanodots for H2O2 Self-Supplying Chemodynamic Therapy | Journal of the American Chemical Society.' n.d. Accessed 15 June 2024. https://pubs.acs.org/doi/abs/10.1021/jacs.9b03457.

Thwala, Lungile Nomcebo, Sphumelele Colin Ndlovu, Kelvin Tafadzwa Mpofu, Masixole Yvonne Lugongolo, and Patience Mthunzi-Kufa. 2023. 'Nanotechnology-Based Diagnostics for Diseases Prevalent in Developing Countries: Current Advances in Point-of-Care Tests.' *Nanomaterials* 13 (7): 1247. https://doi.org/10.3390/nano13071247.

Valdiglesias, Vanessa, Gözde Kiliç, Carla Costa, Natalia Fernández-Bertólez, Eduardo Pásaro, João Paulo Teixeira, and Blanca Laffon. 2015. 'Effects of Iron Oxide Nanoparticles: Cytotoxicity, Genotoxicity, Developmental Toxicity, and Neurotoxicity.' *Environmental and Molecular Mutagenesis* 56 (2): 125–48. https://doi.org/10.1002/em.21909.

Virlan, Maria Justina Roxana, Daniela Miricescu, Radu Radulescu, Cristina M. Sabliov, Alexandra Totan, Bogdan Calenic, and Maria Greabu. 2016. 'Organic Nanomaterials and Their Applications in the Treatment of Oral Diseases.' *Molecules* 21 (2): 207. https://doi.org/10.3390/molecules21020207.

Wang, Edina C., and Andrew Z. Wang. 2014. 'Nanoparticles and Their Applications in Cell and Molecular Biology.' *Integrative Biology* 6 (1): 9–26. https://doi.org/10.1039/c3ib40165k.

Wassenaar, Pim N. H., and Eric M. J. Verbruggen. 2021. 'Persistence, Bioaccumulation and Toxicity-Assessment of Petroleum UVCBs: A Case Study on Alkylated Three-Ring PAHs.' *Chemosphere* 276 (August):130113. https://doi.org/10.1016/j.chemosphere.2021.130113.

Wong, Stella W. Y., and Kenneth M. Y. Leung. 2014. 'Temperature-Dependent Toxicities of Nano Zinc Oxide to Marine Diatom, Amphipod and Fish in Relation to Its Aggregation Size and Ion Dissolution.' *Nanotoxicology* 8 (sup1): 24–35. https://doi.org/10.3109/17435390.2013.848949.

Wongrakpanich, Amaraporn, Imali A. Mudunkotuwa, Sean M. Geary, Angie S. Morris, Kranti A. Mapuskar, Douglas R. Spitz, Vicki H. Grassian, and Aliasger K. Salem. 2016. 'Size-Dependent Cytotoxicity of Copper Oxide Nanoparticles in Lung Epithelial Cells.' *Environmental Science: Nano* 3 (2): 365–74. https://doi.org/10.1039/C5EN00271K.

Xu, Jingru, and Edward Kai-Hua Chow. 2023. 'Biomedical Applications of Nanodiamonds: From Drug-Delivery to Diagnostics.' *SLAS Technology* 28 (4): 214–22. https://doi.org/10.1016/j.slast.2023.03.007.

Xu, Zhi Ping Gordon. 2022. 'Strategy for Cytoplasmic Delivery Using Inorganic Particles.' *Pharmaceutical Research* 39 (6): 1035–45. https://doi.org/10.1007/s11095-022-03178-1.

Xuan, Lihui, Zhao Ju, Magdalena Skonieczna, Ping-Kun Zhou, and Ruixue Huang. 2023. 'Nanoparticles-induced Potential Toxicity on Human Health: Applications, Toxicity Mechanisms, and Evaluation Models.' *MedComm* 4 (4): e327. https://doi.org/10.1002/mco2.327.

Yang, Tingting, Jingming Zhai, Dong Hu, Ruyue Yang, Guidan Wang, Yuanpei Li, and Gaofeng Liang. 2022. '"Targeting Design" of Nanoparticles in Tumor Therapy.' *Pharmaceutics* 14 (9): 1919. https://doi.org/10.3390/pharmaceutics14091919.

Zhang, Tianlu, Liming Wang, Qiang Chen, and Chunying Chen. 2014. 'Cytotoxic Potential of Silver Nanoparticles.' *Yonsei Medical Journal* 55 (2): 283. https://doi.org/10.3349/ymj.2014.55.2.283.

Zhang, Xiao, Guanghui Ma, and Wei Wei. 2021. 'Simulation of Nanoparticles Interacting with a Cell Membrane: Probing the Structural Basis and Potential Biomedical Application.' *NPG Asia Materials* 13 (1): 1–18. https://doi.org/10.1038/s41427-021-00320-0.

Chapter 5.2

Toxicological Assessment of the Nanomaterials Used in Biomedicine

Anshika Kushwaha
Anuj Kumar Tomar
and Paulraj Rajamani*
School of Environmental Sciences, Jawaharlal Nehru University, New Delhi, India

Abstract

The advancement in nanotechnology has resulted in a widespread application of nanomaterials in diverse areas, heightening the possibility of human exposure. The term "nanomaterials" encompasses various nanosized particles with unique physicochemical properties, categorized based on shape into nanofibers, nanowires, nanorods, and others. Despite the growing development of nanomaterial applications in industry and medicine, their possible detrimental impacts on health remain largely unexplored. In medicine, nanomaterials are used in drug delivery, diagnostics, imaging, implants, tissue engineering, and surgery. They can enter the human body through the skin, food, or inhalation, with occupational and environmental inhalation being the primary entry routes. Nanoparticles have been found in a number of organs and tissues, including the brain, where they can cross biological barriers. Elimination occurs mainly through urine, bile, and feces. Prosperous animal studies indicate potential health risks, including genotoxicity, carcinogenicity, and reproductive and developmental health. This chapter explores various exposure routes of nanomaterials and their effects on human health risk assessment. We delve into nanomaterials' chemical and physical characteristics that influence their toxicity, elucidate mechanisms of interaction with biological entities, and analyze consequential health implications. Furthermore, we classify bionanomaterials and highlight their biomedical uses, such as tissue engineering, imaging, and medication delivery. The chapter examines factors influencing nanotoxicity, such as particle size, surface chemistry, and dose, and discusses the toxicokinetics of nanomaterials, including distribution, metabolism, excretion, and absorption. We also investigate nanomaterials' genotoxicity and possible carcinogenicity, providing a comprehensive understanding of their risks and benefits. This knowledge will guide future research efforts toward developing safer and more sustainable nanotechnology.

---

* Corresponding Author's Email: paulraj@mail.jnu.ac.in

In: Emerging Nanomaterials in Biomedical Science
Editors: Eepsita Priyadarshini and Ajith Manayil Parambil
ISBN: 979-8-89530-600-0

Keywords: biomedicine, nanotechnology, nanomaterials toxicity, ecotoxicity, health risks

1. Introduction

Rapid advancements in nanotechnology have expanded nanomaterial applications, increasing the risk of human exposure. These materials have unique properties, including nanofibers, nanorods, and nanowires. However, despite their industrial and medical use, their potential health effects remain largely underexplored (Bundschuh et al. 2018). Nanotechnology offers vast medical potential, with innovations such as diagnostic devices, artificial organ design, bone grafting, nanorobotic biosensors, microscopic drug delivery systems, genomics, etc. Key nanocarriers for drug delivery (De Jong and Borm 2008) consist of carbon nanotubes (CNTs), liposomes, dendrimers, iron oxide, quantum dots (QDs), etc. Nanotechnology promises precise drug delivery, crucial for minimizing side effects in cancer treatment. Additionally, the ability of nanomaterials to transport multiple substances simultaneously offers an exciting advantage and a potential solution for various therapies. However, despite these benefits, there are increasing worries about the possible harm nanoparticles could do to the environment and human health (Martínez et al. 2021).

These substances may be ingested, inhaled, or absorbed through the skin and eyes, with inhalation being the primary route in occupational and environmental settings. Nanoparticles (NPs) may penetrate biological boundaries and have been detected in various bodily tissues and organs, especially the brain (Hussain et al. 2009; Zhao and Castranova 2011). The primary routes of elimination for these particles are through urine, bile, and feces. Systematic toxicological assessment of nanomaterials is frequently neglected during product development. The delivery of nanoparticulate carriers via intravenous or subcutaneous injection for medical applications is especially concerning, as these methods bypass normal absorption processes.

These carriers can cause toxicity and interact with biological macromolecules. Insoluble NPs may penetrate biological membranes and build up in tissues or organs (Zhao and Castranova 2011). In the medical field, various animal studies have indicated that nanomaterials pose potential health risks, including genotoxicity, carcinogenicity, reproductive, developmental, and environmental consequences. Therefore, toxicological assessment is crucial in the development of nanomedicine.

This chapter aims to explore different exposure routes of nanomaterials and provide a comprehensive assessment of the associated risks. It will guide future research efforts toward developing safer and more sustainable nanotechnology practices.

Biomedical nanotechnology, which focuses on designing and developing NPs for various medical uses, has rapidly advanced in recent years. This is evident from numerous nanotechnology-based products now commercialized in clinical and preclinical testing (Aula et al. 2015). Key applications of nanomaterials include:

- Drug delivery: For therapeutic, prophylactic, gene delivery, and address challenges like poor solubility, altering pharmacokinetics for efficacy and safety, and site-specific and intracellular delivery (De Jong and Borm 2008).
- Disease therapy: Localized induction of stimuli in target tissues (Mamot et al. 2012).

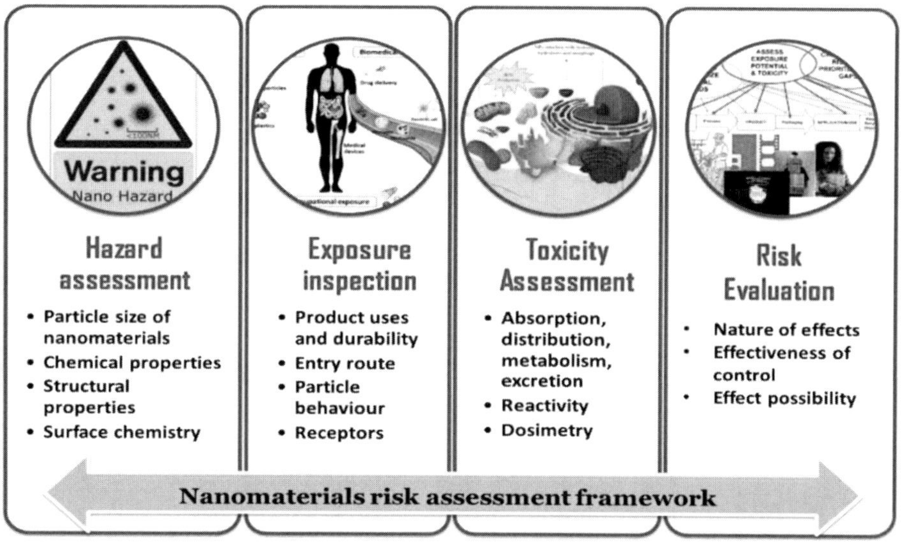

Figure 1. Framework for Assessing Risks Associated with Nanomaterials. Applications of Nanomaterials in Biomedicine.

- Imaging: Utilizing nanomaterials as magnetic resonance contrast agents or fluorescence markers to monitor bio-distribution and detect pathologies (Cho et al. 2010).
- Theranostics: Combining therapy and imaging for personalized medicine.
- Tissue engineering: Supporting the repair or replacement of damaged tissues (Dvir et al. 2011).
- Detoxification: Removing toxins from the body (Hoshino et al. 2012).

The advancements in drug delivery include improving the dissolution of poorly soluble drugs, achieving desired pharmacokinetics, and enabling intracellular delivery of challenging drugs like nucleic acids and proteins. NPs also enhance vaccine efficacy by delivering antigens and adjuvants to antigen-presenting cells, boosting immune responses.

Nanomaterials are used as screening and translational tools in drug discovery to understand disease mechanisms and develop improved versions of existing drugs. Diagnostic imaging applications include using nanomaterials for magnetic resonance imaging, near-infrared fluorescence imaging, and more (Aula et al. 2015), with several products already approved for commercialization, such as Ferumoxsil and Ferrixan (Cho et al. 2010).

2. Classification of the Biomedical Nanomaterials

Nanomaterials are particles with a size range of 1 to 100 nm, come in various types, and are produced through diverse nano-manufacturing technologies. This broad term covers particles with diverse characteristics, including composition, shape, charge, and size variations, which complicates classification. Nanomaterials can be made of carbon, metals, biological materials, polymers, ceramics, or composites and are categorized by shape (e.g., nanofibers, nanowires, nanotubes) and chemical composition (organic and inorganic).

Table 1. Types of Nanomaterial used in Biomedicine and its properties (De Matteis and Rinaldi 2018)

	Types of Biomedical nanomaterials			
Properties	Carbon-Based Nanomaterials	Metal-Based Nanomaterials	Biological Nanomaterials	Nanopolymers
Composition	Carbon-based sources	Metal-based nanoparticles	Derived from proteins, nucleic acids, and carbohydrates	Nanostructured polymer molecules
Shape	Fullerenes are hollow spherical and ellipsoidal particles; nanotubes are cylindrical particles.	Can be spherical, rod, elongated, flat, cube, sheet, triangular etc.	Can be spherical, rod, elongated, flat, cube, sheet, triangular etc.	Lamellar formations are examples of self-assembling structures, while dendrimers and nanofibers are examples of non-self-assembling structures.
Surface area	CNT- large surface area	Substantial surface area relative to their volume	High	Nano size provides high surface area
Examples	CNT and Fullerene nanoparticles	QDs, nanogold, nanosilver, selenium NPs talc, calcium-phosphate and metal oxides (Mg, Fe, Zn, Ni, Al, Cu, Si, Bi, Ti, Mo etc),	Liposomes, Phospholipids etc	Polymeric nanoparticles eg, Poly(styrene), poly(ethylene), poly(ethylene glycol), poly(methyl aerylate), poly(tetrafluoroethylene), poly(lactic-co-glycolic acid) (PLGA) etc), Chitosan, Dendrimers (branched polymers)
Degradability	Non-biodegradable	Mostly Non-biodegradable, some are biodegradable-metal oxides of (Mg, Fe, Zn) and calcium-phosphate	Generally biodegradable and biocompatible, with degradation products being non-toxic or minimally toxic.	Natural materials are biodegradable, whereas synthetic materials are not. Poly(glycolic acid and lactic acid) (PLGA)
Application	CNT is utilized in the biomedical disciplines as tissue scaffolds, medication delivery, bio separators, DNA and protein sensors, and biocatalysts.	Anisotropic metal NPs- specific biomarkers for clear imaging contrasts. Iron oxide particles Used in MRI Able to kill tissue at targeted sites and are used in multifunctional applications, Used as biosensors. QDs- MRI and PET GNPs- disease detection and drug delivery SNPs- antibacterial properties in wound and burn treatment, contraceptives, water disinfectants, and room sprays (highest commercial applications). QDs (fluorescence imaging) are combined with antibodies and exhibit a broad spectrum. Biosensing and bioimaging.	For cancer imaging and treatment, peroxalate ester polymers with encapsulated fluorescent dye can detect cancer by emitting a signal when in contact with hydrogen peroxide, produced by pre-cancerous cells.	Utilized as drug carriers in chemotherapy, allowing cytotoxic medicines to be released into cancer cells and facilitating quick drug transport across cancer cell membranes. Effective drug delivery with PEG-PLGA nanospheres; surface modification with polymer films.

Table 2. Toxicological effects of some widely used nanomaterial's in biomedicine

Nanomaterial's	Toxic effect	References
SWCNTs	Comet in V79, Ames in Salmonella typhimurium, MN test, and micronucleus assay were used to identify the genotoxicity of SWCNTs, albeit the latter was not statistically significant.	(Dhawan and Sharma 2010)
MWCNTs	Nanoparticles damage DNA, induce oxidative stress, trigger apoptosis, and disturb cell architecture *in vitro*. *In vivo*, they cause pulmonary toxicity, inflammation, rapid fibrosis, and immune suppression while promoting allergic responses and liver damage. In mice, they increase chromosomal aberrations. In zebrafish embryos, they affect reproduction, cause defects, and are toxic to bacteria.	(Cheng et al. 2009); (Reddy et al. 2010)
QDs	*In vitro* studies, NPs can induce cytotoxicity, inflammation, and oxidative stress in diverse cell cultures. *In vivo* research shows that QDs pass from mice in pregnancy to their young via the placental barrier, adversely affecting the development of mouse oocytes and can penetrate skin barriers damaged by UV radiation. Environmental impact: Phototoxic to *Daphnia magna* in UV-B light.	(Su et al. 2009; L. Wang et al. 2008; Tang et al. 2008)
SNPs	*In vitro*, NPs induce cytotoxicity, chromosome instability, oxidative stress, apoptosis, and interfere with DNA replication. *In vivo*, they induce oxidative stress, alter gene expression, damage the blood-brain barrier, and cause neuronal degeneration. In fish, they are cytotoxic, genotoxic, and impact development. In Drosophila, they cause stress and DNA damage. In *C. elegans*, they reduce reproduction and cause toxicity.	(AshaRani et al. 2009); (Kawata et al. 2009); (Miura and Shinohara 2009)
GNPs	*In vitro*, NPs alter cellular micromotility oxidative stress, induce mitochondrial damage, and trigger autophagy. *In vivo*, NPs accumulate in vital organs, induce acute liver inflammation and apoptosis, impair human sperm motility, and penetrate sperm heads and tails. Rainbow trout studies: Nanoparticles have adverse effects on hepatocytes.	(Pan et al. 2009; Li et al. 2010) (Patra et al. 2020)
Metal oxide NPs	In mammals, NPs cause cytotoxicity, inflammation, membrane damage, oxidative stress, apoptosis, DNA damage, and altered gene expression. They inhibit osteogenic differentiation and disrupt neural functions. Additionally, nanoparticles are toxic to yeast, bacteria, nematodes, and aquatic species like fish and zebrafish.	(Fahmy and Cormier 2009; Chen et al. 2010; Falck et al. 2009)

Nanomedicine includes analytical tools, novel therapeutics, drug delivery systems, nanoimaging, and nanodevices (Zhao and Castranova 2011). However, it faces clinical, regulatory, and toxicological challenges. Medical nanomaterials are categorized into biological mimetics, biomaterials, sensors, nanoscale devices, and diagnostics. Frequently utilized nanomaterials in biomedicine are listed in Table 1 and include nanotubes, fullerenes, QDs, TiO_2, nanogold, nanosilver, dendrimers, nanocrystals, nanoshells, calcium phosphate vaccines, metal colloidal particles, and nanosized electrodes.

3. Factors Influencing Nano-Toxicity

Nanomaterials offer significant benefits in biomedicine, but their possible toxicity is still a major worry. The toxicity of nanomaterials is triggered by various factors, including their biological interactions, physicochemical properties, and environmental conditions (Pan et al. 2007; Walkey et al. 2012).

1. Size: Smaller NPs exhibit a greater ratio of surface area to volume, which increases reactivity and possible toxicity. Smaller particles can penetrate cellular barriers more easily and may lead to higher cytotoxicity.
2. Shape: How nanomaterials interact with biological systems is influenced by their form, including cellular uptake and distribution. Rod-shaped particles might penetrate cells more effectively than spherical particles, potentially causing more damage.
3. Surface charge: The surface charge determines the interaction with cell membranes and proteins. Positively charged NPs are often more toxic because they have the capability to strongly interact with negatively charged cell membranes to damage them and make absorption easier for cells.
4. Surface Chemistry and Functionalization: Surface modifications can enhance biocompatibility or target specificity but also alter toxicity. Coating nanoparticles with polyethylene glycol (PEG) can reduce toxicity by decreasing protein adsorption and immune recognition.
5. Aggregation State: Nanomaterial aggregation affects distribution, cellular uptake, and toxicity. Aggregated particles may have reduced bioavailability and lower toxicity compared to dispersed particles.
6. Solubility and Dissolution Rate: The solubility and rate at which nanomaterials dissolve can affect their bioavailability and toxicity. Soluble nanomaterials might release toxic ions, whereas insoluble particles might persist in tissues and cause prolonged toxicity.
7. Chemical Composition: Nanomaterials' elemental composition and chemical characteristics dictate their reactivity and potential for toxicity. SNPs release silver ions, which are toxic to bacteria and potentially harmful to human cells at high concentrations.
8. Crystallinity and Phase: The crystalline structure and phase can influence nanomaterials' physical stability and reactivity. Crystalline TiO_2 might be more reactive and toxic than its amorphous form.

9. Impurities and Contaminants: The presence of impurities or contaminants in nanomaterials can contribute to toxicity. Residual solvents or catalysts from NPs synthesis might be toxic and need to be removed.
10. Dosage and Concentration: Higher doses or concentrations generally increase the likelihood of toxic effects. At lower doses, nanomaterials might be safe, but at higher doses, they can overwhelm biological defenses and cause toxicity. For example, TiO_2 NPs are relatively safe at low concentrations but can cause oxidative stress and inflammation at high concentrations.
11. Exposure Route: The exposure route determines the initial interaction with biological systems and subsequent distribution and toxicity. Inhaled nanomaterials can deposit in the lungs and enter the bloodstream, potentially causing respiratory and systemic effects. Intravenously injected NPs, on the other hand, may have a more direct impact on the liver, spleen, and circulatory system.
12. Exposure Duration: Long-term exposure may result in distinct toxic effects compared to short-term or acute exposure. Short-term exposure might cause acute inflammation, whereas long-term exposure might lead to fibrosis or cancer. Chronic exposure to CNT, for instance, has been associated with lung fibrosis and cancer in animal research.

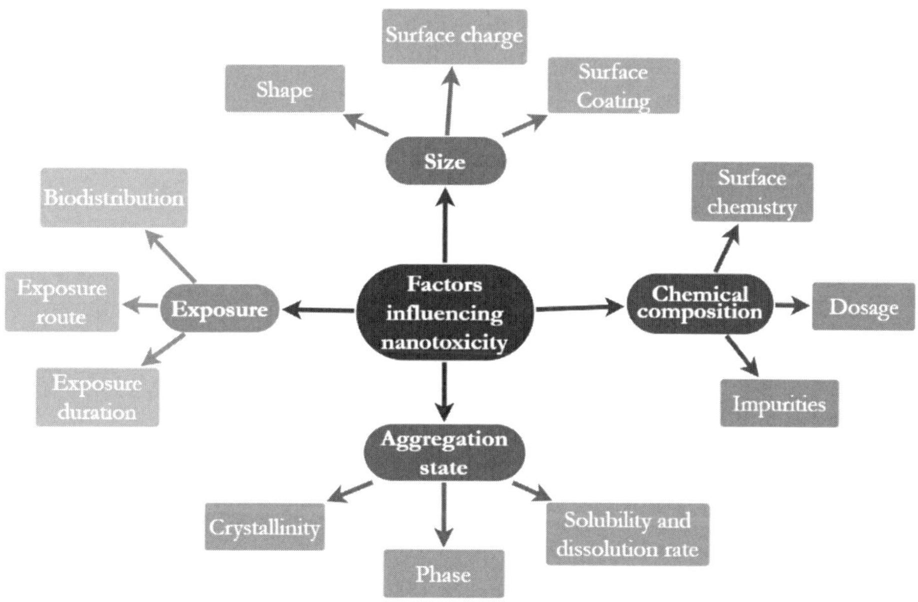

Figure 2. Different attributes influence the toxicity of nanomaterials.

4. Biomedical Toxicokinetics of Nanomaterials

4.1. Absorption

Toxicokinetics' initial stage is absorption. The amount of inhaled or consumed nanomaterials is not the only factor affecting the amount that the human body absorbs; their physicochemical properties also play a role. However, compared to other ways, intravenous and subcutaneous injections of nanomaterial carriers may present a unique and perhaps more significant exposure pathway in biomedicine.

4.1.1. Oral Absorption

The gastrointestinal tract (GIT) is a significant route for nanomaterial absorption, such as food, water, drug carriers, and beverages (Hagens et al. 2007; Lomer et al. 2002). In biomedicine, efforts have focused on enhancing oral drug and vaccine absorption via GIT absorption of micro- and nanomaterials (Hillyer and Albrecht 2001). Airborne nanoparticles can be cleared by the mucociliary escalator and enter the gastrointestinal tract (GIT). These particles are then absorbed by the body, pass through the intestinal lining, and enter the circulatory and lymphatic systems. (Hagens et al. 2007). GIT absorption increases as particle diameter decreases (Desai et al. 1996). The maximum absorption for polystyrene latex particles (Hillyer and Albrecht 2001) was observed at 50–100 nm, with Peyer's patches trapping particles larger than 1 μm. Particle charge also influences GIT absorption, with positively charged NPs being more effectively absorbable than negatively or neutrally charged particles (Bourges et al. 2003; Hussain et al. 2010).

4.1.2. Absorption through Inhalation

The lungs, directly exposed to the environment, are the primary entry point for nanomaterials (Tsuji et al. 2006). Long-term studies show that nanomaterials cause lung injury due to oxidative stress from ROS (De Matteis and Rinaldi 2018). Inhaled nanomaterials absorb biomolecules, altering their aggregation and size. Most nanomaterials are captured by alveolar macrophages (AMs) or penetrate the alveolar epithelial barrier. Small nanomaterials that evade AMs can enter the interstitium, causing prolonged inflammation, fibrosis, and lung damage (Zhang et al. 2014). For example, intratracheal instillation of Fe_2O_3 NPs in mice showed some NPs being phagocytosed by AMs while others penetrated the alveolar epithelium, leading to lung fibrosis and pathological changes. Nano-sized Fe_2O_3 NPs significantly, compared to larger particles, enhance lung epithelial microvascular permeability and cell lysis.

4.1.3. Dermal Absorption

Understanding nanomaterial penetration and toxicity is crucial for biomedicine (Choksi et al. 2010; Tsuji et al. 2006). Nanomaterials in cosmetics are applied directly to the skin, often incorporating metal oxide NPs such as ZnO for UV protection. Sadrieh et al. 2010 found minimal Ti penetration through skin intact on a little pig model, whereas (Gulson et al. 2010) reported human skin absorbs trace amounts of zinc from ZnO NPs in sunscreens. Dermal exposure could be a significant absorption route in biomedicine (Hagens et al. 2007), with medical textiles and wound dressings often containing antibacterial SNP (Lee et al. 2003). Nanomaterials are also utilized for transdermal medication administration (Prausnitzet al. 2004). TiO_2 NPs are increasingly used in various products, indicating a possible risk of human exposure (Long et al. 2007). TiO_2 particles have historically been used as a "negative control" in numerous toxicological studies.

4.1.4. Intravenous and Subcutaneous Absorption

Different exposure paths exist for intravenous and subcutaneous injections of nanomaterial carriers compared to exposures in the workplace and environment. (Allen 2002; Allen and Cullis 2004; Panyam and Labhasetwar 2003; Sahoo and Labhasetwar 2003). Additionally, nanomaterials can be administered straight into the body as contrast agents for imaging and bypassing procedures.

4.1.5. Ocular Absorption

Ocular absorption is the process by which substances applied to the eye are absorbed into ocular tissues. This includes medications, dyes, and nanomaterials delivered through eye drops, injections, or other methods. The absorption depends on the drug's properties, delivery method, and ocular barriers like the cornea and blood-retinal barrier. Effective ocular absorption ensures therapeutic drug levels in eye tissues while minimizing systemic exposure and side effects.

The ocular absorption study examined polylactide NPs' behavior in the retina after intravitreous injection (Bourges et al. 2003), focusing on their distribution and potential for drug delivery. Key findings showed that NPs rapidly reach the retina, localize in RPE cells, and release encapsulated materials while preserving retinal anatomy. The encapsulated dye diffused slowly and uniformly. The study concluded that these nanomaterials support targeted tissue delivery and sustained drug release, suggesting potential applications for chronic disease and genetic therapy in the eye's posterior segment. Further research is needed to assess impacts on retinal function and vision before clinical use.

4.2. Metabolism of the Biomedical Nanomaterials

Hepatocytes play a major role in metabolizing nanomaterials, whose breakdown can cause unpredictable molecular responses (Hussain et al. 2009). Absorbed nanomaterials may change their chemical forms. Inert NPs like gold, silver, fullerenes, and CNT are not metabolized effectively. However, functionalized NPs might be metabolized, such as protein-capped QDs being broken down by proteases and their metallic cores bound by metallothionein for excretion. Enzymes like metalloproteinases in the liver and kidneys help restore cellular metal homeostasis. Liposome-based NPs drug-delivery systems can fuse with cell membranes (Zhao and Castranova 2011). Recent studies show myeloperoxidase can biodegrade single-walled carbon nanotubes (SWCNTs) in neutrophils, potentially influencing inflammatory responses. Further, *in vivo* studies are needed to understand SWCNT metabolism mechanisms.

It is crucial to understand and document the degradation of these nanostructures. *In vivo* rodent models show that oral nanomaterials tend to accumulate in the liver, potentially causing tissue injury. Wang, Langer, and Farokhzad 2012 reviewed the *in vivo* metabolism of nanomaterials, with an emphasis on organ clearance and blood flow in the kidneys, liver, and lungs. Once in the bloodstream through direct application or indirect exposure (inhalation, ingestion, dermal), their small size allows them to reach various body parts. Studies indicate that the lungs, liver, and kidneys are the primary distribution and target sites for nanomaterial exposure, making their clearance patterns essential for understanding there *in vivo* fate (De Matteis and Rinaldi 2018).

4.3. Biological Fate of Biomedical Nanomaterials: Distribution, Accumulation, and Elimination

Translocation refers to the movement of particles from the absorption site to other bodily regions. Upon entering the bloodstream, nanomaterials interact with blood cells, platelets, coagulation factors, and plasma proteins, influencing their metabolism, distribution, and elimination (Aula et al. 2015). For instance, apolipoprotein-A1 binds to SiO_2, asbestos, and TiO_2 particles. Nanomaterials smaller than 200 nm can penetrate red blood cells without undergoing endocytosis and can traverse the blood-brain barrier. A key principle of nanomaterials is that surface area increases as particle size decreases. This significantly influences nano-bio interactions. In the bloodstream, plasma proteins bind to nanomaterials, forming protein coronas that alter their surface properties. The binding affinity of blood proteins to NPs can create a dynamic protein corona (Milani et al. 2012), which can be soft (reversibly bound proteins) or hard (irreversibly bound proteins). These interactions are crucial for determining the nanomaterial's behavior with biological surfaces and receptors. Proteins adsorbed onto nanomaterials may undergo conformational changes, affecting their function and leading to significant implications for protein interactions, cellular signaling, and related mechanisms.

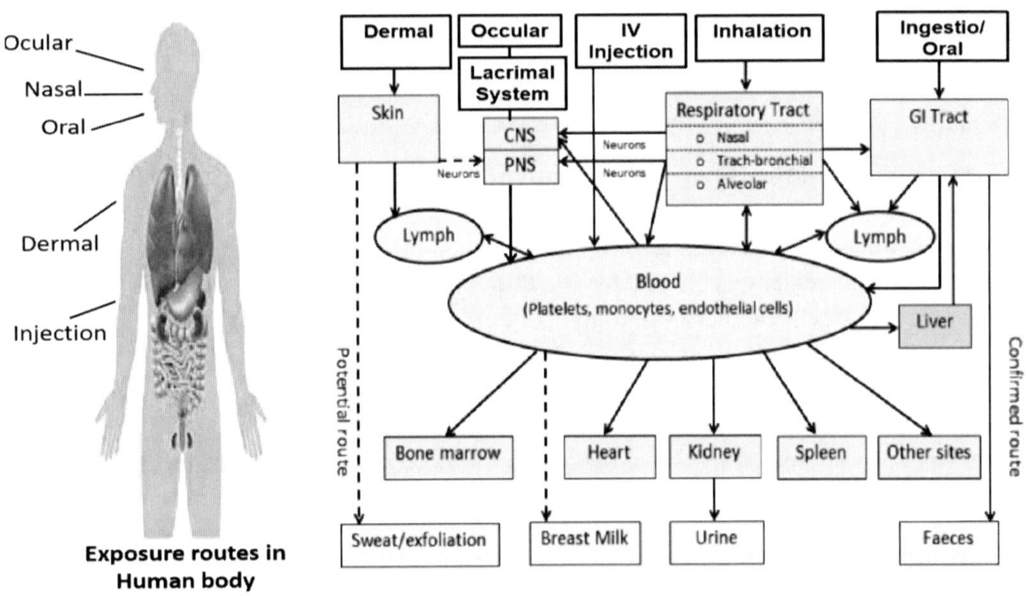

Figure 3. Toxicokinetics route of biomedical nanomaterials (Bakand and Hayes 2016) Dashed arrows: Potential routes; Solid arrows: Confirmed routes; GI: Gastrointestinal; CNS: Central nervous system; IV: Intravenous; PNS: Peripheral nervous system.

The protein corona impacts nanomaterial's interactions with tissues by their dimensions and surface features. These interactions, influenced by the structure and nature of surface-linked proteins, impact cellular internalization. Key factors in nanomaterial-protein interactions include size (Hussain et al. 2009); and surface charge (Lundqvist et al. 2008). Smaller nanomaterials show lower protein binding and longer plasma half-life than larger ones. For example, 80 nm nanomaterials had 6% protein binding, while 240 nm had 34%. Nanomaterials injected intravenously, ranging from 80-150 nm, were rapidly removed, with plasma half-life of 8–30 minutes, while those sized 20–40 nm displayed a 25–30-hour plasma half-life. Inhalation studies showed that carbon NPs reach the bloodstream and various organs within minutes (Zhao and Castranova 2011). Intravenously injected nanomaterials are distributed to the spleen, liver, heart, and brain. Polymeric nanomaterials and lipid NPs show controlled release and translocation to different tissues. Germanium and iron oxide NPs also distribute to organs like the spleen and liver, excreting mainly through urine.

NPs can accumulate in tissues for extended periods (De Matteis and Rinaldi 2018), affecting organs including the brain, liver, kidneys, and thymus. Additional *in vivo* research is needed to comprehensively grasp diverse nanomaterials' metabolism and prolonged impacts.

5. Acute and Chronic Toxicity of the Nanomaterials

Nanomaterials have distinct physico-chemical properties like shape, size, charge, solubility, and extent of agglomeration, influencing their toxicity (Egbuna et al. 2021). Compared to larger particles manufactured from the same materials, they show a different toxicity profile because of their small size and higher specific surface area. Despite their widespread use in biomedicine for drug delivery, their acute and chronic toxicities are not well-studied.

Recent incidents, like the fatal inhalation of nickel NPs, underscore the need for caution. Studies have shown varied toxicity levels of these materials. For instance, SiO_2 NPs can cause liver injury depending on size, and GNPs may induce inflammation and apoptosis (Li et al. 2010), although some studies show minimal toxicity. SNPs are potent antimicrobials but can reduce cell viability in high concentrations (Kawata et al. 2009). Fullerenes and CNTs show promise in drug delivery but have demonstrated potential cytotoxicity and inflammatory responses. Overall, while the potential of nanomedicine is significant, comprehensive studies are required to guarantee the effectiveness and safety of nanomaterials used in medical applications (Zhao and Castranova 2011).

6. Genotoxicity

Genotoxicity refers to DNA damage caused by chemicals or agents, resulting in gene mutations. Nanomaterials possess distinctive physico-chemical properties—like large surface area, small size, shape, and composition—that facilitate cell entry and interaction with cellular components, including the nucleus. This interaction can potentially disrupt cell signaling and function, leading to genotoxic effects (Aula et al. 2015).

Genotoxicity can be primary or secondary. Primary genotoxicity results from the agent's direct contact with nuclear DNA, causing DNA damage. Secondary genotoxicity occurs due to adverse effects like inflammation and oxidative stress (Stone 2009). Genotoxicity assays, such as the comet assay, chromosomal abnormalities (chromosomal aberration and cytokinesis-blocked micronucleus assays), DNA base modifications (8-oHdG detection by ELISA), gene mutations (Ames assay), and changes in gene expression, are used to evaluate different types of DNA damage.

Genotoxic substances can be mutagenic or carcinogenic, causing genetic damage or mutations that contribute to tumor development (Egbuna et al. 2021). *In vivo* genotoxicity tests have used rat and mouse tissues, BAL cells, bone marrow, and Drosophila melanogaster. *In vitro* tests have included the Ames test, MN test, Comet assay, chromosomal aberrations (CA), SCE, gene mutation in mammalian cells (human, rat, and mouse), and cell transformation (De Matteis and Rinaldi 2018).

7. Carcinogenicity

The TiO_2 NPs, widely used in biomedicine, have been shown to cause lung tumors in rodents via intratracheal instillation and breathing, with studies by (Heinrich et al. 1995) (Rittinghausen et al. 1997) supporting these findings. The IARC has classified TiO_2 nanoparticles as potentially carcinogenic to humans (Group 2B) due to available evidence in animals but lacking human epidemiological data (Gheshlaghi et al. 2008). The NIOSH has recommended restrictions on ultrafine and fine TiO_2 exposure at work. CBNs have similarly been linked to lung tumors in rats, earning a Group 2B classification from the WHO/IARC. CNTs, including SWCNTs and MWCNTs, show mixed results; while short-term studies did not report tumors, specific exposures led to mutations and mesothelioma. Notably, long MWCNTs cause inflammatory lesions, unlike short MWCNTs. Overall, TiO_2 NPs and CBNs are carcinogenic in animals, while MWCNTs have shown carcinogenicity with certain injections, necessitating further research to assess their lung cancer risk (Rittinghausen, et al. 1997).

8. Reproductive and Developmental Nanotoxicity

While investigations into the detrimental properties of nanoparticles are growing, there still needs to be more information on the toxicity of nanomaterials to reproduction and development. (Park et al. 2009). There is no conclusive evidence that exposure of humans to nanomaterials causes reproductive or developmental effects. Research has indicated that TiO_2 NPs may cause reproductive defects in Daphnia magna and affect the expression of genes linked to the development of the mice brain. TiO_2 exposure in mother mice resulted in transfer to offspring, affecting their genital and cranial nervous systems (Takeda et al. 2009). CBNs have been linked to decreased sperm production in mice. SWCNTs and MWCNTs have shown substantial developmental toxicity in the embryos of zebrafish. Fullerene C60 caused mortality and abnormalities in mouse embryos. SiO_2 NPs inhibited stem cell differentiation, while GNPs reduced human sperm motility and caused fragmentation. SNPs were harmful to mouse spermatogonia stem cells, and ZnO NPs hurt the hatching process of zebrafish embryos (Chen et al. 2011).

In the accumulation of TiO_2 NPs in the testes and brain, caution is advised regarding their reproductive and developmental toxicity (J. Chen et al. 2011). Additional research is required to compare toxicity between fine and nanosized materials and evaluate the possibility that nanomaterials could get through the placenta and fetal tissue.

9. The Challenges of Using Nanomaterials in Biomedicine

The use of nanomaterials in biomedicine holds significant promise but also presents several challenges. Here's an overview of these challenges and future perspectives (Parvin et al. 2024)

- Toxicity and Biocompatibility: Certain nanomaterials can be toxic to cells and tissues, potentially causing adverse health effects. Understanding and mitigating these toxic effects is essential. Ensuring that nanomaterials do not trigger an immune response or other negative reactions in the body is critical for their successful use in biomedicine.
- Drug Delivery and Release: Precisely controlling the frequency of drug release from nanomaterials can be difficult, affecting the efficacy and safety of treatments. Accurately targeting specific tissues or cells without impacting healthy tissues remains a major challenge.
- Manufacturing and Scalability: Scaling up the production of nanomaterials while maintaining quality and consistency is challenging. High manufacturing costs can limit the widespread adoption of nanomaterials in clinical settings.
- Regulatory and Ethical Issues: Navigating the complex regulatory landscape for nanomaterials is challenging due to many uncertainties and gaps in current regulations. Addressing ethical issues, such as potential long-term effects and privacy concerns, is essential.
- Characterization and Standardization: Thoroughly characterizing nanomaterials, including their size, shape, surface properties, and interactions with biological systems, is crucial but challenging. Developing standardized methods for testing and evaluating nanomaterials is necessary to ensure consistency and reliability across studies and applications.

Conclusion and Future Direction

The rapid advancements in nanotechnology have unlocked significant potential in various fields, especially in biomedicine (Fadeel and Bennett 2010), but they also pose considerable health risks that need thorough evaluation. Nanomaterials like nanofibers, nanowires, and nanorods have unique properties that offer potential uses, particularly targeted drug delivery systems, diagnostic imaging, and tissue engineering. For instance, approved therapies such as Abraxane® leverage the improved permeability and retention (EPR) effect to enhance drug concentration in tumor cells, thereby reducing toxicity in healthy tissues. Despite these benefits, the possible adverse effects of nanomaterials on human health and the environment are significant concerns.

Nanomaterials can enter the human body through the skin, mouth, nose and eyes, with inhalation being predominant in occupational settings. Once inside, they can cross biological boundaries and accumulate in various tissues, including the brain, potentially causing toxicity. Systematic toxicological evaluations are often overlooked during product development, which is alarming given the evidence from animal studies indicating genotoxicity, carcinogenicity, and reproductive harm from these materials.

In biomedicine, nanomaterials are integral to drug delivery, disease therapy, imaging, theranostics, tissue engineering, and detoxification. Their ability to improve drug solubility, pharmacokinetics, and intracellular delivery highlights their transformative potential. However, factors influencing nanotoxicity, including their small size and large surface area, lead to unpredictable interactions within the body, necessitating comprehensive toxicological assessments. The systematic investigation of the toxicokinetics of nanomaterials, encompassing their absorption, metabolism, distribution, and elimination, reveals that they can cause significant cellular and tissue-level changes. Notably, nanomaterials like TiO_2 and CBNs have shown carcinogenic effects in animal studies, and reproductive toxicity remains a critical area of concern (Maynard et al. 2006).

In conclusion, while nanotechnology holds immense promise, especially in advancing medical treatments and diagnostics, balancing this with rigorous safety assessments is crucial. Future research should focus on developing safer nanomaterials, understanding their long-term effects, and establishing comprehensive guidelines to mitigate health risks. This balanced approach will enable the sustainable development of nanotechnology, ensuring its benefits are maximized without compromising human health and environmental safety.

Advances in toxicological testing and nanomaterial design are pivotal for reducing toxicity and enhancing the safety of nanotechnologies. Future research should focus on developing innovative testing methods that accurately mimic biological systems, including high-throughput screening techniques and *in vivo* models, to better predict nanomaterials' extended effects. Additionally, the design of nanomaterials should prioritize safety by incorporating features that minimize adverse biological interactions. This includes optimizing particle size, shape, surface chemistry, and functionalization to reduce toxicological risks while maintaining therapeutic efficacy. By fostering collaboration between toxicologists, material scientists, and regulatory bodies, we can develop comprehensive guidelines and standardized testing protocols that ensure the safe deployment of nanomaterials. Ultimately, these advancements will pave the way for safer, more effective nanomaterials that can revolutionize the biomedical field and other industries while reducing any possible threats to human and environmental health.

References

Allen, Theresa M. 2002. "Ligand-Targeted Therapeutics in Anticancer Therapy." *Nature Reviews Cancer* 2 (10): 750–63. https://doi.org/10.1038/nrc903.

Allen, Theresa M., and Pieter R. Cullis. 2004. "Drug Delivery Systems: Entering the Mainstream." *Science* 303 (5665): 1818–22. https://doi.org/10.1126/science.1095833.

AshaRani, PV V., M. Prakash Hande, and Suresh Valiyaveettil. 2009. "Anti-Proliferative Activity of Silver Nanoparticles." *BMC Cell Biology* 10 (September): 65. https://doi.org/10.1186/1471-2121-10-65.

Aula, Sangeetha, Samyuktha Lakkireddy, Kaiser Jamil, Atya Kapley, A. V. N. Swamy, and Harivardhan Reddy Lakkireddy. 2015. "Biophysical, Biopharmaceutical and Toxicological Significance of Biomedical Nanoparticles." *RSC Advances* 5 (59): 47830–59. https://doi.org/10.1039/C5RA05889A.

Aula, Sangeetha, Samyuktha Lakkireddy, Kaiser Jamil, Atya Kapley, Avn Swamy, and Harivardhan Reddy Lakkireddy. n.d. "Biophysical, Biopharmaceutical and Toxicological Significance of Biomedical Nanoparticles."

Bakand, Shahnaz, and Amanda Hayes. 2016. "Toxicological Considerations, Toxicity Assessment, and Risk Management of Inhaled Nanoparticles." *International Journal of Molecular Sciences*. MDPI AG. https://doi.org/10.3390/ijms17060929.

Bourges, Jean Louis, Sandrine E. Gautier, Florence Delie, Riad A. Bejjani, Jean Claude Jeanny, Robert Gurny, David BenEzra, and Francine F. Behar-Cohen. 2003. "Ocular Drug Delivery Targeting the Retina and Retinal Pigment Epithelium Using Polylactide Nanoparticles." *Investigative Ophthalmology and Visual Science* 44 (8): 3562–69. https://doi.org/10.1167/iovs.02-1068.

Bundschuh, Mirco, Juliane Filser, Simon Lüderwald, Moira S. McKee, George Metreveli, Gabriele E. Schaumann, Ralf Schulz, and Stephan Wagner. 2018. "Nanoparticles in the Environment: Where Do We Come from, Where Do We Go To?" *Environmental Sciences Europe*. Springer Verlag. https://doi.org/10.1186/s12302-018-0132-6.

Chen, Jinyuan, Xia Dong, Yuanyuan Xin, and Meirong Zhao. 2011. "Effects of Titanium Dioxide Nano-Particles on Growth and Some Histological Parameters of Zebrafish (Danio Rerio) after a Long-Term Exposure." *Aquatic Toxicology* 101 (3–4): 493–99. https://doi.org/10.1016/j.aquatox.2010.12.004.

Chen, Ying-Chun, Jong-Kai Hsiao, Hon-Man Liu, I-Yin Lai, Ming Yao, Szu-Chun Hsu, Bor-Sheng Ko, Yao-Chang Chen, Chung-Shi Yang, and Dong-Ming Huang. 2010. "The Inhibitory Effect of Superparamagnetic Iron Oxide Nanoparticle (Ferucarbotran) on Osteogenic Differentiation and Its Signaling Mechanism in Human Mesenchymal Stem Cells." *Toxicology and Applied Pharmacology* 245 (2): 272–79. https://doi.org/10.1016/j.taap.2010.03.011.

Cheng, Crystal, Karin H. Müller, Krzysztof K. K. Koziol, Jeremy N. Skepper, Paul A. Midgley, Mark E. Welland, and Alexandra E. Porter. 2009. "Toxicity and Imaging of Multi-Walled Carbon Nanotubes in Human Macrophage Cells." *Biomaterials* 30 (25): 4152–60. https://doi.org/10.1016/j.biomaterials.2009.04.019.

Cho, Eun Chul, Charles Glaus, Jingyi Chen, Michael J. Welch, and Younan Xia. 2010. "Inorganic Nanoparticle-Based Contrast Agents for Molecular Imaging." *Trends in Molecular Medicine* 16 (12): 561–73. https://doi.org/10.1016/j.molmed.2010.09.004.

Choksi, Adrienne N, Tasneem Poonawalla, and Michael G Wilkerson. 2010. "Nanoparticles: A Closer Look at Their Dermal Effects." *Journal of Drugs in Dermatology : JDD* 9 (5): 475–81.

Desai, Manisha P., Vinod Labhasetwar, Gordon L. Amidon, and Robert J. Levy. 1996. "Gastrointestinal Uptake of Biodegradable Microparticles: Effect of Particle Size." *Pharmaceutical Research* 13 (12): 1838–45. https://doi.org/10.1023/A:1016085108889.

Dhawan, Alok, and Vyom Sharma. 2010. "Toxicity Assessment of Nanomaterials: Methods and Challenges." *Analytical and Bioanalytical Chemistry*. https://doi.org/10.1007/s00216-010-3996-x.

Dvir, Tal, Brian P. Timko, Daniel S. Kohane, and Robert Langer. 2011. "Nanotechnological Strategies for Engineering Complex Tissues." *Nature Nanotechnology* 6 (1): 13–22. https://doi.org/10.1038/nnano.2010.246.

Egbuna, Chukwuebuka, Vijaykumar K. Parmar, Jaison Jeevanandam, Shahira M. Ezzat, Kingsley C. Patrick-Iwuanyanwu, Charles Oluwaseun Adetunji, Johra Khan, et al. 2021. "Toxicity of Nanoparticles in Biomedical Application: Nanotoxicology." *Journal of Toxicology*. Hindawi Limited. https://doi.org/10.1155/2021/9954443.

Fadeel, Bengt, and Alfonso E. Garcia-Bennett. 2010. "Better Safe than Sorry: Understanding the Toxicological Properties of Inorganic Nanoparticles Manufactured for Biomedical Applications." *Advanced Drug Delivery Reviews*. https://doi.org/10.1016/j.addr.2009.11.008.

Fahmy, Baher, and Stephania A. Cormier. 2009. "Copper Oxide Nanoparticles Induce Oxidative Stress and Cytotoxicity in Airway Epithelial Cells." *Toxicology in Vitro* 23 (7): 1365–71. https://doi.org/10.1016/j.tiv.2009.08.005.

Falck, GCM, HK Lindberg, S. Suhonen, M. Vippola, E. Vanhala, J. Catalán, K. Savolainen, and H. Norppa. 2009. "Genotoxic Effects of Nanosized and Fine TiO2." *Human & Experimental Toxicology* 28 (6–7): 339–52. https://doi.org/10.1177/0960327109105163.

Gheshlaghi, Zahra Naghdi, Gholam Hossein Riazi, Shahin Ahmadian, Mahmoud Ghafari, and Roya Mahinpour. 2008. "Toxicity and Interaction of Titanium Dioxide Nanoparticles with Microtubule Protein Acta Biochim Biophys Sin (2008): 777-782 | Toxicity and Interaction of Titanium Dioxide Nanoparticles with Microtubule Protein." *Acta Biochim Biophys Sin* 40: 777. https://doi.org/10.1111/j.1745-7270.2008.

Gulson, Brian, Maxine McCall, Michael Korsch, Laura Gomez, Philip Casey, Yalchin Oytam, Alan Taylor, et al. 2010. "Small Amounts of Zinc from Zinc Oxide Particles in Sunscreens Applied Outdoors Are

Absorbed through Human Skin." *Toxicological Sciences* 118 (1): 140–49. https://doi.org/10.1093/toxsci/kfq243.

Hagens, Werner I., Agnes G. Oomen, Wim H. de Jong, Flemming R. Cassee, and Adriënne J. A. M. Sips. 2007. "What Do We (Need to) Know about the Kinetic Properties of Nanoparticles in the Body?" *Regulatory Toxicology and Pharmacology* 49 (3): 217–29. https://doi.org/10.1016/j.yrtph.2007.07.006.

Heinrich, U., R. Fuhst, S. Rittinghausen, O. Creutzenberg, B. Bellmann, W. Koch, and K. Levsen. 1995. "Chronic Inhalation Exposure of Wistar Rats and Two Different Strains of Mice to Diesel Engine Exhaust, Carbon Black, and Titanium Dioxide." *Inhalation Toxicology* 7 (4): 533–56. https://doi.org/10.3109/08958379509015211.

Hillyer, Julián F., and Ralph M. Albrecht. 2001. "Gastrointestinal Persorption and Tissue Distribution of Differently Sized Colloidal Gold Nanoparticles." *Journal of Pharmaceutical Sciences* 90 (12): 1927–36. https://doi.org/10.1002/jps.1143.

Hoshino, Yu, Hiroyuki Koide, Keiichi Furuya, Walter W. Haberaecker, Shih-Hui Lee, Takashi Kodama, Hiroaki Kanazawa, Naoto Oku, and Kenneth J. Shea. 2012. "The Rational Design of a Synthetic Polymer Nanoparticle That Neutralizes a Toxic Peptide *in Vivo*." *Proceedings of the National Academy of Sciences* 109 (1): 33–38. https://doi.org/10.1073/pnas.1112828109.

Hussain, Saber M., Laura K. Braydich-Stolle, Amanda M. Schrand, Richard C. Murdock, Kyung O. Yu, David M. Mattie, and John J. Schlager. 2009. "Toxicity Evaluation for Safe Use of Nanomaterials: Recent Achievements and Technical Challenges." *Advanced Materials* 21 (16): 1549–59. https://doi.org/10.1002/adma.200801395.

Hussain, Salik, Leen Cj Thomassen, Ioana Ferecatu, Marie-Caroline Borot, Karine Andreau, Johan A Martens, Jocelyne Fleury, Armelle Baeza-Squiban, Francelyne Marano, and Sonja Boland. 2010. "Carbon Black and Titanium Dioxide Nanoparticles Elicit Distinct Apoptotic Pathways in Bronchial Epithelial Cells." *Particle and Fibre Toxicology*. Vol. 7. http://www.particleandfibretoxicology.com/content/7/1/10.

Jong, Wim H De, and Paul Ja Borm. 2008. "Drug Delivery and Nanoparticles: Applications and Hazards." *International Journal of Nanomedicine*. Vol. 3.

Kawata, Koji, Masato Osawa, and Satoshi Okabe. 2009. "*In Vitro* Toxicity of Silver Nanoparticles at Noncytotoxic Doses to HepG2 Human Hepatoma Cells." *Environmental Science & Technology* 43 (15): 6046–51. https://doi.org/10.1021/es900754q.

Lee, H J, S Y Yeo, and S H Jeong. n.d. "Antibacterial Effect of Nanosized Silver Colloidal Solution on Textile Fabrics."

Li, Jasmine J., Deny Hartono, Choon-Nam Ong, Boon-Huat Bay, and Lin-Yue L. Yung. 2010. "Autophagy and Oxidative Stress Associated with Gold Nanoparticles." *Biomaterials* 31 (23): 5996–6003. https://doi.org/10.1016/j.biomaterials.2010.04.014.

Lomer, Miranda C. E., Richard P. H. Thompson, and Jonathan J. Powell. 2002. "Fine and Ultrafine Particles of the Diet: Influence on the Mucosal Immune Response and Association with Crohn's Disease." *Proceedings of the Nutrition Society* 61 (1): 123–30. https://doi.org/10.1079/PNS2001134.

Long, Thomas C., Julianne Tajuba, Preethi Sama, Navid Saleh, Carol Swartz, Joel Parker, Susan Hester, Gregory V. Lowry, and Bellina Veronesi. 2007. "Nanosize Titanium Dioxide Stimulates Reactive Oxygen Species in Brain Microglia and Damages Neurons *in Vitro*." *Environmental Health Perspectives* 115 (11): 1631–37. https://doi.org/10.1289/ehp.10216.

Lundqvist, Martin, Johannes Stigler, Giuliano Elia, Iseult Lynch, Tommy Cedervall, and Kenneth A. Dawson. 2008. "Nanoparticle Size and Surface Properties Determine the Protein Corona with Possible Implications for Biological Impacts." *Proceedings of the National Academy of Sciences* 105 (38): 14265–70. https://doi.org/10.1073/pnas.0805135105.

Mamot, Christoph, Reto Ritschard, Andreas Wicki, Gregor Stehle, Thomas Dieterle, Lukas Bubendorf, Christoph Hilker, Stefanie Deuster, Richard Herrmann, and Christoph Rochlitz. 2012. "Tolerability, Safety, Pharmacokinetics, and Efficacy of Doxorubicin-Loaded Anti-EGFR Immunoliposomes in Advanced Solid Tumours: A Phase 1 Dose-Escalation Study." *The Lancet Oncology* 13 (12): 1234–41. https://doi.org/10.1016/S1470-2045(12)70476-X.

Martínez, Guillermo, Manuel Merinero, María Pérez-Aranda, Eva María Pérez-Soriano, Tamara Ortiz, Belén Begines, and Ana Alcudia. 2021. "Environmental Impact of Nanoparticles' Application as an Emerging Technology: A Review." *Materials*. MDPI AG. https://doi.org/10.3390/ma14010166.

Matteis, Valeria De, and Rosaria Rinaldi. 2018. "Toxicity Assessment in the Nanoparticle Era." In *Advances in Experimental Medicine and Biology*, 1048:1–19. Springer New York LLC. https://doi.org/10.1007/978-3-319-72041-8_1.

Maynard, Andrew D., Robert J. Aitken, Tilman Butz, Vicki Colvin, Ken Donaldson, Günter Oberdörster, Martin A. Philbert, et al. 2006. "Safe Handling of Nanotechnology." *Nature* 444 (7117): 267–69. https://doi.org/10.1038/444267a.

Milani, Silvia, Francesca Baldelli Bombelli, Andrzej S. Pitek, Kenneth A. Dawson, and Joachim Rädler. 2012. "Reversible *versus* Irreversible Binding of Transferrin to Polystyrene Nanoparticles: Soft and Hard Corona." *ACS Nano* 6 (3): 2532–41. https://doi.org/10.1021/nn204951s.

Miura, Nobuhiko, and Yasushi Shinohara. 2009. "Cytotoxic Effect and Apoptosis Induction by Silver Nanoparticles in HeLa Cells." *Biochemical and Biophysical Research Communications* 390 (3): 733–37. https://doi.org/10.1016/j.bbrc.2009.10.039.

Pan, Yu, Annika Leifert, David Ruau, Sabine Neuss, Jörg Bornemann, Günter Schmid, Wolfgang Brandau, Ulrich Simon, and Willi Jahnen-Dechent. 2009. "Gold Nanoparticles of Diameter 1.4 Nm Trigger Necrosis by Oxidative Stress and Mitochondrial Damage." *Small* 5 (18): 2067–76. https://doi.org/10.1002/smll.200900466.

Pan, Yu, Sabine Neuss, Annika Leifert, Monika Fischler, Fei Wen, Ulrich Simon, Günter Schmid, Wolfgang Brandau, and Willi Jahnen-Dechent. 2007. "Size-Dependent Cytotoxicity of Gold Nanoparticles." *Small* 3 (11): 1941–49. https://doi.org/10.1002/smll.200700378.

Panyam, Jayanth, and Vinod Labhasetwar. 2003. "Biodegradable Nanoparticles for Drug and Gene Delivery to Cells and Tissue." *Advanced Drug Delivery Reviews* 55 (3): 329–47. https://doi.org/10.1016/S0169-409X(02)00228-4.

Park, Margriet V. D. Z., Daniëlle P. K. Lankveld, Hank van Loveren, and Wim H. de Jong. 2009. "The Status of *in Vitro* Toxicity Studies in the Risk Assessment of Nanomaterials." *Nanomedicine*. https://doi.org/10.2217/nnm.09.40.

Parvin, Nargish, Vineet Kumar, Sang Woo Joo, and Tapas Kumar Mandal. 2024. "Emerging Trends in Nanomedicine: Carbon-Based Nanomaterials for Healthcare." *Nanomaterials* 14 (13): 1085. https://doi.org/10.3390/nano14131085.

Patra, Biswajit, Rohit Gautam, Eepsita Priyadarsini, Paulraj Rajamani, Surya Narayan Pradhan, Muthupandian Saravanan, and Ramovatar Meena. 2020. "Piper Betle: Augmented Synthesis of Gold Nanoparticles and Its In-Vitro Cytotoxicity Assessment on HeLa and HEK293 Cells." *Journal of Cluster Science* 31 (1): 133–45. https://doi.org/10.1007/s10876-019-01625-5.

Prausnitz, Mark R., Samir Mitragotri, and Robert Langer. 2004. "Current Status and Future Potential of Transdermal Drug Delivery." *Nature Reviews Drug Discovery*. Nature Publishing Group. https://doi.org/10.1038/nrd1304.

Reddy, Anreddy Rama Narsimha, Yellu Narsimha Reddy, Devarakonda Rama Krishna, and Vurimindi Himabindu. 2010. "Multi Wall Carbon Nanotubes Induce Oxidative Stress and Cytotoxicity in Human Embryonic Kidney (HEK293) Cells." *Toxicology* 272 (1–3): 11–16. https://doi.org/10.1016/j.tox.2010.03.017.

Rittinghausen, S., U. Mohr, and D.L. Dungworth. 1997. "Pulmonary Cystic Keratinizing Squamous Cell Lesions of Rats after Inhalation/Instillation of Different Particles." *Experimental and Toxicologic Pathology* 49 (6): 433–46. https://doi.org/10.1016/S0940-2993(97)80131-5.

Sadrieh, Nakissa, Anna M. Wokovich, Neera V. Gopee, Jiwen Zheng, Diana Haines, David Parmiter, Paul H. Siitonen, et al. 2010. "Lack of Significant Dermal Penetration of Titanium Dioxide from Sunscreen Formulations Containing Nano- and Submicron-Size TiO2 Particles." *Toxicological Sciences* 115 (1): 156–66. https://doi.org/10.1093/toxsci/kfq041.

Sahoo, Sanjeeb K., and Vinod Labhasetwar. 2003. "Nanotech Approaches to Drug Delivery and Imaging." *Drug Discovery Today* 8 (24): 1112–20. https://doi.org/10.1016/S1359-6446(03)02903-9.

Stone, Vicki, Helinor Johnston, and Roel P. F. Schins. 2009. "Development of *in Vitro* Systems for Nanotoxicology: Methodological Considerations." *Critical Reviews in Toxicology* 39 (7): 613–26. https://doi.org/10.1080/10408440903120975.

Su, Yuanyuan, Yao He, Haoting Lu, Liman Sai, Qingnuan Li, Wenxin Li, Lianhui Wang, Pingping Shen, Qing Huang, and Chunhai Fan. 2009. "The Cytotoxicity of Cadmium Based, Aqueous Phase – Synthesized,

Quantum Dots and Its Modulation by Surface Coating." *Biomaterials* 30 (1): 19–25. https://doi.org/10.1016/j.biomaterials.2008.09.029.

Takeda, Ken, Ken-ichiro Suzuki, Aki Ishihara, Miyoko Kubo-Irie, Rie Fujimoto, Masako Tabata, Shigeru Oshio, Yoshimasa Nihei, Tomomi Ihara, and Masao Sugamata. 2009. "Nanoparticles Transferred from Pregnant Mice to Their Offspring Can Damage the Genital and Cranial Nerve Systems." *Journal of Health Science* 55 (1): 95–102. https://doi.org/10.1248/jhs.55.95.

Tang, Mingliang, Tairan Xing, Jie Zeng, Huili Wang, Chenchen Li, Shuting Yin, Dan Yan, et al. 2008. "Unmodified CdSe Quantum Dots Induce Elevation of Cytoplasmic Calcium Levels and Impairment of Functional Properties of Sodium Channels in Rat Primary Cultured Hippocampal Neurons." *Environmental Health Perspectives* 116 (7): 915–22. https://doi.org/10.1289/ehp.11225.

Tsuji, Joyce S., Andrew D. Maynard, Paul C. Howard, John T. James, Chiu Wing Lam, David B. Warheit, and Annette B. Santamaria. 2006. "Research Strategies for Safety Evaluation of Nanomaterials, Part IV: Risk Assessment of Nanoparticles." *Toxicological Sciences* 89 (1): 42–50. https://doi.org/10.1093/toxsci/kfi339.

Walkey, Carl D., Jonathan B. Olsen, Hongbo Guo, Andrew Emili, and Warren C.W. Chan. 2012. "Nanoparticle Size and Surface Chemistry Determine Serum Protein Adsorption and Macrophage Uptake." *Journal of the American Chemical Society* 134 (4): 2139–47. https://doi.org/10.1021/ja2084338.

Wang, Andrew Z., Robert Langer, and Omid C. Farokhzad. 2012. "Nanoparticle Delivery of Cancer Drugs." *Annual Review of Medicine* 63 (1): 185–98. https://doi.org/10.1146/annurev-med-040210-162544.

Wang, Lin, Dattatri K. Nagesha, Selvapraba Selvarasah, Mehmet R. Dokmeci, and Rebecca L. Carrier. 2008. "Toxicity of CdSe Nanoparticles in Caco-2 Cell Cultures." *Journal of Nanobiotechnology* 6 (October). https://doi.org/10.1186/1477-3155-6-11.

Zhang, Mingyi, Junjiang Jin, Ya Nan Chang, Xueling Chang, and Gengmei Xing. 2014. "Toxicological Properties of Nanomaterials." *Journal of Nanoscience and Nanotechnology*. https://doi.org/10.1166/jnn.2014.9198.

Zhao, Jinshun, and Vincent Castranova. 2011. "Toxicology of Nanomaterials Used in Nanomedicine." *Journal of Toxicology and Environmental Health - Part B: Critical Reviews*. https://doi.org/10.1080/10937404.2011.615113.

Chapter 5.3

Unraveling *In Vivo* Mechanisms of Nanoparticle Toxicity and Mitigation Strategies

Sonali Pardhiya[1,*]
Paulraj Rajamani[2]
and Rajeev Singh[1,†]
[1]Department of Environmental Science, Jamia Millia Islamia, New Delhi, India
[2]School of Environmental Sciences, Jawaharlal Nehru University, New Delhi, India

Abstract

Due to their distinctive properties, nanoparticles play a diverse role across numerous industries, including food, cosmetics, nutraceuticals, and pharmaceuticals, particularly in drug delivery. However, their increased usage raises concerns about potential toxicological repercussions. The chapter discusses the toxicity concerns associated with nanoparticles, focusing on various mechanisms that induce cellular damage. One primary pathway is generating reactive oxygen species (ROS) influenced by nanoparticle size, surface area, and composition. Smaller nanoparticles with higher surface-to-volume ratios generate more ROS, leading to oxidative stress, cytotoxicity, and DNA damage. Nanoparticles also disrupt the cytoskeleton by direct interaction during internalization and indirectly via ROS production, affecting cell motility, protein expression, and structure. Mitochondrial dysfunction is another critical mechanism, with nanoparticles impairing mitochondrial biogenesis, fusion, and fission, leading to increased ROS production and apoptosis. Nanoparticles induce endoplasmic reticulum stress by disrupting protein folding, triggering inflammation and apoptosis. Additionally, nanoparticles cause DNA damage through direct interaction and oxidative stress, contributing to genotoxicity. Epigenetic changes, such as DNA methylation and histone modification, are also induced by nanoparticles, potentially affecting gene expression and inheritance. This chapter further explores the impact of nanoparticles on vital organs, including the brain, lungs, liver, and kidneys, highlighting the various pathways through which nanoparticles cause cellular and tissue damage. The chapter concludes with approaches to mitigate nanoparticle toxicity, such as modifying size, shape, surface coatings, doping, grafting, and loading, aiming to enhance nanoparticle safety in biomedical applications.

---

[*] Corresponding Author's Email: sonalipardhiya111@gmail.com
[†] Corresponding Author's Email: rsingh5@jmi.ac.in

In: Emerging Nanomaterials in Biomedical Science
Editors: Eepsita Priyadarshini and Ajith Manayil Parambil
ISBN: 979-8-89530-600-0

Keywords: nanoparticle toxicity, reactive oxygen species, human health, nanoparticle toxicity, mitigation

1. Introduction

Due to their unique properties, nanoparticles (NPs) have revolutionized various industries, including food, cosmetics, nutraceuticals, and pharmaceuticals, particularly in drug delivery (Bajpai et al. 2018). These minute particles, frequently less than 100 nanometers in size, provide benefits like improved bioavailability, targeted delivery, and enhanced stability of active compounds. However, the increased use of NPs raises significant concerns regarding their potential toxicological effects (Kumah et al. 2023). Nanoparticle toxicity primarily arises from cellular injury, which can manifest as reversible damage, irreversible impairment, or even cell death. The interaction of nanoparticles with biological systems can trigger oxidative stress, inflammation, and genotoxicity, posing risks to human health and the environment. Understanding nanoparticle toxicity mechanisms is crucial for developing effective mitigation strategies. Approaches such as modifying nanoparticle size and shape, applying coatings, doping, grafting, ligating, and incorporating antioxidants enhance therapeutic potential while minimizing adverse effects. These strategies are essential for creating safe and effective biomedical applications.

2. Toxicity Concerns of NPs

2.1. Nanotoxicity through the Generation of Reactive Oxygen Species

NP toxicity becomes evident through various pathways, particularly involving reactive oxygen species (ROS) production. ROS are extremely reactive oxygen-based molecules capable of inducing cellular damage when uncontrolled. ROS plays crucial roles in cellular processes such as survival, apoptosis, signaling, and inflammation (Checa and Josep 2020). The production of ROS by NPs is influenced by their size, surface area, chemical composition, and internalization into cells. Smaller NPs, with larger surface-to-volume ratios, have more reactive sites, facilitating ROS generation (Canaparo et al. 2020). ROS generates oxidative stress, a crucial element in nanotoxicity, and contributes to changes in cell motility, cytotoxicity, dysregulated cell signaling, DNA damage, apoptosis, cancer proliferation, and metastasis (Abdal Dayem et al. 2017).

Intermediate-sized NPs interact with cell membranes to enter cells through endocytosis, which can be caveolin- or clathrin-dependent, or independent (Skotland et al. 2021, Villanueva-Flores et al. 2020). Small NPs often exhibit cytotoxicity by inducing oxidative stress, leading to cytoskeletal disruption, mitochondrial damage, and DNA impairment. Larger NPs, however, are internalized via micropinocytosis (Figure 1). It is important to consider that NP material and cell lines significantly influence internalization processes.

Oxidative stress from NP exposure is categorized into primary and secondary types. Primary oxidative stress involves direct ROS generation through mechanisms like Titanium dioxide (TiO_2) photocatalysis or metal ion reactions. Secondary oxidative stress arises from NP-induced mitochondrial dysfunction, disrupting the electron transport chain and increasing ROS production. The accumulation of NPs in mitochondria leads to depolarization of the mitochondrial membrane and inhibition of the electron transport chain, enhancing ROS-mediated cytotoxicity. Additionally, NPs activate immune cells via ROS-dependent mechanisms mediated by NOX activation, further contributing to oxidative stress (Møller et al. 2014b).

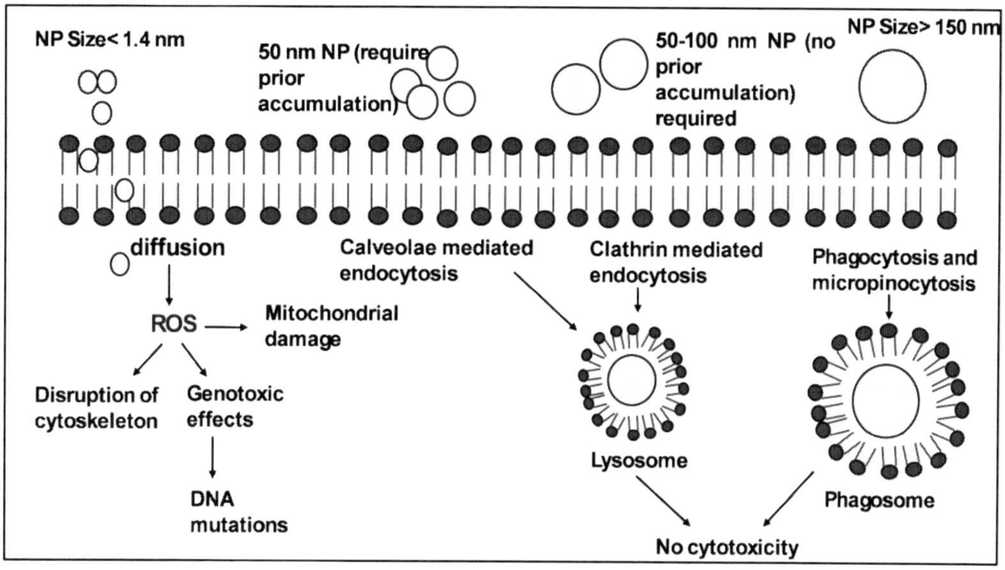

Figure 1. Impact of NP size on cellular internalization mechanisms and cytotoxicity.

2.2. Cytoskeletal Damage by NPs

NPs interact with the cell cytoskeleton directly during internalization and indirectly by generating ROS. These interactions can lead to adenosine triphosphate (ATP) (the molecule that supplies energy to drive and support numerous processes in living cells) reduction, lipid alterations, and changes in cytoskeletal protein expression (Shim et al. 2012). Initially, NPs engage with the cytoskeleton via clathrin- and caveolae-mediated endocytosis, involving actin, plasma membrane receptors, clathrin, and adapter proteins (Wang et al. 2009). Previous investigations have shown that metal oxide NPs, such as TiO_2, can disrupt the cytoskeleton by breaking down actin and tubulin networks (Vuong et al. 2016, Thai et al. 2015). Proteomic studies indicate changes in proteins associated with cell motility after TiO_2 exposure, while carbon black NPs can dysregulate migratory cell proteins, though the affected pathways differ (Vuong et al. 2016). Microarray analysis of BEAS-2B cells (Human Bronchial Epithelial Cell Line) treated with TiO_2 NPs indicated alterations in mRNA (messenger RNA) and miRNA (microRNA- small, single-stranded non-coding RNA molecules) expression related to the cytoskeleton, influencing both cell adhesion and motility (Thai et al. 2015).

NP shape also affects cytoskeletal interactions. For instance, HeLa cells (cervical cancer cell line) showed greater uptake of cylindrical Silicon dioxide (SiO_2) NPs compared to spherical ones, involving Rac1 activation and F-actin stress fiber production (Meng et al. 2011). This suggests different NP shapes induce varying cytoskeletal responses.

Indirectly, NPs can affect the cytoskeleton through oxidative stress. A549 cells (lung cells) exposed to low SiO_2 concentrations showed gene expression changes (Wang et al. 2009). Silver (Ag) NPs stimulate cytokeratin 8 and 18 and gelsolin expression while downregulating actin and tubulin in co-cultured intestinal epithelial cells, indicating cytoskeletal disruption (Georgantzopoulou et al. 2015). In human keratinocyte HaCaT cells, SiO_2 NPs reduced cell survival, caused morphological changes, cell cycle arrest, and altered expression of stress response and cytoskeletal proteins (Liu et al. 2017). Zinc oxide (ZnO) NPs led to actin rearrangement and tubulin network bundling around the nucleus, with abnormal spindles and chromosomes scattered throughout the cytoplasm (Sudhakaran et al. 2019). Organic NPs like dendrimers also impact the cytoskeleton. Breast cancer cells quickly internalize hyperbranched block copolymer micelles via clathrin and macropinocytosis pathways (Zeng et al. 2012). Sixth-generation cationic dendrimers interact with actin filaments, affecting polymerization depending on concentration (Ruenraroengsak and Alexander 2010).

Comprehending both the direct and indirect impacts of NPs on the cytoskeleton is essential for developing safer nanomaterials in biomedical applications and evaluating the potential risks associated with NP exposure.

2.3. Effect of NPs on Mitochondria

Mitochondria, vital for energy production, are key targets of NP toxicity, leading to various disorders such as neurological, cardiac, and immune dysfunctions. Mitochondrial dysfunction is a critical mechanism underlying NP toxicity. For example, carbon black Printex 90 hinders the differentiation of mesenchymal stem cells into osteogenic cells by interfering with mitochondrial biogenesis and dynamics (Shen et al. 2018). Likewise, silicon NPs disrupt mitochondrial fusion, fission, and biogenesis in endothelial cells (Guo et al. 2018a), whereas silver NPs cause neurodevelopmental toxicity through mitochondrial dysfunction (Yamada et al. 2018).

Mitochondrial integrity is crucial for the electron transfer chain (ETC) activity. NPs like polyhydroxylated fullerene and nano-TiO_2 disrupt the ETC, increase ROS, and cause mitochondrial membrane potential (MMP) collapse (Yang et al. 2016, Nichols et al. 2018). This leads to cardiac dysfunction and alters mitochondrial respiratory function. Furthermore, ROS are intricately associated with mitochondrial impairment and programmed cell death. ZnO NPs triggered an excessive ROS generation, activating the mitochondrial-mediated apoptosis pathway in zebrafish embryos (Zhao et al. 2016).

Cadmium sulfide (CdS) quantum dots affect genes related to apoptosis, oxidative stress, autophagy, and mitochondrial function, serving as biomarkers for toxicity (Paesano et al. 2016). NPs like TiO_2 can synergistically enhance the toxicity of other compounds via mitochondria-dependent apoptosis (Wang et al. 2017). Ag NPs cause structural alterations in mitochondria, leading to dysfunction, as observed in mice (Ansari et al. 2016) and fish cells (Bermejo-Nogales et al. 2016). QDs disrupt MMPs, impair respiration, and decrease ATP synthesis (Pasquali et al. 2017). Superparamagnetic iron oxide NPs, used in cancer therapy, induce ROS formation and cancer cell apoptosis by affecting mitochondrial functions (He et al. 2016). Hydroxyapatite NPs inhibit mitochondrial respiratory chain complexes, increasing ROS levels (Xue at al. 2017).

NPs trigger mitochondrial dysfunction through ROS generation, MMP collapse, and impaired biogenesis, leading to apoptosis and various cellular dysfunctions. This highlights the importance of mitochondrial integrity in assessing NP-induced cytotoxicity.

2.4. Effect of NPs on Endoplasmic Reticulum

The endoplasmic reticulum (ER) is crucial for protein folding and assembly, lipid and sterol synthesis, calcium homeostasis, and clearing misfolded or unassembled proteins via ER-associated protein degradation (ERAD), maintaining internal homeostasis. Typically, there is a delicate equilibrium between the protein load in the ER and its folding capacity. Nevertheless, when ERAD is unsuccessful, or stress hinders protein folding, misfolded proteins accumulate, resulting in ER stress.

TiO_2 NPs induce ER stress and inflammation in mouse liver, leading to insulin resistance (Hu et al. 2018). ER stress, typically an adaptive mechanism, can lead to cell death through necrosis or apoptosis if prolonged. Several NPs, like Ag NPs (Zhang et al. 2012), ZnO NPs (Yang et al. 2015), and Si NPs (Guo et al. 2018b), induce ER stress-mediated apoptosis in various cells.

The endoplasmic reticulum (ER) and mitochondria work together to maintain calcium balance. ER stress can increase Ca^{2+} transfer to mitochondria, disrupting mitochondrial homeostasis and triggering apoptosis. Ag NPs, for example, enhance Ca^{2+} transfer, disrupting mitochondrial function and inducing apoptosis in neuroblastoma cells (Li et al. 2018).

2.5. DNA Damage

Research into particle genotoxicity traces back to Doll's seminal work 1955 on asbestos carcinogenicity (Doll 1955). Genotoxic mechanisms are classified into primary and secondary categories. Primary genotoxicity encompasses direct and indirect effects. Direct primary genotoxicity involves particles interacting directly with genetic material upon entering the cytosol or nuclear membrane. This interaction can induce ROS/RNS species, generating free radicals that directly damage DNA or indirectly increase mitochondrial and NADPH oxidase-produced radicals, leading to oxidative stress and subsequent DNA damage. Secondary genotoxicity is associated with inflammatory responses, where the recruited macrophages and PMN (Polymorphonuclear neutrophil) cells also produce reactive radicals, contributing to DNA damage.

Research indicates that nanoparticles like ZnO, SiO_2, and TiO_2 trigger the generation of reactive oxygen and nitrogen species, leading to oxidative damage to DNA (Shukla et al. 2011). NPs have been observed to enter cellular compartments, including the nucleus, where they interact directly with DNA and associated proteins. This size-dependent entry into specific cellular locations underscores their potential for primary genotoxic effects.

Robinson (2008) proposed that NPs activate innate immune responses in phagocytes, increasing cellular oxygen consumption and releasing $O^{\cdot-}$ and H_2O_2 through the NADPH (nicotinamide adenine dinucleotide phosphate hydrogen)-oxidase system. Scientists have proposed three main theories to explain the genotoxicity caused by NPs: surface interactions, the release of ROS/RNS due to transition metal ions, and the activation of membrane receptors, such as the epidermal growth factor receptor, by these metals.

2.6. Effect of NPs on Epigenome

Recently, scientists have acknowledged that disruptions in a cell's epigenome can harm an individual. These epigenetic changes are stable and inheritable, potentially impacting future generations. Concerns arise regarding the impact of NPs on the epigenome. Epigenetic alterations include DNA methylation and hydroxymethylation, histone modifications, chromatin remodeling, RNA methylation, and the involvement of non-coding RNAs (ncRNAs).

NPs have been shown to induce specific alterations in DNA methylation. Studies on various NPs like SWCNTs (single-walled carbon nanotubes)/ MWCNTs (multi-walled carbon nanotubes), TiO_2, Copper oxide (CuO), and SiO_2 have revealed both locus-specific and global DNA hypomethylation effects (Lu et al. 2016). For instance, exposure to SiO_2 NPs reduced global DNA methylation levels by more than 20% in human keratinocyte cell lines, attributed to decreased levels of DNMTs (DNA methyl transferases) (Gong et al. 2010). *In vivo* studies on mice exposed to MWCNTs also indicated global DNA hypomethylation in lung and blood samples (Brown et al. 2016). Conversely, some studies have reported DNA hypermethylation in human lung carcinoma cells after MWCNT exposure, highlighting contrasting outcomes (Li et al. 2016).

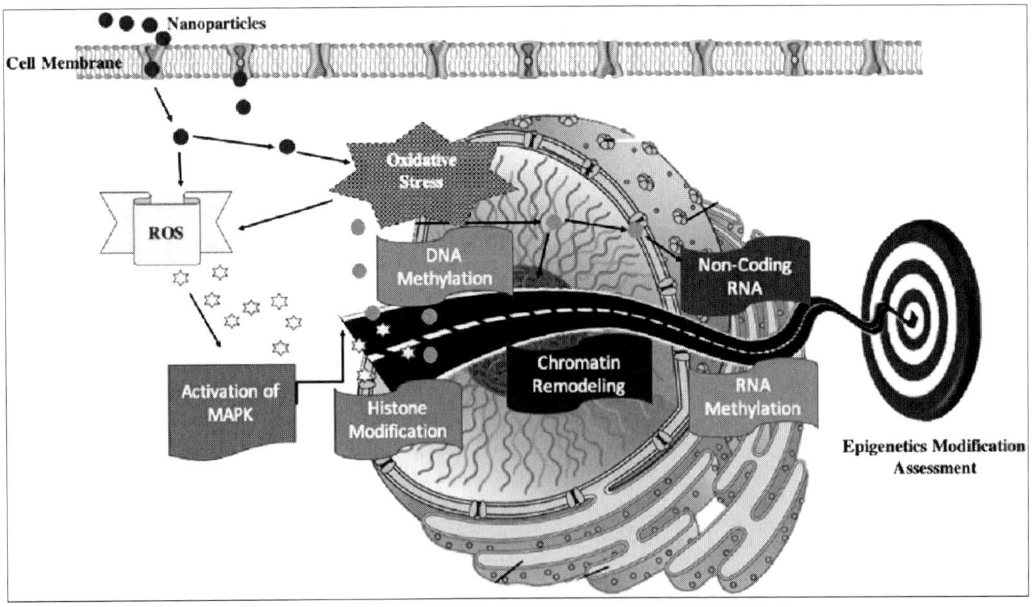

Figure 2. Potential Pathways Illustrating the Impact of Nanoparticles on Epigenome (Reproduced with permission from Shukla et al. Copyright 2021, Frontiers in Genetics).

Additionally, NPs can influence histone modifications, altering chromatin packaging and gene expression control. Studies have demonstrated that NPs cause chromatin condensation and global H3 hypoacetylation in breast cancer cells treated with cadmium telluride quantum dots (Choi et al. 2008).

Furthermore, NPs affect non-coding RNAs associated with epigenetic modifications, although research in this area is limited. Studies have shown altered miRNA expression patterns in mouse lungs exposed to nano-titanium dioxide and carbon black nanoparticles (Halappanavar et al. 2011).

In summary, NPs significantly impact the epigenome, affecting DNA methylation, histone modifications, and non-coding RNA expression (Figure 2), yet further research is crucial to understand these mechanisms fully.

3. Effect of NPs on Vital Organs

3.1. The Effect of NPs on the Brain

NPs can cross the blood-brain barrier due to their small size. Once inside the brain, NPs disrupt cellular processes, triggering microglial activation and the release of pro-inflammatory cytokines such as tumor necrosis factor-alpha (TNF-α), interleukin-1β (IL-1β), and interleukin-6 (IL-6), which exacerbate neuroinflammation and neuronal damage. They also interfere with proteostasis, causing protein misfolding and aggregation, contributing to neurodegenerative disorders (Zhao et al. 2020). Studies have shown that TiO_2-NPs and carbon-NPs can induce oxidative stress, glial proliferation, and markers of neurodegeneration in mice (Zhe et al. 2014, Onoda et al. 2017).

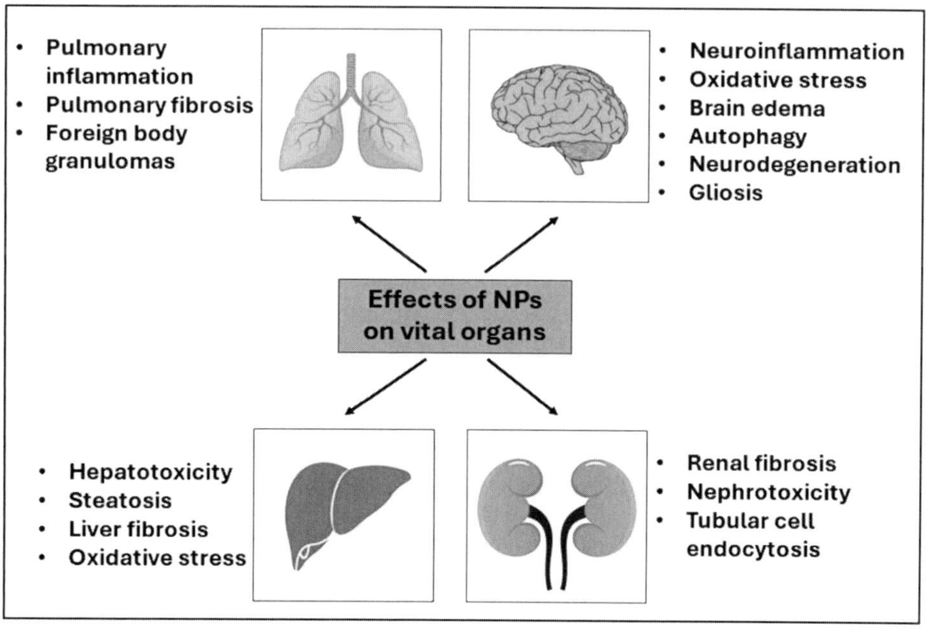

Figure 3. Effect of NPs on vital organs.

3.2. Effect of NPs on Lungs

NPs can induce complex and varied toxicity pathways in the lungs. Key mechanisms include ROS generation, inflammation, apoptosis, necrosis, pulmonary fibrosis, and genotoxicity. ROS production upon NP entry triggers oxidative stress, leading to cellular damage and inflammation. Inflammatory responses are stimulated by NP activation of immune cells like macrophages, releasing cytokines that further damage lung tissue. NP-induced apoptosis and necrosis contribute to tissue injury and impaired lung function. Shvedova et al. studied the impact of silica nanoparticles on mice, revealing their ability to trigger lung inflammation and fibrosis, underscoring the occupational hazards associated with exposure to silica nanoparticles (Shvedova et al. 2009). Research on various NPs such as titanium and cerium highlight their potential lung toxicity, emphasizing the need for further investigation into NP-induced lung damage (Park et al. 2008, Hussain et al. 2012).

3.3. Effect of NPs on Liver

The liver is a major site for the accumulation of NPs, especially metallic ones, impacting both hepatocytes, which make up 60% of liver cells, and non-parenchymal cells, such as sinusoidal endothelial cells, hepatic stellate cells, and Kupffer cells, which constitute 35% of liver cells. This accumulation alters liver cell structure and function (Pardhiya et al. 2022), leading to hepatotoxicity through mechanisms involving inflammation, oxidative stress, and programmed cell death. NP exposure causes increased lipid peroxide levels and decreased antioxidant enzyme activities. NP-induced ROS production further exacerbates liver inflammation and apoptosis, affecting metabolic pathways and immune responses.

3.4. Effect of NPs on Kidney

NPs pose significant risks to the kidneys due to their potential accumulation and toxicity after systemic circulation. Renal vulnerability stems from the organ's high blood flow, exposing renal cells to NPs like silver, titanium dioxide, carbon nanotubes, and quantum dots (Sengul and Asmatulu 2020). Mechanisms of nephrotoxicity involve oxidative stress through ROS, inflammation via cytokine release, fibrosis from activated fibroblasts, and apoptosis triggered by caspase activation.

4. Approaches to Alleviate the Adverse Effects of NPs

Various safety strategies to reduce nanotoxicity without compromising NP performance include altering size and shape, coating, doping, loading, and grafting NPs, ligating groups to NPs, and adding antioxidants. These approaches are summarized below:

4.1. Size and Shape of NPs

The cytotoxicity of nanocarriers is notably affected by their particle size. Smaller NPs provide a larger surface area for interaction with cellular components like proteins, nucleic acids, and lipids per unit mass, potentially increasing toxicity due to enhanced ion release and cellular damage (Długosz et al. 2020). Larger particles accumulate in organs such as the liver and spleen, while smaller NPs are more likely to localize in the kidneys and lungs. Wang et al. observed this pattern in an animal model, noting that 20 nm gold nanoparticles accumulated less in the spleen and kidneys compared to 40 nm particles. In contrast, accumulation was higher in the lungs for the smaller particles (Wang et al. 2021). Similarly, studies by Caster et al. and Seiffert et al. showed that smaller NPs were more toxic to specific organs and induced more proinflammatory responses than larger counterparts (Caster et al. 2017, Seiffert et al. 2015).

The shape of NPs also plays a critical role in their toxicity profile. Spherical NPs are generally considered safer than non-spherical shapes of the same composition, possibly due to their better cellular uptake characteristics. For instance, rod-shaped Ferric oxide (Fe_2O_3) nanoparticles were reported to induce stronger cytotoxic responses in murine macrophages compared to spherical counterparts (Lee et al. 2014). Similarly, rod-shaped Cerium oxide (CeO_2) nanoparticles exhibited greater toxicity than octahedral or cubic shapes (Forest et al. 2017).

Choosing the right NP shape is crucial for optimizing their performance in biological systems, considering their interaction with cells and potential toxic effects. However, it's essential to balance shape optimization with other factors to ensure overall effectiveness and safety in biological applications.

4.2. Coating on NPs

Surface coatings reduce toxicity in diverse applications, including drug delivery and medical implants. These coatings act as protective layers applied to the surface of materials or products, shielding them from potential environmental contaminants or degradation. Metallic NPs, due to their propensity to disperse ions throughout the body or release them into the environment, can pose health risks (Sánchez-López et al. 2020). Variables like particle size, shape, surface coatings, and environmental changes can affect the release of metal ions from nanoparticles, impacting cell viability by inducing oxidative stress and cytokine release. The chemical composition of nano-metallic carrier surfaces allows for the development of biocompatible coatings tailored for interactions with human cells. These coatings have two main functions: they protect the metallic nanoparticle cores from oxidation and facilitate the attachment of functional molecules like therapeutic agents, targeting ligands, or contrast agents (Zhang et al. 2022).

Surface modification through coating is a widely used strategy to mitigate NP toxicity, often achieved through reversible non-covalent alterations (Hwang et al. 2018). Coatings modify the dispersion of NPs, which greatly affects their bioavailability and potential toxicity (Wang et al. 2011). Various materials are employed for coating, including polyethylene glycol (PEG), polyvinyl pyrrolidone (PVP), zwitterionic polymers, poly (N-isopropyl acrylamide) (PNIPAM), and polyvinyl alcohol (PVA) (Debayle et al. 2019). Among these, PEG is particularly favored for its ability to reduce cellular uptake, prolong circulation duration, and shield surface charges of NPs (Patlolla et al. 2019).

Studies have demonstrated the effectiveness of surface coatings in reducing NP toxicity. For instance, in animal models, PEG-coated silver NPs showed fewer hepatotoxic effects than their uncoated counterparts (Karakoti et al. 2008). Hydrophobic coatings on zinc NPs also reduced toxicity, potentially impairing bioavailability (Lai et al. 2021). Coatings like PLGA and proteins have been utilized to prevent the accumulation of autophagosomes caused by metal-based NPs, enhancing biocompatibility. Silica coatings have been shown to reduce the toxicity of zinc NPs by preventing the core from dissolving into zinc ions in neutral and acidic environments, all while preserving their antibacterial properties (Chia and Leong 2016).

In summary, surface coatings offer versatile solutions to enhance the safety and efficacy of NPs in biomedical applications by modifying their chemical properties, improving biocompatibility, and reducing toxicity.

4.3. Doping

Doping is a key method for altering the energy band gap of NPs, which can help reduce the generation of ROS and oxidative stress. The energy band gap, which defines the separation between the valence and conduction bands, plays a crucial role in determining the stability and performance of nanomaterials, affecting their conductivity and related processes (Sharma et al. 2023). Doping alters the band gap by introducing foreign atoms into the NP lattice, potentially mitigating surface charge and binding properties. A reduced band gap could decrease the formation of harmful polymorphs and irradiated by-products, enhancing NP safety profiles.

Numerous studies have validated the efficacy of doping in nanomaterials. For example, Hwang et al. proposed doping titanium NPs with iron (Fe) to modify the band gap, reducing ROS production under solar radiation (Luque and Varma 2012). They noted that varying Fe concentrations and synthesis methods influenced ROS generation. Similarly, Ag-doped TiO_2 NPs have the potential to selectively target and kill cancer cells while leaving normal cells unharmed, thus regulating the cytotoxicity of TiO_2 NPs (Ahamed et al. 2017).

In summary, doping offers a strategic avenue to tailor NP properties, particularly the energy band gap, to minimize oxidative stress and improve safety in diverse applications. Careful selection of doping materials and methodologies can effectively modulate NP behavior, reducing potential environmental and biological risks associated with nanomaterial exposure.

4.4. Grafting and Loading

Grafting and loading are two distinct processes used to modify NPs and mitigate their toxicity, particularly in metallic NPs. Grafting involves the covalent attachment of other compounds to the NP surface, whereas loading refers to non-covalently bonding molecules onto NPs. Both methods effectively alter the toxicological profile of NPs by neutralizing their surface.

Reports indicate that surface neutralization through grafting or loading can significantly impact the behavior of NPs (Pounraj et al. 2018). Polymers commonly used for these purposes include fused rings, mirror structures, and cyclo-cages, which feature protective lobed channels. These structures shield the underlying NPs from deeper layer irradiation, enhancing their stability and reducing cytotoxicity. Improving NP performance with organic molecules facilitates active drug delivery, medical imaging, and tissue repair and regeneration (Ermis et al. 2022). For instance, grafting TiO_2-NPs with ligated cyclo-cages exemplifies a safer-by-design approach to nanostructure formation. This adjustment decreases pyrolysis and cytotoxicity in human Caco-2 cell lines (a colorectal adenocarcinoma cell line) while increasing intracellular uptake compared to nanoparticles without ligands. Additionally, it markedly enhances the stability of these nanostructures and minimizes free radical formation (Das et al. 2021). Similarly, grafting small organic molecules onto carbon nanotubes can reduce their cytotoxicity and improve cells' durability while greatly boosting their theranostic activity. (Wang et al. 2015).

Grafting and loading are pivotal techniques for modifying NPs, enhancing their functionality, and minimizing their potential adverse effects in various biomedical applications.

4.5. Ligation

Certain NPs can be functionally altered by binding with different compounds through various methods. These modifications are essential for mitigating toxicity risks by preventing the unintended breakdown and metabolism of the NPs. They form a protective layer around the NPs, which boosts their therapeutic effectiveness *in vivo* by targeting specific diseases and enhancing bioavailability (Zangabad et al. 2017).

Several ligated entities are available for modifying NPs, including Lactoferrin, glutathione, bovine serum albumin, and vitamin E TPGS (tocopheryl polyethylene glycol succinate). The ligation process is particularly recommended when NPs are prone to biodegradation in the body, ensuring sensitivity and safety (Huerta-Ángeles et al. 2018). Petković et al. studied titanium NPs bound with Zn-acetate as a targeting agent, finding that they produced fewer mono-oxidated radicals in HeGP2 hepatoma cell lines (Petković et al. 2011). This safer approach to developing ligated NPs significantly enhanced their stability in commercial sunscreens and improved their therapeutic effectiveness. Furthermore, Akhtar et al. found that substituting the cationic amino terminal on polyamidoamine NPs with an anionic carboxyl terminal reduced toxicity (Akhtar et al. 2019).

In summary, modifying NPs through ligation with specific moieties offers a promising strategy to mitigate toxicity, improve stability, and enhance therapeutic outcomes in various biomedical applications.

4.6. Addition of Antioxidant

NPs have been linked to inflammation, DNA damage, denaturation of proteins and lipids, and even mutagenesis because they can trigger oxidative stress. To counteract these effects, incorporating antioxidants has been proposed as a potential therapeutic approach. Antioxidants are compounds that neutralize highly reactive free radicals, thereby stopping the chain of biological damage caused by oxidative stress (Li et al. 2013).

Antioxidants can be incorporated into NPs through various mechanisms. One method is to coat NPs with antioxidants, chemically or physically attaching them to the NP surface, which helps prevent ROS formation and subsequent oxidative stress. Another approach is to incorporate antioxidants directly into the NP formulation during synthesis, ensuring they remain active throughout the NP's lifecycle. Additionally, pre-treating cells or the environment with antioxidants before exposing them to NPs can further reduce oxidative damage (Lai et al. 2021, Møller et al. 2014a).

Numerous studies have demonstrated the efficacy of antioxidants in reducing NP toxicity. For example, in animal models exposed to NPs, vitamins E and C have demonstrated the ability to decrease oxidative stress and inflammation. Curcumin, a natural antioxidant, has also proven effective in mitigating lung damage caused by NPs through its antioxidant properties (Tomás-Zapico and Coto-Montes 2005). Moreover, antioxidants such as N-acetylcysteine and glutathione have been recognized for their potential in shielding against NP-induced toxicity. N-acetylcysteine acts as a precursor to glutathione, an essential intracellular antioxidant. Both substances have been demonstrated in various cell and animal studies to alleviate oxidative stress and inflammation caused by NPs (Kwak et al. 2003).

While co-administration of antioxidants with NPs holds promise in reducing toxicity, several considerations are crucial. The selection of the appropriate antioxidant is paramount to achieving reduced NP toxicity without compromising therapeutic efficacy. Furthermore, careful attention to antioxidant dosage is essential, as inadequate doses may fail to effectively neutralize ROS, while excessive doses could lead to unintended consequences or inefficacy against oxidative stress (Li et al. 2020). Thus, optimizing the strategy for antioxidant co-administration with NPs is vital for maximizing therapeutic benefits while minimizing potential risks.

Conclusion and Perspective

The study of NPs has evolved significantly since the early 2000s, with initial research identifying oxidative stress caused by mechanisms like TiO_2 photocatalysis and metal ion reactions. By 2008, it was proposed that NPs activate innate immune responses, leading to increased ROS production. Between 2011 and 2015, studies revealed that metal oxide NPs disrupt cytoskeletal networks and NP shape affects cellular interactions. By 2016, research demonstrated proteomic and genomic alterations due to NPs, including DNA methylation changes. From 2017 to 2018, studies emphasized mitochondrial dysfunction and its role in neurotoxicity and proteostasis disruption. By 2010 coating techniques, such as PEG and silica, emerged to reduce toxicity, while doping methods were developed to modify energy band gaps and reduce oxidative stress. Grafting and loading techniques were advanced to enhance NP stability and functionality, and ligation strategies were refined to improve targeting and safety. Current trends focus on personalized approaches, multifunctional coatings, advanced synthesis techniques, and increasing attention to regulatory and safety considerations.

NPs are revolutionizing biomedicine by offering innovative therapeutic solutions to enhance human health. However, ensuring their safety within biological systems remains a significant challenge. Certain NPs, especially metal oxides, cause organ toxicity via ROS generation, neuroinflammation, and cellular interactions, varying by shape and size.

There are various methods to prevent NPs from interacting with cells, thereby minimizing their effects on biological systems. Modifying the physicochemical properties of NPs can significantly mitigate NP-related toxicity, ensuring a justified risk-benefit assessment. The primary protective strategy involves using biodegradable or biocompatible materials for NP development. Additional modifications, such as adjusting size, shape, morphology, and surface functionalization (e.g., surface coating, PEGylation, and stealthing), as well as ligation and co-administration of chelating agents and antioxidants, can further reduce toxicity. This chapter underscores the various toxicological pathways affecting essential human organs and organelles and summarizes factors that can diminish the adverse effects of NPs.

Advancements in NP research have significantly expanded our understanding of their toxicological mechanisms and have led to the development of various strategies to mitigate adverse effects. Future directions will likely focus on refining these mitigation strategies to enhance NP safety and efficacy. This includes further exploring the molecular mechanisms underlying NP-induced toxicity, developing more sophisticated and biocompatible coatings, and optimizing the size and shape of NPs for specific applications. Additionally, advancements in computational modeling and high-throughput screening will enable more efficient prediction and assessment of NP toxicity. By integrating these approaches, future research aims to maximize the therapeutic potential of NPs while minimizing their risks, thereby supporting their safe and effective use in diverse biomedical applications.

References

Abdal Dayem, Ahmed, Mohammed Kawser Hossain, Soo Bin Lee, Kyeongseok Kim, Subbroto Kumar Saha, Gwang-Mo Yang, Hye Yeon Choi, and Ssang-Goo Cho. "The role of reactive oxygen species (ROS) in the biological activities of metallic nanoparticles." *International journal of molecular sciences* 18, no. 1 (2017): 120.

Ahamed, Maqusood, MA Majeed Khan, Mohd Javed Akhtar, Hisham A. Alhadlaq, and Aws Alshamsan. "Ag-doping regulates the cytotoxicity of TiO2 nanoparticles via oxidative stress in human cancer cells." *Scientific reports* 7, no. 1 (2017): 17662.

Ansari, Mohammad Azam, Arun Kumar Shukla, Mohammad Oves, and Haris M. Khan. "Electron microscopic ultrastructural study on the toxicological effects of AgNPs on the liver, kidney and spleen tissues of albino mice." *Environmental toxicology and pharmacology* 44 (2016): 30-43.

Bajpai, Vivek K., Madhu Kamle, Shruti Shukla, Dipendra Kumar Mahato, Pranjal Chandra, Seung Kyu Hwang, Pradeep Kumar, Yun Suk Huh, and Young-Kyu Han. "Prospects of using nanotechnology for food preservation, safety, and security." *Journal of food and drug analysis* 26, no. 4 (2018): 1201-1214.

Bermejo-Nogales, Azucena, Marta Fernández, María Luisa Fernández-Cruz, and J. M. Navas. "Effects of a silver nanomaterial on cellular organelles and time course of oxidative stress in a fish cell line (PLHC-1)." *Comparative Biochemistry and Physiology Part C: Toxicology & Pharmacology* 190 (2016): 54-65.

Brown, Traci A., Joong Won Lee, Andrij Holian, Virginia Porter, Harley Fredriksen, Minju Kim, and Yoon Hee Cho. "Alterations in DNA methylation corresponding with lung inflammation and as a biomarker for disease development after MWCNT exposure." *Nanotoxicology* 10, no. 4 (2016): 453-461.

Canaparo, Roberto, Federica Foglietta, Tania Limongi, and Loredana Serpe. "Biomedical applications of reactive oxygen species generation by metal nanoparticles." *Materials* 14, no. 1 (2020): 53.

Caster, Joseph M., K. Yu Stephanie, Artish N. Patel, Nicole J. Newman, Zachary J. Lee, Samuel B. Warner, Kyle T. Wagner et al. "Effect of particle size on the biodistribution, toxicity, and efficacy of drug-loaded polymeric nanoparticles in chemoradiotherapy." *Nanomedicine: Nanotechnology, Biology and Medicine* 13, no. 5 (2017): 1673-1683.

Checa, Javier, and Josep M. Aran. "Reactive oxygen species: drivers of physiological and pathological processes." *Journal of Inflammation research* (2020): 1057-1073.

Chia, S. L., & Leong, D. T. (2016). Reducing ZnO nanoparticles toxicity through silica coating. *Heliyon, 2*(10).

Choi, Angela O., Shelley E. Brown, Moshe Szyf, and Dusica Maysinger. "Quantum dot-induced epigenetic and genotoxic changes in human breast cancer cells." *Journal of molecular medicine* 86 (2008): 291-302.

Das, Partha Pratim, Vijay Chaudhary, Furkan Ahmad, and Ankit Manral. "Effect of nanotoxicity and enhancement in performance of polymer composites using nanofillers: A state-of-the-art review." *Polymer Composites* 42, no. 5 (2021): 2152-2170.

Debayle, Manon, Elie Balloul, Fatimata Dembele, Xiangzhen Xu, Mohamed Hanafi, Francois Ribot, Cornelia Monzel et al. "Zwitterionic polymer ligands: an ideal surface coating to totally suppress protein-nanoparticle corona formation?." *Biomaterials* 219 (2019): 119357.

Długosz, Olga, Krzysztof Szostak, Anita Staroń, Jolanta Pulit-Prociak, and Marcin Banach. "Methods for reducing the toxicity of metal and metal oxide NPs as biomedicine." *Materials* 13, no. 2 (2020): 279.

Doll, Richard. "Mortality from lung cancer in asbestos workers." *British journal of industrial medicine* 12, no. 2 (1955): 81.

Ermis, Esra, Zeinab Bagheri, Ebrahim Behroodi, Hamid Latifi, Mahban Rahimifard, and Elham Ajorlou. "Red emissive N–S co-doped carbon dots for live imaging of tumor spheroid in the microfluidic device." *Journal of Science: Advanced Materials and Devices* 7, no. 2 (2022): 100404.

Forest, Valérie, Lara Leclerc, Jean-François Hochepied, Adeline Trouvé, Gwendoline Sarry, and Jérémie Pourchez. "Impact of cerium oxide nanoparticles shape on their *in vitro* cellular toxicity." *Toxicology in Vitro* 38 (2017): 136-141.

Georgantzopoulou, Anastasia, Tommaso Serchi, Sébastien Cambier, Céline C. Leclercq, Jenny Renaut, Jia Shao, Marcin Kruszewski et al. "Effects of silver nanoparticles and ions on a co-culture model for the gastrointestinal epithelium." *Particle and fibre toxicology* 13 (2015): 1-17.

Gong, Chunmei, Gonghua Tao, Linqing Yang, Jianjun Liu, Qingchen Liu, and Zhixiong Zhuang. "SiO2 nanoparticles induce global genomic hypomethylation in HaCaT cells." *Biochemical and biophysical research communications* 397, no. 3 (2010): 397-400.

Guo, Caixia, Ji Wang, Li Jing, Ru Ma, Xiaoying Liu, Lifang Gao, Lige Cao et al. "Mitochondrial dysfunction, perturbations of mitochondrial dynamics and biogenesis involved in endothelial injury induced by silica nanoparticles." *Environmental Pollution* 236 (2018a): 926-936.

Guo, Caixia, Ru Ma, Xiaoying Liu, Tian Chen, Yang Li, Yang Yu, Junchao Duan, Xianqing Zhou, Yanbo Li, and Zhiwei Sun. "Silica nanoparticles promote oxLDL-induced macrophage lipid accumulation and apoptosis via endoplasmic reticulum stress signaling." *Science of the total environment* 631 (2018b): 570-579.

Halappanavar, Sabina, Petra Jackson, Andrew Williams, Keld A. Jensen, Karin S. Hougaard, Ulla Vogel, Carole L. Yauk, and Håkan Wallin. "Pulmonary response to surface-coated nanotitanium dioxide particles includes induction of acute phase response genes, inflammatory cascades, and changes in microRNAs: A toxicogenomic study." *Environmental and molecular mutagenesis* 52, no. 6 (2011): 425-439.

He, Chengyong, Shengwei Jiang, Haijing Jin, Shuzhen Chen, Gan Lin, Huan Yao, Xiaoyong Wang et al. "Mitochondrial electron transport chain identified as a novel molecular target of SPIO nanoparticles mediated cancer-specific cytotoxicity." *Biomaterials* 83 (2016): 102-114.

Hu, Hailong, Li Li, Qian Guo, He Zong, Yuheng Yan, Yao Yin, Yu Wang et al. "RNA sequencing analysis shows that titanium dioxide nanoparticles induce endoplasmic reticulum stress, which has a central role in mediating plasma glucose in mice." *Nanotoxicology* 12, no. 4 (2018): 341-356.

Huerta-Ángeles, Gloria, Martina Brandejsová, Jaroslav Novotný, Kateřina Kopecká, Jana Šógorková, Daniela Šmejkalová, and Vladimír Velebný. "Grafting of steroids to hyaluronan towards the design of delivery systems for antioxidants: The role of hydrophobic core." *Carbohydrate polymers* 193 (2018): 383-392.

Hussain, Salik, Faris Al-Nsour, Annette B. Rice, Jamie Marshburn, Brenda Yingling, Zhaoxia Ji, Jeffrey I. Zink, Nigel J. Walker, and Stavros Garantziotis. "Cerium dioxide nanoparticles induce apoptosis and autophagy in human peripheral blood monocytes." *ACS nano* 6, no. 7 (2012): 5820-5829.

Hwang, Ruth, Vahid Mirshafiee, Yifang Zhu, and Tian Xia. "Current approaches for safer design of engineered nanomaterials." *Ecotoxicology and environmental safety* 166 (2018): 294-300.

Karakoti, Ajay S., Satyanarayana VNT Kuchibhatla, Donald R. Baer, Suntharampillai Thevuthasan, Dean C. Sayle, and Sudipta Seal. "Self-Assembly of Cerium Oxide Nanostructures in Ice Molds." *Small* 4, no. 8 (2008): 1210-1216.

Kumah, Elizabeth Adjoa, Raoul Djou Fopa, Saeed Harati, Paul Boadu, Fatemeh Vida Zohoori, and Tannaz Pak. "Human and environmental impacts of nanoparticles: a scoping review of the current literature." *BMC Public Health* 23, no. 1 (2023): 1059.

Kwak, Mi-Kyoung, Nobunao Wakabayashi, Jennifer L. Greenlaw, Masayuki Yamamoto, and Thomas W. Kensler. "Antioxidants enhance mammalian proteasome expression through the Keap1-Nrf2 signaling pathway." *Molecular and cellular biology* 23, no. 23 (2003): 8786-8794.

Lai, Racliffe Weng Seng, Hye-Min Kang, Guang-Jie Zhou, Mana Man Na Yung, Yan Ling He, Alan Man Ching Ng, Xiao-yan Li, Aleksandra B. Djurisic, Jae-Seong Lee, and Kenneth Mei Yee Leung. "Hydrophobic surface coating can reduce toxicity of zinc oxide nanoparticles to the marine copepod Tigriopus japonicus." *Environmental Science & Technology* 55, no. 10 (2021): 6917-6925.

Lee, Jang Han, Jae Eun Ju, Byung Il Kim, Pyo June Pak, Eun-Kyung Choi, Hoi-Seon Lee, and Namhyun Chung. "Rod-shaped iron oxide nanoparticles are more toxic than sphere-shaped nanoparticles to murine macrophage cells." *Environmental Toxicology and Chemistry* 33, no. 12 (2014): 2759-2766.

Li, Chen-Wen, Lan-Lan Li, Sheng Chen, Jian-Xiang Zhang, and Wan-Liang Lu. "Antioxidant nanotherapies for the treatment of inflammatory diseases." *Frontiers in Bioengineering and Biotechnology* 8 (2020): 200.

Li, Junyi, Meiping Tian, Li Cui, John Dwyer, Nigel J. Fullwood, Heqing Shen, and Francis L. Martin. "Low-dose carbon-based nanoparticle-induced effects in A549 lung cells determined by biospectroscopy are associated with increases in genomic methylation." *Scientific reports* 6, no. 1 (2016): 20207.

Li, Lin, Jiahui Cui, Zi Liu, Xuejiao Zhou, Zengqiang Li, Yang Yu, Yuanyuan Jia, Daiying Zuo, and Yingliang Wu. "Silver nanoparticles induce SH-SY5Y cell apoptosis via endoplasmic reticulum-and mitochondrial pathways that lengthen endoplasmic reticulum-mitochondria contact sites and alter inositol-3-phosphate receptor function." *Toxicology Letters* 285 (2018): 156-167.

Li, Wei, Xiujuan Zhang, Xiaojun Hao, Jiansheng Jie, Baishun Tian, and Xiaohong Zhang. "Shape design of high drug payload nanoparticles for more effective cancer therapy." *Chemical communications* 49, no. 93 (2013): 10989-10991.

Liu, Wei, Tao Hu, Li Zhou, Desheng Wu, Xinfeng Huang, Xiaohu Ren, Yuan Lv et al. "Nrf2 protects against oxidative stress induced by SiO2 nanoparticles." *Nanomedicine* 12, no. 19 (2017): 2303-2318.

Lu, Xiaoyan, Isabelle R. Miousse, Sandra V. Pirela, Stepan Melnyk, Igor Koturbash, and Philip Demokritou. "Short-term exposure to engineered nanomaterials affects cellular epigenome." *Nanotoxicology* 10, no. 2 (2016): 140-150.

Luque, Rafael, and Rajender S. Varma, eds. *Sustainable preparation of metal nanoparticles: methods and applications*. Royal Society of Chemistry, 2012.

Ma, Yan-Bo, Chun-Jiao Lu, Muhammad Junaid, Pan-Pan Jia, Li Yang, Jing-Hui Zhang, and De-Sheng Pei. "Potential adverse outcome pathway (AOP) of silver nanoparticles mediated reproductive toxicity in zebrafish." *Chemosphere* 207 (2018): 320-328.

Meng, Huan, Sui Yang, Zongxi Li, Tian Xia, Justin Chen, Zhaoxia Ji, Haiyuan Zhang et al. "Aspect ratio determines the quantity of mesoporous silica nanoparticle uptake by a small GTPase-dependent macropinocytosis mechanism." *ACS nano* 5, no. 6 (2011): 4434-4447.

Møller, Peter, Daniel Vest Christophersen, Ditte Marie Jensen, Ali Kermanizadeh, Martin Roursgaard, Nicklas Raun Jacobsen, Jette Gjerke Hemmingsen et al. "Role of oxidative stress in carbon nanotube-generated health effects." *Archives of toxicology* 88 (2014a): 1939-1964.

Møller, Peter, Pernille Høgh Danielsen, Dorina Gabriela Karottki, Kim Jantzen, Martin Roursgaard, Henrik Klingberg, Ditte Marie Jensen et al. "Oxidative stress and inflammation generated DNA damage by exposure to air pollution particles." *Mutation research/Reviews in mutation research* 762 (2014b): 133-166.

Nabi, Showkat Ul, Sofi Imtiyaz Ali, Muzafar Ahmad Rather, Wajid Mohammad Sheikh, Mehvish Altaf, Hemant Singh, Peerzada Tajamul Mumtaz, Narayan Chandra Mishra, Sheikh Uzma Nazir, and Showkeen Muzamil Bashir. "Organoids: A new approach in toxicity testing of nanotherapeutics." *Journal of Applied Toxicology* 42, no. 1 (2022): 52-72.

Nichols, Cody E., Danielle L. Shepherd, Quincy A. Hathaway, Andrya J. Durr, Dharendra Thapa, Alaeddin Abukabda, Jinghai Yi, Timothy R. Nurkiewicz, and John M. Hollander. "Reactive oxygen species damage drives cardiac and mitochondrial dysfunction following acute nano-titanium dioxide inhalation exposure." *Nanotoxicology* 12, no. 1 (2018): 32-48.

Onoda, Atsuto, Ken Takeda, and Masakazu Umezawa. "Dose-dependent induction of astrocyte activation and reactive astrogliosis in mouse brain following maternal exposure to carbon black nanoparticle." *Particle and fibre toxicology* 14 (2017): 1-16.

Paesano, Laura, Alessio Perotti, Annamaria Buschini, Cecilia Carubbi, Marta Marmiroli, Elena Maestri, Salvatore Iannotta, and Nelson Marmiroli. "Markers for toxicity to HepG2 exposed to cadmium sulphide quantum dots; damage to mitochondria." *Toxicology* 374 (2016): 18-28.

Pardhiya, Sonali, Usha Singh Gaharwar, Rohit Gautam, Eepsita Priyadarshini, Jay Prakash Nirala, and Paulraj Rajamani. "Cumulative effects of manganese nanoparticle and radiofrequency radiation in male Wistar rats." *Drug and Chemical Toxicology* 45, no. 3 (2022): 1395-1407.

Park, Eun-Jung, Jongheop Yi, Kyu-Hyuck Chung, Doug-Young Ryu, Jinhee Choi, and Kwangsik Park. "Oxidative stress and apoptosis induced by titanium dioxide nanoparticles in cultured BEAS-2B cells." *Toxicology letters* 180, no. 3 (2008): 222-229.

Pasquali, Francesco, Caterina Agrimonti, Luca Pagano, Andrea Zappettini, Marco Villani, Marta Marmiroli, Jason C. White, and Nelson Marmiroli. "Nucleo-mitochondrial interaction of yeast in response to cadmium sulfide quantum dot exposure." *Journal of hazardous materials* 324 (2017): 744-752.

Patlolla, Anita K., S. Anitha Kumari, and Paul B. Tchounwou. "A comparison of poly-ethylene-glycol-coated and uncoated gold nanoparticle-mediated hepatotoxicity and oxidative stress in Sprague Dawley rats." *International journal of nanomedicine* (2019): 639-647.

Petković, Jana, Bojana Žegura, Magdalena Stevanović, Nataša Drnovšek, Dragan Uskoković, Saša Novak, and Metka Filipič. "DNA damage and alterations in expression of DNA damage responsive genes induced by TiO2 nanoparticles in human hepatoma HepG2 cells." *Nanotoxicology* 5, no. 3 (2011): 341-353.

Pounraj, Saranya, Prathap Somu, and Subhankar Paul. "Chitosan and graphene oxide hybrid nanocomposite film doped with silver nanoparticles efficiently prevents biofouling." *Applied Surface Science* 452 (2018): 487-497.

Robinson, John M. "Reactive oxygen species in phagocytic leukocytes." *Histochemistry and cell biology* 130 (2008): 281-297.

Ruenraroengsak, Pakatip, and Alexander T. Florence. "Biphasic interactions between a cationic dendrimer and actin." *Journal of drug targeting* 18, no. 10 (2010): 803-811.

Sánchez-López, Elena, Daniela Gomes, Gerard Esteruelas, Lorena Bonilla, Ana Laura Lopez-Machado, Ruth Galindo, Amanda Cano et al. "Metal-based nanoparticles as antimicrobial agents: an overview." *Nanomaterials* 10, no. 2 (2020): 292.

Seiffert, Joanna, Farhana Hussain, Coen Wiegman, Feng Li, Leo Bey, Warren Baker, Alexandra Porter et al. "Pulmonary toxicity of instilled silver nanoparticles: influence of size, coating and rat strain." *PloS one* 10, no. 3 (2015): e0119726.

Sengul, Ayse Busra, and Eylem Asmatulu. "Toxicity of metal and metal oxide nanoparticles: a review." *Environmental Chemistry Letters* 18, no. 5 (2020): 1659-1683.

Sharma, Anshul, Nitin Sharma, Sachin Singh, and Kamal Dua. "Review on theranostic and neuroprotective applications of nanotechnology in multiple sclerosis." *Journal of Drug Delivery Science and Technology* 81 (2023): 104220.

Shen, Yulai, Lu Wu, Dongdong Qin, Yankai Xia, Zhu Zhou, Xuemei Zhang, and Xin Wu. "Carbon black suppresses the osteogenesis of mesenchymal stem cells: the role of mitochondria." *Particle and fibre toxicology* 15 (2018): 1-17.

Shim, Wooyoung, Man Jeong Paik, Duc-Toan Nguyen, Jin-Kyu Lee, Yuwon Lee, Jae-Ho Kim, Eun-Hee Shin et al. "Analysis of changes in gene expression and metabolic profiles induced by silica-coated magnetic nanoparticles." *ACS nano* 6, no. 9 (2012): 7665-7680.

Shukla, Ritesh K., Ashish Badiye, Kamayani Vajpayee, and Neeti Kapoor. "Genotoxic potential of nanoparticles: structural and functional modifications in DNA." *Frontiers in Genetics* 12 (2021): 728250.

Shukla, Ritesh K., Vyom Sharma, Alok K. Pandey, Shashi Singh, Sarwat Sultana, and Alok Dhawan. "ROS-mediated genotoxicity induced by titanium dioxide nanoparticles in human epidermal cells." *Toxicology in vitro* 25, no. 1 (2011): 231-241.

Shvedova, A. A., E. R. Kisin, D. Porter, P. Schulte, V. E. Kagan, B. Fadeel, and V. Castranova. "Mechanisms of pulmonary toxicity and medical applications of carbon nanotubes: two faces of Janus?" *Pharmacology & therapeutics* 121, no. 2 (2009): 192-204.

Skotland, Tore, Tore Geir Iversen, and Kirsten Sandvig. "Cellular uptake of nanoparticles: Involvement of caveolae?" *Precision Nanomedicine* 4 (2021): 782-786.

Sudhakaran, S., S. S. Athira, and P. V. Mohanan. "Zinc oxide nanoparticle induced neurotoxic potential upon interaction with primary astrocytes." *Neurotoxicology* 73 (2019): 213-227.

Thai, Sheau-Fung, Kathleen A. Wallace, Carlton P. Jones, Hongzu Ren, Raju Y. Prasad, William O. Ward, Michael J. Kohan, and Carl F. Blackman. "Signaling pathways and microRNA changes in nano-TiO2 treated human lung epithelial (BEAS-2B) cells." *Journal of Nanoscience and Nanotechnology* 15, no. 1 (2015): 492-503.

Tomás-Zapico, Cristina, and Ana Coto-Montes. "A proposed mechanism to explain the stimulatory effect of melatonin on antioxidative enzymes." *Journal of pineal research* 39, no. 2 (2005): 99-104.

Villanueva-Flores, Francisca, Andrés Castro-Lugo, Octavio T. Ramírez, and Laura A. Palomares. "Understanding cellular interactions with nanomaterials: Towards a rational design of medical nanodevices." *Nanotechnology* 31, no. 13 (2020): 132002.

Vuong, Ngoc Q., Patrick Goegan, Susantha Mohottalage, Dalibor Breznan, Marianne Ariganello, Andrew Williams, Fred Elisma, Subramanian Karthikeyan, Renaud Vincent, and Premkumari Kumarathasan. "Proteomic changes in human lung epithelial cells (A549) in response to carbon black and titanium dioxide exposures." *Journal of proteomics* 149 (2016): 53-63.

Wang, Xiang, Tian Xia, Susana Addo Ntim, Zhaoxia Ji, Sijie Lin, Huan Meng, Choong-Heui Chung et al. "Dispersal state of multiwalled carbon nanotubes elicits profibrogenic cellular responses that correlate with fibrogenesis biomarkers and fibrosis in the murine lung." *ACS nano* 5, no. 12 (2011): 9772-9787.

Wang, Xinan, Yun Liu, Juan Wang, Yaguang Nie, Shaopeng Chen, Tom K. Hei, Zhaoxiang Deng, Lijun Wu, Guoping Zhao, and An Xu. "Amplification of arsenic genotoxicity by TiO2 nanoparticles in mammalian cells: new insights from physicochemical interactions and mitochondria." *Nanotoxicology* 11, no. 8 (2017): 978-995.

Wang, Xin-Meng, Xing-Wang Wu, Xiao-Ying Zhao, Chen-Wei Wang, and Jiang-Ning Zhou. "Exposure-time-dependent subcellular staging of gold nanoparticles deposition and vesicle destruction in mice livers." *Nanomedicine: Nanotechnology, Biology and Medicine* 34 (2021): 102393.

Wang, Ye, Abel Santos, Andreas Evdokiou, and Dusan Losic. "An overview of nanotoxicity and nanomedicine research: principles, progress and implications for cancer therapy." *Journal of Materials Chemistry B* 3, no. 36 (2015): 7153-7172.

Wang, Zhenjia, Chinnaswamy Tiruppathi, Richard D. Minshall, and Asrar B. Malik. "Size and dynamics of caveolae studied using nanoparticles in living endothelial cells." *ACS nano* 3, no. 12 (2009): 4110-4116.

Xue, Yang, Qingqing Chen, and Jiao Sun. "Hydroxyapatite nanoparticle-induced mitochondrial energy metabolism impairment in liver cells: *in vitro* and *in vivo* studies." *Journal of Applied Toxicology* 37, no. 8 (2017): 1004-1016.

Yamada, Shigeru, Daiju Yamazaki, and Yasunari Kanda. "Silver nanoparticles inhibit neural induction in human induced pluripotent stem cells." *Nanotoxicology* 12, no. 8 (2018): 836-846.

Yang, Li-Yun, Jia-Ling Gao, Tian Gao, Ping Dong, Long Ma, Feng-Lei Jiang, and Yi Liu. "Toxicity of polyhydroxylated fullerene to mitochondria." *Journal of hazardous materials* 301 (2016): 119-126.

Yang, Xia, Huali Shao, Weirong Liu, Weizhong Gu, Xiaoli Shu, Yiqun Mo, Xuejun Chen, Qunwei Zhang, and Mizu Jiang. "Endoplasmic reticulum stress and oxidative stress are involved in ZnO nanoparticle-induced hepatotoxicity." *Toxicology Letters* 234, no. 1 (2015): 40-49.

Zangabad, Parham Sahandi, Mahdi Karimi, Fatemeh Mehdizadeh, Hedieh Malekzad, Alireza Ghasemi, Sajad Bahrami, Hossein Zare, Mohsen Moghoofei, Amin Hekmatmanesh, and Michael R. Hamblin. "Nanocaged platforms: modification, drug delivery and nanotoxicity. Opening synthetic cages to release the tiger." *Nanoscale* 9, no. 4 (2017): 1356-1392.

Zeng, Xianghui, Yuning Zhang, and Andreas M. Nyström. "Endocytic uptake and intracellular trafficking of bis-MPA-based hyperbranched copolymer micelles in breast cancer cells." *Biomacromolecules* 13, no. 11 (2012): 3814-3822.

Zhang, Naiding, Guiya Xiong, and Zhenjie Liu. "Toxicity of metal-based nanoparticles: Challenges in the nano era." *Frontiers in Bioengineering and Biotechnology* 10 (2022): 1001572.

Zhao, Nanxia, Nicola L. Francis, Hannah R. Calvelli, and Prabhas V. Moghe. "Microglia-targeting nanotherapeutics for neurodegenerative diseases." *APL bioengineering* 4, no. 3 (2020).

Zhao, Xuesong, Xin Ren, Rong Zhu, Zhouying Luo, and Baixiang Ren. "Zinc oxide nanoparticles induce oxidative DNA damage and ROS-triggered mitochondria-mediated apoptosis in zebrafish embryos." *Aquatic Toxicology* 180 (2016): 56-70.

Ze, Yuguan, Renping Hu, Xiaochun Wang, Xuezi Sang, Xiao Ze, Bi Li, Junju Su et al. "Neurotoxicity and gene-expressed profile in brain-injured mice caused by exposure to titanium dioxide nanoparticles." *Journal of Biomedical Materials Research Part A: An Official Journal of The Society for Biomaterials, The Japanese Society for Biomaterials, and The Australian Society for Biomaterials and the Korean Society for Biomaterials* 102, no. 2 (2014): 470-478.

Section 6

Chapter 6.1

Navigating Regulatory Terrain: Empowering Translation Technologists in Biomedical Innovation

Akshay Joshi*

Centre for Biomedical Engineering (CBME), Indian Institute of Technology Delhi (IIT D), New Delhi, India

Abstract

The modern era of startup founders in MedTech has been made possible by training in a wide range of skill sets obtained through academic and professional programs, leading the way in nanomaterial biomedical device innovation. Navigating regulatory uncertainty serves as a crucial "valley of death" for academia spin-off business ventures. This chapter provides an overview of the fundamentals of risk management, marketing strategies, and regulatory compliance for biomedical entrepreneurs based on the distinct perspectives of startup regulatory consultants and founders. The legal frameworks and procedures for product approval in the US and India are covered, along with techniques and approaches for organizing and carrying out preclinical and clinical research. Additionally, IP tactics to protect creations and exploit market opportunities will be briefly covered, alongside risk management, quality assurance, and IP strategies. These conclusions are supported by actual product success stories. The purpose of this chapter is to instill confidence in readers regarding navigating uncharted territories of innovation in the biomedical sciences, transforming breakthroughs into market-ready products.

Keywords: regulatory affairs, risk management, intellectual property management, quality assurance, translation MedTech innovations

1. Introduction

In the fast-paced world of the MedTech industry, the combination of various skill sets from academic and professional training programs has completely transformed the business landscape for startup founders. These forward-thinking individuals are spearheading breakthroughs in nanomaterial biomedical devices, pushing the limits of healthcare technology.

---

* Corresponding Author's Email: bmt232625@iitd.ac.in

In: Emerging Nanomaterials in Biomedical Science
Editors: Eepsita Priyadarshini and Ajith Manayil Parambil
ISBN: 979-8-89530-600-0

Nevertheless, one of their primary hurdles involves maneuvering through the intricate regulatory environment, commonly known as the "valley of death," which presents barriers from initial research to the process of bringing a product to market. Biomedical entrepreneurs face a multitude of challenges, especially in managing risks, complying with regulations, and developing effective marketing strategies. This chapter provides insights specifically tailored to the perspectives of startup founders, aiming to demystify these elements. It delves into the fundamental legal frameworks and product clearance procedures in major markets like the US and India, providing practical guidance on navigating these frameworks. Furthermore, it highlights the significance of managing risks, ensuring quality, and implementing intellectual property (IP) strategies to protect innovations and optimize market opportunities. Practical applications and outcomes of these strategies are illustrated through real success stories. After reading this chapter, readers will understand the key components needed to turn innovative biomedical breakthroughs into marketable products. Our main goal is to equip aspiring biomedical entrepreneurs with the necessary knowledge and confidence to push the boundaries in the biomedical sciences, ultimately driving progress in patient care and healthcare technology.

1.1. The Importance of Regulatory Compliance

In the healthcare industry, adhering to regulations is essential for promoting patient safety, upholding quality standards, and building confidence. It includes abiding by the rules, legislation, and accreditation specifications established by different regulatory organizations. By putting strong safety procedures and standards in place, compliance dramatically lowers risks, including medical errors and infections linked to healthcare. Additionally, it maintains quality standards by encouraging evidence-based procedures and ongoing improvement through adherence to accreditation requirements and clinical recommendations. Conversely, non-compliance may result in harsh financial fines, legal ramifications, and harm to one's reputation. Proactive compliance ensures ethical accountability, fosters patient trust, and safeguards the organization's brand in addition to maintaining financial stability (Sharma, Gamta, and Luthra 2023).

1.2. Challenges in Regulatory Compliance

Healthcare laws are always changing, making it necessary to continuously review and update compliance strategies. Adjusting to these changes and guaranteeing adherence will take a lot of resources and attention to detail. Compliance efforts face several obstacles due to a lack of technology, human, and financial resources. The absence of essential resources such as technology, staffing, and training can make compliance management difficult. They need robust systems to manage and secure the enormous volumes of sensitive data that healthcare businesses handle in compliance. The complexity of data management increases the burden of compliance. It can be challenging to overcome reluctance to change and establish an organizational culture of compliance to implement new compliance procedures and guidelines necessary for adherence. Effective compliance management requires all stakeholders to work together and communicate clearly. Dealing with diverse regulatory frameworks operating in multiple jurisdictions makes compliance management more difficult. Companies need to keep an eye on and respond to changes in regulations while keeping their compliance procedures uniform. To ensure successful compliance management in the healthcare industry, these issues underscore the necessity of strategic resource allocation, technological investment, cultural transformation, interdisciplinary collaboration, and regulatory change monitoring. (Guerra-Bretaña and Flórez-Rendón 2018).

2. Regulatory Landscapes in the US and India

To guarantee medical equipment's safety, effectiveness, and caliber, separate authorities oversee the regulatory environments for biomedical devices in the US and India. To succeed in these markets, biomedical entrepreneurs must thoroughly understand these regulatory regimes. Medical devices are largely categorized in the United States and India according to the danger they pose to patients and users; however, other variables taken into account in this categorization process include the device's intended purpose and the degree of control required to guarantee safety and efficacy. The Central Drugs Standard Control Organization (CDSCO) is India's main medical device regulating authority. The Food and Drug Administration (FDA) is in charge of regulating medical devices in the US. More stringent review procedures may be applied to innovations or new technologies that are not materially comparable to devices already on the market. Therefore, although risk is the primary factor in the classification of medical devices in both the United States and India, the regulatory framework also takes into account the device's intended use and technical specifications to guarantee thorough safety and efficacy assessments. (Ramdas Maske and Ware 2023).

2.1. Regulatory Framework in the US

Depending on the level of risk, the FDA divides devices into three classes: Class I (low risk), Class II (moderate risk), and Class III (high risk) (Table 1). The classification determines the regulatory pathway as well. General Controls, such as correct labeling, registration, and adherence to Good Manufacturing Practices (GMP), are usually necessary for Class I devices. General controls and special controls, such as post-market surveillance, performance criteria, and labeling specifications, are necessary for Class II devices. The 510(k) premarket notification procedure is applied to numerous Class II devices that show significant equivalency to already available devices in the market (predicate), lawfully marketed products. The FDA also oversees clinical study Investigational Device Exemptions (IDEs), ensuring that novel devices are thoroughly evaluated before approval (Jarow and Baxley 2015).

2.2. Regulatory Framework in India

The Medical Device Rules, 2017, which are by international standards, regulate the regulatory framework for medical devices in India. These rules are designed to facilitate the safe and effective use of medical devices in healthcare. Based on the complexity of the technology, intended use, and risk, this framework categorizes medical devices, which are classified into four categories according to their level of risk (Table 2).

Table 1. Medical device classification as per US regulations

Class	Potential harm	Controls required	Submission Type
I	Minimal harm	General Controls	510(k) Exempt or 510(k)
II	Higher risk than class I	General Controls and Special controls	510(k) Exempt or 510(k) or De Novo
III	They sustain/ support life. Implanted or present potentially larger harm	General and Pre-Market approvals (PMA) that involve Technical, Non-clinical Laboratory Studies, Clinical investigations data	Pre-Market Approval

Table 2. Comprehensive overview of medical devices in India and regulations associated

Device Class	Risk Level	Regulatory Requirements	Examples
Class A	Low Risk	Manufacturing license from the State Licensing Authority (SLA). Adherence to fundamental safety and performance standards. Self-certification or conformity assessment by a Notified Body may be necessary.	Surgical dressings, Tongue depressors, Hand-held surgical instruments, etc.
Class B	Low to Moderate Risk	Manufacturing license from SLA. More comprehensive conformity assessment. Third-party assessment may be required for compliance.	Hypodermic needles, Suction apparatus, Blood pressure monitoring devices, etc.
Class C	Moderate to High Risk	License from Central Licensing Authority (CLA). Clinical data required to demonstrate safety and efficacy. Rigorous regulatory assessment.	Cardiac stents, Bone fixation plates, Dialysis equipment, etc.
Class D	High Risk	License from CLA. Submission of substantial clinical evidence. Thorough evaluation including preclinical and clinical studies.	Ventilators, Implantable defibrillators, Heart valves, etc.

3. Planning and Conducting Preclinical and Clinical Studies

Preclinical and clinical investigations are critical for assessing biomedical devices' biological interactions and potential dangers. Preclinical research, which includes both *in vitro* and *in vivo* studies, detects problems early, allowing for necessary changes before human testing. Clinical trials confirm device function in humans, assuring both safety and efficacy. These studies are critical for protecting public health by ensuring that only safe goods reach patients, thereby creating trust between healthcare providers and patients. Moreover, they contribute to the advancement of scientific knowledge by conducting cutting-edge research and disseminating findings that can lead to new insights and technologies. They play a crucial role in promoting innovation within their field, driving the development of novel solutions that address emerging challenges. Their efforts assist in the continual improvement of devices, ensuring they meet evolving standards of efficacy and safety. Additionally, they facilitate entry into global markets by navigating regulatory requirements and establishing partnerships that enhance product accessibility. By fostering confidence among stakeholders—including regulatory authorities, healthcare professionals, and patients—they help build trust in new technologies, ultimately leading to improved health outcomes and patient care.

3.1. Preclinical Studies

Preclinical studies are an important step in the biomedical device development process, providing foundational data that informs subsequent clinical trials and regulatory applications. These studies are critical for determining the safety and efficacy of medical devices before they are tested on humans.

a) *In vitro* investigations involve testing the gadget in a controlled laboratory environment with cell cultures or tissue samples. These investigations evaluate biocompatibility, cytotoxicity, and how the technology interacts with biological systems. *In vitro* testing enables researchers to detect potential concerns early in development, guiding future design iterations.
b) *In vivo* experiments on animals provide crucial information about the device's performance in living beings. These investigations look at physiological reactions, tissue compatibility, and any side effects connected with the device. Animal studies assist researchers in understanding how the device works in a biological system and provide information about its safety profile.

Preclinical studies have various key applications: 1) Risk identification assists developers in identifying possible device concerns, such as toxicity, inflammation, or adverse reactions, allowing them to resolve these issues before human testing. 2) Preclinical data informs the optimization of device design, materials, and manufacturing processes to improve safety and efficacy. 3) Before conducting human trials, regulatory agencies require preclinical data to assess the device's safety. Adequate preclinical data is required for regulatory approval. 4) Preclinical investigations reduce risks to human subjects by ensuring that only technologies with a reasonable expectation of safety and efficacy progress to clinical trials. (Thomasstecklerr editors, n.d.)

3.2. Clinical Studies

Clinical trials give crucial proof of a medical device's safety and effectiveness, which helps to guide regulatory decisions and informs healthcare practice (Figure 1). These trials are critical for establishing the clinical benefits of devices and guaranteeing patient safety in real-world environments. Some key considerations for clinical studies involve 1) Patient selection; clinical study participants should represent the device's target population and meet certain inclusion and exclusion criteria. 2) Rigorous study design, including randomization, blinding, and adequate control groups, is critical for ensuring valid and impartial results. 3) Clinical studies collect information on safety endpoints, efficacy endpoints, and other relevant outcomes to assess the device's performance. 4) Clinical research must adhere to Good Clinical Practice (GCP) principles and regulatory criteria to assure data integrity and ethical behavior (Neugebauer et al. 2017).

Phase I (Pilot Study)	Phase II (Feasibility Study)	Phase III (Pivotal Study)	Phase IV (Post-Market Surveillance)
1)Small sample of healthy volunteers or patients. 2)Evaluate device safety profile. 3)Identify early efficacy and ideal operating parameters. 4)Ensure device safety.	1)Larger number of patients. 2)Assess safety and efficacy. 3)Collect more data on device performance. 4)Fine-tune device settings 5)Detect negative consequences of long-term use.	1)Large-scale studies. 2)Confirm device efficacy and safety. 3)Compare performance to established treatment. 4)Offer reliable data in controlled environments.	1)Post-approval studies. 2)Assess real-world performance. 3) Detect long-term harmful effects. 4)Provide ongoing safety and efficacy data.

Figure 1. Phases of clinical studies with some key outcome measures (Adapted from Kandi and Vadakedath 2023).

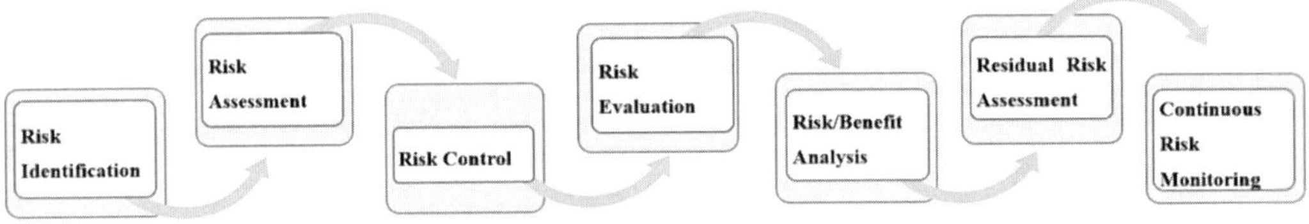

Figure 2. Stages of Risk Management Process for Medical Devices (Adapted from Khinvasara, Ness, and Tzenios 2023).

4. Risk Management and Quality Assurance

Risk management and quality assurance are essential components of the medical device research and manufacturing process, guaranteeing their safety, effectiveness, and adherence to regulatory standards. These procedures aim to identify, evaluate, reduce, and monitor hazards related to the use of devices while ensuring consistent quality throughout the whole lifespan of the product. Quality assurance measures guarantee the consistent manufacturing of medical devices to fulfill safety and performance standards, instilling confidence in regulatory authorities, healthcare professionals, and patients.

4.1. Risk Management Strategies

Effective risk management systems seek to detect, assess, mitigate, and monitor the hazards associated with the use of medical devices.

Here is a comprehensive summary of risk management measures for biomedical devices:

Each stage of the risk management process is depicted, with brief descriptions (Yang 2024).

1) Risk Identification: Use hazard and error analysis to identify potential device risks.

2) Risk Assessment: Determine the severity and likelihood of the identified risks.
3) Risk Control: Take steps to limit risks to acceptable levels.
4) Risk evaluation entails reviewing and analyzing the efficacy of risk-management techniques.
5) Risk/Benefit Analysis: Compare the device's benefits and risks to determine whether the benefits outweigh the dangers.
6) Residual Risk Assessment: After control measures are implemented, assess the remaining risks.
7) Continuous Risk Monitoring: Track risks throughout the device's lifecycle, including post-market surveillance and feedback collecting.

4.2. Risk Mitigation Practices

Fundamentally, some aspects of strong design controls established by utilizing failure modes and effects analysis ensure that the device meets intended use and regulatory standards. Detailed instructions for use can mitigate risks associated with device misuse. Incorporating redundancy and backup systems in the device design can prevent failures. Regular maintenance and calibration schedules are essential to ensure the device operates correctly over time. Continuously monitoring the device's post-market performance through adverse event reporting, user feedback, and ongoing clinical studies can also help identify and address potential risks.

Case Study: Pharmallama - India's First Pre-Sorted Medicine Packs.

Pharmallama integrates advanced design features such as automated sorting systems and barcode technology to minimize errors in medication packing. However, they failed to maintain legal standards, leading to their being banned by the government until further notice.

4.3. Quality Assurance Practices

Robust quality assurance methods are critical for ensuring regulatory compliance, enhancing patient outcomes, and building stakeholder trust. Here's a detailed look at the main quality assurance procedures for biomedical devices:

1. Quality Management System (QMS): A Quality Management System (QMS) is the foundation for quality assurance in biomedical device manufacturing. ISO 13485 is an internationally recognized standard in the medical device industry, outlining specific requirements for establishing and maintaining a robust QMS. This standard is designed to ensure that organizations meet regulatory requirements consistently and produce safe and effective medical devices. ISO 13485 compliance is often essential for market entry and certification in global markets, as it focuses on risk management, process control, and continuous improvement throughout the product lifecycle. Key features include A) Document control, which is the process of managing documents and records to ensure that they are correct, up-to-date, and easily accessible. This comprises SOPs, job instructions, quality manuals, etc. B) Process Control involves creating procedures to oversee medical equipment design, development,

manufacturing, and distribution. This comprises specifying process parameters, putting controls, and assuring consistent results. Management responsibility includes ensuring that top management is committed to the QMS, allocating the appropriate resources, and frequently monitoring its efficacy.

2. Design Controls are essential for assuring biomedical equipment's safety and efficacy. These practices include A) Design planning, which is creating a detailed design and development plan that specifies the stages of product development, responsibilities, and timescales. B) Design Input defines and describes the device's requirements, such as user demands, regulatory requirements, and performance parameters. C) Design Output is documentation of the final design specifications, including drawings, specifications, and manufacturing procedures. D) Design verification and validation entails testing and evaluating the device to confirm that it fulfills the design inputs and is suitable for its intended purpose. Verification ensures that the design outputs match the inputs, whereas validation ensures that the device fulfills the user's needs and intended use.
3. Supplier Quality Management: Managing the quality of materials and components obtained from suppliers is critical. The key practices include A) Supplier qualification, which is evaluating and selecting suppliers based on their ability to meet quality and regulatory standards; B) Conduct regular audits to ensure suppliers meet quality and regulatory requirements. C) Inspecting materials and components upon receipt to ensure they fulfill quality standards.
4. Production and Process Control, where maintaining consistent product quality during manufacturing is critical. This involves A) Process validation, which entails checking manufacturing processes to ensure that they consistently create goods that match predefined requirements. B) In-process controls monitor and control processes during production to detect and remedy errors in real-time C) Calibration and maintenance of equipment should be performed regularly to ensure accuracy and reliability.
5. Corrective and Preventive Action (CAPA): The CAPA system is critical for resolving and preventing quality issues. Key steps include A) Nonconformance Management: The process of identifying and documenting nonconformities, such as departures from specifications, as well as examining the main reasons. B) Corrective action is addressing recognized issues and preventing them from recurring. C) Preventive actions identifying possible issues ahead of time and putting steps in place to avert them.
6. Internal Audits and Continuous Improvement: Internal audits are critical for confirming the QMS's performance and identifying improvement opportunities. The key practices include A) Audit planning, which creates an audit program that addresses all aspects of the QMS. B) Audit execution entails conducting audits to determine compliance with QMS requirements, identifying nonconformities, and assessing the effectiveness of corrective actions. C) Conduct regular reviews of audit findings and the QMS's overall performance to achieve continual improvement via management (Li et al. 2015).

5. Intellectual Property (IP) Strategies

Having a solid Intellectual Property (IP) strategy is essential in the biomedical device industry. It allows companies to safeguard their innovations, maintain a competitive edge, and seize market opportunities. Efficient IP management requires identifying, acquiring, and protecting different forms of intellectual property rights. Patents play a crucial role in safeguarding the intellectual property of biomedical devices. They provide inventors with exclusive rights for a set period in return for publicly revealing their inventions. Patents encompass a wide range of innovative creations, such as device structures, manufacturing processes, and new, useful, and non-obvious methods. Trade secrets are crucial for safeguarding confidential business information that gives companies a competitive advantage. This includes proprietary formulas, manufacturing processes, and research data. Trade secrets offer the advantage of not needing public disclosure, allowing for indefinite protection as long as the information remains confidential. Trademarks provide legal protection for brand names, logos, and other distinguishing identifiers that set a company's products apart in the marketplace. They play a crucial role in establishing brand recognition and fostering customer trust (Mayfield, n.d.).

5.1. Patent Strategies

Legal frameworks and strategic considerations involved in securing patents for medical devices in the US and Indian markets require careful attention. Provided is a comprehensive analysis of patent strategies for both markets. There are two main types of patents for medical devices in the US: utility patents, which cover inventions and discoveries, and design patents, which protect ornamental designs, regulated by the United States Patent and Trademark Office (USPTO) per the US Patent Act. Conducting prior art searches is crucial to ensure the invention's uniqueness. Make use of databases such as USPTO, Google Patents, and other relevant sources. Prepare thorough claims that precisely outline the scope of the invention and encompass a wide range of embodiments to encompass all possible uses and configurations. Ensure the patent application has a thorough description, complete with drawings and examples, to meet the enablement requirement. File a Patent Cooperation Treaty (PCT) application to seek protection in multiple countries simultaneously, just like a savvy business manager would do.

India offers utility patents for the functional features of medical devices and design patents for aesthetic aspects, similar to the United States. These patents are administered by the Indian Patent Office (IPO) under the Indian Patent Act of 1970. To ensure novelty, applicants should conduct comprehensive searches through IPO databases and international patent databases. A priority date can be established by filing a provisional application, with a complete specification submitted within 12 months. Patent prosecution involves interactions with IPO examiners, responding to office actions, and making necessary amendments to claims. To maintain patent validity for the full 20-year term from the filing date, annual renewal fees must be paid promptly. As a business manager, it is crucial to monitor and address any potential infringements. This includes taking necessary actions such as litigation or settlements to enforce patent rights. Prepare for potential patent challenges and develop effective strategies to defend against invalidity or patent infringement claims (Gutman, Capraro, and Chen 2016).

5.2. IP Management and Commercialization

Having a solid grasp of IP management and commercialization strategies is essential for fully capitalizing on the value of biomedical device innovations. To effectively manage intellectual property, having a well-rounded collection of patents, trademarks, trade secrets, and copyrights is important. Regularly conducting audits can help identify any areas that may need improvement and assess the value of existing assets. By implementing a strategic filing approach, priority can be given to high-value innovations and ensure that they are protected in various locations. Being proactive in monitoring the market for potential IP infringements and taking swift enforcement actions are crucial for safeguarding IP rights. To expand market reach and generate revenue, it is important to consider commercialization, licensing intellectual property to third parties, forming strategic alliances, ensuring regulatory compliance, and conducting thorough market analysis. Understanding various sources of income, effectively communicating the unique benefits, and allocating resources to promote and establish a strong reputation is essential in attracting both investors and customers. Through effective management of intellectual property assets, strategic partnerships, and successful market adoption, companies have the opportunity to turn biomedical device innovations into profitable products. This helps them gain a competitive edge and ensures long-term growth and success. ("IP Portfolio Management Strategies for Medical Device Manufacturers," n.d.).

6. Real-World Product Story

Serigen Mediproducts: Pioneering Silk-based Solutions for Wound Healing ("Pioneering Silk-Based Solutions for Wound Healing" 2006).

Dr. Anuya Nisal, a chemical engineer with significant academic credentials, intended to make her scientific study commercially viable. After completing a degree in Polymer Engineering from Pune, followed by a Master's in Materials Science in the US and a PhD in Chemical Engineering from IIT Bombay, she went on to join the National Chemical Laboratory (NCL) in Pune, where she contributes her expertise in advanced materials research and polymer science. At NCL, she worked with Dr. Ashish Lele, a well-known polymer scientist, to investigate novel applications for silk, a versatile biomaterial common in India. This collaboration resulted in the formation of Serigen Mediproducts, which focuses on transforming silk-based technologies into commercial products.

Serigen's journey started with the creation of silk scaffolds for bone regeneration. Serigen developed scaffolds with mechanical strength and porosity that were conducive to natural bone deposition by transforming silk into diverse forms ranging from hard bone-like structures to soft cartilage. This discovery addressed the constraints of current approaches, such as autografting, by creating a synthetic alternative that combines the performance of autografts with the convenience of synthetic materials. Their unique strategy for controlling the scaffold's breakdown rate-maintained compatibility with the body's regeneration process. Now, they have Seriderm, an advanced wound dressing, which solves the issues of complicated wounds such as diabetic foot ulcers and second-degree burns. Serigen, with patents granted in India, Europe, Japan, and the United States, continues to pioneer novel biomedical solutions that improve patient care and advance healthcare technology.

7. Future Directions for Biomedical Entrepreneurs

There are thrilling opportunities driven by swift technological advancements, growing healthcare demands, and constantly evolving regulatory landscapes. Personalized medicine is an exciting area of development that aims to tailor devices and treatments to each patient's specific profile and health conditions. This approach has the potential to revolutionize therapies, making them more efficient and precise. As a result, patients can experience fewer side effects and achieve better outcomes. Emerging technologies hold immense potential to revolutionize the MedTech industry. Entrepreneurs who fully embrace these technologies will be at the forefront of innovation, creating intelligent and highly adaptable medical devices. The integration of nanotechnology in biomedical devices opens up exciting possibilities. Utilizing nanoscale materials and devices presents an incredible chance to attain unmatched accuracy in both diagnosis and treatment, as they allow for direct interaction with biological systems at a molecular level. These advancements promise to transform fields like cancer treatment, regenerative medicine, and minimally invasive surgeries. Sustainability is becoming increasingly important in the field of medical device development. Entrepreneurs are increasingly being challenged to incorporate sustainable practices and materials into their businesses, placing equal importance on environmental well-being and technological progress. Providing guidance on the dynamic regulatory environment and advocating for policies that encourage innovation can significantly help entrepreneurs simplify the process of launching their products. From a business perspective, the future looks incredibly bright for biomedical entrepreneurs. We have a chance to revolutionize global health by creating innovative, personalized, and eco-friendly technologies that can make a real difference.

Conclusion

Successfully navigating the biomedical device industry necessitates a thorough grasp of regulatory compliance, risk management, quality assurance, and intellectual property strategies. Understanding these areas is crucial for biomedical entrepreneurs to achieve successful product development, market-entry, and sustained growth. Ensuring safety and efficacy is paramount in the development of biomedical devices, with strict guidelines provided by regulatory authorities such as the US FDA and India's CDSCO. A solid grasp of these frameworks is essential for effectively navigating regulatory landscapes, minimizing risks, and streamlining product approvals. Entrepreneurs encounter a multitude of challenges, such as navigating intricate regulatory landscapes, adapting to diverse regional demands, and ensuring extensive documentation. Successfully navigating these challenges requires staying informed about regulatory updates, actively collaborating with regulatory authorities, and implementing strong compliance strategies. Ensuring that devices meet all safety and efficacy standards is crucial for protecting public health and fostering trust among stakeholders. When examining the regulatory frameworks of the U.S. and India, it becomes evident that there are notable differences in their approaches to medical device approval. The FDA prioritizes rigorous testing and documentation, ensuring high standards are met.

On the other hand, the CDSCO's guidelines aim to tailor global standards to the specific needs of the local context. A clear understanding of these differences allows entrepreneurs to customize their strategies for each market, guaranteeing compliance and gaining a competitive edge. Effective planning and execution of preclinical and clinical studies play a crucial role in the development process. Preclinical studies follow Good Laboratory Practices (GLP) to detect potential issues at an early stage. Clinical studies, on the other hand, adhere to Good Clinical Practices (GCP) to confirm the safety and effectiveness of the device in human subjects. These studies help with regulatory approvals and contribute to scientific knowledge and innovation. Ensuring device safety and performance requires a strong focus on effective risk management and quality assurance. ISO 14971:2019 and ISO 13485:2016 offer comprehensive risk management and quality system frameworks, guaranteeing that potential risks are thoroughly identified, assessed, and addressed at every product lifecycle stage. By implementing these standards, entrepreneurs can effectively prevent adverse events and guarantee compliance with regulatory requirements. Ensuring intellectual property protection is essential for safeguarding innovations and achieving market success. Implementing effective IP management practices is crucial for companies to safeguard their technological advancements and stay ahead of the competition. This involves conducting regular audits, strategically filing patents, and diligently enforcing IP rights. Case studies showcasing successful products demonstrate the practical application of these principles. They emphasize the importance of following regulatory guidelines, implementing strong risk management practices, ensuring quality assurance, and strategically managing intellectual property to achieve successful commercialization. These stories are valuable learning tools for aspiring biomedical entrepreneurs, showcasing the significance of taking a comprehensive approach to innovation. By taking a comprehensive approach, we guarantee regulatory approval and market success and make significant contributions to the progress of medical science and the improvement of patient care.

Acknowledgments

The author is thankful to Team Regulatory Information & Facilitation Center, Venture Centre, Pune Maharashtra, for giving me Internship exposure to medical device regulations. Also, I am grateful to Team Serigen Mediproducts via the Venture Centre startup tour I got to discover the inspiring story.

References

"IP Portfolio Management Strategies for Medical Device Manufacturers." n.d. www.hrfmlaw.com.

"Pioneering Silk-Based Solutions for Wound Healing." 2006. https://www.venturecenter.co.in/startups-and-success-stories/success-stories/serigen-mediproducts.

Guerra-Bretaña, Rosa Mayelin, and Andrea Lucía Flórez-Rendón. 2018. "Impact of Regulations on Innovation in the Field of Medical Devices." *Research on Biomedical Engineering*. Brazilian Society of Biomedical Engineering. https://doi.org/10.1590/2446-4740.180054.

Gutman, Siegmond Y., Joe Capraro, and Tom Chen. 2016. "Introduction to Patent Strategies for Medical Device Inventions." *Surgery (United States)* 160 (5): 1139–44. https://doi.org/10.1016/j.surg.2016.06.055.

Jarow, Jonathan P., and John H. Baxley. 2015. "Medical Devices: US Medical Device Regulation." *Urologic Oncology: Seminars and Original Investigations*. Elsevier Inc. https://doi.org/10.1016/j.urolonc.2014.10.004.

Kandi, Venkataramana, and Sabitha Vadakedath. 2023. "Clinical Trials and Clinical Research: A Comprehensive Review." *Cureus*, February. https://doi.org/10.7759/cureus.35077.

Khinvasara, Tushar, Stephanie Ness, and Nikolaos Tzenios. 2023. "Risk Management in Medical Device Industry." *Journal of Engineering Research and Reports* 25 (8): 130–40. https://doi.org/10.9734/jerr/2023/v25i8965.

Li, Tzu Wei, Pei Weng Tu, Li Ling Liu, and Shiow Ing Wu. 2015. "Assurance of Medical Device Quality with Quality Management System: An Analysis of Good Manufacturing Practice Implementation in Taiwan." *BioMed Research International* 2015. https://doi.org/10.1155/2015/670420.

Mayfield, Denise L. n.d. "Medical Patents and How New Instruments or Medications Might Be Patented."

Neugebauer, Edmund A. M., Ana Rath, Sunya Lee Antoine, Michaela Eikermann, Doerthe Seidel, Carsten Koenen, Esther Jacobs, et al. 2017. "Specific Barriers to the Conduct of Randomised Clinical Trials on Medical Devices." *Trials*. BioMed Central Ltd. https://doi.org/10.1186/s13063-017-2168-0.

Ramdas Maske, Tulshidas, and Nilesh Ware. 2023. "THE COMPARISON OF MEDICAL DEVICE REGULATIONS IN INDIA WITH USA, EU: AN QUALITATIVE STUDY." www.ijnrd.org.

Sharma, Akash, Vriti Gamta, and Gaurav Luthra. 2023. "Ensuring Patient Safety and Trust: The Critical Importance of Regulatory Compliance in Healthcare." *Journal of Pharmaceutical Research International* 35 (18): 1–15. https://doi.org/10.9734/jpri/2023/v35i187390.

Thomasstecklerreditors, Antonnbespalov Martinnc Michel. n.d. "Handbook of Experimental Pharmacology 257." http://www.springer.com/series/164.

Yang, Kiki. 2024. "Risk Management in Medical Devices: An Application of ISO 14971." In *2024 IEEE International Symposium on Product Compliance Engineering (ISPCE)*, 1–3. IEEE. https://doi.org/10.1109/ISPCE61193.2024.10541258.

Chapter 6.2

Ethical Implications of Nanomaterials in Biomedicine: Regulatory Challenges and Perspectives

Priyanka Sarsaiya
Neha Jha
and Paulraj Rajamani*
School of Environmental Science, Jawaharlal Nehru University, New Delhi, India

Abstract

Emerging nanotechnological applications are not limited to a particular area but have grown tremendously in the bio-medical field, which encompasses disease prevention, diagnosis, and therapy. Since their development, concerns about their effects and ethical considerations have been raised. As their applications include their role in disease prevention, *in vitro* and *in vivo* diagnostic tests, and therapy, it raises substantial ethical and regulatory concerns. In this chapter, we focus on exploring the ethical implications and regulatory challenges associated with the application of nanomaterials in biomedicines. Nanomedicine involves a wide range of social, ethical, and regulatory issues, including distributive justice and human and civil rights. When involving humans, there is a need to respect patient autonomy and beneficence, as nanomedicine could have possible unknown effects. Thus, the key ethical issues include clinical trials for which patient autonomy and informed consent and the patient should be made aware of unknown risks. Ensuring the consent and social acceptance of this new technology in the field of medicine can be very challenging, as insufficient knowledge or education about nanomedicine can lead to a backlash from the public and stakeholders. The strictness of the guidelines and principles or their lack thereof will be considered as the implementation of the existing regulations and tackling the gaps within them. Since the application of nanomedicine is growing rapidly, there is a subsequent increase in databases, which should be maintained with transparency. We will also discuss how government and industry play a greater role in the acceptance of nanomedicine in the present and future.

Keywords: biomedicine, clinical trials, ethical issues, nanomedicine, regulatory framework

---

* Corresponding Author's Email: paulrajr@mail.jnu.ac.in

In: Emerging Nanomaterials in Biomedical Science
Editors: Eepsita Priyadarshini and Ajith Manayil Parambil
ISBN: 979-8-89530-600-0

1. Introduction

Nanotechnology finds its application in the field of nanomedicine. Nanotechnology has revolutionized several areas within medicine, providing innovative solutions for diagnosis, treatment, and prevention of diseases. Nanomedicine uses nanotechnology to innovate in healthcare and is used in diagnostics, medical imaging, therapeutics, vaccines, and regenerative medicine (Jiang and Carmichael 2019). The nanometric scale of these materials, closely matching that of biological processes, allows them to penetrate potential barriers and interact directly at the molecular level, engaging with components such as DNA and proteins. It is common to modify their surfaces to enhance the biocompatibility of these particles, increase their circulation time in blood, and ensure they bind specifically to targets. Nanotechnology encompasses the advanced manipulation of matter at the atomic and molecular scale (Florczyk 2007). Nanomaterials present challenges due to their diverse composition, size, form, surface characteristics, and degree of disintegration. The hardest challenge posed by nanotechnology's emergence is managing the uncertainties and risks involved in this domain. This raises concerns regarding policy issues and risk assessment. Uncertainty looms large regarding health and environmental risks and potential government regulations. Ensuring the safe and sustainable use of nanomedicine requires rigorous scientific investigation and the establishment of clear regulatory standards that can adapt to the evolving landscape of this innovative field (Malloy 2011; Van de Voorde and Jeswani 2021). Therefore, thoroughly assessing the risks of nanoparticles is crucial for their long-term use.

The societal impact of nanotechnology is substantial and widespread, bringing up ethical concerns such as privacy violations, economic exploitation, security, autonomy, environmental pollution, and global justice. To address these knowledge gaps, urgent attention is required from both the scientific and regulatory communities, necessitating inclusive participation from all segments of society in its development and commercialization. Collaboration between natural sciences (mathematics, engineering, physics, chemistry, and biology) and human sciences (sociology, philosophy, law, and ethics) is crucial for developing ethical norms for nanotechnology.

Throughout history, novel scientific and technological advancements have often been met with societal apprehension. Notable examples include the emergence of synthetic chemicals, nuclear power, biotechnology, and genomics in the twentieth century. Nanotechnology faces similar scrutiny due to its potential risks. Toxicological studies on nanomaterials have consistently demonstrated that nanoparticles can adversely affect both human health and the ecosystem (Van de Voorde and Jeswani 2021).

Consequently, the ethics of employing these technologically generated nanomaterials in biomedicine involves addressing various concerns, including ethical, safety, and regulatory challenges inherent in the development, regulation, and clinical application of nanomedicine (Jain 2008, 330-31). While this field promises advantages such as targeted drug delivery, organ cultivation, improved imaging and diagnostics, gene therapy, and regenerative medicine, it also faces similar challenges, such as safety and toxicity, regulation and standards, cost and scalability, and ethical considerations. Specifically, it raises questions about patient autonomy (informed consent) and beneficence (balancing benefits and risks). The study of the moral and social implications of nanotechnology, known as nanoethics, is a nascent and contentious field (Allhoff and Lin 2009). An overview of nanotechnology in biomedicine applications is represented in Figure 1.

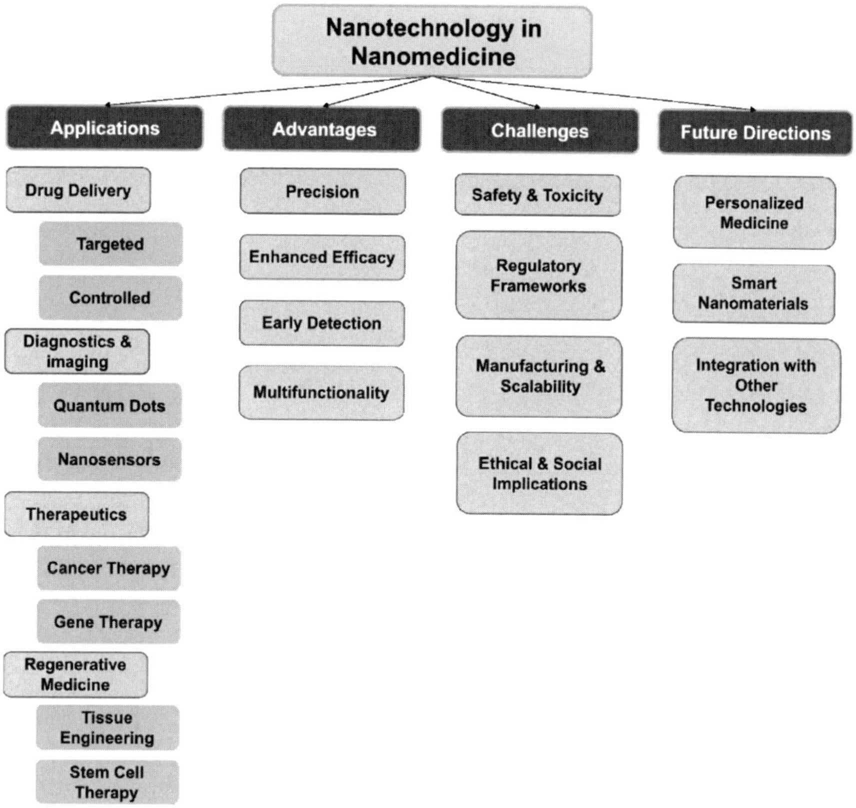

Figure 1. Overview of Nanotechnology in Biomedicine.

Some key applications of nanomedicine include:

1.1. Drug Delivery Systems

Nanoparticles are engineered for targeted drug delivery to specific cells or tissues, improving therapeutic efficacy and minimizing side effects through mechanisms such as ligand-receptor interactions, magnetic guidance, and pH-sensitive release. Nanocarriers can also provide controlled release of drugs, maintaining optimal therapeutic levels over extended periods and enhancing patient compliance (Mukherjee 2014).

1.2. Diagnostics and Imaging

Quantum dots, which emit bright and stable fluorescence, are valuable for imaging and tracking biological processes at the cellular level. Nanosensors, incorporating nanoscale materials, can detect biomarkers with high sensitivity and specificity, enabling early diagnosis of diseases like cancer and infectious diseases (Wang et al. 2006).

1.3. Therapeutics

Nanotechnology has emerged as a promising approach in cancer therapeutics (Chaturvedi et al. 2019). In cancer therapy, nanoparticles are used for hyperthermia, photodynamic therapy, and enhancing the delivery of chemotherapeutic agents. Nanocarriers deliver genetic material to specific cells for gene therapy, offering potential treatments for genetic disorders and other diseases.

1.4. Regenerative Medicine

In tissue engineering, nanomaterials support the growth of cells and tissues by providing scaffolds that mimic the natural extracellular matrix, promoting tissue regeneration. In stem cell therapy, nanoparticles enhance the delivery and efficacy of stem cells, improving outcomes in regenerative medicine.

The integration of nanotechnology into medicine offers several advantages, such as precision in targeting disease cells or tissues and enhanced efficacy because nanoscale materials can improve the solubility, stability, and bioavailability of therapeutic agents, enhancing their effectiveness (Arora et al. 2012). It can also help in early detection of diseases with the help of nanosensors, thus improving treatment outcomes. Moreover, nanoparticles can be designed to perform multiple functions simultaneously, such as targeting, imaging, and therapy, providing comprehensive solutions for complex medical conditions.

2. Challenges and Considerations

Nanomedicine development faces more obstacles than other nano products. Extensive preclinical research, rigorous clinical trials, and clear identification of the intended clinical use of the drug are necessary for its successful translation. The low clinical translation efficiency of nanotherapeutics highlights the existence of safety, biological barriers, and regulatory challenges (Zhang et al. 2020, 2) While nanotechnology holds great promise for medicine, it also presents several challenges that must be addressed, as given below.

2.1. Safety and Toxicity

Understanding nanomaterials' potential toxicity and long-term effects on human health and the environment is critical. Rigorous safety assessments and regulations are necessary to ensure safe application. Although nanoparticles (NPs) have numerous biological applications, there is limited knowledge regarding the toxicity risks associated with their use in biological contexts. These risks are influenced by various parameters, including dose, surface area, charge, shape, and structure. Further research is necessary to ensure these interactions are safe for human use (Merugu et al. 2023).

2.2. Regulatory Frameworks

Developing appropriate regulatory frameworks that address nanomaterials' unique properties and behavior is essential for their safe and effective use. Addressing regulatory gaps specific to nanotechnology and advocating transparent governing principles and practices. There is a need for transparent legislation with strong regulations and strict guidelines to avoid future problems (Bhat, et al. 2023).

2.3. Manufacturing, Scalability, and Cost

Ensuring consistent and scalable production of nanomaterials with precise control over their properties is a significant challenge. Consistent surface properties, such as chemical modification, electrical charge, and parameters like size and polydispersity, must be maintained during the scale-up process (Souto 2020). Clinical translation of any medicine requires large-scale synthesis with high uniformity and consistency. The high costs associated with the scale-up process also hinder the development of novel nanomedicines (Zhang et al. 2020, 10).

2.4. Ethical and Social Implications

Addressing ethical considerations, such as informed consent, privacy, and equitable access, is crucial for the responsible development and deployment of nanomedicine. The ethical challenges surrounding the application of nanotechnology in the field of medicine are complex and multifaceted, comprised of several areas of concern, such as informed consent of patients involved in treatment and participation in clinical trials. Ethical challenges arise in conducting comprehensive preclinical and clinical studies to accurately assess potential risks and benefits. It often deals with the collection of sensitive data and personal information, which must be secured and safeguarded to maintain privacy and any unauthorized access or misuse (Hunt and Mehta 2013, 183-245).

3. Ethical Considerations and Patient Consent in Nanomedicine

The promising possibilities of nanomedicine must be weighed against potential risks. For example, nanoparticles' ability to interact with DNA raises questions about genetic privacy and the possibility of unintended genetic modifications. Social acceptance involves public perception and trust in nanomedicine, which can be influenced by transparent research practices, clear regulations, and effective communication of risks and benefits. Addressing these ethical and social aspects is essential for developing and integrating nanomedicine into healthcare. By ensuring thorough regulation, conducting extensive safety research, and engaging with the public and ethical bodies, nanomedicine can achieve its full potential while minimizing risks and addressing societal concerns. Ethical practices and informed consent are paramount to safeguarding patient rights and maintaining public trust in nanomedicine. Ethical consideration refers to the moral principles and values that guide decision-making procedures.

Whereas regulatory consideration refers to the legal and compliance aspects that must be adhered to when developing, testing, and implementing new technologies, products, or services and legal requirements that guide decision-making and actions in various fields, including research, healthcare, and technology development. These considerations ensure that activities are conducted with integrity, respect for individuals and society, and comply with laws and regulations (Rai et al. 2021).

In addition to the precautionary principle, which is the law in many nations while being seldom followed, we also need to implement a global responsible care principle to protect users, consumers, and employees. Although there are some societal issues that nanoscience and new technologies might help with, workplace safety and environmental protection considerations must come first. Currently, there are many questions regarding the sustainability of nanotechnologies, mostly because financial interests rather than logic and science are driving those (Rai et al. 2021). The nanoscale world has a dual nature, offering solutions for health and the environment while also posing threats. However, ensuring nanosafety is key to addressing these challenges minimizing harm, and harnessing maximum benefits (Gatti et al. 2015, 242-43). The major ethical considerations include:

3.1. Safety and Risk Assessment

One of the foremost ethical concerns in nanomedicine is the safety of nanomaterials and their potential risks to patients. Due to their small size, Nanoparticles can interact with biological systems in unique and unpredictable ways. Ethical considerations include:

- Long-term Effects: The long-term health impacts of nanomaterials are not fully understood. Thorough and ongoing research is necessary to monitor these effects.
- Toxicity: Nanoparticles may exhibit toxic properties, which is not associated with their bulk counterparts. Ethical protocols must prioritize minimizing harm and preventing toxic exposure.
- Environmental Impact: Nanomaterials' production, use, and disposal may have environmental consequences. Ethical practices must encompass environmental stewardship to prevent ecological damage.

3.2. Equity and Access

Nanomedicine has the potential to widen the gap between different socio-economic groups if access to advanced treatments is not equitable. Ethical considerations include:

- Accessibility: Ensuring that nanomedicine is available to all, regardless of economic status, is crucial. Policies must be in place to prevent disparities in access to cutting-edge treatments.
- Cost: The high cost of developing nanomedicine could translate to expensive treatments. Balancing innovation with affordability is a key ethical challenge.

3.3. Privacy and Data Security

The integration of nanotechnology with digital health tools raises concerns about privacy and data security. Ethical considerations include data protection and Informed Consent for data use.

3.3.1. Patient Consent in Nanomedicine/Informed Consent

Obtaining informed consent is a crucial ethical and legal obligation in medical practice, ensuring that patients understand and agree to the treatments they receive. To provide informed consent, the information must be comprehensible. Given the knowledge gaps, uncertainty, and complexity surrounding the long-term implications of therapeutic instruments and nanomedical diagnostics, it might be challenging to provide enough information for informed consent regarding a suggested treatment, diagnosis, and therapy (Van de Voorde and Jeswani, 2021).

In nanomedicine, the informed consent process must address specific challenges:

- Complexity and Understanding: The science behind nanomedicine can be complex and difficult for patients to understand. Consent forms and discussions should be tailored to provide clear, comprehensible information about the treatment, its benefits, risks, and alternatives.
- Disclosure of Risks: Given the novel nature of nanomedicine, it is crucial to disclose known risks as well as uncertainties regarding long-term effects. Patients should be aware of the experimental status of certain treatments.
- Voluntariness: Patients must give their consent freely, without any pressure or undue influence, ensuring they feel empowered to decline participation in nanomedical treatments or trials.

Also, special consideration should be given to certain populations and situations that require additional ethical scrutiny when obtaining consent. These include vulnerable groups such as children, the elderly, or those with cognitive impairments. Their capacity to understand and voluntarily provide consent should be carefully assessed. Clinical Trials and emergency cases also need ethical guidelines intervention.

4. The Role of Government in Nanomedicine

The role of government in nanomedicine is multifaceted and crucial for advancing this field. These agencies assess safety, efficacy, and quality, often collaborating with international counterparts. The key aspects that these government bodies undertake are regulation and safety, research funding and support, standardization and guidelines, and public health preparedness. Some of the key government regulatory bodies across the globe are:

4.1. Food and Drug Administration (FDA)

The FDA in the United States regulates nanomedicines and other products containing nanomaterials. They guide nanotechnology-based products' safety, efficacy, and quality assessment (Administration 2014).

4.2. Medicines and Healthcare Products Regulatory Agency (MHRA)

In the United Kingdom, the MHRA oversees the regulation of nanomedicines, ensures compliance with safety standards, and evaluates clinical data (Foulkes et al. 2020).

4.3. Other National Agencies

Various countries have their regulatory bodies overseeing nanomedicine. For example, Health Canada (Canada) and the National Medical Products Administration (NMPA) in China. In Japan, medicine regulation is controlled jointly by the Ministry of Health, Labour and Welfare (MHLW) and the Pharmaceuticals and Medical Devices Agency (PMDA) (Foulkes et al. 2020). The Department of Science and Technology is undertaking an initiative to address nanotechnology regulation in India. However, it is not the specific nanomedicine regulatory body (Bhatia and Chugh 2017).

5. Research and Development Ethics

Scientists and engineers play important roles in nanoethics to ensure responsible and ethical behavior in the field of nanotechnology. They employ nanotechnology to generate novel, innovative materials, technologies, and systems with unique features. When matter is controlled at such a small scale of atoms and molecules, it exhibits entirely new phenomena and features. As a result, scientists and engineers are at the forefront of research and development and should acknowledge their moral obligations. They should engage in open communication about the risks and benefits of existing products, as well as the consequences of their recent findings and projects. This transparent discourse should extend to fellow scientists, policymakers, and the broader public, facilitating informed decision-making and responsible innovation (Van de Voorde and Jeswani, 2021).

5.1. Research Integrity and Transparency

Scientists must adhere to high research standards, assuring transparency, honesty, and data integrity in their work (McGinn 2010). Scientists must stay informed in their field. This entails being aware of scientific breakthroughs (novel nanomaterials and procedures) and new rules and regulations in other nations. While exploring the beneficial applications of nanotechnology, researchers should also consider potential adverse effects. Furthermore, non-scientists and nanotechnologists should actively share their knowledge with the general public, contributing to increased scientific literacy. Ultimately, informed public opinion can be shaped through advocacy campaigns aimed at establishing better regulations and evidence-based policies (Van de Voorde and Jeswani 2021).

5.2. Risk Assessments

Technological risk is a long-standing issue in technology ethics; it includes ethical, epistemological, and theoretical-decision dilemmas. Scientists assess the dangers connected with nanomaterials, considering their effects on human health, the environment, and ecosystems. Engineers attempt to reduce the risks connected with nanoparticles by putting safety elements into designs. As a result, risk minimization is a key goal of technological development, and codes of ethics typically hold engineers accountable for designing safe products and lowering the risks involved with their operations. While risk reduction is a key goal in technological development, achieving absolute security is often challenging and may sometimes conflict with ethical considerations. Factors such as cost, usability, and durability influence product design (Van de Voorde and Jeswani, 2021). It is critical to conduct risk assessments before authorizing the commercialization of nanotechnology products. Risk assessment significantly aids in predicting the behavior and fate of engineered nanomaterials across various environmental contexts (Patel and Pathak 2021, 414).

6. Future Perspectives and Challenges

6.1. Developing Robust Guidelines for Nano-Ethics

Developing robust guidelines for nano-ethics is essential to address the unique ethical challenges posed by the advancement of nanotechnology. This process requires interdisciplinary collaboration among scientists, ethicists, policymakers, and other stakeholders to ensure that diverse perspectives are considered, resulting in balanced and inclusive guidelines. A key component is the assessment and management of risks associated with nanomaterials and nanoproducts. Guidelines should provide frameworks for identifying, evaluating, and mitigating potential risks to human health and the environment. Transparency in research and development processes is crucial for maintaining public trust. Therefore, guidelines must emphasize clear communication about the benefits and risks of nanotechnology and actively involve the public in decision-making processes. Furthermore, robust guidelines must be supported by appropriate regulatory frameworks, including standards for the safe production, use, and disposal of nanomaterials, with regular monitoring and enforcement to ensure compliance. Businesses that prioritize ethics in compliance not only keep out of trouble with the law but also contribute to a more ethical and sustainable business climate (Vall, 2023).

Promoting ethical research practices is fundamental to nano-ethics, and guidelines should outline principles for responsible conduct in nanotechnology research, including data integrity, intellectual property, and conflicts of interest. Additionally, ensuring equitable access to the benefits of nanotechnology is a key ethical consideration. Guidelines should address social justice issues to prevent disparities in the distribution of nanotechnology's benefits and burdens. Sustainability must also be a core principle, with guidelines promoting practices that minimize environmental impact and encourage the development of sustainable nanotechnological solutions. In summary, developing robust guidelines for nano-ethics involves a comprehensive, interdisciplinary approach that emphasizes risk management, transparency, regulatory oversight, ethical research, equity, and sustainability, all of which are essential for guiding the responsible development and application of nanotechnology. (Allhoff and Lin 2009).

6.2. Engaging Stakeholders in Regulation

Stakeholders play a crucial part in shaping the development, governance, and societal impact of nanomaterials and nanotechnology. These individuals or groups with vested interests in nanotechnology include experts (industry experts, government experts), civil organizations, and the public. They play an influential role in formulating policies and decisions and risk management related to nanotechnology. Ethical implications and societal benefits are central to their considerations. Public organizations represent the interest of citizens and would advocate for transparency, awareness, and ethical use of nanotechnology, ensuring the safety and well-being of society (Larsson et al. 2019, 13-17). The regulatory agencies, such as the FDA, EPA, etc., also collaborate with other stakeholders to ensure safe and sound nanotechnological use (National Nanotechnology Initiative 2014).

Engaging stakeholders in the regulation of nanotechnology is important. The key groups identified as critical to the decision-making process in nanotechnology are: (1) managers and policymakers who oversee and regulate nanotechnology initiatives; (2) scientists, researchers, and engineers who drive innovation and technical advancements; and (3) stakeholders, including industry representatives, environmental advocates, and public interest groups, who contribute diverse perspectives and assess potential impacts on society and the environment. The grant's targeted focus on nanotechnology is highly valuable for engaging stakeholders and the public, as it fosters informed dialogue, promotes transparency in technological advancements, and supports broader awareness of nanotechnology's potential applications and societal impacts. It advocates for concurrent social science research alongside toxicology studies on nanoparticles, highlighting this as a modest yet crucial investment complementing ongoing environmental health and safety research. Addressing ethical, social, and regulatory dimensions pre-emptively in nanomedicine is crucial to mitigate potential public resistance akin to controversies surrounding genetically modified foods in Europe (Mills and Fleddermann 2005). Effective education of the public and stakeholders regarding nanomedicines' benefits and risks is recommended.

Conclusion

As nanomedicine continues to evolve, ethical considerations and patient consent must remain at the forefront of its development and application. Ensuring nanomedical treatments' safety, equity, and privacy while facilitating a robust informed consent process is essential for maintaining ethical standards and fostering public trust in this transformative field. By addressing these ethical challenges thoughtfully and proactively, the medical community can harness the full potential of nanomedicine to improve patient outcomes and advance healthcare. To enable the government and businesses to continue researching and developing new technologies, the public must support government investment, purchase consumer goods, and accept the net benefits of nanotechnology. Teams of stakeholders, including regulators, businesses, academia, and others, will need to lead efforts to disseminate information concerning the dangers of nanotechnology (Allhoff and Lin 2009, 92).

References

Administration, Drug. 2014. "*Guidance for Industry Considering Whether an FDA-Regulated Product Involves the Application of Nanotechnology Contains Nonbinding Recommendations*." http://www.fda.gov/RegulatoryInformation/Guidances/ucm257698.htm.

Allhoff, Fritz, and Patrick Lin, eds. 2009. *Nanotechnology & society: Current and emerging ethical issues*. Dordrecht: Springer Netherlands. 1-206. https://doi.org/10.1007/978-1-4020-6209-4.

Arora, Pooja, Annu Sindhu, Neeraj Dilbaghi, Ashok Chaudhury, Govindasamy Rajakumar, and Abdul Abdul Rahuman. 2012. "Nano-regenerative medicine towards clinical outcome of stem cell and tissue engineering in humans." *Journal of Cellular and Molecular Medicine* 16, no. 9: 1991–2000. https://doi.org/10.1111/j.1582-4934.2012.01534.x.

Bhat, Mansoor Ahmad, Kadir Gedik, and Eftade O. Gaga. 2023. "Environmental Impacts of Nanoparticles: Pros, Cons, and Future Prospects." In *Synthesis of Bionanomaterials for Biomedical Applications*, Edited by Munir Ozturk, Arpita Roy, Rouf A. Bhat, Fazilet V. Sukan, and Fernanda M. P. Tonelli. Elsevier. 493–528. https://doi.org/10.1016/B978-0-323-91195-5.00002-7.

Bhatia, Pooja and Chugh, Archana. 2017. "A multilevel governance framework for regulation of nanomedicine in India" *Nanotechnology Reviews* 6, no. 4. 373-382. https://doi.org/10.1515/ntrev-2016-0083.

Chaturvedi, Vivek K, Anshuman Singh, Vinay K. Singh, and Mohan P. Singh. 2019. "Cancer Nanotechnology: A New Revolution for Cancer Diagnosis and Therapy." *Current Drug Metabolism* 20, no. 6: 416–29. https://doi.org/10.2174/1389200219666180918111528.

Florczyk, Stephen J. 2007. "Ethical Issues in Nanotechnology." In *Journal of Long-Term Effects of Medical Implants*: Edited by Subrata Saha, 17, no. 3: 271–80. https://doi.org/10.1615/jlongtermeffmedimplants.v17.i3.90.

Foulkes, Rachel, Ernest Man, Jasmine Thind, Suet Yeung, Abigail Joy, and Clare Hoskins. 2020. "The regulation of nanomaterials and nanomedicines for clinical application: Current and future perspectives." *Biomaterials science* 8, no. 17, 4653-4664. https://doi.org/10.1039/D0BM00558D.

Gatti, Antonietta M., and Stefano Montanari. 2015. *Case studies in nanotoxicology and particle toxicology*. Academic Press. 242-43. https://doi.org/10.1016/C2013-0-18692-8.

Hunt, Geoffrey, and Michael Mehta. 2013. *Nanotechnology: "Risk, Ethics and Law."* Routledge, 183-245. https://doi.org/10.4324/9781849772051.

Jain, Kewal K. 2008. *The handbook of nanomedicine*. 3rd ed. New York: Humana Press, 330-31. https://doi.org/10.1007/978-1-4939-6966-1.

Larsson, Simon, Magnus Jansson, and Åsa Boholm. 2019. "Expert Stakeholders' Perception of Nanotechnology: Risk, Benefit, Knowledge, and Regulation." *Journal of Nanoparticle Research* 21 (3). https://doi.org/10.1007/s11051-019-4498-1. 13-17.

Malloy, Timothy F. 2011. "Nanotechnology Regulation: A Study in Claims Making." *ACS Nano* 5, no.1: 5–12. https://doi.org/10.1021/nn103480e.

McGinn, Robert. 2010. "Ethical Responsibilities of Nanotechnology Researchers: A Short Guide." *NanoEthics* 4, no.1 : 1–12. https://doi.org/10.1007/s11569-010-0082-y.

Merugu, Ramchander, Shiva Sharma, and Manisha Rastogi. 2023. "An Overview of Nanotechnological Applications in Biomedical Engineering." *Materials Today: Proceedings* 76:412–17. https://doi.org/10.1016/j.matpr.2022.12.028.

Mills, K., and C. Fleddermann. 2005. "Getting the Best from Nanotechnology: Approaching Social and Ethical Implications Openly and Proactively." *IEEE Technology and Society Magazine* 24 (4): 18–26. https://doi.org/10.1109/MTAS.2005.1563498.

Mukherjee, Biswajit. 2014. "Editorial (Thematic Issue: 'Nanosize Drug Delivery System')." *Current Pharmaceutical Biotechnology* 14, no. 15: 1221. https://doi.org/10.2174/138920101415140804121008.

Patel, Jayvadan K., Yashwant V. Pathak. 2021. *Emerging Technologies for Nanoparticle Manufacturing*, Springer Nature Switzerland. 414. https://doi.org/10.1007/978-3-030-50703-9.

Rai, Mahendra, Mrunali Patel, and Rashmin Patel, eds. 2021. *Nanotechnology in Medicine: Toxicity and Safety*. Wiley. https://doi.org/DOI:10.1002/9781119769897.

Souto, Eliana B., João Dias-Ferreira, Ranjita Shegokar, Alessandra Durazzo, and Antonello Santini. "Ethical issues in research and development of nanoparticles." In *Drug Delivery Aspects*. 157-168. Elsevier, 2020.

Triennial Review of the National Nanotechnology Initiative. 2014. *Triennial Review of the National Nanotechnology Initiative*. National Academies Press. https://doi.org/10.17226/18271.

Vall, Shin. 2023. "The Importance of Ethical Considerations in Regulatory Compliance." *Pharmaceutical Regulatory Affairs: Open Access*. Vol. 12:3. University of Connecticut, https://doi.org/10.37421/2167-7689.2023.12.366.

Van de Voorde, Marcel, and Gunjan Jeswani. 2021. *Ethics in Nanotechnology: Emerging Technologies Aspects*. Walter de Gruyter GmbH & Co KG. 1-256. https://doi.org/10.1515/9783110701883-202.

Wang, Lin, Meghan B O'Donoghue, and Weihong Tan. 2006. "Nanoparticles for multiplex diagnostics and imaging." *Nanomedicine* 1, no. 4: 413–26. https://doi.org/10.2217/17435889.1.4.413.

Zhang, Chenyang, Liang Yan, Xin Wang, Shuang Zhu, Chunying Chen, Zhanjun Gu, and Yuliang Zhao. 2020. "Progress, challenges, and future of nanomedicine." *Nano Today* 35: 101008. https://doi.org/10.1016/j.nantod.2020.101008.

About the Editors

Dr. Eepsita Priyadarshini, PhD

Dr. Eepsita Priyadarshini is an accomplished researcher with a PhD in biological sciences from CSIR-IMMT and earned her Post-doctorate from Jawaharlal Nehru University, New Delhi. She worked as a DST-Women Scientist at Jawaharlal Nehru University, New Delhi, specializing in development of nanocoatings to prevent biofilm infection on medical implants She is currently working as Assistant Professor, School of Biosciences and Technology, Galgotias University, U.P., India. Her expertise spans a broad spectrum from nanomaterial fabrication to their environmental and biomedical applications, embracing cutting-edge techniques and *in vitro* studies. Dr. Eepsita academic journey is marked by notable accolades. She was conferred with Chancellor's Gold Medal and University Gold Medal for outstanding performance during her MSc. She also received the prestigious Newton-Bhaba PhD fellowship to conduct her research at University College London, United Kingdom. Dr. Eepsita has significantly contributed to the scientific community as a reviewer and editor for numerous esteemed journals as ACS Applied Biomaterials, Frontiers in Biomaterials, Materials Letter, Sensors and Actuators, and Chemical Engineering Journal.

Dr. Ajith Manayil Parambil, PhD

Dr. Ajith Manayil Parambil earned his PhD from Jawaharlal Nehru University, New Delhi, in 2023, specializing in aqueous contaminant sensing and remediation using carbon dots and novel composites. His expertise spans logic gates, device fabrication, and biocomputation. Dr. Parambil has enriched his research experience at Kaohsiung Medical University, Taiwan. With an MPhil. from Madras University and an MSc. from Gandhigram University, he excelled academically, being a gold medalist and first-rank holder. He secured numerous accolades, including the AWSAR National Award for Best Science Research Story and the "Innovation in Science Pursuit for Inspired Research (INSPIRE)" Junior & Senior Research Fellowship Fellowship from the Department of Science and Technology (DST), Government of India. He has also been a recipient of the Young Scientist International Travel Grant from the Science and Engineering Research Board (SERB), DST, Government of India, as well as the Taiwan Experience Education Program (TEEP) Visiting Researcher Scholarship from the Ministry of Education (MOE), Taiwan, and the Global Talent Internship Scholarship in 2021 from the Ministry of Science and Technology (MOST), Taiwan. Dr. Parambil has significantly contributed to the academic discourse with publications in esteemed journals such as Small (Wiley), Chemical Engineering Journal (Elsevier), and Journal of Material Chemistry B (RSC). He is a postdoctoral research fellow at the Nanotechnology Centre, Centre for Energy and Environmental Technology, VSB-Technical University of Ostrava, Czech Republic.

Index

C

D

E

F

G

Q

R

U

V

W

X

Z